Flora and Vegetation of The Teide National Park

Victoria Eugenia Martín Osorio
Beatriz Hernández Bolaños
Wolf-Hermann Wildpret Martín
Wolfredo Wildpret de la Torre

Flora and Vegetation of The Teide National Park

Illustrated Guide to Species and Cartography

Springer

Victoria Eugenia Martín Osorio
Department of Botany, Ecology and Plant Physiology
University of La Laguna
San Cristóbal de La Laguna, Tenerife, Spain

Wolf-Hermann Wildpret Martín
Biodiversity, Macroecology & Biogeography Research Group, Faculty of Forest Sciences and Forest Ecology
University of Göttingen
Göttingen, Germany

Beatriz Hernández Bolaños
Department of Botany, Ecology and Plant Physiology
University of La Laguna
San Cristóbal de La Laguna, Tenerife, Spain

Wolfredo Wildpret de la Torre
Department of Botany, Ecology and Plant Physiology
University of La Laguna
San Cristóbal de La Laguna, Tenerife, Spain

ISBN 978-3-031-88399-6 ISBN 978-3-031-88400-9 (eBook)
https://doi.org/10.1007/978-3-031-88400-9

This Springer imprint is published by the registered company Springer Nature Switzerland AG
The registered company address is: Gewerbestrasse 11, 6330 Cham, Switzerland

Acknowledgments

We would like to thank all the people who have made it possible to produce this work, which we have been preparing and updating for so many years through our studies and research work. It has been an arduous task in collecting data due to the inaccessibility of some very rugged areas of the national park. To do so, we have had the help of several collaborators and researchers who have travelled all over the territory to take samples of the flora and vegetation. Some have stayed in the National Park carrying out their professional work, others have made the leap to other more distant territories to carry out scientific work in line with the tasks they carried out in the National Park. Some have chosen teaching work on the island of Tenerife. In general, the years of intense work have provided a unique time that will be unforgettable in our personal and professional lives. Thanks to Lucas Afonso Jiménez, Yolanda De Rioja Santos, Abel Jacobo Hernández Luis, Francisco Manuel Mesa Luis, Mª Ángeles Montañés, François Tapia e Ico Sanchez-Pinto. Thanks also to the collaborators Javier Martínez-Carbajal and Gerardo García Casanova who have selflessly provided their high-quality, professional photos, some of them of great sentimental value. Thanks to Beatriz Hernández Bolaños for her professionalism, her perseverance, and her dedication to work. The work carried out on the Geobotanical Information System has been impeccable and reflects the responsible nature and seriousness of the work, without always forgetting the respectful and cordial personal treatment. Professor Wildpret has always been a great help in transmitting knowledge based on experience and botanical vocation. His endless workdays also had their risky side; there were situations of loss of vision and contact on some paths that were difficult to follow due to the aggressiveness of the volcanic territories. The incorporation of Wolf-Hermann Wildpret Martín has been of great value for his knowledge of new technologies and bioclimatic content, and he has been able to integrate into a multidisciplinary team.

We would also like to thank all the team of Environmental Agents of the National Park, Wardens, and the Management team and technicians in general who have always collaborated and facilitated the work to carry out this research in the largest natural laboratory in Macaronesia.

Competing Interests The authors have no competing interests to declare that are relevant to the content of this manuscript.

Contents

About the Authors

Victoria Eugenia Martín Osorio is a Professor at the University of La Laguna, on the island of Tenerife, Canary Islands. She studied a degree in Biology in the Plants specialty at the University of Málaga and obtained a PhD in Environmental Analysis, Impact Studies and Land Management with the research work entitled "Study of the Flora and Vegetation of the Grazalema Natural Park, Cadiz, Málaga." At the University of Málaga she carried out research work in international projects such as the LUCDEME "Fight Against the Desertification of the Mediterranean" and the "Cartography of the Habitats of the Natura 2000 Network of the European Union" for Andalusia. She later obtained a position as a professor in the Department of Botany, Ecology and Plant Physiology at the University of La Laguna where she has carried out teaching work in different faculties such as the Faculty of Education, Faculty of Sciences, and in the sections of Environmental Sciences and Biology. In the Canary Islands, she also participated in the project of the "Cartography of the Habitats of the Natura 2000 Network of the European Union" for the Canary Archipelago and the elaboration of the "Canary Islands Biodiversity Strategy" for the Canary Islands Government.

Wolfredo Wildpret de la Torre is Professor Emeritus of the University of La Laguna, Doctor Honoris Causa from the Leibniz University of Hannover, Germany, and winner of the Canary Islands Prize for Research and Innovation from the Canary Islands Government. President of the Board of Trustees of Teide National Park from 1985 to 1998 and Member of said Board representing the University of La Laguna to the present date. He was the founder of the Faculty of Biology of the University of La Laguna and has dedicated his entire professional life to the study of the flora and vegetation of the Canary Islands.

Beatriz Hernández Bolaños is a Geographer and Doctor from the University of La Laguna. Her doctoral thesis focused on the study of the flora and vegetation of the Teide National Park. She has collaborated in several research projects on the endangered flora of the Canary Islands.

Wolf-Hermann Wildpret Martín is a Biologist from the Albert-Ludwigs University of Freiburg, Germany, and has studied a master's degree in Plant Ecology and Biodiversity in the same University. He has collaborated in several research projects related to the conservation of flora and vegetation in Germany and the Canary Islands. Currently he belongs to Biodiversity, Macroecology & Biogeography Research Group, Faculty of Forest Sciences and Forest Ecology, University of Göttingen, Göttingen, Germany.

Chapter 1
Introduction to the Flora and Vegetation of the Teide National Park

1.1 Introduction to the Study of the Flora and Vegetation of the Teide National Park

Professor Victoria Eugenia Martín Osorio, the main author of the text of this book, during one of the countless sessions of comments on the scientific content of the book, suggested that I try to write a historical outline of the different scientific aspects and events of more or less relevant people who visited or lived for a short or longer time in the Cañadas del Teide. Most of them only had the objective of visiting the mentioned space briefly for 2 or 3 days. Not all of them managed, due to different circumstances, to reach the upper edge of the summit of the mythical and famous Teide volcano stratum. But almost all of them published their impressions and observations on the climb, most of them from La Orotava to the longed-for summit. These scientific-cultural impressions undoubtedly contain a personal historical and human value of great interest.

For several weeks I was reflecting on the proposal. Finally, rereading the good part of the literature that is available to us on the subject, I decided to write with enthusiasm this introduction. My growing historical vocation, in my period of professional retirement, has contributed to risking trying to synthesize the most important part of the adventures, sometimes dangerous, with the achievements or attempts of people or groups that for different reasons did not achieve or failed in the attempt of this prodigious experience.

The island of Tenerife is the largest and highest of the Canary Islands, with an area of 2034 km^2. With a height of 3715 m above sea level, the Teide Peak is the highest point in Spain and the Atlantic Ocean. It is the third highest island volcano on the planet after the Hawaiian Mauna-Kea (4207 m above sea level) and Mauna-Loa (4170 m above sea level). It is the highest mountain in the Macaronesia biogeographic region, which includes five archipelagos in the Atlantic Ocean, from north to south: the Azores, Madeira, the Savage Islands, the Canary Islands and Cape Verde.

V. E. Martín Osorio et al., *Flora and Vegetation of The Teide National Park*,
https://doi.org/10.1007/978-3-031-88400-9_1

Tenerife rises nearly 8000 m above the Miocene oceanic crust (Watts et al. 1997) and began to grow very quickly due to the contribution of basaltic lavas through fissural eruptions, which gave rise to a shield volcano that comprises almost 90% of the island. Remnants of this basaltic shield emerge at the corners of Anaga and Teno and in the Roque del Conde. These subaerial rocks have ages between 12 and 3.3 Ma according to existing radiometric dates "K-Ar and Ar-Ar" (Ancochea et al. 1999; Thirlwall et al. 2000).

Named a World Heritage Site by UNESCO in 2007, the Las Cañadas caldera, on the island of Tenerife (Canary Islands), the home of the active volcanic complex Teide-Pico Viejo (TPV), is one of the most beautiful geological landscapes of our planet and one of the best exposed and most interesting volcanic calderas in the world (Martí 2019). In the centre of this Tenerife surface there is a wide depression of cauldron-shaped form called "Las Cañadas" with a surface of about 130 km^2. In it is located the majestic Pico del Teide. This caldera can be considered as a living and natural museum of the geographical, geological, biological, paleontological and especially volcanic activity, currently in a phase of fumarolic rest, since its last eruption in the Narices del Teide, from June 9 to September 8, 1798, on the western flank of Pico Viejo and later, outside the National Park, on the Chinyero volcano in November 1909.

It is difficult to decipher what the biodiversity of this space was like before the arrival of humans on the island. Geological history shows us that the island, which is about 12 million years old, has been subject to significant volcanic activity and as is often the case, this will have meant systematic changes to its terrestrial and marine biodiversity, with the destruction or disappearance of taxa and the subsequent reconstruction of new populations through the classic natural processes of speciation and colonisation.

It is estimated that for some centuries before the present time there was no significant human influence on the island's land surface. The gradual arrival of humans, most of them from the neighbouring north-west of Africa, accompanied by domestic animals, mainly goats, sheep, pigs, rabbits and perhaps dogs, together with the introduction of seeds of food plants for their sustenance, initiated the slow and progressive deterioration and transformation of a landscape unknown to humanity until that time. The primitive vegetation cover would have been until then only subjected to violent destruction due to a series of historical eruptions, some of enormous volume, and the erosive climatic processes of very rainy and stormy periods causing the geomorphology of the current island landscapes, some of spectacular beauty and grandeur. The constant arrival by air of clouds of Saharan dust, the so-called "calima" that since ancient times has not stopped transporting bacteria, unicellular algae, spores, seeds, plant and animal propagules, even being historic, the abundant spectacular and long-suffered invasions of the Saharan locust.

From those distant times until the beginning of the fourteenth century, the arrival of some European and North African expeditions was sporadic. Some reported on the territorial presence of human population, their culture, as well as some geological data and the most spectacular examples of their native fauna and flora.

Juba II, king of Numidia, is responsible for the organization of an expedition to the archipelago who brought the first information about the existence of some islands, populated by human beings, geographically located opposite the African continent. This king was, according to García García (2009), an intellectual scientist who masterfully dominated a wide range of fields of knowledge and who took the archipelago out of the realm of myth, placing it in the geographical reality of the first century of the current era.

This important expedition seems to have been consulted by some of the navigators who sporadically visited the islands until the arrival of the French, Portuguese, Andalusians and Galicians who began, not without violent situations, the conquest of the Canary archipelago that ended with the submission of the inhabitants of the island of Tenerife two years after the discovery of the American continent by Christopher Columbus in 1492 on his ocean voyage from the island of La Gomera to the Caribbean.

A description of how the first Normans could have found the plant landscapes at the beginning of the fifteenth century, upon their arrival on the islands, can be found in Wildpret and Martín Osorio (2006). From the chronicle of Le Canarien (Pico Graña et al. 2003) written by the monks Pierre Bontier and Jean Le Verrier, some data can be obtained on the vegetation and plant resources seen by the Normans, collected through their raids, persecutions, betrayals, Christianization, capture and massacres of aborigines to which they subjected the different populations of human beings of some of the islands of the archipelago visited and even those among themselves. The clerics recount in a superficial way some impressions and this allows us to have a brief vision of the Canarian landscape reality. It is assumed that the conquerors were more committed to their internal struggles, their betrayals and their desire to dominate the new territories than to being interested in the animal or plant life that had inhabited them since ancient times (Wildpret and Martín Osorio 2006).

The description of the vegetation that they make from their boats on the island of Tenerife is the following: there is a large mountain towards the centre, the highest of all the Canary Islands, whose slope extends in all directions over most of the island, and all around there are deep ravines full of thick groves and beautiful springs, dragon trees and many trees of different types and species (Fig. 1.1) (Wildpret and Martín Osorio 2006).

The islands as an attractive scientific stopover due to their volcanic nature, their peculiar fauna and vegetation and the presence of the legendary peak of Teide in Tenerife aroused the interest of European naturalists who, in a continuous and growing increase since the beginning of the seventeenth century, began to create formidable historical information in all fields of culture and incipient scientific research. In this regard, I recommend, among others, reading the very well-documented book by the historian Alfredo Herrera Piqué entitled "The Canary Islands, Scientific Stopover in the Atlantic, Travellers and Naturalists in the eighteenth Century" (Herrera Piqué 1987).

In this introduction we have decided to limit ourselves to making a few brief comments about the people, most of them scientists, of some note, who in their respective visits to the island managed to climb to the summit of Teide in those

Fig. 1.1 Landscape of dragon trees *Dracaena draco* and canary palm trees *Phoenix canariensis*

difficult times of the eighteenth century. The century of enlightenment, when the possibility of breaking with a backward and retrograde culture was opened to the world to make way for the great beginning of free thought. Great figures of eighteenth-century European culture contributed decisively to promoting new scientific progress. Without a doubt, one of the great scientific figures of this century was the doctor, naturalist and professor of Botany at the University of Uppsala, Carl Linnaeus (Rasmunt 23.05.1707-Uppsala 10.03.1778). He was, from our point of view, one of the most relevant botanists in the field of Natural Sciences. His binomial system, still valid, is basic for describing, recognizing or classifying any living being and even past fossils. This great naturalist and to a lesser extent his son were the pioneers in classifying and publishing the material contributed by naturalists visiting the island of Tenerife or the material received through shipments of plants collected directly on the island.

In the Dutch garden of Amsterdam "*Hortus Cliffortianus*" Linnaeus (1737) discovered and described one of the first Canary Island plants grown in Europe, which he named *Campanula canariensis* (=*Canarina canariensis* L.). Around 40 endemic species from the Canary Islands were described by Linnaeus in his magnum opus "*Species plantarum*" (1753), including *Spartocytisus supranubius* L., the most characteristic species of the Teidean flora, the "cardón" *Euphorbia canariensis* L. or the dragon tree *Dracaena draco*. (L.)L. Linneo filius (20.01.1741- Upsala 1.11.1783), after his father's death, continued to describe Canary Island species in his “*Supplementum Plantarum and Generum plantarum*” (Linnaeus fil. 1782) most of which were collected and sent from Tenerife to his teacher, the botanist Aiton of the London Kew Garden, by the Scottish gardener Francis Masson (Ortega et al. 2009).

1.2 Ascensions to the Pico del Teide During the Eighteenth Century

The Frenchman Louis Feuillée (1660–1732), a priest of the Minim order, astronomer and botanist, is often mentioned as the pioneer in collecting “in situ” and classifying the first specimens of plant species collected on the island of Tenerife using the old Tournefort system. At the age of 60, he was chosen in 1724 by the Paris Academy of Sciences to travel to Tenerife. On the 1st of May 1724 he embarked in Marseille bound for Santa Cruz de Tenerife, where he arrived at 3 pm on the 23rd of July. On the 30th of July he travelled from La Laguna, where he lived, to La Orotava and the next day he went down to Puerto de la Cruz to measure the altitude of Mount Teide from sea level to the summit of the volcanic stratum on its coast for the first time. On August 3, the expedition began its ascent to the Teide Peak from La Orotava along the traditional route, reaching the foot of the volcano at four in the afternoon (González de Uriarte et al. 2007). From there, they began the ascent to the Estancia de los Ingleses. They camped there and at four in the morning on the 4th they had breakfast and began their climb to the summit of the volcano. Only the two sons of the French consul, Mr. Porlier, and the young collaborator of Feuillée, Verguin, reached the summit at 8:20 a.m., and with the instruments they had taken, they made the observations and measurements they had planned during the 2 hours they remained on the summit. When they began their descent from the top, they visited the Cueva del Agua with its ice sheets and returned to the town of La Orotava.

On August 10th, they embarked at dusk in Puerto de la Cruz bound for the island of El Hierro and the next morning they arrived at the port of Santa Cruz de La Palma where they remained until the morning of the 12th when they set sail for El Hierro where they arrived at 4 in the afternoon. The main reason for the expedition was to calculate, among other things, the latitude of the island. On the 21st they began their return to Tenerife. “11 days of work on the island of the meridian deserve to be studied” (Herrera Piqué 1987). On the return trip they stopped off on the island of La Gomera so they could visit the capital for a few hours and then continue to Tenerife with a stopover in the Port of Santiago and disembark in Garachico to return by land to Puerto de la Cruz. The next day they returned to the Villa, their place of residence in the Valley.

Feuillée’s ascent of Teide was the fourth known until then and the first of a scientific nature. In the work of Herrera Piqué (1987) there is extensive information about the ascent and stay on the Peak, whose summit he did not reach due to an injury sustained during the ascent. From a historical point of view, the discovery by his collaborator Verguin of the Teide Violet described and drawn by Feuillée as *Viola tenerifera radice fibrosa* (Feuillée 1724) should be highlighted. During the rest of his stay on the island, together with Verguin, they collected phanerogam plants and a lichen on the island. This material, duly pressed, labelled for each species with its drawing and description using the pre-Linnaean Tournefortian multinominal nomenclature, is deposited in the Natural History Museum of Paris. Father Feuillée can also be considered the first naturalist to author a scientific study on the vegetation of the Canary Islands.

Throughout the eighteenth century, more and more expeditions arrived in Tenerife as a necessary stopover on their routes to the discovery and colonisation of distant European territories that were more or less known. The majority stayed for a short time, to obtain supplies, or for a longer time to recover ships wrecked by violent storms on the way and even by attacks by European or North African pirates who at that time roamed the Atlantic Ocean around the archipelagos of the Azores, Madeira and the Canary Islands. Some of the accounts of those travellers contain valuable historical information about their stay in Tenerife. Some of them even contain news of climbs to Mount Teide.

In the spring of 1749, Michel Adanson stopped off in Santa Cruz on his trip to Senegal (Adanson 1759). During his brief stay on the island, he made a brief description of the height of the Teide and stated that it is one of the highest mountains in the universe and that its summit is covered in snow all year round. Three years later, in 1752, Dr. Heberden, an English physician and botanist, at that time resident in La Orotava, calculated the height of the peak's summit at 15,396 English feet, equivalent to 2408 fathoms. This figure coincided with that obtained by trigonometric operations previously carried out by Mr. John Crosse, the English consul in Santa Cruz de Tenerife. In 1754, there is knowledge, but no data, of the ascent of the Teide by the Dutch consul, Mr. Van Steinfort (Herrera Piqué 1987).

On Christmas Eve 1771, the French frigate La Flore arrived in Santa Cruz. The scientists Verdun de la Crenne, the astronomer P. Pingre and the mathematician and geodesist J. Ch. Borda were on board. The scientists were accompanied by the young astronomer Mersais, a student of La Lande, and the draughtsman Ozanne, master draughtsman at the French school of naval guards (Borda et al. 1778). These expeditionary made an excursion to Teide following the traditional route, indicated by Feuillée. The account of the ascent to the peak provides natural references to the Teide crater and the Ice Cave.

Borda returned to Tenerife in 1776 in order to make new calculations, including that of the altitude of the Teide Peak. On this occasion he obtained the result of 1905 fathoms (3712.8 m), the closest altitude to the present one. From La Orotava Borda climbed to the top of Teide accompanied by the Spanish scientists José Varela and Luis Arguedas and several French officers who were part of the aforementioned expedition.

On May 31, 1777, the gardener, plant collector and draughtsman born in Aberdeen, Scotland (August 1724—Montreal 23-12-1805) Francis Masson arrived in Tenerife sent by the botanists of Kew Garden, especially by the botanist of the garden William Aiton. He returned to Madeira on January 1, 1778 (Britten 1884).

During this long period, Masson carried out a significant collection of plants and seeds to increase the collection of specimens, as well as sheets of species, most of which are in the herbarium of the British Museum of Natural History (Ortega et al. 2009). It is known that he made an ascent of Pico del Teide, the account of which, if there was one, has not been able to be consulted.

The French naturalist and botanist Jacques Julien Houttou de Labillardière (Alenzon 28.10.1775-Paris 8.0.1.1834) was part of the D'Entrecasteux expedition in 1791 in search of the missing Jean Françoit Galaup de la Pérouse. On this trip he visited Tenerife to climb Mount Teide. "The last bush I found before reaching the peak was the *Spartium supra nubium*, but there is a herbaceous plant very close to the summit, a violet, its flower had already withered" (Rodríguez and Wildpret 2005).

The eighteenth century ended with the arrival in Tenerife of the Prussian scientist Alexander von Humboldt (Berlin 14.09 1769—Berlin 6.05 1859) and the French doctor and botanist Aimé Jacques Bonpland (1773–1858). They arrived in Tenerife on 17 June 1799, where they stayed for 5 days. On 21 June, from Puerto de la Cruz, they began the climb to Las Cañadas, heading for the summit of Teide. Upon reaching Portillo, Humboldt wrote:. "It is by crossing this narrow landscape between two basaltic hills that one enters the plains of Retama. It took us about two and a half hours to cross it. The goats of Pico feed on the leaves of the Spartium and have been wild in these deserts since time immemorial and have been esteemed as a delicious delicacy" (Humboldt 1820). During their time at the summit and inside the crater, with the help of the numerous devices they took on the expedition, they did all kinds of data collection and mineral collection. On their return, they commented: "On a grass toasted by the heat of the African sun, *Cladonia paschalis* covers arid terrain and shepherds make fire there that spreads over considerable distances. Up to the summit of the Peak, urceolarias and other lichens work on the decomposition of the sludge." It is possible that this botanical information is the first record of the lichens so abundant in the Teidean territory.

Furthermore, I have thought it appropriate to partially transcribe a nice comment by Humboldt (1874) published in the edition of his great work "Cosmos" Essay on a physical description of the world translated into Spanish by Bernardo Giner and José de Fuentes Volume 1 Madrid 1874 page 5:

> If I were allowed to abandon myself to the memories of distant adventures, among the joys that it presents I would describe the summit of the peak of Tenerife, at the moment when a horizontal layer of clouds, dazzling in white, separates the cone of ashes from the lower plain, and suddenly, by ascending currents, allows the view from the very edge of the crater to dominate the vineyards of (la) Orotava and the orange gardens and the thick groups of plane trees on the coast (Humboldt 1874).

As I finish commenting on these details, probably incomplete, of the visits mentioned, in my opinion, of the most important naturalists and scientists of that century, my memory flies to the illustrious naturalist from Tenerife, José de Viera y Clavijo, born in the town of Realejo de Arriba in the Orotava valley, just a few kilometres from Pico del Teide on 28 December 1731, and died in Las Palmas de Gran Canaria on 21 February 1813. In 1799 he wrote the Dictionary of Natural History of the Canary Islands, which was a unique scientific milestone written by a naturalist born in the Canary Islands.

References

Adanson M (1759) Voyage to Senegal, the isle of Gorée and the river Gambia. Londres

Ancochea E, Huertas MJ, Cantagrel JM, Coello J, Fúster JM, Arnaud N, Ibarrola E (1999) Evolution of the Cañadas edifice and its implications for the origin of the Cañadas Caldera (Tenerife, Canary Islands). J Volcanol Geotherm Res 88(1999):177–199

Borda J Ch, Pingre P, Verdun de la Crenne M (1778) Voyage Fait par ordre du Roi en 1771 et 1772 en Diverses Parties de l'Europe, de l'Afrique et de lÁmerique. https://gallica.bnf.fr/ark:/12148/bpt6k1521394b

Britten J (1884) Francis Masson. J Bot (London) 22(1884):114–123

Feuillée L (1724) Voyage aux Isles Canaries, ou Journal des observations Physiques, Mathematiques, Botaniques et Historiques faites par ordre de Sa Majesté París: Museum National f'Histoire Naturelle, ms 38

García García A (2009) Juba II y las Islas Canarias, Ediciones Idea, Colección de Estudio General, Santa Cruz de Tenerife, 2009, 391pp + 21 ilustraciones

González de Uriarte C, Oliver JM, Curell C, Pico B (2007) Louis Feuillée, pionero de las exploraciones científicas en Canarias https://fundacionorotava.org/media/web/files/page40__UriarteOliverCurellPico.pdf

Herrera Piqué A (1987) Las Islas Canarias, escala científica en el Atlántico. In: Viajeros y naturalistas en el siglo XVIII. Editorial Rueda, Madrid

Humboldt A (1820) Voyage aux régions équinoxiales du nouveau continent, fait en 1799, 1800, 1801, 1802, 1803 et 1804, par Al. de Humboldt et A. Bonpland. Tome 6/réd. par Alexandre de Humboldt. https://gallica.bnf.fr/ark:/12148/bpt6k972858

Humboldt A (1874) Cosmos. Ensayo de una descripción física del mundo. Alejandro de Humboldt. Vertido al castellano. Bernardo Giner y José de Fuentes Tomo 1 Madrid

Linnaeus C (1737) *Hortus Cliffortianus*. Amsterdam

Linnaeus C (1753) *Species plantarum* (1753) http://www.botanicus.org/

Linnaeus C fil (1782). https://web.archive.org/web/20110725095053/

Martí J (2019) Las Cañadas caldera, Tenerife, Canary Islands: a review, or the end of a long volcanological controversy. Earth Sci Rev 196. https://doi.org/10.1016/j.earscirev.2019.102889

Ortega JF, Santos Guerra A, Carine M, Jarvis C (2009) Francis Masson y los primeros estudios taxonómicos modernos de la flora macaronésica. El Indiferente 20:2–11. https://www.macaronesian.org/assets/files/file-f4371694886acd.pdf

Pico Graña B, Aznar Vallejo E, Corbella Díaz D (2003) Le Canarien manuscritos, transcripción y traducción. Instituto de Estudios Canarios

Rodríguez O, Wildpret W (2005) Evolución del Paisaje Vegetal del Parque Nacional del Teide. Organismo Autónomo del Parques Nacionales. Ministerio de Medio Ambiente

Thirlwall MF, Singer BS, Marriner GF (2000) 39Ar/40Ar ages and geochemistry of the basaltic shield stage of Tenerife, Canary Islands, Spain. J Volcanol Geotherm Res 103:247–297

Watts AB, Peirce C, Collier J, Dalwood R, Canales JP, Henstock TJ (1997) A seismic study of lithospheric flexure in the vicinity of Tenerife, Canary Islands. Earth Planet Sci Lett 146(1997):431–447

Wildpret W, Martín Osorio VE (2006) Los paisajes vegetales de las Islas Canarias a la llegada de los normandos a principios del siglo XV *in* Le Canarien: Retrato de dos mundos/coord. Eduardo Aznar, Dolores Corbella Díaz, Berta Pico Graña, Antonio Tejera Gaspar, vol 2, 2006 (Contextos), pp 91–122. ISBN 84-88366-59-0

Chapter 2
Physical Environment

2.1 Geographic Framework

The Canary Islands are an Atlantic oceanic archipelago off NW Africa located between 13°20′–18°10′W and 27°37′–29°25′N. It is formed by seven main inhabited islands: Lanzarote, Fuerteventura, Gran Canaria, Tenerife, La Gomera, La Palma and El Hierro. To these are added the small islets and rocky components of the so-called "Chinijo" Archipelago, located in its Eastern sector to the North of Lanzarote; from biggest to smallest these are La Graciosa, Alegranza, Montaña Clara, Roque del Este and Roque del Oeste; the small islet of Lobos between Fuerteventura and Lanzarote, and a series of offshore rocks of small surface area.

The scope of the present manuscript is defined by the territory included within the boundaries of the Teide National Park (Fig. 2.1), created by Decree on January 22, 1954 (BOE, no. 35, of February 4, 1954), reclassified as such by Law 5/1981, of March 25 (BOE, no. 90, of April 15, 1981) in compliance with the provisions of Law 15/1975, of May 2, on Protected Natural Areas; and whose last extension is determined by the *Resolution of October 14, 1999*, of the General Secretariat of the Environment (BOE, no. 310, of December 28, 1999).

The Teide National Park is located on the central part of the island of Tenerife, the largest and highest of the seven islands that make up the Canary Islands which is situated in the Atlantic Ocean (Fig. 2.2), approximately 100 km from the northwestern coast of the African continent (between Morocco and Western Sahara). The "autonomous community" of the Canary Islands belongs to the Spanish Kingdom (one of the member countries of the European Union), from which it is located approximately 1300 km away. This makes these islands one of the Outermost Regions (ORs) of the European Union together with the autonomous regions of Portugal, Madeira and Azores, also a part of the Macaronesian biogeographic region, and a number of other French departments and communities: Reunion, Martinique, Mayotte, Guadalupe, French Guayana and St Martin.

V. E. Martín Osorio et al., *Flora and Vegetation of The Teide National Park*,
https://doi.org/10.1007/978-3-031-88400-9_2

Fig. 2.1 View of the Teide volcano and the vegetal landscape surrounding the National Park. (Photo credit: Wolf-Hermann Wildpret Martín)

The surface area of the Teide National Park has officially increased, after the last extension, to some 18,990 hectares (approximately 9.50% of the total area of the island). However, the boundaries used to delimit the entire protected area after the most recent legal rearrangements take us to a total surface area of 21,318 hectares, since the land belonging to the municipalities of Güímar and Arico, will be acquired by the National Park and subsequently included in the near future.

Furthermore, this territory reaches a truly considerable altitude, thus characterising the high central peak of Tenerife, since it extends from 1600 m above sea level at its lowest point (north of the slopes of Teide itself—"Los Castillos"—municipality of Icod de Los Vinos), to 3714,5 m above sea level ("Pico del Teide"—municipality of La Orotava) (Table 2.1), this being the highest point in the entirety of Spain. Its average altitude is around 2000–2200 m above sea level.

The Teide National Park is crossed by a major road network, which is made up of the following access routes (Fig. 2.3):

- **TF-21 Highway**

  North Access: La Orotava—Portillo de la Villa—Boca Tauce
  South Access: Granadilla—Vilaflor—Boca Tauce

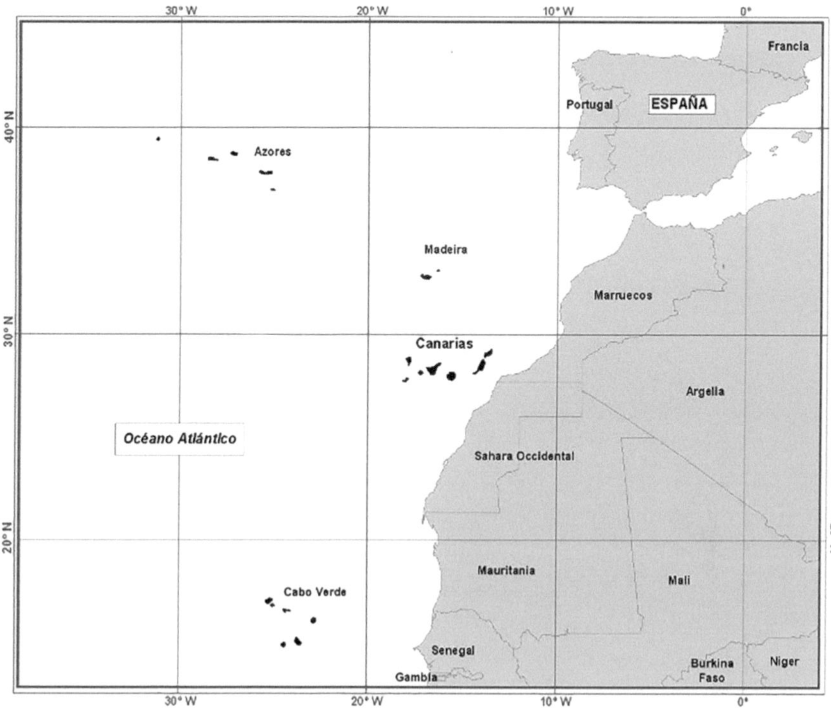

Fig. 2.2 Situation of the Canary Archipelago in the Atlantic-Macaronesian context

Table 2.1 Surface and Maximum Altitude by islands of the Canary Archipelago

Province	Island	Surface (km^2)	Maximum Altitude (m.a.s.l)
Santa Cruz de Tenerife	**Tenerife**	**2034,38**	**3714,5**
	Teide National Park	*189,9*	
	La Palma	708,32	2423
	La Gomera	369,76	1487
	El Hierro	**268,71**	1501
Las Palmas de Gran Canaria	**Gran Canaria**	1560,10	1949
	Fuerteventura	1659,74	807
	Lobos	4,58	127
	Lanzarote	845,94	**671**
	La Graciosa	29,05	266
	Alegranza	10,30	289
	Montaña Clara	1,48	256
Canarias		**7446,95**	**3714,5**

Source: http://www.gobiernodecanarias.org/istac

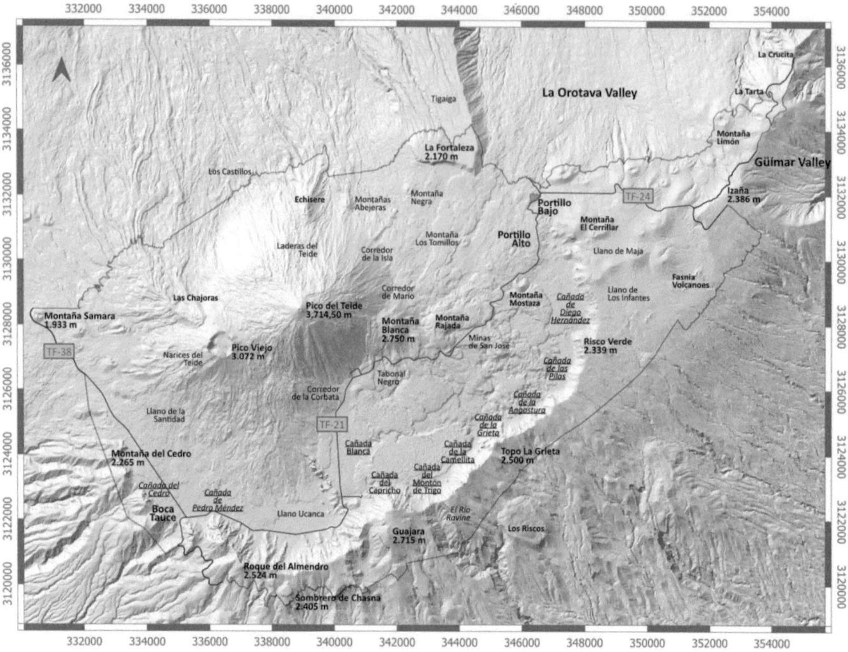

Fig. 2.3 Toponyms and relief of Teide National Park

– **TF-24 Highway**

East Access: La Laguna—Portillo de la Villa

– **TF-38 Highway**

West Access: Chío—Boca Tauce

The municipalities into which the island of Tenerife is administratively divided and which officially have land that forms part of the Teide National Park are twelve (Table 2.2): La Orotava, Guía de Isora, Icod de los Vinos, Fasnia, Granadilla, La Guancha, San Juan de la Rambla, Adeje, Santiago del Teide, Los Realejos, Vilaflor and Garachico. To these should be added Güímar (325 hectares) and Arico (2003 hectares), whose plots are still in the protection zone (Fig. 2.4) and will be incorporated into the National Park in the future.

According to data extracted from the electronic application of the Government of the Canary Islands, Spatial Data Infrastructure of the Canary Islands (IDE Canarias) 2015, our area of study is also affected by the following protection figures:

- **Network of Protected Natural Areas**—Law 12/1994, of December 19, on Natural Areas of the Canary Islands and Law 4/2017, of July 13, of the Soil and Protected Natural Spaces of the Canary Islands https://www.boe.es/eli/es-cn/l/2017/07/13/4/con (Fig. 2.5):

  – ***Teide Natural Monument.***

Table 2.2 Area of administrative participation of each municipality within the Teide National Park

Municipalities	Surface (has) within the Park	%
La Orotava	13748	72.4
Guía de Isora	1683	8.9
Icod de los Vinos	1266	6.7
Fasnia	946	5.0
Granadilla	378	2.0
La Guancha	234	1.2
San Juan de la Rambla	223	1.2
Adeje	201	1.1
Santiago del Teide	143	0.7
Los Realejos	79	0.4
Vilaflor	59	0.3
Garachico	20	0.1

Source: http://www.gobiernodecanarias.org

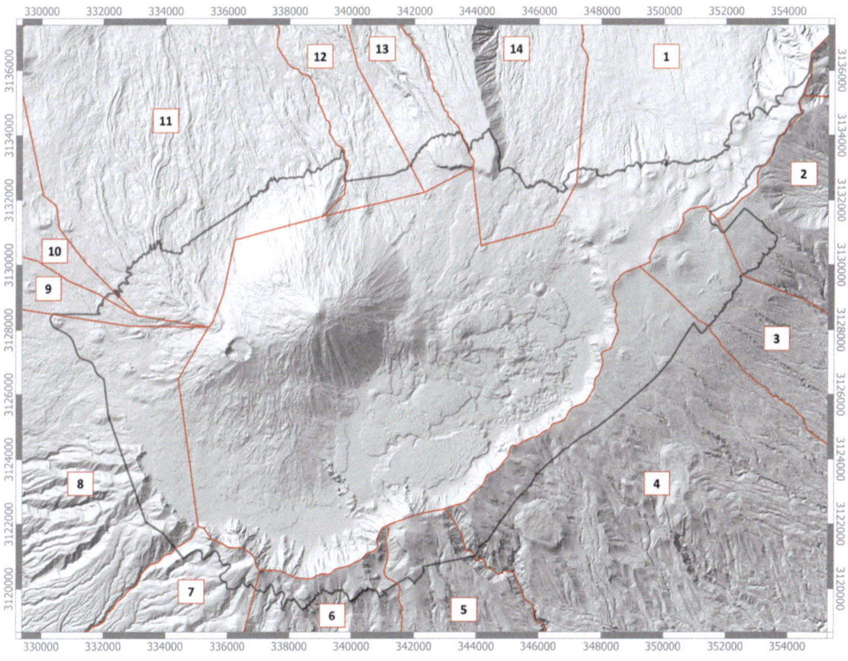

Fig. 2.4 Municipalities with land within the boundaries of the Teide National Park: (1) La Orotava, (2) Güímar, (3) Fasnia, (4) Arico, (5) Granadilla de Abona, (6) Vilaflor, (7) Adeje, (8) Guía de Isora, (9) Santiago del Teide, (10) Garachico, (11) Icod de los Vinos, (12) La Guancha, (13) San Juan de la Rambla and (14) Los Realejos

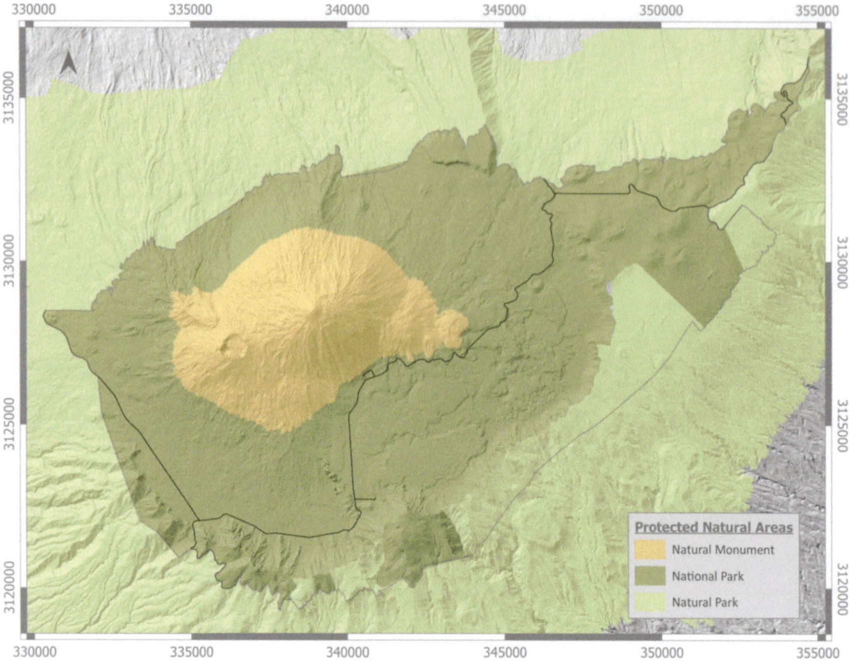

Fig. 2.5 Categories of the Network of Protected Natural Areas of the Canary Islands, according to Law 12/1994, of December 19, which affect the study area

- *Teide National Park*, which is also, by its own definition and according to Law 11/1990, of July 13, on the prevention of Ecological Impact, an ***Area of Ecological Sensitivity.***
- ***Corona Forestal Natural Park.***

- **Special Conservation Zone (ZEC)** part of the Natura 2000 Network in the Canary Islands (Decree 174/2009, of 29 December): affects the entire surface of the National Park.
- **Special Protection Zone for birds (ZEPA)**: according to Community Directive 79/409/CEE and amendments: affects the entire surface of the National Park.
- **Public Utility Mountains (MUP) and High Risk Forest Fire Zones (ZARI)** (Fig. 2.6)—General Directorate of Nature Protection of the Vice-Ministry of the Environment of the Government of the Canary.

Other international recognitions that the Teide National Park also includes:

- ***Natura 2000 Network.***
- ***European Diploma (Council of Europe) since 1989.***
- ***Natural World Heritage Site (UNESCO) since 19 July 2007.***

In addition, since 2002 and on a voluntary basis, this protected area has implemented an Environmental Management System based on the UNE-EN ISO

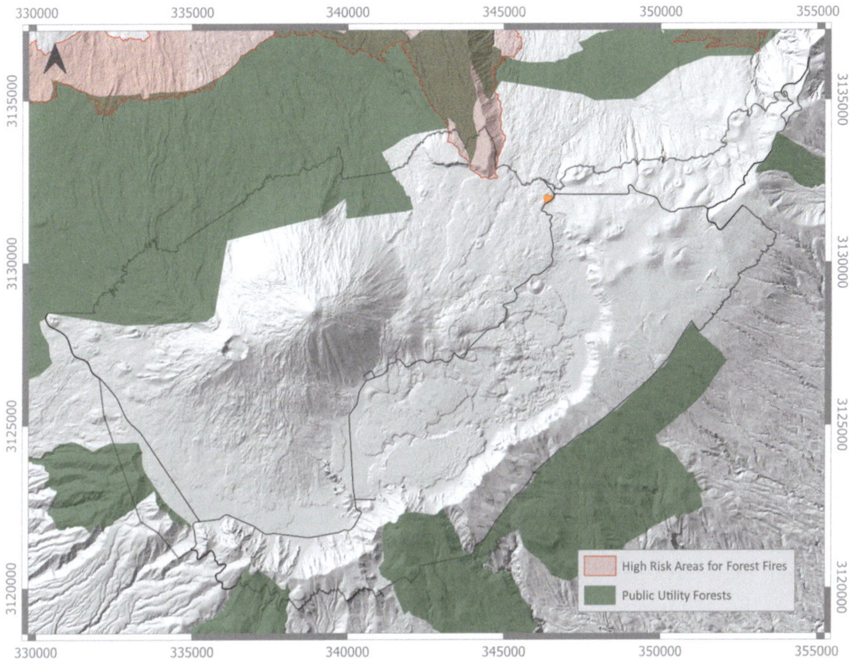

Fig. 2.6 Delimitation of the areas classified as High Risk Forest Fire Zones (ZARI) and as Public Utility Mountains (MUP) that affect our area of study

14001:2004 Standard and Regulation (EC) No. 1221/2009 of the European Parliament and of the Council, of 25 November 2009, on the voluntary participation of organizations in a Community eco-management and audit scheme (EMAS) (Durbán Villalonga and Reverón Gómez 2011). This is the first Spanish Protected Natural Area to obtain the Environmental Management Certificate dated November 8, 2005.

As an instrument for its management, it has the Master Plan for the Use and Management of the Teide National Park according to Decree 153/2002 of October 24 (BOC No. 164, of December 11, 2002). This management became the responsibility of the Autonomous Community of the Canary Islands starting January 1st, 2010, according to Royal Decree 1550/2009, of October 9 (BOE No. 270, of November 9, 2009), specifically of the Ministry of the Environment and Territorial Planning, according to Decree 226/2009, of December 3 (BOC No. 245, of December 16, 2009). Its governing body is constituted through the so-called Board of Trustees of the Teide National Park, which meets in ordinary or extraordinary plenary sessions and is made up of various representatives of the different Public Administrations (of the State and the Autonomous Community of the Canary Islands), competent Federations and/or Associations, representatives of professional-scientific and/or technical groups, etc. The different agendas, debates, opinions and decisions developed during the different sessions are recorded in the minutes book of the Board of Trustees of the Teide National Park.

2.2 Geological and Geomorphological Framework

The different elements that make up the lithological-structural landscape of the Teide National Park are eminently of volcanic origin. As are those that have led to the construction of the island of Tenerife and the entire Canary Archipelago as a whole.

The Archipelago is made up of a set of independent volcanic islands that began to emerge in the form of small "seamounts" (Fig. 2.7), on a flat and rigid ocean floor, whose age is estimated to be between 165 and 175 Ma (Jurassic). These seamounts began to grow below sea level, until, due to the successive accumulation of primary materials, they emerged in the form of islands, thus exceeding the first 3000–4000 m of underwater depth, and thus triggering the beginning of subaerial volcanism phenomena.

There are various theories throughout our history that have attempted to explain the causes that could have led to the formation of this island group. In order to understand them, we must take into account that volcanic processes arise as a result of the manifestation of an enormous energy found within our planet and that, in addition, they form part of a much more global and complex system that explains many of the geological phenomena that develop on the Earth's crust. That is, the continental drift system and plate tectonics.

According to this model, the outermost layer of the Earth (lithosphere) is fragmented into a set of plates of very different sizes, which are in continuous movement, one with respect to the other, and whose dynamism is determined by the "floating" effect that they exert on the "plastic" materials that constitute the first layer of the Earth's mantle (asthenosphere). The majority of the phenomena of tectonic and magmatic activity on the entire planet (earthquakes and volcanoes) are located at the limits of these plates or in their contact zones. However, within this scheme, the Canary Islands are a special case, since if it is true that they are located

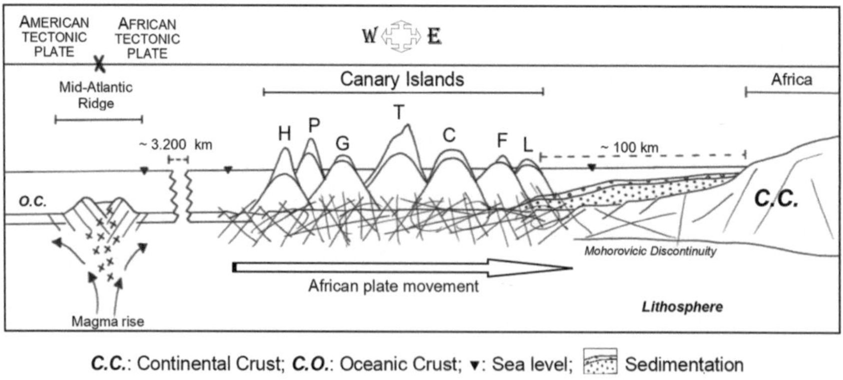

Fig. 2.7 Approximate diagram of the geodynamic situation of the Canary Islands. H: El Hierro, P: La Palma, G: La Gomera, T: Tenerife, C: Gran Canaria, F: Fuerteventura, L: Lanzarote

in a transition area between oceanic crust and continental crust, they are not close to any zone considered as "between plates". On the contrary, they are located inside the Afro-Atlantic plate (intraplate position), about 3220 km from the so-called mid-Atlantic submarine ridge (boundary between the American and African plates).

But especially, the location of the islands is a geodynamic zone subject to intense tectonic stresses, an area of complex movements of tearing, distension and compression, hence the existence of folds and an important network of fractures that appear in the seabed between islands and in the underlying crust.

A large amount of radiometric dating (K/Ar and Ar/Ar) measurements has scientifically demonstrated the progressive decrease in the age of the islands that make up the Canary Archipelago in a west-east direction, a characteristic that is fundamental in the hot spot models for other oceanic islands in an intraplate situation. These studies have also differentiated three phases in the construction of the Canary Island buildings, which also occur in Hawaii, and which have become the modern basis of volcanic stratigraphy in oceanic hot spot islands (Carracedo 2006):

- **Juvenile growth phase**: Phase of greatest volcanic production (90% of the total material emitted), in which the island is located in the same vertical of the hot spot and takes the form of a shield (cases of La Palma and El Hierro).
- **Eruptive rest phase**: Due to the displacement of the lithospheric plate and, therefore, the gradual distancing of the island, the influence of the hot spot begins to diminish. In addition, the volume of the building itself progressively makes it difficult for magma to reach the surface (case of La Gomera).
- **Post-erosive rejuvenation phase**: Volcanism restarts, although in a much smaller form, both in volume and frequency of eruptions. In this case, Tenerife would be in the earliest phase, Gran Canaria in the late phase, and Fuerteventura and Lanzarote in a senile phase of advanced erosive dismantling, although in the latter, there have been recent eruptive episodes of some importance.

Formation of Las Cañadas and the Teide—Pico Viejo Building

The formation of the island of Tenerife, as a volcanic and oceanic island, has been quite complex and particular, especially if we refer to the episodes that occurred in the last 3–4 Ma, a period in which its central part has not only achieved enormous topographical prominence, even at the level of the entire Archipelago, but has also been the trigger of very diverse and spectacular volcanic processes, very rare in this type of buildings, of submarine origin.

This makes the island of Tenerife, not only on a European scale but also worldwide, a rather attractive and fascinating field of study, especially for scientists in the branches of volcanology, stratigraphy, petrology and geology and geomorphology in general.

Thanks to the evolution that geochronological (14C, K/Ar, 39Ar/40Ar) and geostratigraphic studies have experienced in particular, regarding the establishment of absolute and/or relative ages of the different materials and formations of the summit, a model has been developed that has come closer and closer to the true reconstruction of the different phases of formation of the centre and culmination of the island. Even so, even today, there are gaps to be resolved, such that research articles continue to be published one after the other continuously, especially in recent years.

There have been several hypotheses about the formation of Las Cañadas throughout history. Today, the one that seems to be most accepted is the one that states that the culminating part of the island of Tenerife is the result of the achievement of four large constructive episodes that originated large-scale composite volcanic buildings: "Cañadas I", "Cañadas II", "Cañadas III" and "Conjunto Teide-Pico Viejo". These did not occur immediately, but rather were interspersed temporally, sometimes with important events of a destructive nature (vertical collapse or lateral sliding phenomena), sometimes with episodes that were no longer so titanic where morphostructural activity gave way to the development of works of a more morphosculptural nature, which continued with greater or lesser persistence over time.

Cañadas I Building (3.5 to 2.7 Ma)

The oldest materials found in The Wall (also La Pared) of Las Cañadas (the semicircular cliff encircling the National Park and extending along the several "Cañadas" in Fig. 2.3) date back to about 3.5 Ma, so it is assumed that the construction phase of the Cañadas I Building began from that period. There is evidence of spatial overlap between this and the pre-existing structures (Massifs), although according to the geochronological data obtained by several authors, in the case of the Roque del Conde complex, there is a gap of approximately 3 Ma between the youngest materials of this massif and the oldest of Las Cañadas. At the same time, it is known that there was contemporaneous activity between the latter and the Anaga Massif (a vein network of similar age).

The outcrops or remains of this first large central building can be observed in different sectors of La Pared and, outside the Las Cañadas depression, at certain levels of some ravines or inside certain galleries.

Cañadas II Building (2.4 to 1.3 Ma)

Following the landslide that caused the disappearance of a good part of the structure of the Cañadas I Building, new episodes of volcanic construction began to occur, thus constituting the new skeleton of the Cañadas II Building, now located more to the S and W of the previously demolished building. Supposedly, it did not extend to the E and SE, possibly due to the existence of pre-existing structures or remains of residual relief belonging to the formation phase itself. This is known because the lower levels of the following eruptive phase (Cañadas III) rest directly on the upper levels of Cañadas I. These materials, which initially began to rest on those corresponding to Cañadas I, are 2.3 to 2.2 Ma old, which gives us a difference between the last episode of construction of the first building and the beginning of Cañadas II of about 300,000–400,000 years, thus obtaining a period lacking volcanostratigraphic evidence (supposed period of eruptive calm).

The volcanic elements generated in this second phase of formation of the Cañadas building can be found in different outcrops in the S and W part of the island, in the western half of the current Cañadas Wall and to the N, in the Tigaiga massif. As we said before, no remains have been found in eastern sectors. Within this same episode, two very different moments of activity have been interpreted, each of them with different outcrop units.

Cañadas III Building (1.20 to 0.15 Ma)

This episode of construction of the Cañadas building is separated from the previous one by a period of approximately 100,000 years, during which a possible cessation or impoverishment of the eruptive activity is probably responsible for the existence of a fairly generalised level of discordance between the youngest strata of the Cañadas II building and the oldest ones of Cañadas III.

Teide Formation—Pico Viejo

There is hardly any information about the first eruptions of Teide that supposedly began to fill the valley of Icod or Icod—La Guancha, since this first phase was completely covered by the recent eruptions that have finished drawing the current profile of this volcano.

The construction phase of Teide ended about 30,000 years ago. From this moment on, there seems to be an interruption in the leading volcanic patterns, leading to a domatic volcanism peripheral to the volcano. Until then, Teide would have become a stratovolcano about 3500 m high, with an open crater to the E whose diameter could have reached 900 m.

The last eruption of Teide is dated to about 1150 ± 140 years ago and corresponds to the formation of the terminal cone and the emission of the phonolitic flows known as **black lavas** (Fig. 2.8). This has often been related to the eruption witnessed by Christopher Columbus on the island of Tenerife on his passage through

Fig. 2.8 Black lava on the Teide volcano May 31, 2009. (Photo credit: Victoria Eugenia Martín Osorio)

the Canary Islands in 1492, but the truth is that, according to the dating carried out, it should be attributed to the eruption of the Boca Cangrejo Volcano, located on the Abeque or NW Ridge near the town of Santiago del Teide. After the cessation of activity of the Teide edifice about 30,000 years ago, since its materials do not seem to overlap, the eruptions of the Pico Viejo volcano begin, which emerges adventitiously from the stratovolcano. After these episodes, the volcano also remains practically inactive, except for the last eruptions, which seem to have more relation to manifestations of the NW Ridge.

With the culmination of the **Pico Viejo volcano**, the main phase of formation of the central complex of Tenerife is completed. From this moment on, the most intense volcanism occurs in areas of the Ridge (especially in the NW) and around a peripheral line to the Teide—Pico Viejo complex (Fig. 2.9).

A summary of the volcanological characteristics of the Teide National Park is presented below: Types of Magmas (Table 2.3), Types of Pyroclastic Materials (Table 2.4) and Types of Lava Materials (Table 2.5).

Fig. 2.9 Pico Viejo volcano with 3072 m of altitude. 3 August 2006. (Photo credit: Victoria Eugenia Martín Osorio)

Table 2.3 Summary diagram of the volcanic characteristics of Teide National Park

TEIDE NATIONAL PARK	
VOLCANIC ORIGIN	
Alkaline magmas ~ 70–100 km deep: upper layers of the Earth's mantle - Canary Islands -	
Basic magmas Poor in Silica (SiO_2) - *Canary Islands* -	**Intermediate magmas** Rich in Silica (SiO_2) - *Tenerife, Las Cañadas* -
Primary magma	Magma differentiated from an initial magma (mechanisms of magmatic differentiation)
Fluid magma	Viscous magma
900–1200 °C	700–900 °C
EFFUSIVE CHARACTER Low explosivity	EXPLOSIVE CHARACTER Low effusivity
Long-distance lava flows	Long-distance pyroclastic flows
Heavy material—High density	Lighter material—Low density
Eruption type: Fissure—Strombolian	Eruption Type: Plinian
Origin: NW Ridge and NE Ridge	Origin: Central area of the island
MAGMATIC ROCKS	
Volcanic Rocks (Flows—Pyroclasts)	**Plutonic Rocks** (Dikes—Pitons)
Magma consolidation on the surface	Magma consolidation in depth
Rapid cooling	Slow cooling
Minimum degree of crystallization	High degree of crystallization
Porphyritic texture of the rocks *Phenocrystals + Matrix*	Holocrystalline Texture: *Complete crystallization* Granular Texture: *Aggregate of crystals*
They may contain bubbles or vacuoles	They are compact rocks, without voids or bands

(continued)

Table 2.3 (continued)

BASALT	**PHONOLITE**	**TRACHYTE**	**OBSIDIAN**	**GRANITE**	**SYENITE**	**GABRO**
Dark matrix Microcrystalline	Light matrix Microcrystalline	Glassy matrix	–	Holocrystalline granule	Holocrystalline granule	Holocrystalline granule
Olivine Pyroxene	Pyroxene Feldspars	Feldspars	Vitreous Non-crystal	Quartz, Feldspar Mica	Light-coloured minerals—Feldspars	Dark-coloured minerals

Table 2.4 Classification of products and basic morphologies of volcanic activity

PYROCLASTICS **MATERIALS** Highly fragmented aerial projection products generated from magma in the explosive episodes of an eruption, and which may remain molten or be in a solid state at the outlet			
TRANS-PORT MECHANISM	FALLING PYROCLASTS: Individual movement of particles Ballistic trajectories Fragment by fragment deposition and according to the energy of the eruption and the intensity and direction of the wind. Selective deposition or Granoselection: ↓ weight ⇔ ↑ displacement distance.		FLOW PYROCLASTS (500–800 °C) Trajectories according to topography Pyroclastic flow: At ground level, few mobilized lithics. Poorly classified deposit: poorly defined strata. Pyroclastic surge: Large amount of mobilized lithics. Turbulent flow. Well-classified deposit.
ACCORDING TO SIZE AND GENESIS	<2 mm	**ASHES** Powdery and/or sandy in nature High displacement capacity Highly explosive eruptions	**Kinerite tuff—kinerite**
	2–30 mm	**LAPILLI** Picón—basic; Jable—acidic (PUMICE) Irregular and very porous shapes Eruptions of low—intermediate explosiveness Frequent in the formation of volcanic apparatus	**Tuff—pumice**
		ACCRETIONARY LAPILLI Rounded fragments formed from a condensation nucleus (crystal, water, dust, etc.) Covered with concentric layers of ash Hydro or phreatomagmatic eruptions	
	>30 mm	**SLAG** Very irregular Curly contours	**Agglomerate—gap**
		BOMBS Morphology ↔ Spins in the air + Impact when deposited Fusiform appearance, more or less rounded Clean bombs: Without fragments of other kinds Dirty bombs: With fragments of other kinds *Pele's Tears or Goddess of Fire (Hawaii)* *Flaminated bombs* *Fusiform or spindle-shaped bombs* *Spheroidal bombs, Geminated bombs* *Bread crust bombs, Cauliflower bombs* **BLOCKS** Angular or subangular shapes	

Source: Hernández Bolaños

Table 2.5 Classification of products and basic morphologies of volcanic activity

LAVA MATERIALS Magmatic material that comes to the surface in the form of a stream or lava flow, and that flows downhill along the slope and according to the pre-existing topography. They lose their volatile content through degassing processes
"*Pahoe—hoe*" lavas: Very fluid lavas that form a smooth crust when cooled.
"*Aa*" lava: Low-intermediate viscosity castings that present a very broken and irregular surface.
Block lavas: Lavas with higher viscosity. They present a large number of blocks (chaotic appearance)
Morphology of Lava Flows: Superficially: *Pahoe-hoe* lavas, (Fig. 2.10): Flow arches (cordate lavas), small lava escapes from the interior (gut lavas), or slab-shaped breaks (slab flows). *Aa* lavas: the so-called malpaíses. Internally: Network of retraction cracks that give rise to various types of disjunction: prismatic or columnar (due to cooling), slab-shaped (differences in flow speed), radial or rosette-shaped (channelling of lava flow) Occasionally: Superposition of flows, hornitos (expulsion of large gas bubbles), volcanic tubes, jameos, etc.

Source: Hernández Bolaños

Fig. 2.10 *Pahoe—hoe* lavas. Los Roques, January 2005. (Photo credit: Victoria Eugenia Martín Osorio)

2.3 Climatic and Topographic Context of Teide National Park

Tables 2.6 and 2.7 present a summary of the context of the main topographic and climatic conditions in the Teide National Park.

Table 2.6 Climatic and topographic context of the National Park

TEIDE NATIONAL PARK	
Regional and Insular context **Tenerife, Canary Islands** Altitude of Teide National Park: 1.600–3.718 m a.s.l.	
<u>Atmospheric stratification</u>: *Upper layer:* Hotter and drier air (Teide National Park). *Temperature inversion* (~ 800–1500 m a.s.l.): warmer air layer resting on top of cooler air, causing a temperature increase above a certain height; upper limit of the "sea of clouds", which influences adjacent areas but doesn't usually enter the National Park. *Lower layer:* colder and more humid air. <u>Altitudinal climatic contrasts</u>: differences in climate conditions between the coastal layers, medium altitudes and high altitudes. <u>Lack of direct marine influences</u> on the high mountain climate of Tenerife, lending to an increased climatic harshness.	
Principal characteristics of the Teide National Park climatic elements (based on the analysis of the data of Izaña Observatory weather station)	
Temperatures	Mean Annual Temperature: ~9 °C Mean Maximum Temperature: ~20–25 °C Mean Minimum Temperature: ~10 °C Maximal Temperature: 28,6 °C Minimal Temperature: −10 °C Daily temperature oscillation: ~10 °C (August) Coldest months: January and February (mean ~ 4,1 °C) Coldest month: January Hottest months: July and August (mean ~ 18,4 °C) Hottest month: July Annual temperature amplitude: ~14,3 °C Months with the highest probability of frost: October—June Months with the highest number of frost days: December and January Frost-free months: July, August and September
Precipitations	Related mostly with marine polar air intrusions and "cold drops" Mean annual precipitation: ~540,6 mm Variable precipitation regime Precipitation intensity: ~40 mm/day - Wettest period: November—January Driest Period: June—August Number of days with precipitations: 43 days Number of days with snow: 11–12 days Until 1.500 m a.s.l. (northern slopes of the National Park): positive precipitation gradient with altitude. (<600–700 mm per year) Above 1.500 m a.s.l.: negative precipitation gradient with altitude (~300–500 mm per year)

(continued)

Table 2.6 (continued)

TEIDE NATIONAL PARK	
Humidity	Mean annual relative humidity: ~43% (winter) y ~ 23% (summer) Most humid months: Novembre—February
Irradiance	Usually, clear skies Clean and clear atmosphere Intense solar irradiance Highest number of sun hours: Mai—August Duration of maximum mean solar irradiance: 3.448,5 hours/year—80% of the theoretically highest possible (highest value for Spain)
Winds	Dry dominant winds from the northwest (*antitradewinds*) Maximum frequency: ~50–60% (June) Minimum frequency: July—August Western or southwestern fluxes: arrival of Atlantic disturbances with precipitations and faster winds Southern fluxes: Intrusions of Saharan winds
General characterisation of Teide National Park's Climate	Climate with characteristics: Of high mountains Continental Temperate Mediterranean

Table 2.7 Geographic context

TEIDE NATIONAL PARK	
Geographic context **Topographic differences: altitude, orientation and exposition.** Topographic configuration of the Teide National Park	
Differences between the interior and exterior of the Las Cañadas depression Differences between eastern and western Cañadas Existence of great reliefs: Teide-Pico Viejo edifice and the Pared (Wall) of Las Cañadas	
Main climatic differences within the Teide National Park	
Exterior of the Las Cañadas depression	**Northern and northeastern sides**: Influence of the "sea of clouds" in the lowest areas of the National Park. Occasionally, said sea of clouds manages to flow into the Las Cañadas depression More exposed to dominating winds Lesser risk of frost Higher precipitations: ~600–700 mm annually Area of the *Dorsal*, around Izaña, without large reliefs, is very exposed to the winds, much colder, even than in the summits of the Pared de Las Cañadas **Southern sides of the Pared de las Cañadas**: Much warmer (exposition)

(continued)

Table 2.7 (continued)

TEIDE NATIONAL PARK	
Interior of the Las Cañadas Depression	Local climate with more continental characteristics Increase in temperature harshness Maximum temperature: 30,6 °C (1915) Minimum temperature: −16,1 °C (1915) Increase in daily temperature amplitude: 15 °C (annual mean), 18,9 °C (max.—July) y 14,7 °C (min.—January) Diurnal overheating of the interior of Las Cañadas, especially in its central part: base of the southern slopes of the Teide-Pico Viejo edifice (2.150–2.500 m a.s.l.)—scarce influence of winds on high solar irradiance Western sector of Las Cañadas (2.030–2.300 m a.s.l.): daily temperature amplitude less stark due to increased influence of the winds Oriental sector of Las Cañadas (2.080–2.250 m a.s.l.): Higher influence of the northern and northeastern winds, and therefore, with colder temperatures Temperatures <0 °C are very frequent Increased likelihood of frost, even during the summer Thermic effect akin of that of a valley: strong nocturnal cooling, colder valley bottom compared to the higher summits of the adjacent slopes *Llano de Ucanca* (2020–2080 m a.s.l.): During the night is the coldest sector, with very high temperature amplitudes and sparse influence of winds Lower precipitation regime: ~300–500 mm annually
Altitudinal differences due to the existence of great reliefs	Generation of mountain winds between valley bottoms and the summits Modification of predominating wind patterns due to obstructions *Annual temperature gradient:* ~0,43–0,46 °C/100 m: *Winter*: 0,53–1,23 °C/100 m *Summer*: 0,13–0,30 °C/100 m *S & E slopes:* 0,43 °C/100 m *N & NE slopes:* 0,46-0,48 °C/100 m Possibility of *föhn effect* forming depending on the mountain side At the base and in the interior slopes of the Pared de Las Cañadas (2.200–2.400 m a.s.l.), less solar irradiance hits (orientation predominantly N & NW); fresher area during the day compared to the rest Summits of the Pared de Las Cañadas (2300–2700 m a.s.l.), a decrease in temperatures occurs due to the higher elevation and higher incidence of the predominating winds

2.4 Bioclimatic Study

Conceptual and methodological frameworks

According to Rivas-Martínez et al. (1999, 2007, 2011), bioclimatology is the ecological science which studies the relationship between the climate and the distribution of living beings and the communities they form across the planet. In the case of the plant kingdom (vascular plants), its main objective consists in determining the relationship between the mean climatic values, principally of temperature and precipitation, with the geographic distribution ranges of species and of vegetation; incorporating later information about biogeocoenosis and phytosociology (structured in series, geoseries and geopermaseries of vegetation).

With the intention of being able to be used at a global scale, Prof. Dr. Salvador Rivas-Martínez elaborated, across the years, his so-called Worldwide Bioclimatic Classification System (Rivas-Martínez et al. 2011), which in the last decades suffered numerous modifications and adaptations.

In this worldwide Bioclimatic Classification System, the world is divided in:

- **5 Macrobioclimates**: The uppermost classification unit. Related with the great types of climate, biomes and biogeographic regions of the planet: *Tropical, Mediterranean, Temperate, Boreal* and *Polar.*

For determining the different units of bioclimate the monthly values of the following parameters are used:

- **Precipitation.**
- **Temperature**.

Out of which the following are calculated:

- **Bioclimatic indices,** principally the following ones:
  - *Simple Continentality Index (Ic).*
    - Describes the amplitude in temperature between the mean temperatures of the hottest and the coldest month of the year in degrees centigrade. Higher amplitudes, continental climates; lower amplitudes, oceanic climates.
  - *Thermicity index (It)—Compensated thermicity index (Itc).*
    - Describes the intensity of cold while accounting for mean annual temperature. *Itc* compensates for particularly strong oceanic or continental climates.
  - *Ombrothermic index (Io)—Ombrothermic index of summer months (Ios).*
    - Describes the relationship between annual precipitation and temperature, therefore accounting for evapotranspiration rates and water availability to the vegetation. Low values for dry climates, higher values for more humid climates.

In Rivas-Martínez et al. (2011), a dichotomous key for determining the macrobioclimate, bioclimate and bioclimatic variants of a territory are provided. As well as the formulas and tables for performing detailed bioclimatic analysis of any part of the world. Here we present the most basic and important calculations for performing a diagnosis of the bioclimate of the Canary Islands and the Teide National Park in particular.

2.4.1 Bioclimatic Indices

Simple continentality index (Ic) (Table 2.8)

$$\mathbf{Ic = Tmax - Tmin}$$

Tmax: *Mean temperature of the hottest month of the year in °C.*
Tmin: *Mean temperature of the coldest month of the year in °C.*

Annual ombrothermic Index (Io)

$$\mathbf{Io = (Pp / Tp)^{*}\ 10}$$

Pp: *Annual positive precipitation:* based on the idea that mostly precipitation in liquid form is relevant for plant growth, it indicates the sum of precipitation in months with mean temperature >0 °C. In case of all months having mean temperature >0 °C it is equivalent to the annual precipitation.

Tp: *Annual positive temperature:* sum of mean monthly temperatures of those months where mean temperature >0 °C. If all months have a mean temperature 0 °C, then it can also be obtained by multiplying the mean annual temperature by 12. Regardless, the result is then multiplied by 10 to express it in tens of degrees.

Table 2.8 Types, subtypes and levels of simple continentality, which have been recognised for Earth

Types	Subtypes	Levels	Values
1. Hyperoceanic (0–11)	Ultrahyperoceanic	1.1a.Strong	0–2,0
		1.1b. Weak	2,0-4,0
	Euhyperoceanic	1.2a.Strong	4,0-6,0
		1.2b. Weak	6,0-8,0
	Subhyperoceanic	1.3a. Strong	8,0–10,0
		1.3b. Weak	10,0–11,0
2. Oceanic (11–21)	Semihyperoceanic	2.1a. Strong	11,0–12,0
		2.1b. Weak	12,0–14,0
	Euoceanic	2.2a. Strong	14,0–15,0
		2.2b. Weak	15,0–17,0
	Semicontinental	2.3a. Weak	17,0–19,0
		2.3b. Strong	19,0–21,0
3. Continental (21–66)	Subcontinental	3.1a. Weak	21,0–24,0
		3.1b. Strong	24,0–28,0
	Eucontinental	3.2a.Weak	28,0-37,0
		3.2b. Strong	37,0–46,0
	Hypercontinental	3.3a. Weak	46,0-56,0
		3.3b. Strong	56,0–66,0

From Rivas-Martínez et al. (2011)

Summer ombrothermic Index (Ios) (Table 2.9)

$$\mathbf{Ios = Ps / Tps}$$

Ps: *Summer precipitation:* like Pp, sum of the mean precipitation of the summer months: June, July and August in the northern hemisphere; December January and February in the southern hemisphere.

Tps: *Positive summer temperature:* positive temperature (as Tp) of the summer months, expressed in tens of degrees.

- $\boldsymbol{Ios_1}$: *Ombrothermic Index* of the hottest month of the summer trimester.
- $\boldsymbol{Ios_2}$: *Ombrothermic Index* of the hottest bimester of the summer trimester.
- $\boldsymbol{Ios_3}$: *Ombrothermic Index* of the summer trimester.
- $\boldsymbol{Ios_4}$: *Ombrothermic Index* of the summer trimester and the month directly preceding it. Relevant for border regions between temperate and mediterranean Macrobioclimates (Rivas-Martínez et al. 2011).

Thermicity Index (It)

$$\mathbf{It = (T + m + M) \times 10.}$$

Table 2.9 Ombrothermic values for the Ombric types, horizons (subtypes) recognised for Earth

Ombric types	Ombric horizons	Abr.	Io, Iom
1. Ultrahyperarid	1a. Lower ultrahyperarid	uhai	0,0-0,1
	1b. Upper ultrahyperarid	uhas	0,1-0,2
2. Hyperarid	2a. Lower hyperarid	hai	0,2-0,3
	2b. Upper hyperarid	has	0,3-0,4
3. Arid	3a. Lower arid	ari	0,4-0,7
	3b. Upper arid	ars	0,7-1,0
4. Semiarid	4a. Lower semiarid	sai	1,0-1,5
	4b. Upper semiarid	sas	1,5-2,0
5. Dry	5a. Lower dry	sei	2,0-2,8
	5b. Upper dry	ses	2,8-3,6
6. Subhumid	6a. Lower subhumid	sui	3,6-4,8
	6b. Upper subhumid	sus	4,8-6,0
7. Humid	7a. Upper humid	hui	6,0-9,0
	7b. Lower humid	hus	9,0–12,0
8. Hyperhumid	8a. Lower hyperhumid	hhi	12,0–18,0
	8b. Upper hyperhumid	hhs	18,0–24,0
9. Ultrahyperhumid	9. Ultrahyperhumid	uhu	>24,0

From Rivas-Martínez et al. (2011)

T: *Mean annual temperature* (°C).

m: *Mean temperature of the minima of the coldest month of the year* (°C).

M: *Mean temperature of the maxima of the coldest month of the year* (°C).

Compensated thermicity Index (Itc) (Table 2.10)

$$\mathbf{Itc = It \pm C}$$

It: *Thermicity Index.*

C: *Compensation value.* In the extratropical zones of the world (north of parallel 23°*N* and south of parallel 23°S) the compensated thermicity index is used to equilibrate the excesses of cold during the winter in especially continental areas, or the excessive mildness of the winters in especially oceanic areas. This allows us to analyse, in a more comparable way, the continentality indices of different zones.

Values for the different *compensation values* (C_0–C_4) can be found in Rivas-Martínez et al. (2011), for the territory of the Canary Islands the only one needed is C_0 in areas with IC < 8, as there are no areas with IC > 18. For the calculation:

- If Ic < 8, then, Itc = It − C.
- If Ic = 8–18, then Itc = It.
- If Ic > 18, then Itc = It + C.

Table 2.10 Thermotypic horizons and corresponding *It, Itc* and *Tp* values for the mediterranean macrobioclimate

Thermotypic horizons	Abr.	It, Itc	Tp: Ic > 21, Itc < 120
Lower InfraMediterranean	lim	515 − 580	>2600
Upper InfraMediterranean	uim	450 + 515	2400–2600
Lower ThermoMediterranean	ltm	400 − 450	2250–2400
Upper ThermoMediterranean	utm	350 − 400	2100–2250
LowerMesomediterranean	lmm	285 − 350	1800–2100
UpperMesomediterranean	umm	220 − 285	1500–1800
LowerSupramediterranean	lsm	150 − 220	1200–1500
UpperSupramediterranean	usm	(120) − 150	900–1200
LowerOromediterranean	lom	–	675–900
UpperOroMediterranean	uom	–	450–675
LowerCryoroMediterranean	lcom	–	100–450
UpperCryoroMediterranean	ucom	–	1–100

From Rivas-Martínez et al. (2011)

2.4.2 *Bioclimatic Belts*

Within the worldwide bioclimatic classification system of Rivas-Martínez et al. (2011), bioclimatic belts correspond to climatic conditions which follow one after another along an altitudinal or latitudinal gradient. Their delimitation depends on the thermo- (*It, Itc* and *Tp*) and ombrotypic (*Io*) values. Each of those bioclimatic belts is associated with a typical kind of vegetation: the *vegetation belts.*

Bioclimatic belts are defined through the calculation of:

Thermotypes + Ombrotypes

Thermo- and Ombrotypic horizons: Within the different belts, subdivisions between the lower and upper halves of their thermic and ombric elements are differentiated. These subdivisions improve their relationship with vegetation.

The **bioclimatic diagnosis** or bioclimatic belts of a territory is constructed based on the information of the:

Name of bioclimate + denomination of the thermotype +denomination of the ombrotype

This diagnosis can be with or without information on the specific horizons within the belts.

Bioclimatic belts of the Teide National Park based on station data

With the intention of making a small approximation to the bioclimatic study based on Rivas-Martínez et al. (2011) of the Teide National Park, we have proceeded to make a search, localisation and recopilation of data of climatic stations within the area of the National Park. This takes into account, as in previous sections, previous work for the island of Tenerife in general and for the summits of Teide in particular by Del Arco-Aguilar et al. (1994, 2006). In order to infer quality data and to be able to assert climate-vegetation relationships, we limited this exercise to data of stations which were within the National Park as to avoid inferences made from data of stations further away.

With this purpose, in the year 2002, the office of the Teide National Park, handed us a dataset of data, especially temperature and precipitation of the meteorological stations distributed within the park. Thanks to this dataset, in that year, the study of a total of 32 meteorological stations within the bounds of the National Park was performed. Twenty-nine of these stations belonged to the National Meteorological Institute of Spain, today called AEMET (State Agency of Meteorology), while three were property of TRAGSA, the company which was responsible for providing management and maintenance services for the National Park.

As a first step, the exact position of the stations within the National Park and the exact coverage of data for monthly precipitation and temperature values was performed.

As previously stated by Del Arco Aguilar y Wildpret de la Torre (1994), we could also state that despite a fairly good coverage of meteorological stations, both within and immediately adjacent to the National Park (except on the northwestern sector, where none are found above 1600 m a.s.l), there is a big gap of available data on every single station. Except for Izaña, there were serious difficulties on finding data strings with a duration of more than 2 or 3 consecutive years. It was even harder to find such a dataset for which both temperature and precipitation data was available. This is a problem also pertaining to the main and most reliable station in Izaña (2.367 m a.s.l), which although having data entries since 1916, also presents some data gaps.

In view of this, of the 32 analysed stations a selection process was undertaken to choose stations based on two main criteria: their strategic location within the National Park and the number of years with meteorological data.

With regards to the distribution of stations, we aimed at having at least one station in each of the main sectors of the National Park, although without being able to get one for the north face of the Teide-Pico Viejo edifice.

In order to be able to meet the criteria stated before, we were forced to take stations with extremely short data ranges (2–3 years in some cases) while being aware of the risks this entails.

I any case, strictly speaking, the meteorological station in Izaña is the only reliable one for performing proper climatic analysis. The difficulty lies within extrapolating the results of the Izaña station analysis to the rest of Las Cañadas due to its very rugged topography and sudden changes in climate and microclimate.

In the end the selection fell on the following stations (Fig. 2.11):

- A total of seven stations for the thermicity index calculation **(It/Itc)** (Table 2.11). In order from west to east:
  - **Station C412C (181)**, *Montaña Samara* (*W*-Las Cañadas): 2 years (1991–1992).
  - **Station C406D (117)**, *Boca Tauce A* (*SW*-Las Cañadas): 3 years (1990–1992).
  - **Station C406G (121)**, *Parador Nacional de Turismo* (Centre-south part of Las Cañadas): 3 years (1990–1992).
  - **Station 374**, *Cable car tower* (upper southern slopes of Teide): 5 years (1994–1998).
  - **Station 370**, *Cañada de La Grieta* (*SE*-Las Cañadas): 3 years (1995–1997).
  - **Station 372**, *Pista Media Luna* (eastern half of Las Cañadas): 3 years (1995–1997).
  - **Station C430E (217)**, *Main Izaña station* (summits of La Dorsal-*NE*):
    - ***Station C430E (217a)*:** 37 years (1957–1994).
    - ***Station C430E (217b)*:** 3 years (1970–1972).

The aim of this subdivision of the Izaña station was to perform a validity check of analysis results of a short timeframe station when compared with a long timeframe station in the same place.

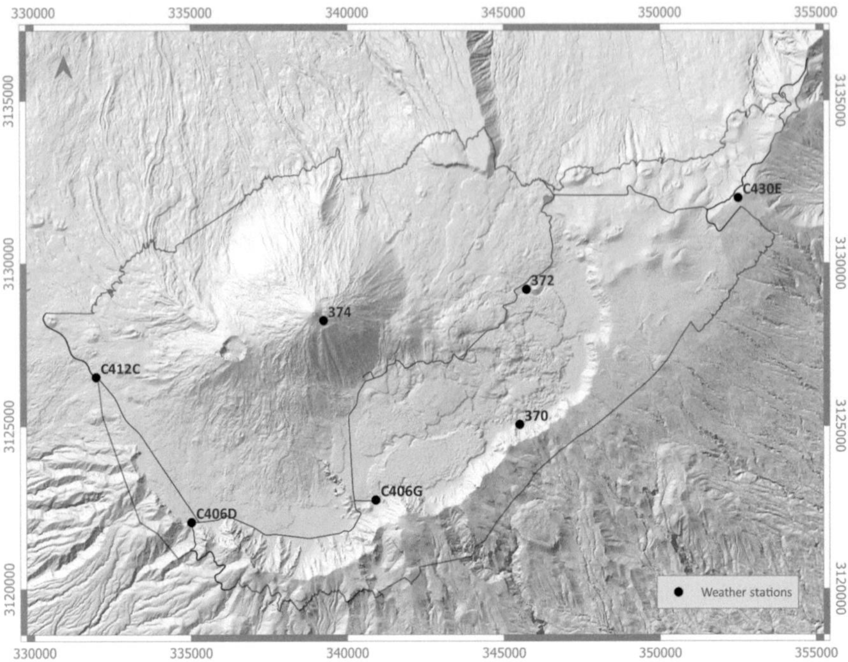

Fig. 2.11 Location of the meteorological stations chosen for the bioclimatology analysis. In orange: Botanical Garden Wolfredo Wildpret de la Torre

- A total of four stations for ombrothermic index calculations. In order from west to east:
  - **Station C412C (181)**, *Montaña Samara* (*W*-Las Cañadas): 2 years (1991–1992).
  - **StationC406D (117)**, *Boca Tauce A* (*SW*-Las Cañadas): 3 years (1990–1992).
  - **Station C406G (121)**, *Parador Nacional de Turismo* (Centre-south part of Las Cañadas): 3 years (1990–1992).
  - **Station C430E (217)**, *Main Izaña station* (summits of La Dorsal-*NE*):
    - ***Station C430E (217a):***
    - 37 years (1957–1994).
    - ***Station C430E (217b):***
- ***3 years (1970–1972).***
- Based on the key in (Rivas-Martínez et al. 2011), the high summit of the island of Tenerife is *Mediterranean pluviseasonal-oceanic.*
- Based on the continentality index, all stations can be described within the *Oceanic* type and the *euoceanic* subtype. The only exceptions were the highest stations, which are of semihyperoceanic character. This is a pattern which can be observed in many mountain summits across the world. The analysed stations changed in oceanity level between strong and weak.v

Table 2.11 Continentality, ombrothermic and thermic indices calculated based on data from the different meteorological stations analysed within the Teide National Park

Code INM Tragsa	Stations	Altitude (m a.s.l.)	Situation in the National Park	Period	N° years	Ic	Io	It
C412C 181	Montaña Samara	1900	*W*-Las Cañadas	1991–1992	2	**14,6**	**3,2**	**226**
C406D 117	Boca Tauce A	2030	*SW*-Las Cañadas	1990–1992	3	**15,5**	**3,2**	**229**
C406G 121	Parador Nacional	2160	Parte centro-meridional de la depresión de Las Cañadas	1990–1992	3	**14,7**	**3,3**	**216**
No INM 374	Torreta del Teleférico	3250	Altos de las laderas meridionales del Teide	1994–1998	5	**11,5**	**–**	**41** Tp = 555
No INM 370	Cañada de La Grieta	2100	*SE*-Las Cañadas	1995–1997	3	**15,3**	**–**	**213**
No INM 372	Pista Media Luna	2240	Mitad oriental de Las Cañadas	1995–1997	3	**12,3**	**–**	**250**
C430E*a* 217*a*	Izaña	2371	Altos de la Dorsal-*NE*	1957–1994	37	**14,2**	**3,7**	**187**
C430E*b* 217*b*	Izaña	2371	Altos de la Dorsal-*NE*	1970–1972	3	**15,0**	**5,8**	**141**

Therefore, we can state that station 374 on the crater of Mount Teide was the most oceanic one, while station C406D was the most continental one.

- With respect to thermotypes:
  - The westernmost sector of the interior of the Las Cañadas depression (stations C412C and C406D) seems to stay within the *Upper Mesomediterranean* thermotype.
  - The central part of the interior of Las Cañadas (stations C406G y 370) are somewhat colder and already seem to be of the *Lower Supramediterranean* thermotype.
  - The easternmost sector of this same area (station 372) seems to be again *Upper Mesomediterranean.*
  - The summits of La Dorsal around Izaña (station C430E*a*), is as the central part of Las Cañadas of *Lower Supramediterranean* thermotype.
  - The upper areas of the Teide edifice (station 374) are of *Upper Oromediterranean* thermotype.
- With respect to ombrotypes:
  - The westernmost sector of the Las Cañadas interior and its central part (stations C412C, C406D y C406G) seem to fall under the *Upper Dry* ombrotype.
  - For the easternmost part of Las Cañadas and the upper half of Mount Teide, no ombrotype could be defined due to a lack of precipitation data.
  - The area of the Dorsal around Izaña (station C430E), is of *Subhumid* ombrotype.

For some of the stations a new dataset was obtained with monthly data from 2008 to 2015 which we combined with data from the previous periods. This data was limited to stations C406D, C406G and C412C. Sadly this data was also fraught with missing values so that we could only enrich previous analysis by up to 3 years for station C406D and 2 to 1 years for C412C and C406G (Fig. 2.13). Additionally, we compiled new data from AEMET for the Izaña station for the 1990–2010. This latter data for Izaña is the most robust to date, as it represents the only modern uninterrupted time series for the Teide National Park that we know of (Fig. 2.12).

Although a very preliminary result, the comparison of periods between the Izaña station for the periods 1957–1994 to the 1990–2010 period shows a clear aridification of the climate with a shift from Lower Subhumid to Upper Dry ombrotypes. This can be seen in the data as a drop in ***Io*** from 3,7 (Table 2.11) to 2,98 (Fig. 2.12). As the ***Itc*** remains very similar (187–188,25) this shift has to be mediated by a descent of measured precipitations. Further analysis to estimate the impacts of global climate change on the temperature and precipitation regimes of the Teide National Park are necessary. Regardless of data quality, one of the main attributes of Mediterranean climates, the summer drought, can be very well seen in all ombrothermic diagrams performed as the drop in precipitations during the summer is very evident.

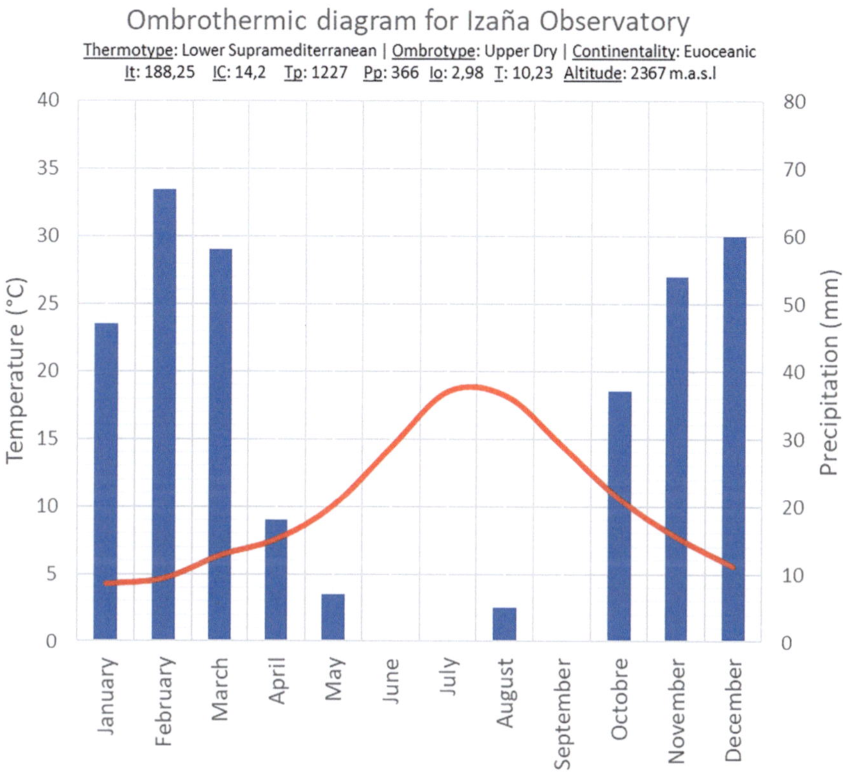

Fig. 2.12 Ombrothermic diagram for the station at Izaña Observatory for the 1990–2010 period. Red line indicates the evolution of mean monthly temperature (°C) and blue bars the mean monthly precipitation (mm)

As we stressed already, except for the Izaña station, all their results of this bioclimatic classification (Table 2.12) should be taken with caution, as a rigorous bioclimatic analysis needs of a robust backbone of monthly meteorological data to be able to extrapolate long time climatic averages to which ultimately the vegetation adapts to. Some reassurance can be found when comparing the analysis of the short-term Izaña station to its long-term counterpart, which shows just variation in the ombrotype and thermotype horizons without bigger changes. Nonetheless, we believe some of these patterns and results to be representative enough to explain part of the differences we do observe in vegetation and climate within the Teide National Park and in which will delve in the next chapters. A better coverage of the Teide National Park with high quality meteorological data is fundamental to be able to observe the trends in current climatic developments and to make better predictions about the future of the vegetation of the National Park with respect to ongoing climate change. This is especially relevant for all schemes looking to preserve endangered or critically endangered taxa native to the National Park.

Table 2.12 Bioclimatic diagnosis obtained from the analysis of the different meteorological stations analysed in the Teide National Park

Code INM Tragsa	Station	Altitude (m a.s.l)	Period	N° Years	Bioclimatic diagnosis
C412C 181	Montaña Samara	1900	1991–1992	2	*Mediterranean pluviseasonal-oceanic Upper Mesomediterranean Upper Dry*
C406D 117	Boca Tauce A	2030	1990–1992	3	*Mediterranean pluviseasonal-oceanic Upper Mesomediterranean Upper Dry*
C406G 121	Parador Nacional	2160	1990–1992	3	*Mediterranean pluviseasonal-oceanic Lower Supramediterranean Upper Dry*
No INM **374**	Torreta del Teleférico	3250	1994–1998	5	*Mediterranean pluviseasonal-oceanic Upper Oromediterranean* –
No INM **370**	Cañada de La Grieta	2100	1995–1997	3	*Mediterranean pluviseasonal-oceanic Lower Supramediterranean* –
No INM **372**	Pista Media Luna	2240	1995–1997	3	*Mediterranean pluviseasonal-oceanic Upper Mesomediterranean* –
C430E*a* 217*a*	Izaña	2371	1957–1994	37	*Mediterranean pluviseasonal-oceanic Lower Supramediterranean Lower Subhumid*
C430E*b* 217*b*	Izaña	2371	1970–1972	3	*Mediterranean pluviseasonal-oceanic Upper Supramediterranean Upper Subhumid*

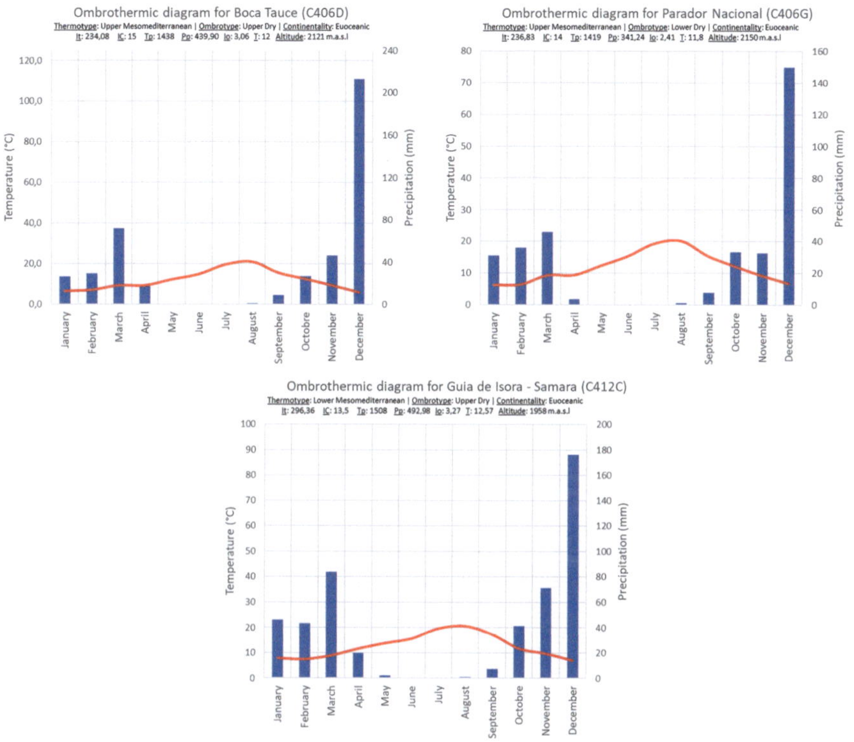

Fig. 2.13 Ombrothermic diagram for the stations at Boca Tauce, Parador Nacional and Montaña de Samara for the period indicated in Table 2.11 and some additional years where no values were missing for the period 2008–2015. Red line indicates the evolution of mean monthly temperature (°C) and blue bars the mean monthly precipitation (mm)

References

Carracedo JC (2006) El Volcán Teide (Volcanología, interpretación de paisajes e itinerarios comentados). Servicio de Publicaciones de la Caja General de Ahorros de Canarias, n° 376, Consejo Superior de Investigaciones Científicas, Santa Cruz de Tenerife, 431 pp Incluye Mapa Geológico del Teide y las Dorsales de Tenerife

Del Arco Aguilar M, Wildpret de la Torre W (1994) Estudio bioclimático del Pico del Teide. Tenerife (Islas Canarias). Instituto de Estudios Canarios 38:7–14. La Laguna

Del Arco M, Pérez de Paz PL, Acebes JR, González Mancebo JM, Reyes-Betancort JA, Bermejo JA, De Armas S, González-González R (2006) Bioclimatology and climatophilous vegetation of Tenerife (Canary Islands). Ann Bot Fennici 43:167–192

Durbán Villalonga M, Reverón Gómez J (coord.) (2011) El Parque Nacional del Teide: Inscripción en la Lista del Patrimonio Mundial. Organismo Autónomo de Parques Nacionales, Madrid

Rivas-Martínez S, Sánchez-Mata D, Costa M (1999) North American Boreal and Western Temperate Forest Vegetation. Itinera Geobot 12:5–316

Rivas-Martínez S et al (2007) Mapa de series, geoseries y geopermaseries de vegetación de España. (Memoria del mapa de vegetación potencial de España). Parte I. Itinera Geobot 17:1–436

Rivas-Martínez S, Coautores et al (2011) Mapa de series, geoseries y geopermaseries de vegetación de España [Memoria del mapa de vegetación potencial de España]. Itinera Geobot 18(1, 2):5–800

Chapter 3
Flora of Teide National Park

3.1 Floristic Diversity

The flora species selected in this work have in common the criterion of being the most abundant taxa sampled per grid within the boundaries of the Teide National Park. They include endemic Canary Island species, island endemic species and endemic species of the Teide National Park, as well as a number of species with a wider distribution.

In total, 102 taxa have been identified, which are grouped in 69 endemic species and 33 non-endemic species. Endemic species have been classified according to their geographical distribution with the following criteria:

- **36 Endemisms of Tenerife**
  - *17 Endemisms of the National Park (NP)*
- **14 Canary Island endemics**
  - 8 HPGTC, 3 PTC, 1 GTC, 1 HPT, 1 PGTC
- **11 Endemisms** of the islands of La Palma and Tenerife
- **1 Endemism** of the islands of El Hierro and Tenerife
- **7 Macaronesian** endemisms
  - 3 HPGTCFL, 3 HPGTC, 1 GTCFL

The acronyms correspond to the following islands:

H: El Hierro, **P**: La Palma, **G**: La Gomera, **T**: Tenerife, **C**: Gran Canaria, **F**: Fuerteventura, **L**: Lanzarote

V. E. Martín Osorio et al., *Flora and Vegetation of The Teide National Park*,
https://doi.org/10.1007/978-3-031-88400-9_3

3.2 Floristic Catalogue

For the correct botanical taxonomic referencing, the Canary Islands Natural Inventory Bank (BIOCAN), Plants of the World Online, from the Royal Botanic Gardens Kew (POWO), The Global Biodiversity Information Facility (GBIF), the International Plant Names Index (IPNI) and the World Checklist of Vascular Plants (WCVP) database have been used (Table 3.1).

BIOTA contains the database of all wild species in the Canary Islands Archipelago and details the following categories according to the origin of the taxon:

Certain Native (CN)

Includes native taxa, both endemic and non-endemic. In both cases, this category will be assigned to those taxa for which there is evidence, or it is suspected that their presence in the Canary Islands is natural and not associated with human activity.

Probable Native (PN)

Includes those taxa for which there are indications that they have hardly been introduced by human action. (for example, taxa that have a restricted distribution and are associated with a natural habitat on the islands).

Certainly Introduced Non-invasive (CIN)

Includes those taxa whose introduction (accidental or deliberate) is recorded in the literature and others considered cosmopolitan and widely distributed outside the Canary Islands. Introduced taxa for which their invasive nature is unknown are also included.

Introduced Certain Invasive (ICI)

Includes those taxa that are introduced or established in a natural or semi-natural habitat and that represent an agent of change and a threat to native biological diversity, either due to their invasive behaviour or the risk of genetic contamination.

Introduced Certain Potentially Invasive (ICP)

Includes those taxa that are introduced or established in a natural or semi-natural habitat and that could become invasive in the Canary Islands, especially those that have demonstrated this character in other countries or regions with ecological characteristics like those of the Canary Islands.

Table 3.1 Flora of the Teide National Park

Scientific name	Distribution	Origin	Endemicity
Abies pinsapo Boiss.		CI	
Adenocarpus viscosus (Willd.) Webb & Berth. **subsp. *viscosus***	T	CN*	
Aeonium aizoon (Bolle) T.H.M.Mes (=*Greenovia aizoon* Bolle)	T	CN*	
Aeonium aureum (C.Sm. *ex* Hornem.) T.H.M.Mes (=*Greenovia aurea* (C. Sm. *ex* Hornem.) Webb & Berth.)	HPGTC	CN*	
Aeonium smithii (Sims) Webb & Berth.	T	CN*	
Aeonium spathulatum (Hornem.) Praeger **var. *spathulatum***	HPGTC	CN*	
Andryala pinnatifida Aiton **subsp. *teydensis*** (Sch. Bip.) Rivas-Mart. & al.	T	CN*	NP
Anogramma leptophylla (L.) Link	HPGTCFL	CN	
Arabis caucasica Willd.	HPGTC	PN	
Arenaria leptoclados (Rchb.) Guss.	HPGTCFL	PN	
Argyranthemum adauctum (Link) Humphries **subsp. *dugourii*** (Bolle) Humphries	T	CN*	
Argyranthemum tenerifae Humphries	T	CN*	
Aria edulis (Willd.) M. Roem.	PT	PN	
Arrhenatherum calderae A. Hansen	PT	CN*	
Asplenium adiantum-nigrum L.	HPGTCF	CN	
Asplenium onopteris L. **var. *onopteris***	HPGTCFL	CN	
Athamanta montana (Webb *ex* Christ) Spalik & Wojew. & S.R.Downie (=*Todaroa montana* Webb *ex* Christ)	HPGTC	CN*	

Table 3.1 (continued)

Scientific name	Distribution	Origin	Endemicity
Bencomia exstipulata Svent.	PT	CN*	
Bethencourtia palmensis (C.Sm.) Choisy (=*Senecio palmensis* (C. Sm. *in* Buch) Link)	PT	CN*	
Bituminaria bituminosa (L.) C. H. Stirt.	HPGTCFL	PN	
Bromus madritensis L.	HPGTCFL	PN	
Bromus tectorum L.	PTC	PN	
Bufonia paniculata Dubois	PGTC	CN	
Bystropogon canariensis (L.) L'Hér. **var.** ***canariensis***	HPGTC	CN*	
Bystropogon origanifolius L'Hér. **var.** ***origanifolius***	HPGTC	CN*	
Carex demissa Hornem. **subsp.** ***demissa***	T	PN	
Carex paniculata L. **subsp.** ***calderae*** (A. Hansen) Lewej. & Lobin	T	CN*	NP
Carlina xeranthemoides L. f.	T	CN*	
Cedrus atlantica (Endl.) Manetti *ex* Carrière	T	IS	
Cerastium sventenii Jalas	HPT	CN*	
Chamaecytisus proliferus (L. f.) Link **subsp.** ***angustifolius*** (Kuntze) G. Kunkel	GTC	CN*	
Cheirolophus teydis (C. Sm. *in* Buch) G. López	PT	CN*	
Cistus osbeckiifolius Webb *ex* Christ **subsp.** ***osbeckiifolius***	T	CN*	
Cistus symphytifolius Lam. **var.** ***symphytifolius***	HPGTC	CN*	
Cystopteris diaphana (Bory) Blasdell **subsp.** ***diaphana***	HPGTC	CN	

Table 3.1 (continued)

Scientific name	Distribution	Origin	Endemicity
Cystopteris fragilis (L.) Bernh.	HPGTC	CN	
Dactylis metlesicsii Schönfelder & Ludwig	T	CN*	NP
Descurainia bourgeauana (E. Fourn.) O. E. Schulz	PT	CN*	
Descurainia gonzalezii Svent.	T	CN*	
Descurainia lemsii Bramwell	T	CN*	
Dittrichia viscosa (L.) Greuter	HPGTCFL	PN	
Echium auberianum Webb & Berth.	T	CN*	NP
Echium virescens DC.	T	CN*	
Echium wildpretii Pearson *ex* Hook f. **subsp. *wildpretii***	T	CN*	NP
Echium × hermanni Martín Osorio & Wildpret (*Echium wildpretii × auberianum*)	T	CN*	NP
Ephedra nebrodensis Tineo	PT	CN	
Erica canariensis Rivas-Martínez, Martín Osorio & Wildpret	HPGTCFL	CN*	Macaronesia
Erigeron calderae A. Hansen	T	CN*	NP
Erysimum scoparium (Brouss. *ex* Willd.) Wettst.	PTC	CN*	
Ferula linkii Webb	HPGTC	CN*	
Festuca agustinii Linding.	HPGTC	CN*	Macaronesia
Festuca myuros L.	HPGTCFL	PN	
Filago pyramidata L.	HPGTCFL	PN	
Galium aparine L.	HPGTCFL	PN	
Gastridium phleoides (Nees & Meyen) C. E. Hubb.	HPTF	PN	
Helianthemum juliae Wildpret	T	CN*	NP

Scientific name	Distribution	Origin	Endemicity
Hemionitis guanchica (Bolle) Christenh. (=*Cheilanthes guanchica* Bolle)	HPGTCFL	CN	
Hypericum grandifolium Choisy	HPGTCFL	CN*	Macaronesia
Juniperus cedrus Webb & Berthel. **subsp. *cedrus***	PGTC	CN*	
Lamarckia aurea (L.) Moench	HPGTCFL	PN	
Laphangium teydeum Wildpret & Greuter	T	CN*	NP
Lotus campylocladus Webb & Berth. **subsp. *campylocladus***	T	CN*	
Melica canariensis W. Hempel	GTCFL	CN*	Macaronesia
Mentha longifolia (L.) Huds. **var. *teydea*** Bolle	T	CN*	NP
Micromeria lachnophylla Webb & Berth.	T	CN*	
Micromeria lasiophylla Webb & Berth. **subsp. *lasiophylla***	T	CN*	NP
Monanthes brachycaulos (Webb *in* Webb & Berth.) Lowe **var. *nivata***	T	CN*	
Nepeta teydea Webb & Berth. **var. *albiflora*** Svent.	T	CN*	NP
Nepeta teydea Webb & Berth. **var. *teydea***	PT	CN*	
Pimpinella cumbrae Link	PT	CN*	
Pinus canariensis C. Sm. *ex* DC. *in* Buch	HPGTC	CN*	
Pinus radiata D. Don	HPGTC	IS	
Piptatherum coerulescens (Desf.) P. Beauv.	HGTCL	PN	
Piptatherum miliaceum (L.) Coss.	HPGTCFL	PN	
Plantago webbii Barnéoud	PTC	CN*	
Poa infirma Kunth	PGTCFL	PN	

Scientific name	Distribution	Origin	Endemicity
Polycarpaea tenuis Webb *ex* Christ	PT	CN*	
Pterocephalus lasiospermus Link *ex* Buch	T	CN*	
Ranunculus cortusifolius Willd.	HPGTCFL	CN*	Macaronesia
Rhamnus integrifolia DC.	T	CN*	
Rhaponticum canariense (Webb & Berth.) DC. (=*Stemmacantha cynaroides* (C. Sm. *in* Buch) Dittrich)	T	CN*	NP
Rosa cannadas-teydensis P Vargas, M Nogales & M Luceño	T	CN*	NP
Rumex maderensis Lowe	HPGTC	CN*	Macaronesia
Sagina procumbens L. **subsp.** ***stolonifera*** Martín Osorio, Rivas Martínez & Wildpret	PT	CN*	
Salix canariensis C. Sm. *ex* Link	HPGTC	CN*	Macaronesia
Scrophularia glabrata Aiton	PT	CN*	
Sideritis eriocephala Marrero *ex* Negrín & P. Pérez	T	CN*	
Sideritis oroteneriffae Negrín & P. Pérez	T	CN*	
Sideritis soluta Clos. **subsp.** ***soluta***	T	CN*	
Silene berthelotiana Webb	HT	CN*	
Silene nocteolens Webb & Berth.	T	CN*	NP
Silene vulgaris (Moench) Garcke **subsp.** ***commutata*** (Guss.) Hayek	HPGTCFL	PN	
Spartocytisus supranubius (L. f.) Christ	PT	CN*	
Spergula pentandra L.	HPTC	PN	

Scientific name	Distribution	Origin	Endemicity
Tolpis lagopoda C. Sm. *in* Buch	PTC	CN*	
Tolpis webbii Sch. Bip. *ex* Webb & Berth.	T	CN*	
Tragopogon porrifolius L.	PGTC	PN	
Umbilicus gaditanus Boiss.	HPGTCFL	PN	
Valeriana calcitrapae L.	HPGTC	PN	
Viola cheiranthifolia Humb. & Bonpl.	T	CN*	NP
Viola guaxarensis M.Marrero, Docoito Díaz & Martín Esquivel	T	CN*	NP
Wahlenbergia lobelioides (L. f.) Link **subsp. *lobelioides***	HPGTCFL	CN	

Acronyms correspond to the first letter of the following islands: *H* El Hierro, *P* La Palma, *G* La Gomera, *T* Tenerife, *C* Gran Canaria, *F* Fuerteventura, *L* Lanzarote. Acronyms on origin: *CN* Certain Native, *CN** Endemism, *PN* Probable native; *CI* Certain Introduced, *NP* Teide National Park endemic

Introduced Certainly Lack of Data (ICF)

Includes those taxa that are introduced or established in a natural or semi-natural habitat, for which, at the moment, there is not enough information available to consider them invasive or potentially invasive, but there are indications that they could form part of one of these categories.

Introduced Probable (IP)

Includes those taxa for which there are indications that indicate the difficulty of their presence in the Canary Islands being due to natural processes not associated with human activity. These are taxa not recognized as introduced in the bibliographic sources, but with a clear association with anthropic environments. It also includes those taxa that are suspected to be introduced in the Canary Islands because they are invasive in other places outside the archipelago, because of their wide distribution on the islands and beyond, or because they have natural forms of dispersion linked to human activity.

In the Teide National Park, 51 taxa have also been cited that we have not included in the description of the floristic cards because they are not abundantly represented in the sampled grids (Table 3.2). Therefore, the total flora cited so far is 153 taxa.

Table 3.2 Flora of the Teide National Park

Species	Family	Distribution
Agrostis castellana Boiss. & Reut.	*Poaceae*	HPGTCFL
Aira caryophyllea L.	*Poaceae*	HPGTCFL
Bromus rigidus Roth	*Poaceae*	HPGTCFL
Bromus rubens L. **subsp. *kunkelii*** (H. Scholz) H. Scholz	*Poaceae*	HPGTCFL
Asphodelus ramosus L. **subsp. *distalis*** Z. Díaz & Valdés	*Asphodelaceae*	HPGTCFL
Asplenium septentrionale (L.) Hoffm.	*Aspleniaceae*	PT
Avena barbata Pott *ex* Link	*Poaceae*	HPGTCFL
Avena sativa L. ***subsp. sativa***	*Poaceae*	TCFL
Avena sterilis L.	*Poaceae*	HPGTCFL
Buglossoides arvensis (L.) I. M. Johnst.	*Boraginaceae*	TCFL
Calendula arvensis L.	*Asteraceae*	HPGTCFL
Centaurium tenuiflorum (Hoffmanns. & Link) Fritsch	*Gentianaceae*	HPGTCFL
Cheilanthes maderensis Lowe	*Sinopteridaceae*	HPGTCFL
Chenopodium album L.	*Chenopodiaceae*	HPGTCFL
Cupressus arizonica E. L. Greene	*Cupressaceae*	–
Cupressus macrocarpa Hartw.	*Cupressaceae*	PTC
Cupressus sempervirens L.	*Cupressaceae*	H.G.
Epilobium tetragonum L. **subsp. *tetragonum***	*Onagraceae*	P.G.T.C.
Erodium botrys (Cav.) Bertol.	*Geraniaceae*	HPGTCFL
Erodium cicutarium (L.) L'Hér.	*Geraniaceae*	HPGTCFL

Species	Family	Distribution
Euphorbia peplus L. **var. *peplus***	*Euphorbiaceae*	HPGTCFL
Euphorbia segetalis L.	*Euphorbiaceae*	PTL
Filago lutescens Jord.	*Asteraceae*	HTCL
Filago pygmaea L.	*Asteraceae*	HTCL
Funaria hygrometrica Hedw.	*Funariaceae*	HPGTCFL
Gastridium ventricosum (Gouan) Schinz & Thell.	*Poaceae*	PGTCFL
Geranium molle L.	*Geraniaceae*	HPGTCFL
Lactuca serriola L.	*Asteraceae*	HPGTCFL
Lolium rigidum Gaudin	*Poaceae*	HPGTCFL
Melica ciliata L. **subsp. *magnolii*** (Green & Godr.) K.Richt	*Poaceae*	GTCFL
Myosotis ramosissima Rochel	*Boraginaceae*	HPTC
Ochlopoa annua (L.) H. Scholz	*Poaceae*	HPGTCFL
Origanum vulgare L.	*Lamiaceae*	HPGTC
Oxalis corniculata L.	*Oxalidaceae*	HPGTCFL
Phagnalon saxatile (L.) Cass.	*Asteraceae*	HPGTCFL
Pinus halepensis Mill.	*Pinaceae*	HGTCFL
Poa trivialis L.	*Poaceae*	TL
Polypogon fugax Nees *ex* Steud.	*Poaceae*	HPGTCFL
Polypogon monspeliensis (L.) Desf.	*Poaceae*	HPGTCFL
Polypogon viridis (Gouan) Breistr.	*Poaceae*	HPGTCFL
Prunus domestica L.	*Rosaceae*	TC
Reseda luteola L.	*Resedaceae*	HPTCFL
Scirpus holoschoenus L. **subsp. *globiferus*** (L. f.) Husn.	*Cyperaceae*	PGTC
Senecio vulgaris L.	*Asteraceae*	HPGTCFL

Species	Family	Distribution
Setaria viridis (L.) P. Beauv.	*Poaceae*	TC
Silene inaperta L.	*Caryophyllaceae*	TC
Sonchus oleraceus L.	*Asteraceae*	HPGTCFL
Torilis leptophylla (L.) Rchb. f.	*Apiaceae*	HPGTFL
Trachynia distachya (L.) Link	*Poaceae*	HPGTCFL
Trisetaria panicea (Lam.) Paunero	*Poaceae*	HPGTCFL
Veronica anagallis-aquatica L.	*Plantaginaceae*	PGTC

3.3 Threatened Flora

Species which are under any level of protection because of their threat status in the Teide National Park differ on their protection based on the applicable legislation. Based on the national Spanish catalogue of threatened species, there are five species in the National Park considered as "Endangered" and two as "Vulnerable".

List of species in the Spanish catalogue of Endangered Species (Real Decreto 139/2011, de 4 de febrero):

Endangered	Vulnerable
Bencomia exstipulata *Helianthemum juliae* *Rhaponticum canariense*	*Silene nocteolens* *Juniperus cedrus*

Parallely there is a Ley 4/2010, de 4 de junio, Canary Island Catalogue of Protected Species in which following categories are defined:

1. Threatened Species:
   - Endangered
   - Vulnerable
2. Species of Interest for Canary Island Ecosystems (SICIE)
3. Special Protection Species

In the Teide National Park, following species are listed in the previous categories:

Endangered	Vulnerable	Interest for Canary Island Ecosystems
Bencomia exstipulata *Helianthemum juliae* *Rhaponticum canariense* *(=Stemmacantha cynaroides)*	*Dactylis metlesicsii* *Silene nocteolens*	*Cerastium sventenii* *Erigeron calderae* *Laphangium teydeum* *Aria edulis (=Sorbus aria)*

Finally six taxa (Table 3.3) are species with the **IUCN Threat Category Classifications** EN (Endangered) or CR (Critically Endangered): *Bencomia exstipulata* (Fig. 3.1), *Helianthemum juliae*, *Silene nocteolens*, *Cerastium sventenii*, *Rhaponticum canariense (=Stemmacantha cynaroides)* and *Juniperus cedrus* subsp *cedrus* of which four are exclusive to the protected area or the high summit (Bañares et al. 1998; Bañares et al. 2004; Moreno 2008).

Table 3.3 Species with the IUCN Threat Category Classifications

Nº	Scientific name	Family	Distribution	Threat Category	Nº Subgrids 250x250 m
1	***Bencomia exstipulata***	Rosaceae	P.T.	CR	3
2	***Helianthemum juliae***	Cistaceae	T.	CR	5
3	***Silene nocteolens***	Caryophyllaceae	T.	CR	64
4	***Cerastium sventenii***	Caryophyllaceae	H.P.T.	EN	39
5	***Rhaponticum canariense*** *(=Stemmacantha cynaroides)*	Asteraceae	T.	EN	10
6	***Juniperus cedrus* subsp. *cedrus***	Cupressaceae	P.G.T.C.	EN	288

Fig. 3.1 ***Bencomia exstipulata*** in its only natural habitat, Roque de Chavao. (Photo credit: Javier Martínez Carbajal)

3.4 Floristic Richness per Subgrid

Of the total amount of sampled subgrids, a visual representation of the floral richness (species number) within the subgrids has been performed. The richest subgrids contains a total of 41 taxa, being located within the Barranco del Río, a ravine with a flowing water stream within the National Park.

Based on the performed sampling and observing the distribution map of the species richness of the Teide National Park (Fig. 3.2), the subgrids with the highest number of species correspond to those located around the Pared de Las Cañadas, Roques de García and La Fortaleza, coinciding with the most rocky and steep areas, as well as geologically oldest. This contrasts with those more geologically younger areas which are still to be colonised, and which are located on the slopes of the Teide stratovolcano and the lava flows of the Narices del Teide. There is a total of 175 subgrids where no taxon was detected during this sampling.

The ten most abundant species by number of subgrids present and which appear in more than 1600 subgrids (Fig. 3.3) are, in descending order: ***Spartocytisus supranubius***, ***Pterocephalus lasiospermus***, ***Argyranthemum tenerifae***, ***Tolpis webbii***, ***Arrhenatherum calderae***, ***Nepeta teydea* var. *teydea***, ***Adenocarpus viscosus***, ***Descurainia bourgeauana***, ***Erysimum scoparium*** and ***Scrophularia glabrata***. In the floristic sheets of each species, a map is represented with the relative abundance of specimens per grid, the scale being <25%, 25–50%, 50–75% and 75–100%.

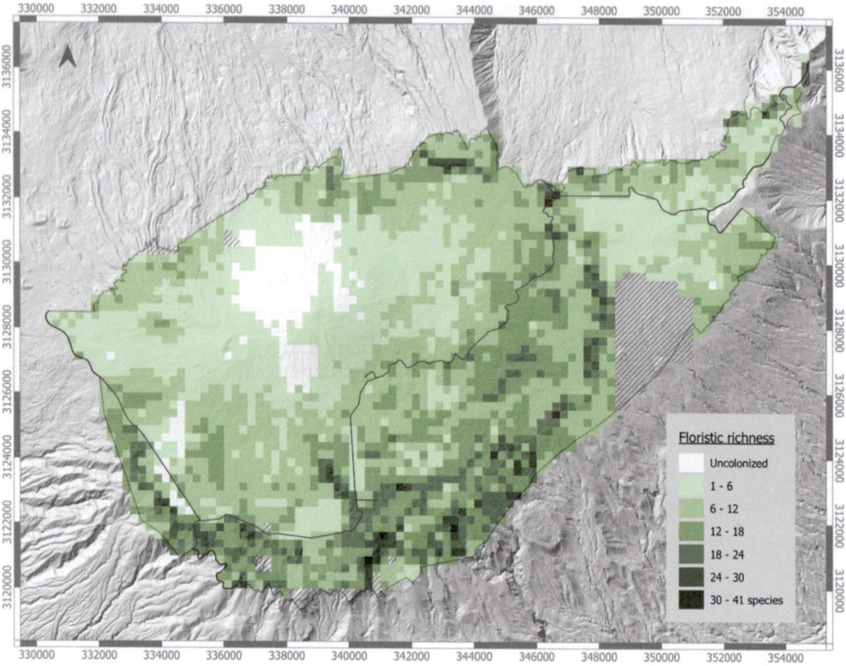

Fig. 3.2 Species richness distribution in the Teide National Park

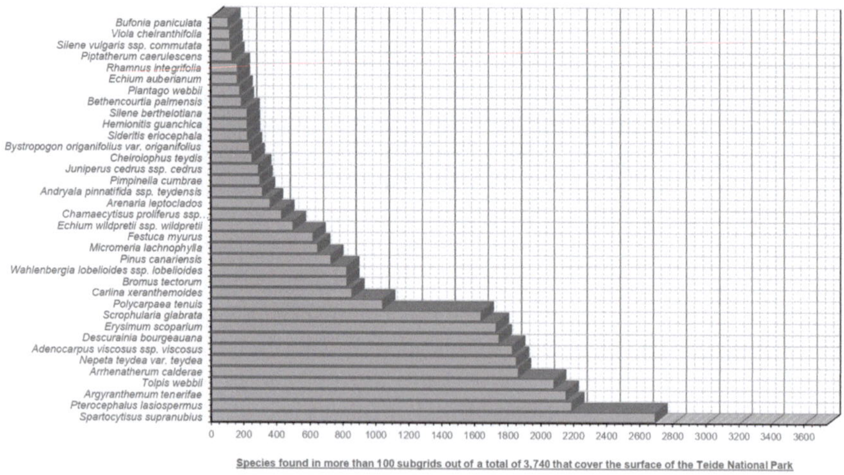

Fig. 3.3 Most represented species in the National Park by 250 × 250 m grids

The families most represented within the Teide National Park are *Fabaceae*, *Asteraceae*, *Brassicaceae*, *Dipsacaceae*, *Lamiaceae* and *Poaceae*. Within them, *Fabaceae* and *Asteraceae* are both the ones with the widest distribution and highest species richness across subgrids.

3.5 Floristic Catalogue

3.5.1 Abies pinsapo *Boiss.*

Biblioth. Univ. Génève, sér. 2, 13, 402, 406 (1838)

Introduced species and new record for the Teide National Park

Not endemic, Introduced in the Canary Islands
Family: *Pinaceae*
Common name: Abeto o Pinsapo (ES); Spanish fir (EN)
Biotype: Tree (phanerophyte)
Flowering: April–May
Fruiting: August–October
Pollination: wind pollination (anemogamy)
Dispersal: wind dispersal (anemochorous)
Chromosome N°: 2n = 24

Description

Tree up to 30 m tall, pyramidal in shape and straight trunk, with a greyish-white rhytidome; branches generally triverticillate, almost horizontal; resinous buds. **Leaves** 9–12 mm, rigid, acute, seated around the entire stem. **Male inflorescences** (Fig. 3.4) in cones of 8–12 × 6–7 mm; **females in cones** of 10–18 × 4–5 cm, erect, with a tectrix bract much shorter than the fruiting scale and this rounded at the apex. **Seeds** of 6–8 mm, with a triangular wing of 15–17 mm.

Fig. 3.4 (**a**) Parador Nacional del Teide, 18 June 2017. (Photo credit: Victoria Eugenia Martín Osorio)

Distribution

Endemism of the south of the Iberian Peninsula (provinces of Cádiz and Málaga) (Fig. 3.5) that lives forming thick forests between 900 and 1600 m a.s.l. A western Mediterranean element considered a relic of the tertiary coniferous forests and closely related to North African species.

In the Canary Islands it is cultivated or planted in garden areas (it bears fruit in autumn). In the Teide National Park (Fig. 3.6) two specimens have been found in the gardens of the Parador Nacional and one specimen has recently been removed from the mountaineers association's hut.

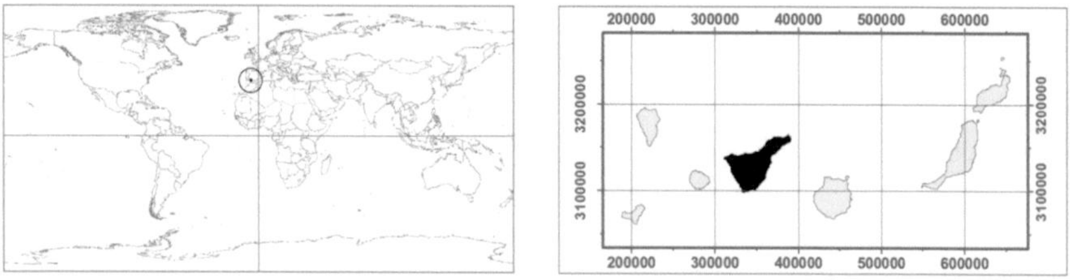

Fig. 3.5 It is located on the island of Tenerife in the surroundings of the Parador Nacional de las Cañadas del Teide

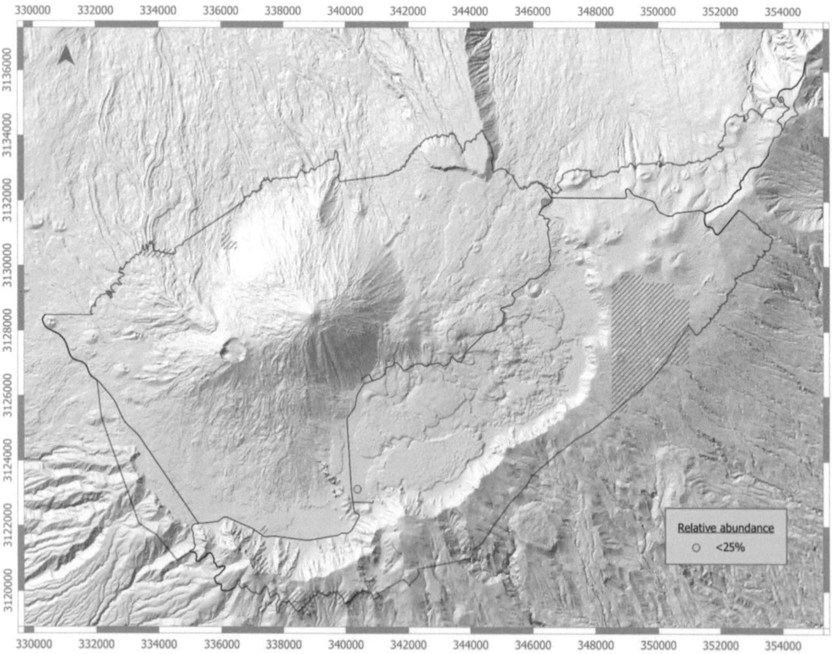

Fig. 3.6 *Abies pinsapo* distribution in National Park, N° Subgrids (250 × 250 m): **1**/3740

Habitat and Phytosociology

It lives in calcareous, dolomite and serpentine mountains between 1000 and 1800 m a.s.l. Always facing north and with a humid—hyperhumid ombroclimate.

It belongs to the class ***Querco-Fagetea sylvaticae***, which groups together the communities of climatophilous, temporo-hygrophilous and edaphoxerophilous planifolia and mixed forests—but not the fluvial edaphohygrophilous ones, with an optimal temperate climate, although they are also found in high-altitude Mediterranean rainy areas; and it is a characteristic species of the ***Paeonio broteri-Abietion pinsapo*** alliance, which groups together the associations corresponding to the humid-hyper-humid Meso-Supra-Mediterranean calcareous-dolomitic or ultramafic Andalusian fir forests (pinsapares) with a distribution in Ronda, Estepona and Grazalema.

Conservation Status

In the two subgrids in which this species was found, no damage of any kind was observed.

Threat Category

VU (Bañares et al. 2004)
VU (Moreno 2008)

Germplasm Banks

Banco de Germoplasma de E.T.S. Ingenieros Agrónomos de Madrid.

Exsiccata (Herbarium Specimens)

Tenerife: Inventario de la Flora y Vegetación del Parque Nacional del Teide (2001–2003) V.E. Martín Osorio, W. Wildpret de la Torre & B. Hernández Bolaños: Parador de Las Cañadas, 4.4.2003, Martín Osorio & Wildpret (TFC 44876).

Uses, Exploitation and/or Medicinal Properties

Normally, due to the characteristics of its wood (soft and resinous), there is no evidence of it having been used for lumber, however, in the past, it was used to some extent in the manufacture of paper. It is mainly used as an ornamental element in gardens.

3.5.2 Adenocarpus viscosus *(Willd.) Webb & Berthel.* *ssp.* viscosus

Phyt. Canar., 3 (2/A): 32 (1836–1850)

Adenocarpus frankenioides Chois! *in* DC. Mém. leg., pg. 217; Prodr., vol. 2, pg. 158
Cytisus foliolosus L'Hérit. Stirp. nov., fasc. 7, pg. 184, table 96. Non Hort.; Kew. Plantae facies optima
Genista viscosa Willd. *Sp. pl.*, vol. 3, pág. 937
Spartium microphyllum Link. *in* Buch, Phys., Beschr., Can. Ins., pg. 155, pro parte; Buch, ibid., pg. 184, quoad plantam alpinam Nivariae et Palmae, non Cav.
Aden *-: gland, –* ***karpos****: fruit, viscosus: sticky*

Endemism, Certain native
Family: *Fabaceae*
Common name: Codeso del pico o de cumbre (ES); Canary Island flatpod (EN)
Biotype: Tall shrub (chamaephyte/nanophanerophyte).
Flowering: April–July
Fruiting: June–August
Pollination: insect pollination (entomogamy)
Dispersal: Ballistic (ballistochory)
Chromosome No: 2n = c. 24

Description

Genus *Adenocarpus*: Divaricate **branches**. Trifoliate **leaves**, often crowded, folded, petiolate, ovate to linear-lanceolate, predominantly revolute, generally in fascicles. Abundant **flowers**, **yellow**, (Fig. 3.7) racemose towards the apex of the inflorescence, lower part of pedicels bracteate, bibracteolate in the centre, bracts deciduous. Tubular calyx, bilabiate, upper lip deeply bifid, lower elongate, trifid, usually curved. Long petal, banners cordate-ovate, shortly stipitate, patent; alae (wings) ovate, obtuse, towards the transverse nail folded carinal longitudinal indeflex. Stamens 10, monadelphous, splitting longitudinally by complete anthesis. Style filiform curved inwards. Stigma capitate. **Legume** narrowly oblong, flattened, apex subround, with glandular papillae, vigorous stipitate glandular nerve, very often obtuse with four to ten seeds. **Seeds** tightly packed without caruncle. Embryo coated in gelatinous perisperm, consistently encapsulated. Radicle curved inwards, subobtuse, longitudinally cotyledons. Cotyledons ovate or rounded, flattened.

Species *viscosus*: Thick **Shrub**, slimy to the touch. **Leaves** very dense, fasciculate, linear, revolute, pubescent and rough, membranous bracts, lanceolate, carinate and glandulous; **calyx** viscous-glandular, partly lanceolate; petal subglabrous. **Legumes** rudimentary glandulous and sparsely haired.

Fig. 3.7 (**a**, **b**) *Adenocarpus viscosus* Majua, May 2009. (Photo credit: Victoria Eugenia Martín Osorio)

Fig. 3.8 Endemism of the island of Tenerife and doubtful record in La Gomera

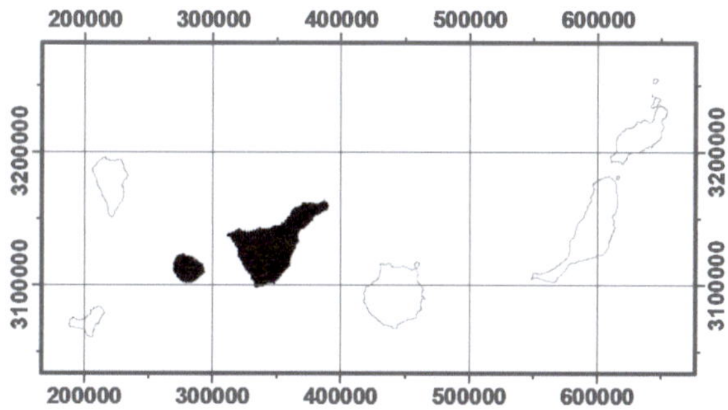

Distribution

In the Canary Islands present in the highest summits of Tenerife and in La Gomera, although records there are doubtful (Fig. 3.8).

Widely distributed in the Teide National Park (Fig. 3.9) above 1600 m a.s.l. up until 3125 m a.s.l.in the southern slopes of the stratovolcano, where its density is also higher. In the northern slopes of the volcano and northeastern of the ridges it is scarcer.

Habitat and Phytosociology

It is present in the understory of the pine tree forests and characterises high mountain summit shrublands of Tenerife.

Species characteristic of the class ***Chamaecytiso-Pinetea canariensis*** which groups coniferous forests, broom shrublands and serial dwarf pioneer shrublands typical of Tenerife and La Palma, as well as other oromediterranean shrublands of the Teide edifice; climatophilous, edaphoxerophilous, pyrophilous or serial, endemic to the midlands and summits of central and western Canary Islands, typical of dry-humid Meso-Oromediterranean bioclimates of the Western Canary biogeographical subprovince.

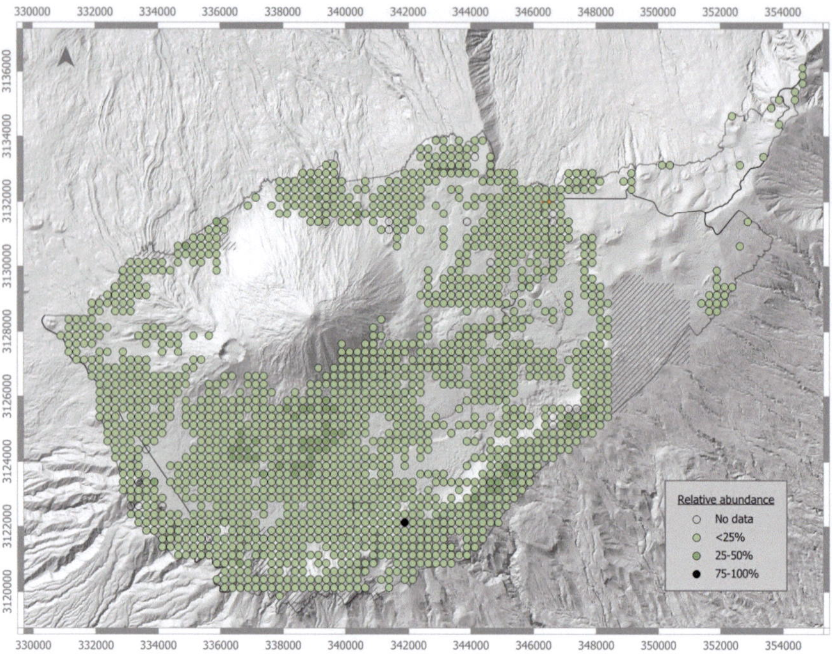

Fig. 3.9 *Adenocarpus viscosus* subsp. *viscosus* distribution map. N° Subgrids (250 × 250 m): **1822**/3741

It also characterises or participates in the following phytosociological units:

Adenocarpo viscosi-Chamaecytisetum angustifolii, *Codesar-Escobonal* which constitute the sun loving forest fringe of Meso-Supramediterranean dry-humid pine forest of Tenerife. In the Cañadas del Teide this formation is located quasi as an arch surrounding the national park.

Adenocarpo viscosi-Juniperetum cedri, typical of the steep slopes and cliffs of the Wall of Las Cañadas (edaphoxerophilous series of Tenerife, lithosolic Mediterranean, pluviseasonal oceanic Supramediterranean dry-subhumid euoceanic of the dwarf forests of *Juniperus cedrus* and *Adenocarpus viscosus* with *Echium wildpretii* and *Pterocephalus lasiospermus*).

Spartocytisetum supranubii, climatophilous and edaphoxerophilous series of Tenerife, Mediterranean pluviseasonal oceanic, upper Mesomediterranean-Oromediterranean dry-subhumid euoceanic of the high altitude shrublands of *Spartocytisus supranubius* and *Adenocarpus viscosus* with *Descurainia bourgeauana* and *Nepeta teydea*.

It can also be found as part of the following syntaxa:

Spartocytisetum supranubii-descurainietosum lemsii
Violo cheiranthifoliae-Adenocarpetum viscosi
Arrhenathero calderae-Plantaginetum webbii

Cheirolopho teydis-Carlinetum xeranthemoides
Tolpido webbii-Micromerietum lachnophyllae
Arrhenathero calderae-Adenocarpetum viscosi

Conservation Status

In 89% of the subgrids (1625) where the plant was found there were no visible damages. Only in 10% (175), there were damages (by herbivory or high levels of mortality, no juvenile recruitment, etc.) but never on more than 25% of the individuals. There was only damage found on more than 25% of individuals on less than 1% (13) of total subgrids.

Threat Status

nt (Barreno et al. 1984)
LR pm (Bañares et al. 1998)

Germplasm Banks

Banco de Germoplasma de E.T.S. Ingenieros Agrónomos de Madrid.
Jardín Botánico Canario Viera y Clavijo.

Exsiccata (Herbarium Specimens)

Tenerife: T. Charcos de Erjos-Los Silos, 9.07.1995, V. Lucía Sauquillo & C. Acosta (TFC 38148); T. Erjos-Los Silos, 22.06.1997, V. Lucía Sauquillo, A. Reyes & A. Díaz (TFC 40829 + Dupl.). Inventario de la Flora y Vegetación del Parque Nacional del Teide (2001–2003) V.E. Martín Osorio, W. Wildpret & B. Hernández Bolaños: Roques al Sur del Llano Ucanca (41D 35 d), 14/05/2003, A. Montañés Rguez. (TFC 44300); Ladera N del Teide—Base de Pico Cabras (29D 7 d), 18/06/2003, Y. de Rioja Santos (TFC 45442).

Traditional or Medicinal Uses

> "*Fodder plant, as aperitif, purgative; flatpod 'sugar' or 'melocha': manna; toxic. Used as infusion and for.*" (Translated from: Pérez de Paz and Hernández Padrón 1999)

3.5.3 Aeonium aizoon *(Bolle) T.H.M.Mes*

H.'t Hart & U.Eggli (eds.), Evol. & Syst. Crassulac.: 41 (1995)

Greenovia aizoon Bolle in Bonplandia (Hannover) 7: 242 (1859)
Greenovia quadrentalis Webb. *ex* Christ *in Bot. Jahrb.*, 9: 114 (1888)
Sempervivum aizoon Christ. Bot. Jahrb., 9: 162 (1888)

Endemism, Certain native
Common name: Bea—Bejeque (ES)/Treeleek (EN)
Family: *Crassulaceae*
Biotype: Subshrub (chamaephyte)
Flowering: Mai–June
Fruiting: June–July
Pollination: insect pollination (entomogamy)
Dispersal: wind dispersal (anemochory)
Chromosome No: NA

Description

Succulent, rosulate, perennial, low growing chamaephyte species, very branched and with numerous rosettes and tomentose (Fig. 3.10). **Branches** slender, horizontal, up to 15 cm long, wrinkly with old and young evergreen leaves. Rosettes flattened, with abundant leaves 5–7 cm in diameter.

Leaves green, oblong-spatulate, 3–4 cm long, 1–1.5 cm wide, truncate and apiculate at the apex, slim at the apex and broader towards the base, densely glandular-pubescent on the upper-, underside and margins.

Fig. 3.10 Rocky outcrops near La Crucita, Nov. 2004. (Photo credit: Victoria Eugenia Martín Osorio)

Flowering stem 10–13 cm high, glandular-pubescent, densely leafy, especially in the lower part with similar leaves that progressively decrease in size.

Inflorescence flattened, 10–40 flowers, around 2-3 cm long and 4–8 cm wide, 3–5 branches or dichotomous branches, glandular-pubescent. Buds oval-spheroidal. Flowers about 20 splitted, 12 mm in diameter, shiny yellow. Calyx green, glandular-pubescent, 4 mm long, cut from the middle downwards with acute oblong-lanceolate segments. Petals broadly oblanceolate, the outermost four lineal, pointy, glabrous, 6–7 mm long. Stamens 5–6.5 mm long, yellow, glabrous, with filiform filaments. Carpels green, glandular-pubescent, 4–5 mm long, style approximately equal to the ovary.

Remarks:

It endures the dry season with open rosettes, different to the rest of species in Sect. *Greenovia*. In shaded conditions it tends to produce long branches equipped with leafy whorls which mark the annual growing segments. It is easily multiplied by seeds and cuttings. It hybridises with ***Aeonium aureum*** **(C. Sm.** ***ex*** **Hornem.) T. Mes** (=*G. aurea* (Chr. Sm. *ex* Hornem) Webb & Berth.).

Distribution

Endemism of the island of Tenerife, (Fig. 3.11) between 1300 and 2200 m a.s.l. It is abundant on the slopes of the Güímar Valley, Igueste de Candelaria, Ayosa, etc. In the Teide National Park, (Fig. 3.12) it is only found in the area of La Crucita and around Montaña Limón.

Habitat and Phytosociology

Rupiculous species which usually appears in cavities or cracks preferentially shaded and oriented to the north and east. Species characteristic and bioindicator of the ***Greenovion aureae*** alliance, which groups comophytic and chasmophytic western canarian communities, well characterised by endemic succulent species of the genus (Sect.) *Greenovia* (=*Aeonium*), typical of the belts: upper Thermomediterrean, Meso- and Supramediterranean dry-humid. Within the alliance it defines and is

Fig. 3.11 Endemism of the island of Tenerife

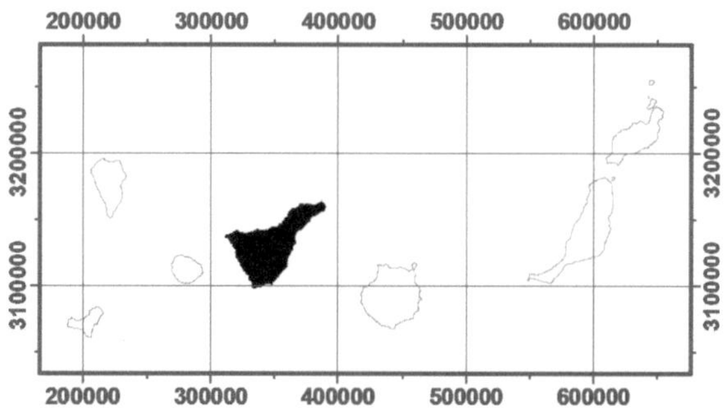

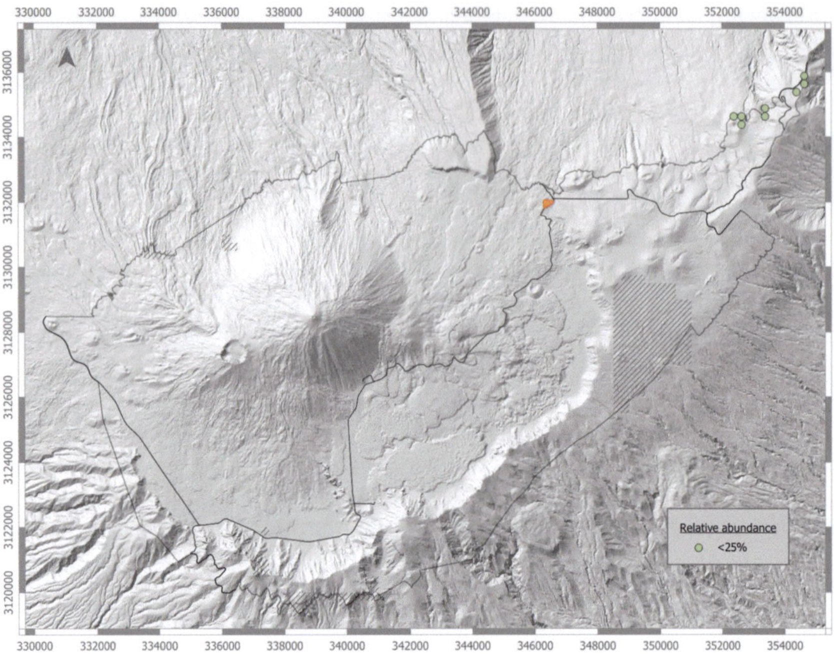

Fig. 3.12 *Aeonium aizoon* distribution. N° Subgrids (250 × 250 m): **8**/3741

characteristic of the community ***Greenovietum aizoi***. It colonises nearly vertical basalts exposed to the northeast.

Conservation Status

In 75% (6) of the subgrids where it was found no damage of any kind was detected. In 12.5% (1), damages were detected (herbivory, high mortality without regeneration, etc.), but those did not impact over 25% of individuals. In another subgrids, damages on more than 25% of individuals were detected.

> "*Its conservation status is relatively good. However, this is a species constrained to a very limited area and which populations seem very dispersed in highly selective places which in most cases lack adequate protection.*" (Translated from: Gómez Campo et al. 1996)

Threat Status

V (Barreno et al. 1984)
Anexo II: Protegidas (Gobierno de Canarias 1991)
V (Walter and Gillett 1998)
VU D2 (VVAA 2000)
VU (Bañares et al. 2004)
VU D2 (Moreno 2008)

Germplasm Banks

E.T.S. Ingenieros Agrónomos de Madrid.

Exsiccata (Herbarium Specimens)

Tenerife: 3.03.1971, P.L. Pérez de Paz (TFC 718); Ladera de Güímar, 16.04.1981, M. del Arco Aguilar, W. Wildpret & P.L. Pérez de Paz (TFC 13721).

3.5.4 Aeonium aureum *(C.Sm.* ex *Hornem.) T.H.M.Mes*

H.'t Hart & U.Eggli (eds.), Evol. & Syst. Crassulac.: 41 (1995)

Greenovia aurea (C. Sm. *ex* Hornem.) Webb & Berthel. *in Hist. Nat. Iles Canaries 3(2; 1): 199 (1841)*
Sempervivum aureum Chr. Smith. *ex* Hornemann *Suppl. Hort. Hafniensis* 61 (preface dated March 15, 1819)
Sempervivum calyciforme Haworth. *Suppl. Pl. Succ.*, 69 (preface dated May 1819)
Greenovia aurea Webb & Berth. *Phyt. Canar.*, 1: 199 (1840)
Greenovia rupifraga Webb. *in Bot. Mag.* 4087 (1844); *ex* Christ *in Bot. Jahrb.*, 9: 161 (1888)
Sempervivum rupifragum Christ. *Bot. Jahrb.*, 9: 161 (1888)

Endemism Certain native
Common name: Bea dorada—Bejeque—Talla (ES), Golden Treeleek (EN)
Family: *Crassulaceae*
Biotype: Subshrub (chamaephyte)
Flowering: March–April
Fruiting: Mai–June
Pollination: insect pollination (entomogamy)
Dispersal: wind dispersal (anemochory)
Chromosome N°: n = 18

Description

Rosulate species with many fibrous roots. Stem-less or with a short stem, with a solid rosette of leaves, from the lower axils, from which they rise short, horizontal and gradually (which is rare in cultivation). Rosette 8–25 cm in diameter, (Fig. 3.13) when with abundant width growth with erect inner leaves; at other times half closed and urn-shaped, sometimes closed, egg-shaped and lined with pinkish dead leaves.

Leaves sessile, spatulate or cuneate, in the base wide, in the apex rounded, truncate or broadly cuneate, minimally apiculate or retuse, 5–10 cm in length, 3–6 cm wide, glaucous, fleshy with thin margins, margins hyaline, often more or less sericeous.

Fig. 3.13 Rocky outcrops near La Crucita, June 2004. (Photo credit: Victoria Eugenia Martín Osorio)

Flowering stem erect, 30–45 cm tall, glandular-pubescent, very leafy with crammed spatulate or obovate leaves, decreasing in size while ascending; divided from the top and spreading inwards, often in horizontal branches, single or twice dichotomously branched, 10–25 cm in length, plumose, with acute lanceolate bracts and subsessile secondary flowers, or on pedicels up to 5 mm long. Buds oblate-spherical.

Flowers golden, 20–35 parted, 2–2.5 cm in diameter. Calyx green, with glandular hairs, 4–5 mm long, divided from the middle down in acute linear-lanceolate teeth. Petals linear-lanceolate, apiculate, 7–8 mm long. Stamens yellow, 5 mm long, filaments filiform. Without scalces. Carpels thin and clustered, green, hairy towards the base, 5 mm in length, ovaries with a dense ring with a central hollow, styles thin, erect, 2 mm long.

Remarks:

It hybridises with ***Aeonium aizoon* (Bolle) THM Mes** (=*Greenovia aizoon* Bolle).

Distribution

Endemic species of the western and central Canary Islands (Fig. 3.14). In the island of Tenerife it is distributed above 1500 m a.s.l., being able to reach 2150 m a.s.l. within the National Park, (Fig. 3.15) where it was only found around La Crucita and in the SW slopes of La Tarta.

Habitat and Phytosociology

Rupiculous species, although it can appear as an epiphyte on trees or on rooftops of certain buildings.

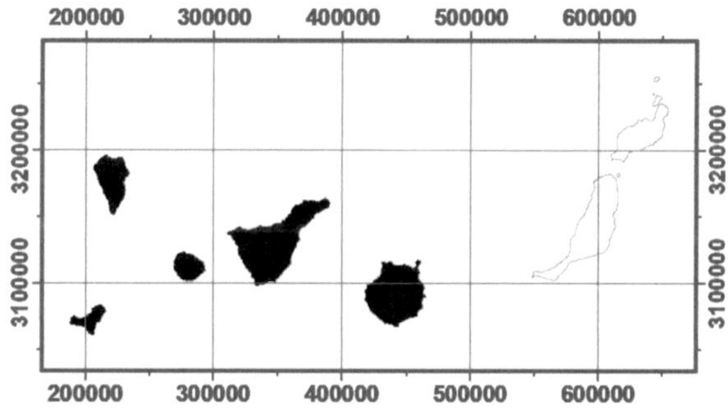

Fig. 3.14 Endemism of the central and western islands

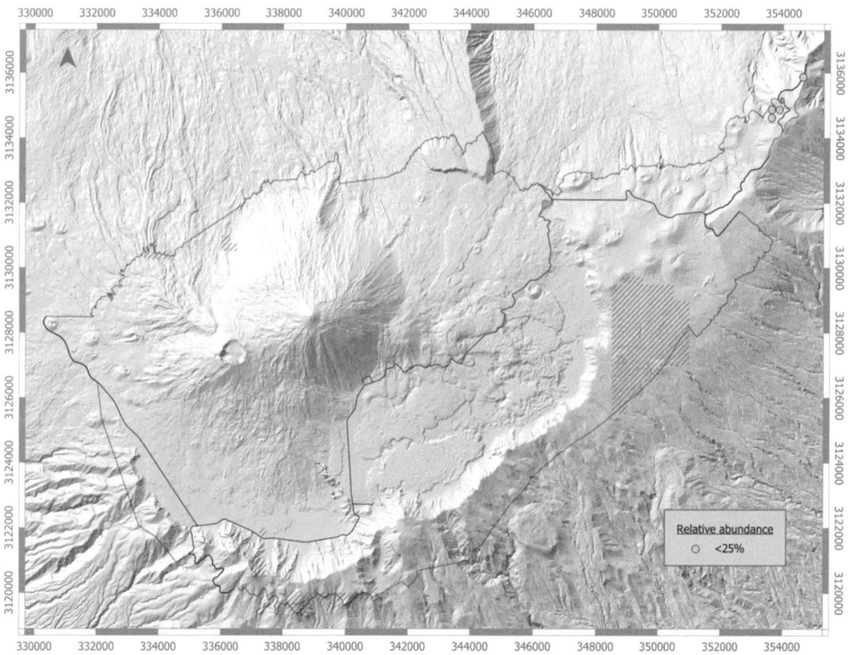

Fig. 3.15 *Aeonium aureum* distribution. N° Subgrids (250 × 250 m): **4**/3741

It is a characteristic species and bioindicator of the ***Greenovion aureae*** alliance, which groups comophytic and casmophytic communities from the western Canary Islands, well characterized by endemic succulent species of the genus *Greenovia* (=*Aeonium*), typical of the belts: upper Thermomediterranean, Mesomediterranean and Supramediterranean dry-humid.

Within that alliance it defines the community ***Greenovietum aureae***, of which it is characteristic. This is typical of higher altitudes (900–2000 m) southward and eastern sides of the island of Tenerife. On southward facing rocks and steep basaltic lava columns.

Conservation Status

In all subgrids this species was found its conservation status was high, no damages of any kind were recorded.

Threat Status

nt (Barreno et al. 1984)
Anexo II: Protegidas (Gobierno de Canarias 1991)

Germplasm Banks

E.T.S. Ingenieros Agrónomos de Madrid.
Jardín Botánico Canario Viera y Clavijo.

Exsiccata (Herbarium Specimens)

Tenerife: mayo 1969, Bramwell 1287 (TFC 230); Sobre Galería Los Cochinos-Los Silos, 6.04.1996, V. Lucía Sauquillo & M. Acosta (TFC 39279).

3.5.5 Aeonium smithii *(Sims) Webb & Berthel.*

Phyt. Canar., 3(2/I): 187–188 (1840)

Aeonium smithii Webb & Berth. *Phyt. Canar.*, 1: 187 (1840)
Sempervivum foliosum Chr. Smith. *ex* Buch *in* Abhandl. Königl. Akad. Wissensch; Berlin, 1816–17: 366 (1819) (name without description); *ex* Otto *in* Horae. *Phys. Berol.* 38 (1820)
Sempervivum hispicaule Haworth. *Suppl. Pl. Succ.*, 69 (1819)
Sempervivum smithii Sims. *Bot. Mag.* 1980 (1818)
Aionion *(greek): "ageless"* ***smithii****: Dedicated to Christen Smith (Norwegian botanist)*

Endemism, Certain native

Vernacular Name: Bejeque/Pastel de risco (ES) Smith's Treeleek (EN)
Family: *Crassulaceae*
Biotype: Subshrub (chamaephyte)
Flowering: Mai–June
Fruiting: summer
Pollination: insect pollination (entomogamy)
Dispersal: wind dispersal (anemochory)
Chromosome N°: n = 18

Description

Short, evergreen **shrub**. Few **branches**, succulent, irregular and fairly knotted, glabrous except under the terminal leaves rosettes, where they are densely hairy along 8–15 cm, stiff 4–8 mm long hairs.

Leaves broad spatulate, (Fig. 3.16) 4–5 cm long and 2.5–4 cm wide, subpeciolate, very sharply apiculate, sparsely hairy, surface rather irregular, reddish, glandules of linear in appearance, which are very long and conspicuous in their lower side, pretty undulated and irregular margins, unevenly ciliate with fine patent hairs and adorned on the upper side with hyaline hemispherical half rosaries as long as the cilia.

Flower stems 30.5 cm long, erect, hairy, densely leafy, with crowded and diminishing leaves. Short inflorescence of about 8 dichotomous branches, each with around six flowers. Flower buds broader than longer, with twisted petals. Flowers

Fig. 3.16 Cañada de Diego Hdez., June 2004 Beatriz Hernández

8–12 parted, big, flat, golden-yellow, 2 cm in diameter. Calyx 6 cm long, glandular-hairy, green, with linear glands, cut towards its bottom half into acuminate ovate-lanceolate segments. Petals obovate-lanceolate, acute, glandular-hairy on both sides, 9 mm long. Stamens yellow, epipetals of around 4.5 mm, episepals 5.5 mm long, filaments sharp; others broadly ovoid, glandular. No scales. Hairy ovaries, 3.5 mm long, very gibbous, styles filiform, erect, 2.5 mm long. **Seeds** miniscule and very numerous.

Remarks:

It also reproduces by cuttings and can hybridise with *A. spathulatum* (Hornem.) Praeger. Great value from a scientific point of view due to its morphological singularity, which differentiates it from other members of the genus and brings it towards Sect. *Greenovia* based on floral peculiarities (lack of nectaries f.ex.).

Distribution

Endemism of the island of Tenerife, (Fig. 3.17) dispersed across its southern side from 200 to 2375 m a.s.l. Important localities include Tamaimo, Vilaflor, Güímar, etc.

Within the national Park it is basically found in the interior cliffs of the Wall de Las Cañadas: Diego Hernández, Las Pilas, La Angostura, Guajara, Ucanca and El Cedro (Fig. 3.18).

Habitat and Phytosociology

Rock dwelling species found mostly within forested areas. It grows from relatively low levels of the Inframediterranean belt up until the dominance of pine forests and broom shrublands. Although it tolerates conditions at high altitudes it is sensible to the cold.

It is a characteristic species and bioindicator of the ***Greenovio aureae-Aeonietea***, class of permanent vegetation, which groups evergreen permanent rocky and casmo-comofitic communities, endemic of the Canario-Madeirense biogeographic subregion; constituted primarily by succulent chamaephytes and rosulate-succulent of Crassulaceae (*Aeonium*, *Aichryson*, *Monanthes*), which develop in rocky

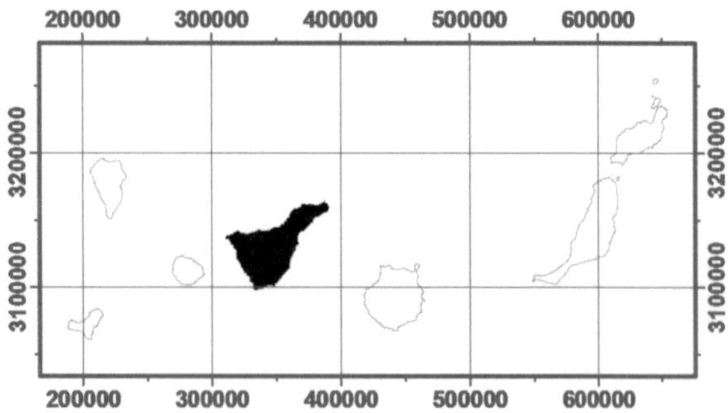

Fig. 3.17 Endemism of the island of Tenerife

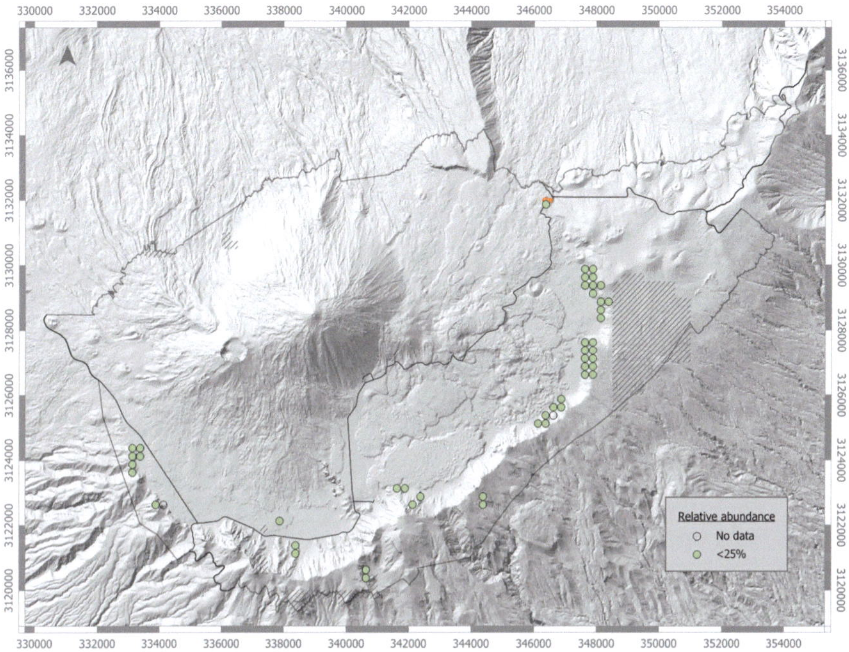

Fig. 3.18 *Aeonium smithii* distribution. N° Sub-grids (250 × 250 m): **49**/3740

surfaces with some sediment, walls, cracks and stony indentations in the Canary and Madeira archipelagos in the Infra-Supramediterranean hyperarid-humid bioclimates. It is characteristic of the ***Cheilantho guanchicae-Aeonietum smithii*** community, an endemic community of the mountains of the Circo de las Cañadas del Teide.

Conservation Status

> "*Its conservation status is relatively good. Most of the localities where it grows are devoid of human impact due to their inaccessibility. But other known localities suffer risks due to soil erosion (Montaña Hoya) or destruction of corresponding badlands (Güímar). Furthermore, older authors comment on its fragility to frost events in Las Cañadas. Poaching of individuals constitute an important threat as it is a very notable and aesthetic plant with remarkable flowers.*" (Translated from: Gómez Campo et al. 1996)

Based on the sampling done in the National Park, in 96% (47) of the subgrids where it was found, no damages were observed. Only in 2% (1) damages were observed (grazing, high mortality with lack of juveniles, etc.) but this didn't reach more than 25% of individuals affected.

Threat Status

V (Barreno et al. 1984)
Anexo II: Protegidas (Gobierno de Canarias 1991)
LR pm (Bañares et al. 1998)

Germplasm Banks

Banco de Germoplasma de E.T.S. Ingenieros Agrónomos de Madrid.
Jardín Botánico Canario Viera y Clavijo.

Exsiccata (Herbarium Specimens)

Tenerife: Cañada de Diego Hernández-Las Cañadas, 31.07.1971, F. Estévez & A. Acuña (TFC 15990 + Dupl.); Las Cañadas, 30.5.1981, W. Wildpret de la Torre, E. Beltrán Tejera, A. Losada Lima & A. García Gallo (TFC 33144). Inventario de la Flora y Vegetación del Parque Nacional del Teide (2001–2003) V.E. Martín Osorio, W. Wildpret & B. Hernández Bolaños: Llano de Ucanca (41D 4 d), 22/04/2003, Y. de Rioja Santos (TFC 45427).

3.5.6 Aeonium spathulatum *(Hornem.) Praeger var.* spathulatum

Praeger (1925): 210

Aeonium bentejui Webb.*ex* Christ *in Bot. Jahrb.*, 9: 109 (1888)
Aeonium strepsicladum Webb & Berth. *Phyt. Canar.*, 3(2/1): 187 (1840)
Aldasorea strepsicladum Hort. *ex* Haage & Schmidt Gen. Cat. 203 (1930)
Sempervivum barbatum Chr. Smith. *ex* Buch *in* Abhandl. Königl. Akad. Wissensch; Berlin 1816–17: 366 (1819) (name without description); *ex* Otto *in* Horae *Phys. Berol.* 37 (1820) (not of Hornemann *Suppl. Hort. Hafniensis* 61 (= *caespitosum*, see p. 193), Weeb & Berth. Phyt. *Canar.*, 1: 188 (= *A. caespitosum* × *spathulatum*, see, p. 197), nor of De la Soie *in Bull. Soc.*; Murithienne, 3: 16 (name without description))
Sempervivum bentejui Christ. *Bot. Jahrb.*, 9: 160 (1888)
Sempervivum lineolare Haworth. *Suppl. Pl. Succ.* 69 (preface dated May 1819)
Sempervivum spathulatum Hornemann. *Suppl. Hort. Hafniensis* 60 (preface dated March 15, 1819)
Sempervivum strepsicladum Webb & Berth. *Phyt. Canar.*, 1: 31 (1840)
Sempervivum villosum Lindley. *Bot.* register 1553 (1832) (not of Aiton, Haworth, nor Lamotte, see p. 116)
***Aionion** (greek): "ageless"* ***spathulatum**: spatulate or like a spoon (referring to the shape of the leaves)*

Endemism, Certain native
Family: Crassulaceae
Common name: Sayón/Bejequillo canario (ES) Spoonleaved Treeleek (EN)
Growth form: Subshrub (chamaephyte)
Flowering: March–Mai
Fruiting: Mai–June

Pollination: insect pollination (entomogamy)
Dispersal: wind dispersal (anemochory)
Chromosome N°: n = 18

Description

Erect **subshrub**, 30–60 cm high, very branched, hard and treelike. Thin, Woody branches, suberect or twisting, blackish, rough, the youngest very leafy and finely pubescent while growing, later glabrous except at the apex. **Leaves** obovate-spatulate or obovate-rhomboid, apiculate, 1–3 cm long and 7–8 mm wide, sparsely stalked, finely pubescent, somewhat sticky and thick, underside round and marked with longitudinal glands, brownish-green in colour, (Fig. 3.19) leaf margins with weak and irregular rosary shaped cilia, young leaves form an ovoid bud. **Floral stem** to the most part large, erect, 15–20 cm long, slender, at the beginning leafy with similar leaves, at flowering time mostly bare. Glandular-pubescent and sticky. Terminal **inflorescence**, compact, flattened, 5–10 cm wide and long, few bracts, almost linear, broader towards the apex, glandular pubescent. Flowers 8–10 parted, Golden-yellow coloured, 1 cm in diameter. Pedicels slender, 5–7 mm in longitude. Calyx 2 mm in longitude, glandular-pubescent, split 2/3 along the middle to the base in acute lanceolate segments. Petals 4.5–5 mm long, oblong, blunt, golden. Stamens 4–5 mm long, golden, filaments filiform. Without scales. Carpels golden, not very slender, ovaries and styles 2 mm in length.

Fig. 3.19 Teide Botanical Garden, June 2004. (Photo credit: Beatriz Hernández)

Distribution

Found on all Canary Islands except Lanzarote and Fuerteventura. In the Teide National Park, it reaches 2175 m a.s.l. and is found north-western of the Cerro Las Chajoras, in areas of Montaña Abejera, La Fortaleza, upper parts of La Orotava valley, Barranco del Río and Roques de El Cedro (Figs. 3.20 and 3.21).

Habitat and Phytosociology

Fissure and rupicolous species. It is a characteristic species and bioindicator of the ***Greenovio aureae-Aeonietea***, class of permanent vegetation, which groups evergreen permanent rocky and casmo-comofitic communities, endemic of the Canario-Madeirense biogeographic subregion; constituted primarily by succulent chamaephytes and rosulate-succulent of Crassulaceae (*Aeonium*, *Aichryson*, *Monanthes*), which develop in rocky surfaces with some sediment, walls, cracks and stony indentations in the Canary and Madeira archipelagos in the Infra-Supramediterranean hyperarid-humid bioclimates. Participant as well in the ***Cheilantho guanchicae-Aeonietum smithii*** community.

Conservation Status

In 98% (44) of the sub-grids it was found in, conservation status was high, meaning, no damages were detected. Only in 2% (1) of the sub-grids, plants showed damages (northern slopes of Teide).

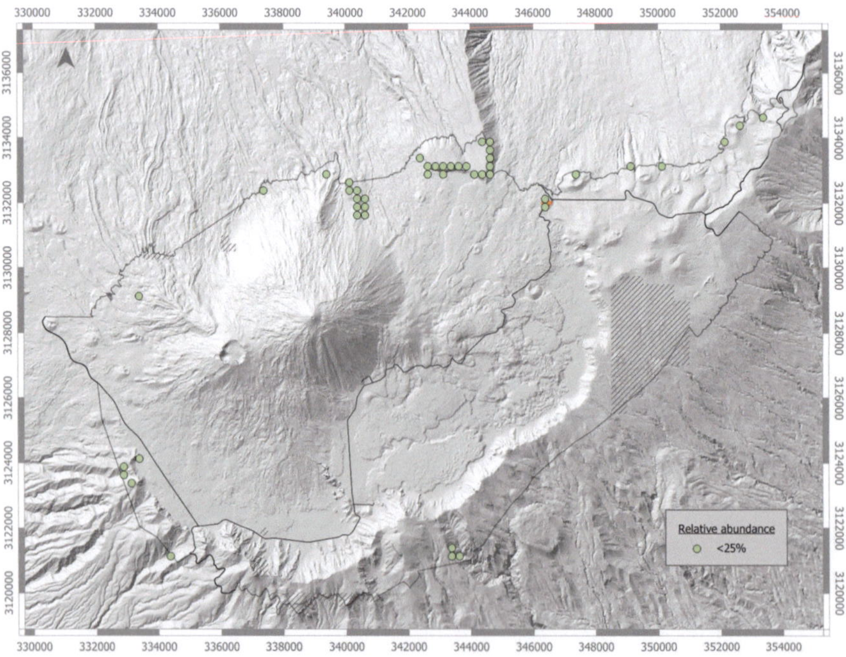

Fig. 3.20 *Aeonium spathulatum* distribution. N° Sub-grids (250 × 250 m): **45**/3740

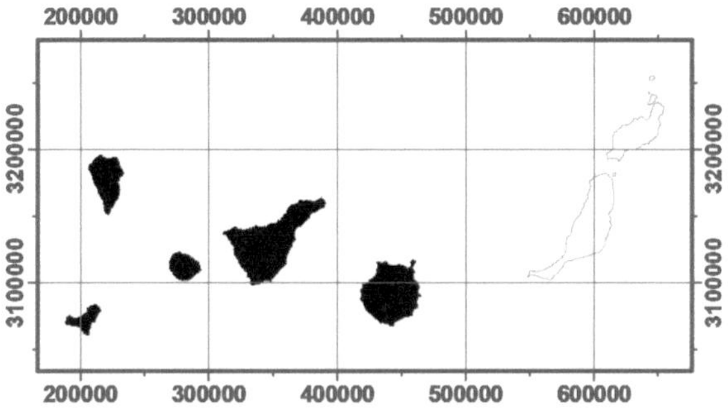

Fig. 3.21 Endemism of the Canary Islands except Lanzarote and Fuerteventura

Threat Status

nt (Barreno et al. 1984)
Anexo II: Protegidas (Gobierno de Canarias 1991)
LR pm (Bañares et al. 1998)

Germplasm Banks

Banco de Germoplasma de E.T.S. Ingenieros Agrónomos de Madrid.
Jardín Botánico Canario Viera y Clavijo.

Exsiccata (Herbarium Specimens)

Tenerife: T. Erjos-Los Silos, 11.05.1997, V. Lucía Sauquillo, C. Acosta, J.A. Reyes Betancort & A. Díaz (TFC 40818). Inventario de la Flora y Vegetación del Parque Nacional del Teide (2001–2003) V.E. Martín Osorio, W. Wildpret & B. Hernández Bolaños: Pinar de la parte Norte (29D), 24/06/2003, Y. de Rioja Santos (TFC 44285); Cañada de Los Guancheros—La Fortaleza (30A 24), 09/07/2001, Y. de Rioja Santos (TFC 44266).

3.5.7 **Andryala pinnatifida** *Aiton ssp.* **teydensis** *(Sch. Bip.) Rivas-Mart. et al*

Rivas-Martínez, S. *et al. (2002). Itinera Geobotanica, 15(2): 433–922*
pinnatus*: pinnate* ***fidus****: split or dented (referring to the form of the leaves)*

Endemism, Certain native
Family: *Asteraceae*
Common name: Estornudera del Teide/Espinera/Hierba peluda (ES)
Growth form: Subshrub (chamaephyte)
Flowering: Mai–July
Fruiting: Mai–July

Pollination: insect pollination (entomogamy)
Dispersal: wind dispersal (anemochory)
Chromosome N°: 2n = 18

Description

Suffruticose, cano-tomentose. **Roots** of 20–30 cm, vertical or twisted, upper parts 4–10 mm in diameter, rounded, fusiform, some looking like fibrillose branches, lignose, white line, hard, brown bark. **Stems** solitary or branched from the base in a single cauliform bunch, round, foliate, with compound or racemose corymb, tomentose, leaves with sessile stellate hairs. **Leaves** on the upper part ashy-green, tomentum of the opposite side not so dense, ashy-albaceous. Tomentum hairs towards the corymb or the raceme articulate, rounded, on the upper side smaller, subglobose apex, glanduliferous, greenish-flavescent, patent, in the upper parts dense and, frequently, as dense in the peduncle and involucre as in the corymb. Hairs of the corymb leaves, especially on the upper side, in flanks and nerves of the opposite side, grayish, not as dense. Leaves arranged in a spiral line from the base until the stem apex, those in the lower section, bigger and denser, shallowly pinnatifide or sinuate-toothed, acute, rachis short, linear-lanceolate-elliptic, narrow, subracemose, capitulum small; teeth triangular-oblong. Lower middle section with attenuate winged petiole; upper part, base dilated and amplexicaule. Involucre greyish, hairs glanduliferous. In the axes of the involucre leaves and in the dichotomous panicles, a kind of reddish gum sometimes flows.

Capitulum always erect, multifloral. Involucre hemispherical campanulate, subuniserial, 4–8 mm high, tomentose. Scales 13 to 2, narrow-patent, inner side glabrous, back side carinate. Outer scales longer, narrow, carina much more hispid, margin hardly scarious, in the inner side carina less hispid and tomentose, margin very scarious. Receptacles 2–4.4 mm in diameter, alveolate-reticulate, margin of alveoli slightly raised, toothed, some setae very simple. **Flowers** golden, (Fig. 3.22) hermaphrodyte, liguliform, 5-dentate, radial involucres over 2–4 mm. Ligule 1–1.5 mm wide, 5-toothed at the apex, triangular-lanceolate. Cylinder of anthers not

Fig. 3.22 Cañada de El Capricho, July 2003. (Photos credit: Victoria Eugenia Martín Osorio)

very long, apex rounded; filaments glabrous, flat; styles subarched. Achenes 35–112, conformable, elongate-ovate-cylindrical, lower part attenuate, upper part shortly contracted, glabrous, 10-costata. Achenes 35–112, conformable, elongated-ovate-cylindrical, lower part attenuated, upper part slightly contracted, glabrous, 10-ribbed, ribs very prominent, lower part with basilar areoles, circular, shortly subevanescent, upper part in horns or teeth 10, linear-lanceolate, not very prominent. Pappus deciduous, whitish, uniserial, setae slightly unequal.

Remarks:

It was described by Webb & Berthelot as form *teydensis* (*Phyt. Canar.*, 3(2/2): 412–418, 1836–1850). Later, it was elevated by Christ to the rank of variety, and even later, to the rank of subspecies by Rivas-Martínez et al. (2002).

Distribution

Endemism which is only found in the summits of the island of Tenerife, where it reaches 2500 m a.s.l. Within the National Park it is widely distributed in the aster reaches of Las Cañadas, especially in zones of the interior and exterior of the Wall, from Montaña Guajara until La Fortaleza and southern slopes of Izaña and Siete Fuentes. In a more disperse manner, in south-eastern areas of the interior of the Circo de las Cañadas (Figs. 3.23 and 3.24).

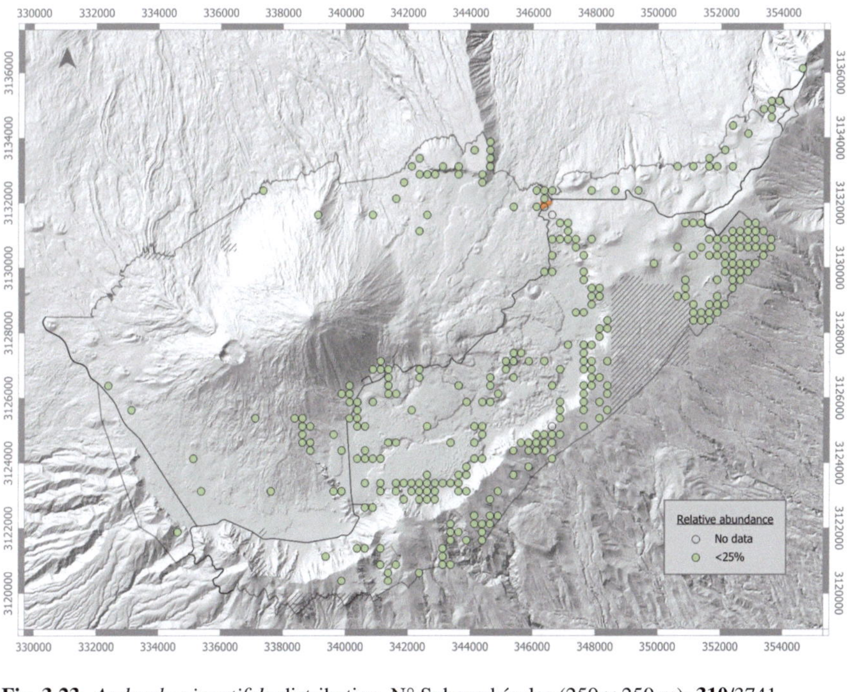

Fig. 3.23 *Andryala pinnatifida* distribution. N° Subcuadrículas (250 × 250 m): **310**/3741

Fig. 3.24 Endemism of the Teide National Park

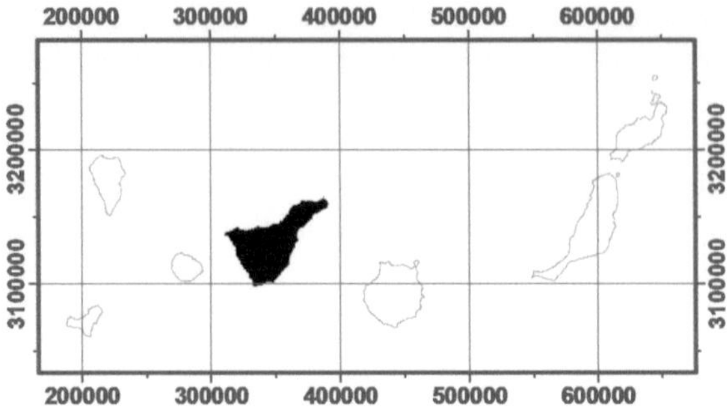

Habitat and Phytosociology

It lives preferably in rocky and/or stony areas. Species characteristic of the class ***Chamaecytiso-Pinetea canariensis***. It can also be found in the following phytosociological units: ***Chamaecytiso-Pinetalia canariensis***, ***Spartocytisetalia supranubii***

Conservation Status

In 91% (281) of subgrids where this taxon was found, conservation status was high. Only in 7% (22), damages were observed (herbivory, high mortality without regeneration, etc.), but those did not affect more than 25% of individuals. Only in 2% (6) of subgrids, damages in more than 25% of individuals was detected.

Threat Status

VU D2 (Bañares et al. 1998)

Exsiccata (Herbarium Specimens)

Tenerife: Pista de Lomo Blanco-Vilaflor, 1600 m.s.n.m., 13.06.1979, P.L. Pérez de Paz & C. Hernández Padrón (TFC 32410 + Dupl.); Ctra. Gral. San Andrés, 20.07.1979, P.L. Pérez de Paz & C. Hernández Padrón (TFC 32411); Inventario de la Flora y Vegetación del Parque Nacional del Teide (2001–2003) V.E. Martín Osorio, W. Wildpret & B. Hernández Bolaños: Camino de la Degollada de Guajara, 25/06/2003, A. Montañés Rodríguez (TFC 44282).

3.5.8 **Anogramma leptophylla** ***(L.) Link***

Fil. Sp. Cultae: 137 (1841)

Asplenium leptophyllum (L.) Sw. *in* Observ. Bot.: 403 (1791)
Osmunda leptophylla (L.) Savigny *in* J.B.A.M.de Lamarck, Encycl. 4: 657 (1798)

Grammitis leptophylla (L.) Sw. *in* Syn. Fil. (Kiliae): 23, 218 (1806)
Polypodium leptophyllum L. *in* Sp. Pl.: 1092 (1753)

Anogramma: Generic name for the arrangement of sori in a line at the end of nerves.
Leptos- (greek): fragile -phyllos (greek): leaf

Not endemic, Certain Native
Family: *Hemionitidaceae*
Common name: Helecho de tiempo (ES); Jeresey Fern (EN)
Biotype: Annual (therophyte)
Sporulation: March–June
Dispersal: wind dispersal (anemochory)
Chromosome N°: 2n = 52, 58; n = 26, 29

Description

Rhizome minute, covered with linear-lanceolate palea, rusty at the base, hyaline elsewhere. **Fronds** dimorphic, pubescent when young (Fig. 3.25). The outer ones (2) 3–7 (9) cm, sterile: blade 1–2 pinnate, ovate to kidney-shaped, with the last segments flabellate, lobed at the apex. The inner ones (5) 7–15 (25) cm, fertile; blade 2–3 pinnate, ovate-oblong to narrowly triangular, with the last segments ovate-oblong, cuneate at the base, emarginate, or slightly lobed at the apex; last order of veins bifurcated. **Sporangia** exindusidate, arranged on the ends of the veins, confluent at maturity. **Spores** triplet, subglobose-tetrahedral, warty, with 3 prominent ribs in the equatorial zone.

Fig. 3.25 *Anogramma leptophylla*, January 2015. (Photo credit: Victoria Eugenia Martín Osorio)

Distribution

Present throughout Macaronesia and the Mediterranean region; western Iberian Peninsula, Catalonia, the Balearic Islands and other isolated localities in central and eastern Spain, and Andalusia; in Atlantic Europe, tropical Africa, South Africa, the Near East, India, Australia, New Zealand, South and Central America. It also appears in all the Canary Islands (Fig. 3.26). In the Teide National Park (Fig. 3.27) it was found in a very dispersed manner and with little representation throughout the protected area; only in certain isolated points in areas of the ridge, the interior of Las Cañadas and, in La Pared, only in the Cañada de Ucanca, between 2050 and 2300 m a.s.l.

Habitat and Phytosociology

This fern colonizes damp earthy slopes, cracks, rocks and/or walls, in the shade and with frequent dew. It is a characteristic species and bioindicator of ***Selaginello denticulatae-Anogrammion leptophyllae***, an alliance that groups the associations of shady and cool earthy slopes, Thermo-Meso-Mediterranean subhumid-humid hyperoceanic-oceanic, with a wide Mediterranean distribution.

Conservation Status

In virtually all the subgrids in which this species was found (14), no damage of any kind was observed. Only in one of them was its condition poor, with damage to more than 25% of individuals.

Exsiccata (Herbarium Specimens)

Tenerife: Entre el Pitón y la Caldera de Aguamansa, La Orotava, 1500 m, 5.1981, W. Wildpret & M. del Arco Aguilar (TFC 13676); La Laguna, 27.3.1985, W. Wildpret & A. García Gallo (TFC 10313); La Esperanza, 27.3.1985, W. Wildpret & A. García Gallo (TFC 23640); Bco. Tahodio, sobre la presa, Anaga, 18.2.1985, P.L. Pérez (TFC 24730); La Fortaleza, 5.1983, H. Metlesics, W. Wildpret et col. (TFC 28292); Cañada de Diego Hdez., 2.6.1973, C. Gil Rguez. (TFC 28293); Bco. de Añavigo, 25.5.1973, C. Gil Rodríguez (TFC 28294); Cuchillos del Saltadero, 11.1972,

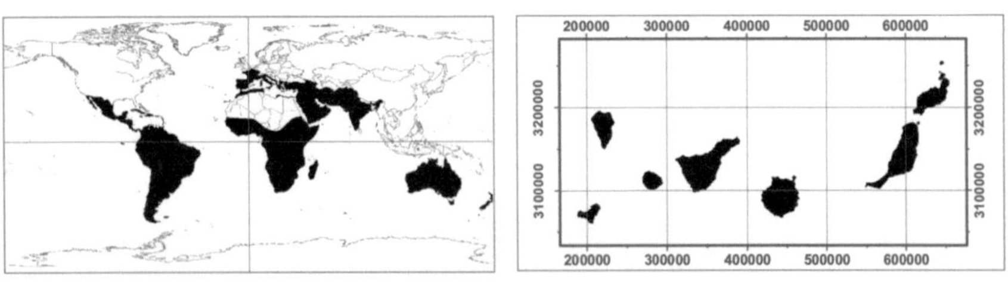

Fig. 3.26 Present on all the Canary Islands

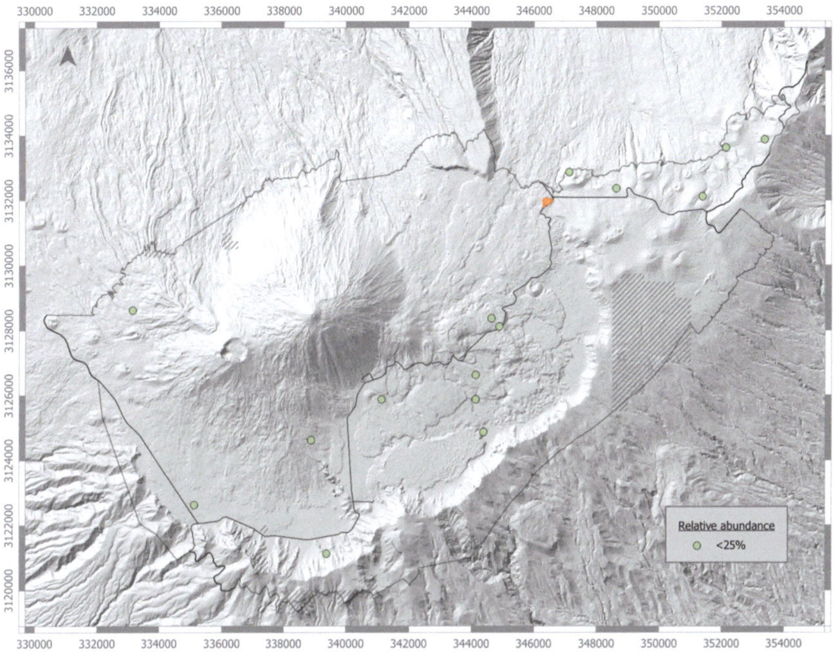

Fig. 3.27 *Anogramma leptophylla* distribution, N° Subgrids (250 × 250 m): **15**/3740

A. Acuña (TFC 28295 + duplic.); La Laguna, 16.3.1979, W. Wildpret & M. del Arco Aguilar (TFC 8631 + duplic.); Mña. Mostaza, 2.1969, A. Santos (TFC 28296); Bco. de la Orchilla, 22.4.1973, E. Beltrán (TFC 28297); Bco. de la Orchilla, 20.5.1973, E. Beltrán (TFC 28298); Las Mercedes, 5.1974, FRG (TFC 28299); Las Mercedes, 3.1970, C. Gil Rguez. (TFC 28300); San Diego, La Laguna, 5.4.1971, C. Gil Rguez. (TFC 28301 + duplic.); Bco. de Añavigo, 5.1973, W. Wildpret, H. Metlesics, E. Beltrán, P.L. Pérez & C. Gil Rguez. (TFC 38302); Mte. de las Mercedes, 6.1974, (TFC 28319); Los Realejos, 4.1974, C.L. García Palmero (TFC 28578); Las Mercedes, 15.5.1969, M. C. Gil (TFC 28582); La Laguna, Rancho Grande, 27.3.1985, W. Wildpret & A. García Gallo (TFC 10313, 10314, 10315); La Laguna, 22.4.1986, A. García Gallo (TFC 22612); Bco. Badajoz en comienzo cascada, 7.3.1986, O. Rguez. Delgado (TFC 27517); Lomo cresta Ladera bajo mirador, 17.3.1986, O. Rguez. Delgado (TFC 27516); Bco. de Badajoz, 14.3.1988, W. Wildpret, O. Rguez. Delgado & Prof. Berg. Arteaga (TFC 27515); Bco. de la Pasada, Agache, 19.5.1981, E. Beltrán & O. Rguez. Delgado (TFC 12846); Los Silos, 21/3/88, E. Beltrán, A. Losado, J.R. Acebes & C. León Arencibia (TFC 32803); Galería superior Valle Ucanca, 2375 m, 24.4.1983, Ene 92, J.S. Socorro, MCL (TFC 34145); Galería 1ª del Bco. del Riachuelo, 2150 m, 12.10.1982, Ene 92, E. Beltrán, J.S. Socorro, W. Wildpret & M.C. León Arencibia (TFC 34146); Güímar, 1.1973, J.L. Rguez. Armas (TFC 35329); La Laguna, Rancho Grande, 11.3.1997, J.R. Acebes (TFC 40497); Pista de los Madroños, Los Silos, 6.4.1997, V. Lucía

Sauquillo, A. Díaz, J.A. Reyes-Betancort & E. Hernández (TFC 40036); Cuenca del Bco. de Tahodio, Anaga, 550 m, 9.4.1999, S. Becher, T. Himstedt & E. Beltrán (TFC 42142); Bco. Hoya de las Limeras, 300 m, UTM 28RCS5950, 14.4.2002, M. Rguez. López & M.L. Rguez. Navarro (TFC 44688). Inventario de la Flora y Vegetación del Parque Nacional del Teide (2001–2003) V.E. Martín Osorio, W. Wildpret & B. Hernández Bolaños: Roques al S del Llano Ucanca (41D 27 d), 22/04/2003, Y. de Rioja Santos (TFC 44862).

Gran Canaria: Bco. Oscuro, Valleseco, 5.1980, P.L. Pérez & C. Suárez (TFC 11095); Bco. Oscuro, El Tablero, 2.6.1979, J.R. Acebes, M. del Arco Aguilar & M.C. León (TFC 29935); Camino vecinal Presa de los Pérez, 800–900 m, 15.10.1983, J.S. Socorro (TFC 18497).

El Hierro: El Cres, 10.4.1979, W. Wildpret, M. del Arco Aguilar, P.L. Pérez & C. Hernández Padrón (TFC 8265); El Golfo, 26.3.1972, E. Barquín Díez (TFC 22909); El Golfo, 1.1992, W. Wildpret, E. Beltrán, E. Barquín Díez & M.C. León (TFC 22948); El Hierro, Ventejea, cumbres, 16.7.1981, P.L. Pérez & C. Hernández Padrón (TFC duplic. 34353).

La Palma: Mazo, 10.4.1974, (TFC 28572); Curva de los Palos Blancos, a Fajana de los Revoltillos, 12.2.1990, M.C. León, E. Beltrán, A. Losada, J.M. González, C. Hernández Padrón, A. Bañares & J.L. (Proyecto DGICYT) (TFC duplic 28847, 28871); Inmediaciones de la Montaña de los Charcos, Fuencaliente, 1200 m, 1.6.1991, P.L. Pérez, J. Leal (TFC 41129); La Fajana, Garafía, 9.4.1998, E. Beltrán, J.L. Rguez.-Armas, J. Leal Pérez & S.A. Medina (TFC 42795).

Lanzarote: Mña. Los Helechos, 2.4.1996, J. Cazorla Godoy, J.A. Reyes-Betancort (TFC 38811); Haría, Bco. Chafarís, 29.1.1997, J.A. Reyes—Betancort & A. Perdomo (TFC 40350); Haría, Jameo cumplido, 6.3.1997, J.A. Reyes—Betancort (TFC 40394); Haría, Jameo de la Gente, 4.3.1997, J.A. Reyes—Betancort (TFC 40455).

3.5.9 Arabis caucasica *Willd.*

Willd., *Enum. Pl. Hort. Berol. suppl.*, 45 (1813)

Arabis alpina subsp. caucasica (Willd.) Briq. *in* Prodr. Fl. Corse 2(1): 48 (1913)
Arabis caucasica Schltdl. First published *in* Enum. Pl. Hort. Berol., Suppl.: 45 (1814)
Arabis albida Stev. nom. illeg.
Arabis alpina Georgi *in* Beschr. Nation. Russ. Reich 3(4): 1137 (1800), nom. illeg.
Arabis caucasica var. *leiopoda* Pau *in* P.Font i Quer, Exsicc. (Iter Marocc.) 1929: n.° 197 (1930), nom. nud.
Cardamine viscosa S.G.Gmel. *ex* DC. *in* Syst. Nat. 2: 216 (1821)
Erysimum albidum (Steven *ex* M.Bieb.) Kuntze *in* Revis. Gen. Pl. 1: 28 (1891)

Not endemic, Probable native
Family: *Brassicaceae*

Common name: Pelusilla (ES); Mountain rock cress (EN)
Biotype: Hemicryptophyte
Flowering: April–June
Fruiting: April–June
Pollination: self-pollination (autogamy)
Dispersal: wind dispersal (anemochory)
Chromosome N°: 2n = 16

Description

Plant 15–35 cm, procumbent, with many rosettes, usually smooth and densely pubescent. Basal leaves greyish green or whitish, with 2–5 obtuse teeth on each side; cauline leaves auriculate-sagittate at base.

Inflorescence axes and pedicels usually glabrous (Fig. 3.28). Sepals 5–8 mm; Petals 9–18 × 5–8 mm, with a broadly patent and abruptly contracted limb.

Silique 40–70 × 1.6–2.4 mm, somewhat constricted between widely spaced seeds; pedicels 10–20 mm; valves with a patent median vein. Seeds not winged or with a narrow, inconspicuous wing.

Remarks:

According to the consulted bibliography, this taxon has several synonyms and even inaccuracies in the citations of authors, so a taxonomic study and review would be necessary. It is widely cultivated in gardens and frequently escapes.

Fig. 3.28 On the path to Guajara, before the Degollada, Apr. 2009. (Photo credit: Beatriz Hernández)

Distribution

Distributed throughout southern Europe, frequently naturalized in other parts of the globe. It appears throughout the Canary Islands except on the islands of Lanzarote and Fuerteventura.

In the Teide National Park it can be found more or less frequently colonizing the interior slopes of the SE of the Pared de Las Cañadas (from Risco Verde to Cañada de Ucanca), between 2100 and 2700 m a.s.l. (Figs. 3.29 and 3.30) In La Palma it grows together with *Viola palmensis* Webb & Berthel.

Habitat and Phytosociology

It is a characteristic species and bioindicator of the ***Greenovio aureae-Aeonietea*** class, perennial vegetation, which groups together permanent perennial comophytic and chasmo-comophytic communities, endemic to the Canary-Madeiran biogeographic subregion; constituted mainly by succulent and rosulate-succulent chamaephytes of *Crassulaceae* (*Aeonium*, *Aichryson*, *Greenovia*, *Monanthes*), which develop on somewhat earthy rock surfaces, walls, cracks and rocky crevices; hyper-arid-humid Infra-Supra-Mediterranean floors. It usually colonizes eminently arid areas.

It may also appear in ***Sileno berthelotianae-Cerastietum sventenii***, a rupicolous association of phonolitic cracks and spurs and trachybasalts oriented to N and NW

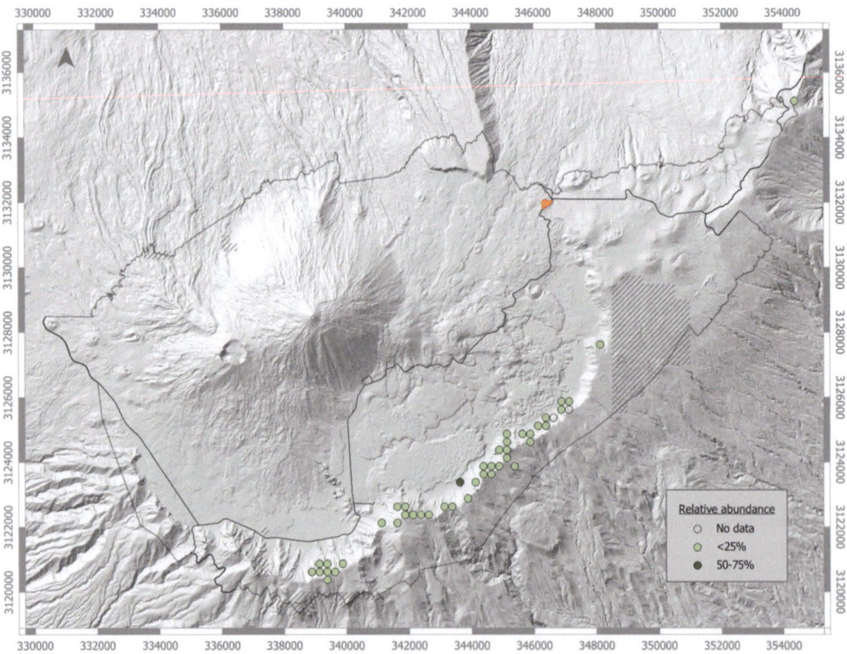

Fig. 3.29 *Arabis caucasica* distribution, N° Subgrids (250 × 250 m): **45**/3.740

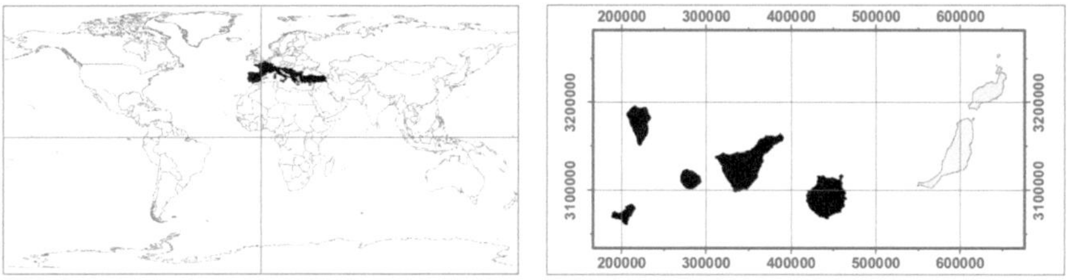

Fig. 3.30 Present on occidental Canary Islands and Gran Canaria

of the dry Supra-Mediterranean floor. It is found at altitudes between 2200 and 3000 m, but always in windward situations within the area of the ***Spartocytisetea supranubii***, ***Spartocytision*** and ***Violion cheiranthifoliae*** communities. This community replaces ***Cheilantho guanchicae-Aeonietum smithii*** at altitudes above 2200 m a.s.l.

Conservation Status

In practically all the subgrids in which this species was found (43), no damage of any kind was observed. Only in one of the sub-grids, its conservation status was estimated to be medium. In the rest (1), this variable was not specified at the time of sampling.

Germplasm Banks

Real Jardín Botánico de Madrid.

Exsiccata (Herbarium Specimens)

Cited as *Arabis albida* Stev.:

Tenerife: Las Cañadas, Topo de la Grieta, 31.5.1973, W. Wildpret, E. Beltrán, L. GR, P.L. Pérez, M. Galván & H. Metlesics (TFC 21725).

La Palma: Inmediaciones Cumbres de La Palma, 2200 m, 8.8.1998, P.L. Pérez (TFC 41616).

Cited as *Arabis alpina* ssp. *caucasica* (Schlecht in Willd.) Briq.:

La Palma: Pico de la Cruz, 6.4.1980, J. Fernández Casas (TFC 10557).

Cited as *Arabis grex. caucasica*:

La Palma: El Canal y Los Tiles, bordes del canal, 1.4.1989, MCL, E. Beltrán, A. Losada, J.M. González, A. Bañares (TFC 28976).

Cited as *Arabis caucasica* Schlecht in Willd.:

Tenerife: Altos de Arafo, Güímar, 27.4.1974, E. Barquín Díez, W. Wildpret, P.L. Pérez, E. Beltrán, J.R. Acebes, B. Méndez, RAL, CLA (TFC duplic. 32580); Galería superior, Valle de Ucanca, 24/4/83, S. Socorro, (TFC 34257); 11.06.1969, 1929, Bramwell (TFC 31); Topo de la Grieta, 31/5/73, W. Wildpret, P.L. Pérez, MG, CG, H. Metlesics & E. Beltrán (TFC 3898); Cumbre de los Andenes, 01.04.1977,

J.R. Acebes (TFC 15048); Valle volcán de las Arenas, cumbres de Arafo, 09.05.1981, W. Wildpret, M. del Arco Aguilar & P.L. Pérez (TFC duplic. 23750); El Canal y Los Tiles, 01.04.1989, E. Beltrán, A. Bañares, J.M. Glez. Mancebo, A. Losada, J.P. Leal & M.C. León Arencibia (TFC duplic. 28838); Las Cañadas del Teide, frente Degollada de Guajara, 2400 m, 28.05.1983, J.S. Socorro, W.Wildpret (TFC 18462). Inventario de la Flora y Vegetación del Parque Nacional del Teide (2001–2003) V.E. Martín Osorio, W. Wildpret & B. Hernández Bolaños: Roques al Sur del Llano Ucanca (41D 36 d), 21.04.2003, A. Montañés Rodríguez (TFC 44844); Roques al Sur del Llano Ucanca (41D 37 c), 21.04.2003, Y. de Rioja Santos (TFC 44861); Parque Nacional del Teide, 05.06.2003, W. Wildpret & V.E. Martín Osorio (TFC 44975); Guajara, Teide, 2622 m.s.n.m., 20.05.2002, A. Hdez. Luis, L. Afonso, Y. de Rioja (TFC 43855).

La Palma: Caldera Morro Negro, 16.06.95, P.L. Pérez (TFC duplic. 38726); Cumbres, s.d., P.L. Pérez, (TFC duplic. 38787); P.N. Caldera de Taburiente, 14/5/99, E. Beltrán et al. (TFC 2 duplic 42218); Pico de la Cruz, 22.05.2002, J.P. Oval, A. Acevedo, V. García (TFC 43678); Puntallana, pista forestal que sube desde el cementerio hacia el Pico de las Nieves, 900 m, 15.02.2002, P.L. Pérez & J.A. Reyes-Betancort (TFC 43769); La Palma, 13.05.1984, A. Charpin & A. Santos (TFC 20275).

3.5.10 Arenaria leptoclados *(Rchb.) Guss.*

Sic. Syn., 2: 824 (1845)

Arenaria serpyllifolia var. *leptoclados* Rchb. *in* Icon. Fl. Germ. Helv. 5: 32 (1841)
Arenaria serpyllifolia subsp. *leptoclados* (Rchb.) Nyman *in* Consp. Fl. Eur.: 115 (1878)

Not endemic, Probable native
New record for Teide National Park
Family: *Caryophyllaceae*
Common name: Morujilla (ES); Slender Sandwort (EN)
Biotype: Annual (therophyte)
Flowering: February–June
Fruiting: February–June
Pollination: wind pollination (anemogamy)
Dispersal: wind dispersal (anemochory)
Chromosome N°: 2n = 40

Description

Annual plant with a 10–12 cm delicate, glandular stem. **Leaves** ovate-acuminate, lanceolate, linear, hirsute-glandular with 3 prominent nerves, the lower ones petiolate. **Flowers** pentamerous, sepals 3–4 mm, lanceolate, long subulate, hirsute; entire petals 1/2–1/3 the length of the sepals (Fig. 3.31). **Subcylindrical capsule**

Fig. 3.31 *Arenaria leptoclados*. (Photo credit: GG Casanova)

dehiscent by 6 entire teeth less than the sepals or equal. **Seeds** 0.4–0.5 mm, reniform, finely echinulate.

Distribution

Distributed throughout Macaronesia, the Mediterranean region, N Africa and SW and Central Europe. In the Canary Islands it appears on all the islands (Figs. 3.32 and 3.33).

In the Teide National Park, it is distributed mainly in the interior of the southeastern Las Cañadas, although it also forms important populations in the surroundings of Montaña Echisere, slopes and lava flows of Montaña or Cerro las Chajoras, between Punta de Los Roques (Cañada del Cedro) and Montaña de Chasogo and the western part of Llano Ucanca; between 1700 and 2750 m a.s.l. Now, more punctually, it is found beyond La Pared towards the southern slopes and in the heights of the Valley of La Orotava and Güímar.

Habitat and Phytosociology

It is a characteristic species of the ***Tuberarietea guttatae*** class, in xerophilous and ephemeral therophytic grasslands that colonize incipient or degraded soils, poor in organic matter.

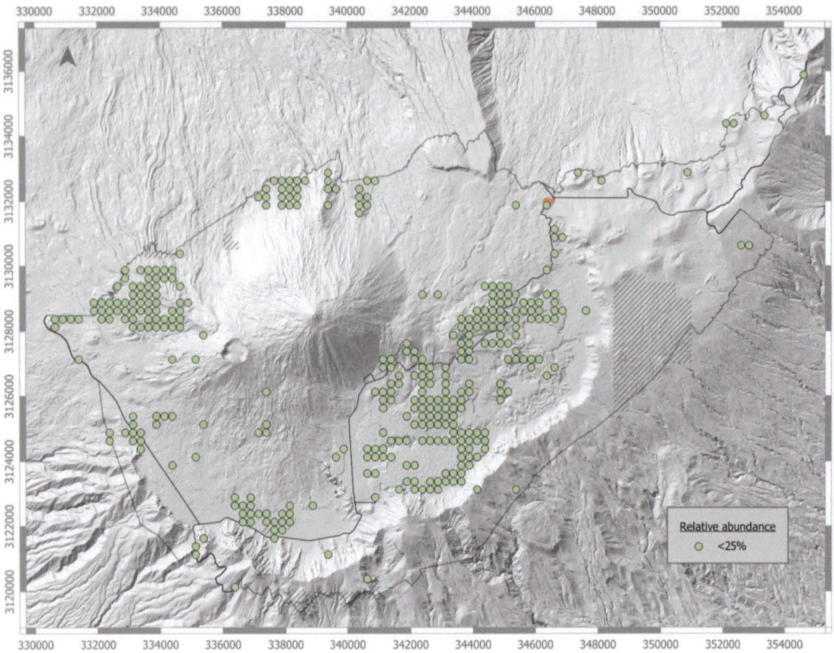

Fig. 3.32 *Arenaria leptoclados* distribution, N° Subgrids (250 × 250 m): **357**/3740

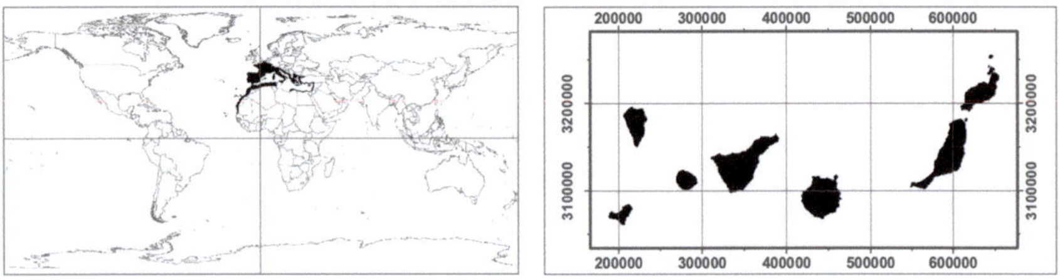

Fig. 3.33 Present on all the Canary Islands

Conservation Status

In 93% (333) of the subgrids in which this species was found, no damage of any kind was observed. Only in one (1) of them, damage was detected (browsing, high mortality without regeneration, etc.), but this did not affect more than 25% of individuals. Damage was only found on more than 25% of individuals in 6% (23) of the sub-grids where this taxon appeared.

Exsiccata (Herbarium Specimens)

Tenerife: Tenerife, Arafo, 11.4.1974, P.L. Pérez (TFC 4617); Ladera de Güímar, 16.4.1981, P.L. Pérez, W. Wildpret, M. del Arco Aguilar (TFC 13735); Pinar sobre el Bailadero, 1450 m, W. Wildpret, P.L. Pérez, M del Arco Aguilar, O. Rguez.

Delgado (TFC 27642); Las Cañadas, 30.5.1981, W. Wildpret, E. Beltrán, A. Losada, A. García Gallo (TFC 33123); Valle del Volcán de las Arenas, Arafo, 1700 m, 9.5.1981, W. Wildpret, M del Arco Aguilar (TFC 33979); Las Cañadas, Bco. del Riachuelo, 2100 m, s.d., J.S. Socorro, W. Wildpret (TFC 18451); Inventario de la Flora y Vegetación del Parque Nacional del Teide (2001–2003) V.E. Martín Osorio, W. Wildpret & B. Hernández Bolaños: Al W de la Montaña del Limón (31A 7), 6.04.2001, L. Afonso Giménez (TFC 44911); Al SW de La Tarta (31A 3 d), 5.04.2001, A. Hernández Luis (TFC 44912); Roques al Sur del Llano Ucanca (41D 27), 22.04.2003, Y. de Rioja (TFC 44860); Área no ajardinada del Centro de Visitantes, orientación NE, 2100 m.s.n.m., W. Wildpret de la Torre (TFC 45420).

Como cf. *Arenaria leptoclados* (Reichenb.)Guss.:

Las Cañadas, Mña. Rajada, 20.6.1982, E. Beltrán, T. Bravo (TFC 23767); inmediaciones Mña. Colorada, Adeje, 16.4.1987, P.L. Pérez, M del Arco Aguilar (TFC 24508).

El Hierro: El Cres. 4.1979, W. Wildpret, M. del Arco Aguilar, P.L. Pérez, C.H. Padrón (TFC 7902); Jinama, 1200 m, 8.7.1981, P.L. Pérez, C.H. Padrón (TFC 34335); Los Jables, 22.3.1998, P.L. Pérez, GCT (TFC 42965) (termófilos 206–201/97)—Ver PLP.

La Palma: Malpaís de la Corona, 3.1.1994, J.A. Reyes-Betancort (TFC 38572); San Bartolomé, 23.2.1995, J.A. Reyes-Betancort (TFC 37166); Mácher, 15.2.1996, J.A. Reyes-Betancort (TFC 39896 + duplic.).

Lanzarote: Haría, Bco. Chafarís, 28.1.1997, J.A. Reyes-Betancort, A. Perdomo (TFC 40347).

Fuerteventura: Valle Ginamar, 1.5.1980, W. Wildpret, I. La Serna Ramos, M. Arco Aguilar, A. García Gallo, CR, CL. (TFC 16674).

3.5.11 Argyranthemum adauctum *(Link) Humphries ssp.* dugourii *(Bolle) Humphries*

Humphries, C.J. (1976). *Bull. of the British Museum (Natural History) Botany, 5(4): 233–234.*

Chrysanthemum canariense v. *tenuisectum* Christ. Loc. cit. "tenuisecta".
Chrysanthemum dugourii (Bolle) Christ. *Bot. Jb.*, 9: 146 (1887).
Preauxia dugourii Bolle. *Bonplandia*, Hannover 7: 297 (1859); Pitard & Proust, Îles Canar., *Fl. Archipel*: 231 (1908).

Endemism, Certain native
Family: *Asteraceae*
Common name: Margarita/Magarza tinerfeña de pinar (ES); Pine forest dill daisy (EN)
Biotype: Subshrub (chamaephyte)
Flowering: February–September

Fruiting: April–September
Pollination: insect pollination (entomogamy)
Dispersal: wind pollination (anemochory)
Chromosome N°: n = 9

Description

A. adauctum: **Shrub** with stems between 70 and 90 cm high, erect, branched throughout or only from the base, densely tomentose to glabrous. Leaves 2–8(10) × 0.5–4(5) cm, obovoid, bi- to tripinnate (or pinnatisect), sessile; primary lobes 4–24, 2–30 × 0.5–15 mm, acute. **Inflorescence** corymbose, 5–20 capitula (Fig. 3.34); bracts filiform linear-oblong; pedunculus 1–8(15) cm long. Involucre 4–12 mm in diameter; bracts in three series, triangular to obovoid-oblong; outer bracts fleshy with a narrow scarious margin; inner bracts scarious, usually with an extension, apex hyaline. Floral rays 10–15 × 2–3.5 mm, white, apex 1–3 parted; floral discs 3–4 mm, corolla lobes yellow, tubes white. Cypselae rays 2.5–4.5 × 1.3–6.5 mm, turbinate, cross section trigonous to semi-cylindric, dorsal surface arched, ventral surface slightly concave or smooth, winged, coalescing always in groups of 2–5; pappus absent or as a narrow marginal crest with a floral abscission at the apex; discs of the cypselae 1.6–2.2 × 0.5–1 mm, obconical, coiled, winged, usually sterile; pappus non-existant.

ssp. ***dugourii*****:** **Stems** 60–80 cm, branched from the base, scabrous to tomentose. **Leaves** 2–6 × 0.3–2 cm, bipinnatisect, scabrous to tomentose, with 6–22 primary lobes, 0.5–2 × ca. 0.2 cm wholly or partially opposite; 2–10 secondary lobes, 0.2–0.6 × ca. 0.15 cm, opposite to alternating, acute. Bracts and cypselae as in ssp. ***canariense*** (lower bracts involucral expanded until the apex; cypsella rays 2.5–4.2 mm; cypsela discs 1.6–2 mm, light brown to chestnut coloured).

Fig. 3.34 Teide Botanical Garden, June 2006. (Photo credit: Victoria Eugenia Martín Osorio)

Remarks:

This subspecies is very variable regarding its leaves and indumentum. It looks like as is frequently mistaken with *A. teneriffae*. However, it differs in its sectional characters, like its pappus-less and winged rays and discs of the cypselae, the spatulate-obtuse sessile flowers and involucre interior bracts. Morphologically it is most similar to the subspecies *gracile*, but its leaves are usually longer and more dissected and tomentose.

Distribution

Endemism of the island of Tenerife. It is distributed from 1500 to 2000 m a.s.l. According to bibliography, it was present in the northeast of Las Cañadas, until the southern front of the slopes of Barranco de Pasa Jirón, El Portillo, Lomo de Retama and pine forest of Vilaflor. This last three localities could not be located during this survey; it was only found in northeastern areas of Izaña (proximities of Montaña Limón and La Tarta) (Figs. 3.35 and 3.36).

Habitat and Phytosociology

Generally, it can be found in pine forest areas. Species characteristic of pine forests included in the ***Cisto symphytifolii-Pinion canariensis*** alliance, which groups the closed climatophilous and edaphoxerophilous meso-macroforests, characterized by

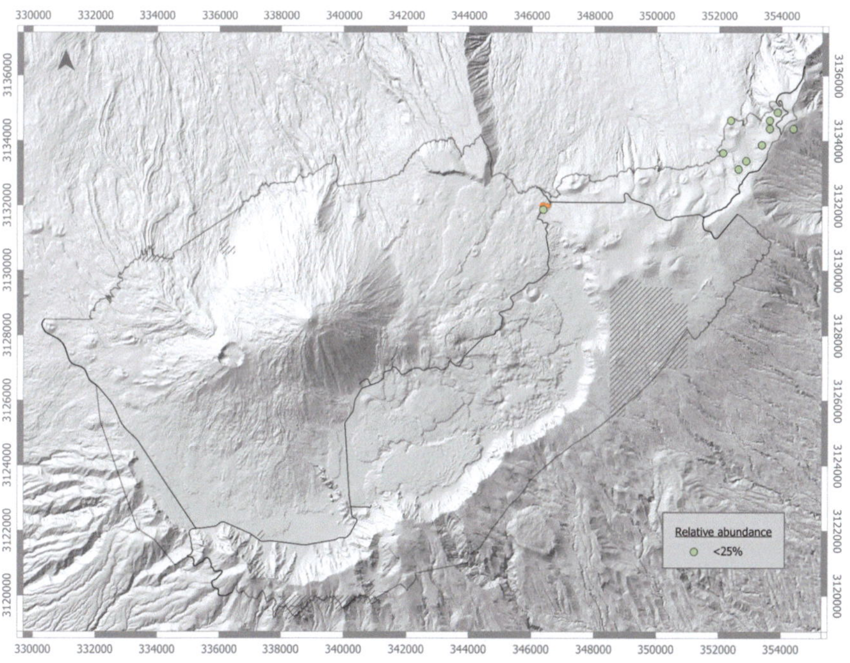

Fig. 3.35 *Argyranthemum adauctum* subsp. *dugourii*. N° Subgrids (250 × 250 m): **10**/3740

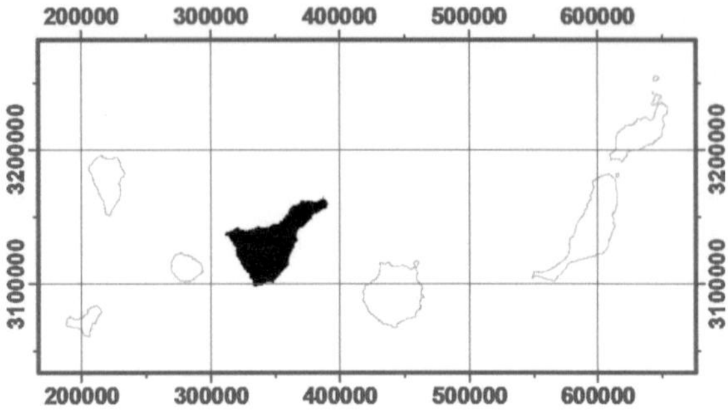

Fig. 3.36 Endemism of the island of Tenerife

canarian pine (*Pinus canariensis*), Meso-Supramediterranean dry-humid, little times foggy, of western canarian distribution. It is present in the ***Sideritido solutae-Pinetum canariensis*** community (climatophilous and edaphoxeripholous Series of Tenerife, mediterranean pluviseasonal oceanic, Termo-Supramediterranean dry-subhumid semihyperoceanic of ***Pinus canariensis*** and ***Sideritis soluta*** forests with ***Chamaecytisus angustifolius***). Occasionally it can be found in the summit broom communities of the ***Spartocytision supranubii*** alliance.

Conservation Status

In 70% (7) of the subgrids in which this taxon was found, its conservation status was high, meaning no damages were found. In 20% (2), damages were detected (herbivory, high mortality without regeneration, etc.), but never on more than 25% of individuals. Only in 10% (1) sub-grids damage was detected on more than 25% of individuals.

Threat Status

R (Barreno et al. 1984)
VU (U.I.C.N., 1994)
VU D2 (Bañares et al. 1998)
R (Walter and Gillett 1998)
VU–D2 (VVAA 2000)
VU (Bañares et al. 2004)
NT (Moreno 2008)

Exsiccata (Herbarium Specimens)

Tenerife: El Sombrerito, 1700 m.s.n.m. (Cult. No. 83), Larsen (C); In Pinetis convallis Tamadaya, Bourgeau 1388 (C; FI; G; K; P, neotypus; W); Misma localidad, 1200 m.s.n.m., 18.06.1855, Perraudière (P); Misma localidad, 17.07.1855, Perraudière (P); Tamadaya, cerca de Arico, 1.03.1855, Bourgeau 1388 (C; FI; G; K; P; W); Fasnia, 17.06.1855, Perraudière (P); El Sombrerito, 1882, Askenasy in herb. Christ (Z); Cerca de Arico, 700 m.s.n.m., 1926, Burchard 312 (G; S; Z); Granadilla,

1954, Wall 554 (S); Peñones, 6.05.1956, Lems 2611 (MICH); Llano de Ucanca, 2100 m.s.n.m., 22.05.1957, Larsen (C); Las Cañadas, El Portillo, entre El Cabezón y El Portillo, 2000 m.s.n.m., 14.04.1969, Bramwell 1315 (RNG); El Sombrerito, 2000 m.s.n.m., 30.04.1969, Bramwell 1411 (RNG) y 2.05.1957, Larsen (C); El Retamar, 2300 m.s.n.m., 10.04.1971, Bramwell & Humphries 3386 (RNG); Vilaflor, 2 km N-W, 10.04.1971, Bramwell & Humphries (RNG); Santiago del Teide-Cumbre Bolico, 14.03.1974, E. Beltrán Tejera (TFC 8488); Mña. Colorada-Adeje, 1700 m.s.n.m., 16.04.1987, P.L. Pérez de Paz & M. del Arco Aguilar (TFC 24506); Tágara, 1500 m.s.n.m., 3.05.1990, M. del Arco Aguilar & M. Salas Pascual (TFC 38397); T. Agua Agria—Vilaflor, 28.05.1993, M. del Arco Aguilar (TFC 36059 + Dupl.).

3.5.12 Argyranthemum tenerifae *Humphries*

Humphries, C. J. (1976). Bulletin of the British Museum (Natural History) Botany, 5(4), 200–202.

Argyranthemum anethifolium Webb. Webb & Berthel., *Phyt. Canar.*, 2: t. 94 (1840), nom invalid.

Argyranthemum anethifolium Webb *ex* Schultz Bip. Webb & Berthel., tom. cit.: 267 (1844), pro parte, non *Pyrethrum anethifolium* Willd.—Pitard & Proust, Îles Canar., *Fl. Archipel*: 230 (1908).

Chrysanthemum anethifolium sensu Masferer. *An. Soc. Esp. Hist. Nat.*, 10: 207 (1881) reimpr. ut. *Rec. Bot. Tenerife*: 131 (1881), non Buch; Burchard *in* Biblthca bot. 98: 200 (1929), excl. specim. a Cumbre de Bolico.- Ceballos & Ortuño, *Veg. Fl. For. Canar. Occid.*: 440 (1951).

Endemism, Certain native
Family: *Asteraceae*
Common name: Margarita del Teide (ES), Teide dill daisy (EN)
Biotype: Subshrub (chamaephyte)
Flowering: March–October
Fruiting: April–October
Pollination: insect pollination (entomogamy)
Dispersal: wind pollination (anemochory)
Chromosome N°: n = 9

Description

Shrub with stems between 30 and 50 cm tall, erect, branched from the base, glabriusculous to scabrous. **Leaves** 2–6 × 0.3–1.6 cm, usually obovoid to oblong, pinnatisect or bipinnatisect, glabriusculous to scabrous, the petiole expanded towards a long parallel rachis; primary lobes 4–10, 0.5–1.6 × 0.1–0.8 cm, opposite; secondary lobes 2–16, 2–6 × 0.5–0.8 mm, alternating to subopposite, acuminate. **Inflorescence**

corymbose with (1) 2–12 capitula (Fig. 3.37); ligule white; peduncles ca. 30 cm; bracts in general filiform to linear-lanceolate, usually sessile, occasionally shortly petiolate. Involucre 7–15 mm in diameter; bracts in three series; outer bracts triangular, scarious with a fleshy flank; inner bracts linear-lanceolate to linear-triangular, acute, laciniate, with narrow scarious margins. Flower rays white, apex emarginate to trifid; flower discs 3–4 mm, corolla lobes yellow. Ray cypselae 4–5 × 2.5–5 mm, trigonous, arched, usually triallose, with two broad lateral wings and a small cuneate ventral one; pappus coroniform, divided towards the back margin, usually reduced to a coriaceous marginal crest, disc of cypselae 4–5 × 1.5–2.5 mm, obconical, cylindrical to laterally compressed with a cuneate ventral wing; pappus coroniform, divided towards the dorsal margin into outer series.

Remarks:

Morphologically very similar and frequently mistaken for *A. adauctum* ssp. *dugourii* but conversely it is easily distinguishable due to its sparse habitus, petiolated leaves, tri-winged cypselae rays and coroniform pappus. In Sventenius (1946), the author cited the taxon as *Chrysanthemum anethifolium* Brouss. With the common name *Margarita del Teide*. Later in the *Argyranthemum* genus revision, Humphries (1976) later synonymized *C. anethifolium* with *Argyranthemum foeniculaceum* (Willd.) Webb *ex* Schultz Bip.

Distribution

It is only found on the island of Tenerife. Although previously only found between 1900 and 2300 m a.s.l., with only some individuals up to 3600 m a.s.l., nowadays it is found through the National Park, with individuals growing up to 3700 m a.s.l. Less abundant towards north-western slopes of Teide and around Llano de Maja, Izaña and Siete Fuentes (Figs. 3.38 and 3.39).

Fig. 3.37 (**a**) Izaña, May 2005. (Photo credit: Beatriz Hernández) (**b**) Siete Cañadas, May 2004. (Photo credit: Beatriz Hernández) (**c**) Pista de El Filo, June 2004. (Photo credit: Victoria Eugenia Martín Osorio)

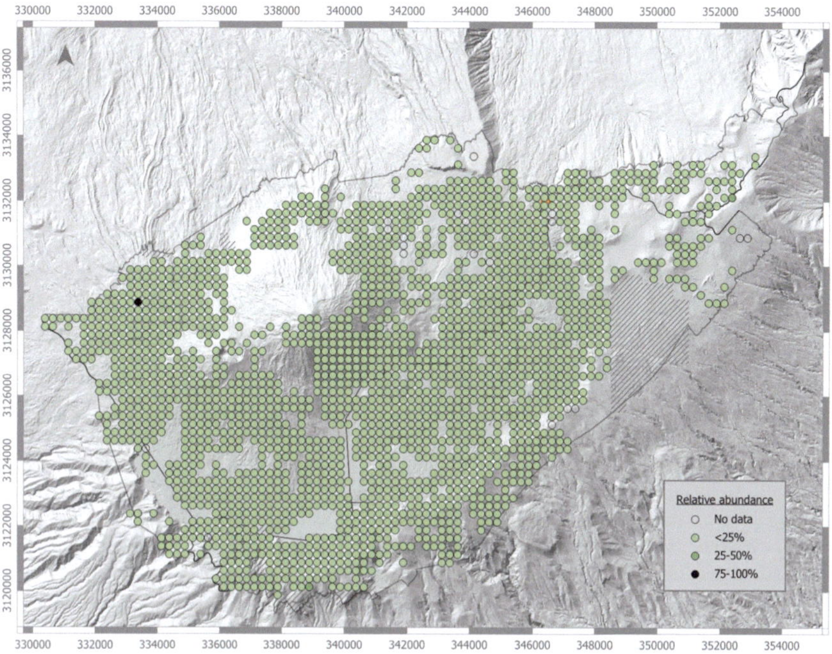

Fig. 3.38 *Argyranthemum tenerifae* distribution. N° Subgrids (250 × 250 m): **2150**/3740

Fig. 3.39 Endemism of the island of Tenerife

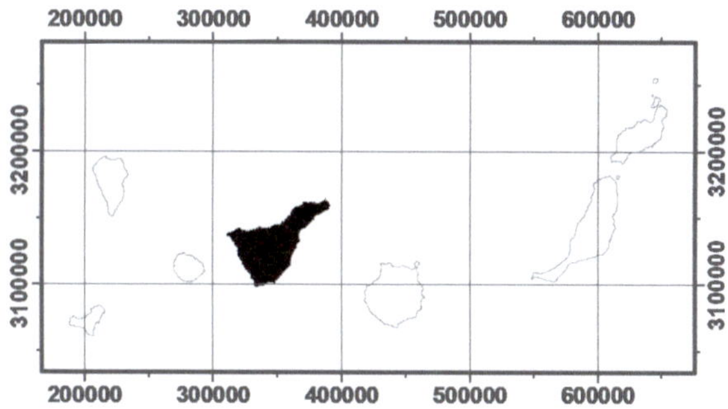

Habitat and Phytosociology

Grows on lava rock fields and in small cracks with soil depositions. It also grows on top of pumitic substrate and in nitrophilous environments such as road margins where the soil has been disturbed.

Characteristic species of the ***Spartocytision supranubii*** alliance, which groups high altitude broom, etc., which are climatophile or edaphoxerophile, sometimes replacement or serial communities of the Canary Island pine forest.

It can also be found as a part of following syntaxa:

Adenocarpo viscosi-Chamaecytisetum angustifolii,
Violetum cheiranthifoliae,

Arrhenathero calderae-Plantaginetum webbii*, *Cheirolopho teydis-Carlinetum xeranthemoides*, *Tolpido webbii-Micromerietum lachnophyllae*, *Arrhenathero calderae-Adenocarpetum viscosi

Conservation Status

In 50% of subgrids (1075) where this taxon was found, no damages of any kind were observed. In 36% (784) damages were detected (herbivory, high mortality without regeneration, etc.), but never on more than 25% of individuals. Only in 13% (283) sub-grids damage was detected on more than 25% of individuals. In the remaining, 1% (8) of sub-grids, assessment is missing.

> *"In a superficial analysis of the flora of the summits of Tenerife, it could appear that this plant has no threats whatsoever, due to the fact that its populations are currently widely distributed across the whole territory of the Teide National Park and its surroundings. Nonethess, its inflorescence attracts the attention of visitors, many of which, despite prohibitions, like to cut and take them. The most limiting factor, however, is the rabbit, which shows a clear preference for this plant causing serious disruptions to its populations. Abundant remains of this plant have also been found in the stomach of mouflons."*
> (Translated from Gómez Campo et al. 1996)

Threat Status

nt (Barreno et al. 1984)
LR ca (Bañares et al. 1998)

Germplasm Banks

Banco de Germoplasma de E.T.S. Ingenieros Agrónomos de Madrid.
Jardín Botánico Canario Viera y Clavijo.

Exsiccata (Herbarium Specimens)

Tenerife: Más abajo del Pico del Teide, 2600 m.s.n.m., 1826, Berthelot *in* herb. Schultz Bip. (FI; P, holotypus); 2200 m.s.n.m., 27.08.1839, Simoniz (Z); Paso de Arenas Negras, 30.05.1846, Bourgeau 849 (BM; FI; G; K; P); 1847, *ex* herb. Webb (G-DC); Región del Pico del Teide, 6.07.1855, Bourgeau (P); Chazua, 5.06.1882, Hillebrand *in* herb. Christ (Z); Montaña Blanca, 2.06.1899, Murray (BM); 2000 m.s.n.m., 25.02.1935, Chaytor (K); Pico del Teide, 6.07.1946, Ceballos & Ortuño 14424 (MA); 1800 m.s.n.m., 2.05.1957, Larsen (C); Montaña Blanca, 2900 m.s.n.m., 22.05.1963, Landbohojsk (C); Montaña Blanca, 12.11.1965, Lems 6268 (MICH); Pico del Teide, 2000 m.s.n.m., 22.07.1963, Brookes (BM); 2000 m.s.n.m., 14.01.1965, Kaae (C); El Portillo, 2000 m.s.n.m., 22.10.1968, Bramwell 265, 265(a) (RNG); Entre El Cabezón y El Portillo, 2000 m.s.n.m., 15.04.1969, Bramwell 1315 (RNG); El Sombrerito, 30.04.1969, Bramwell 1411 (RNG); Chazua, 22.10.1969, Hansen (C); Chazua, de Jussieu (P); El Sombrerito,

10.04.1971, Bramwell & Humphries 3385 (RNG); Las Cañadas, 21.05.1972, E. Beltrán Tejera (TFC 8483); Valle de Ucanca, 26.05.1973, I. La Serna Ramos (TFC 35877); Las Cañadas, 28.05.1979, E. Beltrán Tejera (TFC 7939, 7940, 7941); Las Cañadas, 23.05.1982, R. Leucona (TFC 13123); Montaña Colve o Corchado-Cumbre Agache, 5.01.1984, O. Rodríguez Delgado (TFC 27743); El Portillo-Las Cañadas, 28.05.1993, M. del Arco Aguilar (TFC 36054 + Dupl.); La Fortaleza, W. Wildpret de la Torre, AA, PP, E. Barquín Díez, CG, H. Metlesics & E. Beltrán Tejera (TFC 3888, 3889). Inventario de la Flora y Vegetación del Parque Nacional del Teide (2001–2003) V.E. Martín Osorio, W. Wildpret & B. Hernández Bolaños: Ladera Sur del Teide, hacia Pico Viejo (35D 21), 16/06/2003, F. Mesa (TFC 44203 + dupl.); Al NE de la Cañada de La Angostura (36D 30 a), 25/06/2001, L. Afonso, A. Hernández, Y. de Rioja & F. Tapia (TFC 44870); Roques al Sur del Llano Ucanca (41D 35 d), 14/05/2003, A. Montañés Rodríguez (TFC 44295).

3.5.13 Aria edulis (Willd.) M.Roem.

First published in Fam. Nat. Syn. Monogr. 3: 124 (1847)

Sorbus aria (L.) Crantz *Stirp. Austr. Fasc.*, 2, 46 (1763)
Aria nivea var. *edulis* (Willd.) Koehne *in* Deut. Dendrol.: 250 (1893)
Crataegus aria L. *Sp. Pl.*, 1: 475–476 (1753) [basión.]
Hahnia aria var. *edulis* (Willd.) Dippel *in* Handb. Laubholzk. 3: 375 (1893)
Pyrus aria proles *edulis* (Willd.) Asch. & Graebn. *in* Syn. Mitteleur. Fl. 6(2): 95 (1906)
Pyrus edulis Willd. *in* Enum. Pl. Hort. Berol. 2: 527 (1809)
Sorbus aria subsp. *edulis* (Willd.) Sved., Alm & Örtendahl *in* Index Seminum (UPS, Upsaliensi) 1933: 16 (1934)
Sorbus aria var. *edulis* (Willd.) Wenz. *in* Linnaea 38: 54 (1873)
Sorbus edulis (Willd.) K.Koch *in* Hort. Dendrol.: 176 (1853)

Not endemic, Probable native
Family: *Rosaceae*
Common name: Peralillo de cumbre (ES); Whitebeam (EN)
Biotype: Tree (phanerophyte)
Flowering: April–May
Fruiting: summer-autumn
Pollination: insect pollination (entomogamy)
Dispersal: animal dispersal (zoochory)
Chromosome N°: 2n = 34, 51, 68

Description

Tree up to 20 (25) m. young branches hairy, with abundant lenticels; smooth, grey bark; buds 5–16 mm, conical-ovoid, acute, with 5–9 ciliated scales, glabrescent—tomentose when young—, not slimy.

Leaves 4–12 (16) × 2–9 (13) cm, simple, obovate or oval-lanceolate to suborbicular (width/length ratio = 0.5/1.2), 1–2 times serrated—generally with irregular teeth-, toothed—often doubly-, sometimes with more or less marked lateral lobes—the basal one split up to 8 (12) % of its length and the adjacent upper one split up to 8 (10) %-, with a cuneate or rounded base, not leathery, glabrescent on the upper surface and white-tomentose on the underside; petiole 8–20 mm; stipules 3–5 × 1.5–2.5 mm, linear, without lateral teeth, tomentose on the margin.

Inflorescence corymbiform, with branches densely hairy in flowering and almost glabrous in fruiting (Fig. 3.40). Receptacle 3–5 mm, tomentose. Sepals 2.5–5 × 1.5–2.5 mm, without lateral teeth, tomentose at least on the external surface. Petals 6–7 × 4–5 mm, patent, unguiculate, glabrous or almost so, white, with orbicular, cochleariform and entire limb. Stamens 20; filaments 2.5–7 mm, linear, glabrous; anthers 1–1.5 mm, white, cream or salmon coloured. Carpels 2–3, fused in the lower 2/3; styles 2–3, c. 3 mm, free with a hairy base, persistent. Fruit (8) 10–17 × 8–15 mm, subglobose, sometimes pyriform, red, with (17–) 30–300 (–450) lenticels of 0.2–0.5 (0.8) mm; epidermis unistrata; flesh not heterogeneous, with large islands of tannin cells, visible to the naked eye, and few sclereids -up to 40 μm in diameter, with walls of c. 30 μm, in large groups, without central cavity. **Seeds** (0) 1–3 (4), (4) 5–8 × 2–4.5 mm, with broadly elliptical cross section, smooth, reddish-brown.

Remarks:

The description includes both the forms of regularly serrated leaves -possibly diploid, of sexual reproduction- and those of irregularly serrated or slightly lobed leaves. For example, plants with somewhat lobed leaves are very common, which have often been confused with S. intermedia—a species characterized by its deep lobes.

Fig. 3.40 (**a**) *Guajara*, June 2004, (**b**) *Guajara*, June 2003. (Photo credit: Victoria Eugenia Martín Osorio)

Distribution

It is found in the N, Central and W of Europe, and NW of Africa; in almost all of the Iberian Peninsula, except in most of the provinces of the SW quadrant and Mallorca. In the Canary Islands, it is only present in the islands of La Palma and Tenerife. In the Teide National Park, it has been located in several points of the Pared de Las Cañadas, between 2200 and 2500 m a.s.l.: Topo de La Grieta, Montaña Guajara and Roque Bermejo (east of Roque del Almendro) (Figs. 3.41 and 3.42).

Habitat and Phytosociology

In the Mediterranean region it grows preferably in beech forests, holm oak forests, oak forests, oak groves, maple groves, pine forests, often in clearings and margins, bushes and rocks, and on all types of substrates; between 0 and 2200 m a.s.l.

It is a characteristic species and bioindicator of ***Quercetalia pubescenti-petraeae***, an order that represents the communities of climatophilous and edaphoxerophilous forests of central-southern Europe, broad-leaved and sometimes needle-leaved developed on calcareous, dolomitic, sandy or siliceous substrates rich in bases, in submesophytic or xerophytic soils, which generally have a summer water deficit. They thrive in dry-humid Thermo-Supratemperate bioclimates, frequently sub-Mediterranean and subhumid-hyperhumid Meso-Supra-Mediterranean, from

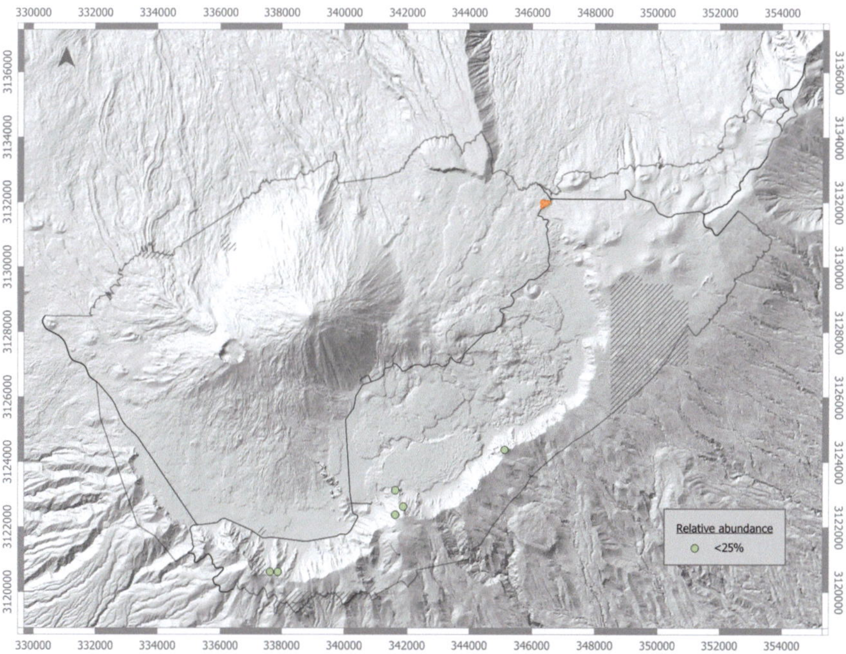

Fig. 3.41 *Aria edulis* distribution, N° Subgrids (250 × 250 m): **6**/3740

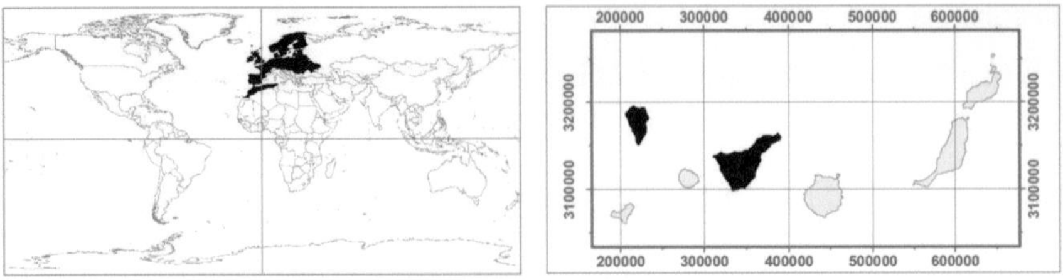

Fig. 3.42 Present on La Palma and Tenerife islands

Euoceanic to subcontinental. In the Teide National Park, it is found in the area of ***Juniperion cedri*** (Fig. 3.43).

Conservation Status

According to the sampling carried out, in none of the six (6) subgrids in which this species was found, damage of any kind was observed.

"*Scarce in Las Cañadas.*" (Sventenius 1946)

Threat Category

Anexo II: Estrictamente protegida (Gobierno de Canarias 1991)
De interés especial (Gobierno de Canarias 2001)

Exsiccata (Herbarium Specimens)

Tenerife: Inventario de la Flora y Vegetación del P.N. del Teide (2001–2003) V.E. Martín Osorio, W. Wildpret & B. Hernández Bolaños: Guajara (42C 2 a), 15/05/2003, Y. de Rioja Santos (TFC 44961).

La Palma: Los Andenes, 7.1983, A. Bañares (TFC 19963).

3.5.14 Arrhenatherum calderae *A. Hansen*

Cuad. Bot. Canar., 14/15: 59–70 (1972).
***calderae**: referring to the crater or the caldera of a volcano, in this case to the one of Las Cañadas, where it was described.*

Endemism, Certain native
Family: *Poaceae*
Common name: Cerrillo de cumbre—Mazorrilla del Teide (ES); Teide oat grass, summit oat grass (EN)
Growth form: hemicryptophyte
Flowering: April–July

Fig. 3.43 *Aria edulis* and Prof. Wildpret on the slopes of Guajara, June 2003

Fruiting: July–August
Pollination: wind pollination (anemogamy)
Dispersal: short-range dispersal (atelechory)

Description

Perennial grass, very caespitose (Fig. 3.44). Stout erect canes, always up to approximately 1 m long, with 3–4 nodes. **Laminar** leaves, always around 20 cm long, 3–4 mm wide, linear-acuminate, flat, glaucous, very glabrous, on the underside and the margins, scabrous, ligula 3–4 mm long, obtuse, denticulate or truncate. Panicle erect, ovate-lanceolate, lax, 8–9 cm long, after anthesis, narrow, appressed. **Spicule** lanceolate, 8–10 mm long, biflorate; lower flower male, upper flower hermaphrodite. Glumes unequal, lower 6.5 mm long (excl. awn) uninerved, upper 8.5 mm long, trinerved, membranous on both sides, glabrous. Lemma 7–8 mm long, on one side apically marginal, on other side protruding from acumen; palea ca. 5 mm long. Awn of lower flower ca. 18 mm long, during anthesis more or less straight, dry geniculate ca. 1 mm inserted above base; awn of upper flower ca. 3.5 mm long, inserted ca. 2 mm below the apex. Stamens 3, anthers 4 mm long, flavescent. Stigma 2 mm long, feathery. Caryopsis oblong-lanceolate, hairy, furrowed inner part 5 mm long.

Distribution

Endemism of the islands of Tenerife and La Palma. In the Teide National Park, according to consulted bibliography, was abundant in the screes of the talus cones of rocky outcrops in the southern arch of Las Cañadas, as well as in of the slopes of the so called El Cerrillar mountain, among other places. Today it is widely distributed across the whole park, reaching on occasions 3668 m a.s.l.; more scarce in the slopes of the Teide-Pico Viejo complex, especially on its northern side, Montaña Blanca and areas of the Izaña ridge (Figs. 3.45 and 3.46).

Fig. 3.44 (**a**) Cañada de Diego Hernández, June 2003 (**b**) Cañada de El Capricho, June 2004. (Photos credit: Victoria Eugenia Martín Osorio)

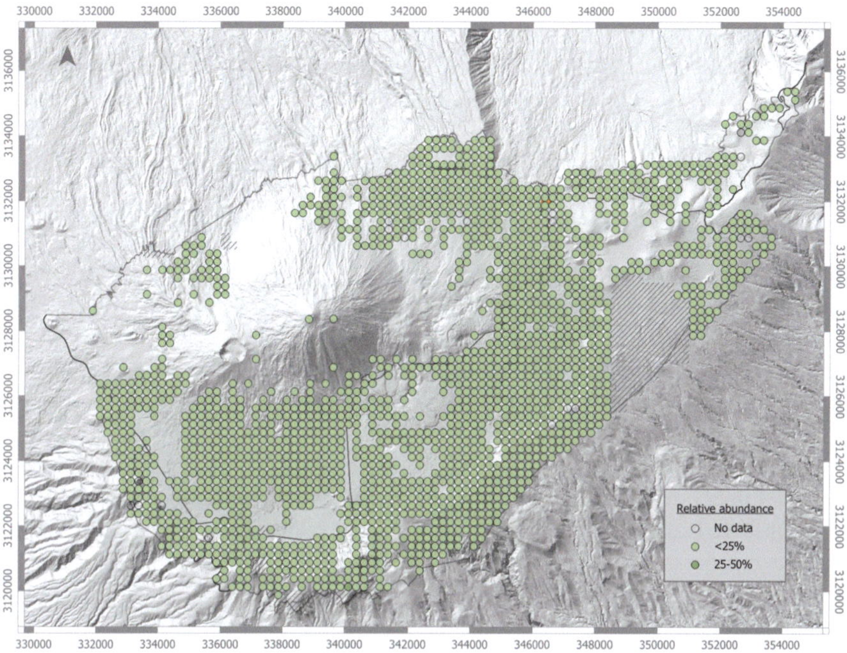

Fig. 3.45 *Arrhenatherum calderae* distribution, N° Subgrids (250 × 250 m): **1858**/3740

Fig. 3.46 Endemism of the island of Tenerife and La Palma

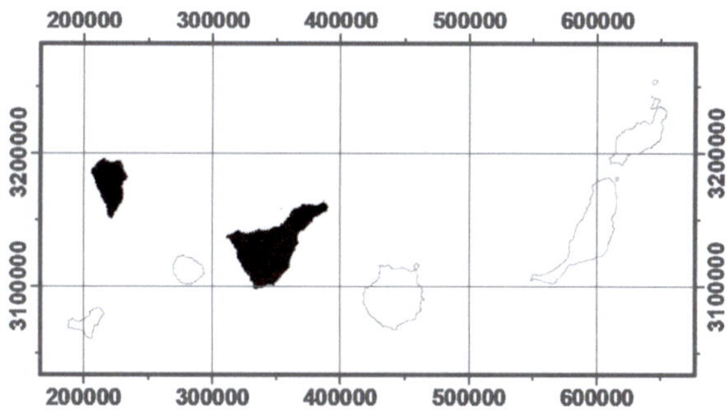

Remarks:

Rodríguez et al. (2009), cite it for the first time in the Cone of the Teide volcano, growing on slopes of 45°, on permeable frost withered substrates of salic materials, concretely on stony soils around the Telesforo Bravo trail, at an altitude of 3668 m a.s.l.

Habitat and Phytosociology

Species characteristic of the ***Plantaginion webbii*** alliance, which groups the nanoshrublands of *Plantago webbii*, primocoloniser of lava flows, lapilli (tephra) and other volcanic deposits, especially recent ones, as well as serial on eroded soils of canary pine forests, juniper forests or Teide broom shrublands; distributed in La Palma and Tenerife, Meso-Supramediterranean, dry-humid.

It also characterises or intervenes in:

Arrhenathero calderae-Plantaginetum webbii
Cheirolopho teydis-Carlinetum xeranthemoides
Tolpido webbii-Micromerietum lachnophyllae
Arrhenathero calderae-Adenocarpetum viscosi

Conservation Status

In 87% (1622) of subgrids where this taxon was found, no damages of any kind were observed. In 11% (214) damages were detected (herbivory, high mortality without regeneration, etc.), but never on more than 25% of individuals. Only in 0.4% (8) of sub-grids damage was detected on more than 25% of individuals. In the remaining, of sub-grids, assessment is missing.

Threat Status

R (Barreno et al. 1984)
LR pm (Bañares et al. 1998)

Germplasm Banks

Banco de Germoplasma de E.T.S. Ingenieros Agrónomos de Madrid.
Jardín Botánico Canario Viera y Clavijo.

Exsiccata (Herbarium Specimens)

Tenerife: La Fortaleza, 19.05.1973, E. Beltrán Tejera (TFC 27044 + Dupl.); La Fortaleza-Las Cañadas, 19.05.1973, E. Barquín Díez & C. León Arencibia (TFC 33576); Las Cañadas, 10.06.1973, J.F. Pérez Francés (TFC 25847, 25852); Altos de Güímar, 1800 m.s.n.m., 6.04.1981, W. Wildpret de la Torre & M. del Arco Aguilar (TFC 9575); El Portillo-Las Cañadas, June 1981, CRP (TFC 24493); Valle de Ucanca, 2250 m.s.n.m., 24.04.1983, S. Socorro (TFC 18466); Bco. del Río, 2100 m.s.n.m., 15.05.1983, S. Socorro (TFC 18465); Los Azulejos-Las Cañadas, 23.05.1983, E. Beltrán Tejera (TFC 18302 + Dupl.); Degollada de Diego Hernández, 20.06.1985, P.L. Pérez de Paz, I. la Serna Ramos & IMM (TFC 24754 + Dupl.); Izaña, 25.05.1987, W. Wildpret de la Torre & O. Rodríguez Delgado (TFC 29208); La Fortaleza-Las Cañadas, 13.06.1989, E. Beltrán Tejera & M.C. León Arencibia (TFC 35262). Inventario de la Flora y Vegetación del Parque Nacional del Teide

(2001–2003) V.E. Martín Osorio, W. Wildpret & B. Hernández Bolaños: Ladera suroriental de Izaña (31C 17), 26/04/2001, L. Afonso, A. Hernández, Y. de Rioja & F. Tapia (TFC 44915); Jardín Botánico del Parque Nacional del Teide, 23/05/2003, V.E. Martín Osorio & W. Wildpret de la Torre (TFC 44945); Ladera Sur del Teide, hacia Pico Viejo (35D 31 c), 16/06/2003, F. Mesa (TFC 44204 + dupl.); Cañada del Capricho, 02/07/2003, V.E. Martín Osorio & W. Wildpret de la Torre (TFC 44200); Por debajo del Roque del Mal Abrigo, al W—NW (31A 21), 17/05/2005, V.E. Martín Osorio, W. Wildpret de la Torre & B. Hernández Bolaños (TFC 44978 + dupl.); Ruta de Siete Cañadas (30D 32 y 30D 42), 15/06/2005, V.E. Martín Osorio & W. Wildpret de la Torre (TFC 44985 + dupl.); Sendero Telesforo Bravo, Cono del Teide, 2009 Rodríguez, Mª.L., B. Fariña, M. Rodríguez & J. Escolástico Moreno (TFC 49466).

3.5.15 Asplenium adiantum-nigrum *L.*

Sp. Pl.: 1081 (1753)

Chamaefilix adiantum-nigrum (L.) Farw. *in* Amer. Midl. Naturalist 12: 273 (1931)
Tarachia adiantum-nigrum (L.) C.Presl *in* Abh. Königl. Böhm. Ges. Wiss., ser. 5, 6: 82 (1851)

Not endemic, Certain Native to the Canary Islands
Family: *Aspleniaceae*
Vernacular name: Black Adianto or Culipollo (ES); Black Maidenhair (EN)
Biotype: hemicryptophyte
Sporulation: all year round
Dispersal: wind dispersal (=anemochory)
Chromosome N°: 2n = 144; n = 72 (allotetraploid)

Description

Rhizome more or less elongated, sometimes branched, with lanceolate to oblong-lanceolate palea, attenuated into a long filiform tip, dark brown, with wider cells and thinner walls than in *A. onopteris*.

Fronds 10–30 cm, densely fasciculate (Fig. 3.47); petiole 1/2–3/5 of the length of the blade, blackish brown to purplish black, sometimes green near the blade, glabrous, paleaceous at the base; blade 6–15 × 3–7 cm, 2 (3) pinnate, more or less triangular, not acuminate at the apex, dark green, membranous to coriaceous, generally shiny; pinnae 8–15 pairs, ovate-lanceolate, acute, not caudate; the lower ones, straight or slightly curved antrorse; pinnules narrowly ovate to obovate, sometimes cuneate-flabellate, cuneate below, with short, acute teeth.

Sori oblong-linear. **Spores** (30) 33–39 (45) µm, ovoid, dark brown to black-brown, with perisporium equinulate-reticulate and warty.

Fig. 3.47 Guajara, June 2003. (Photo credit: Victoria Eugenia Martín Osorio)

Remarks:

According to the consulted bibliography, it has been frequently confused with *A. cuneifolium* and *A. onopteris*, since it is an allotetraploid, the result of the crossing of these two species and subsequent duplication. Quite polymorphic in relation to the substrate; various authors have admitted different subspecies and varieties.

Distribution

It is distributed throughout Europe, Macaronesia, Asia, Africa and North America. Towards the South its presence is increasingly rare, being restricted to the most mountainous areas (tropical Africa and Australia). Present throughout the Canary Islands, except on the island of Lanzarote (Figs. 3.48 and 3.49).

In the Teide National Park, it appears in a dispersed form in areas of the Dorsal, some sectors of the Pared de Las Cañadas and the northwestern slopes of Pico Viejo (Cerro or Montaña Las Chajoras), between 2100 and 2400 m a.s.l.

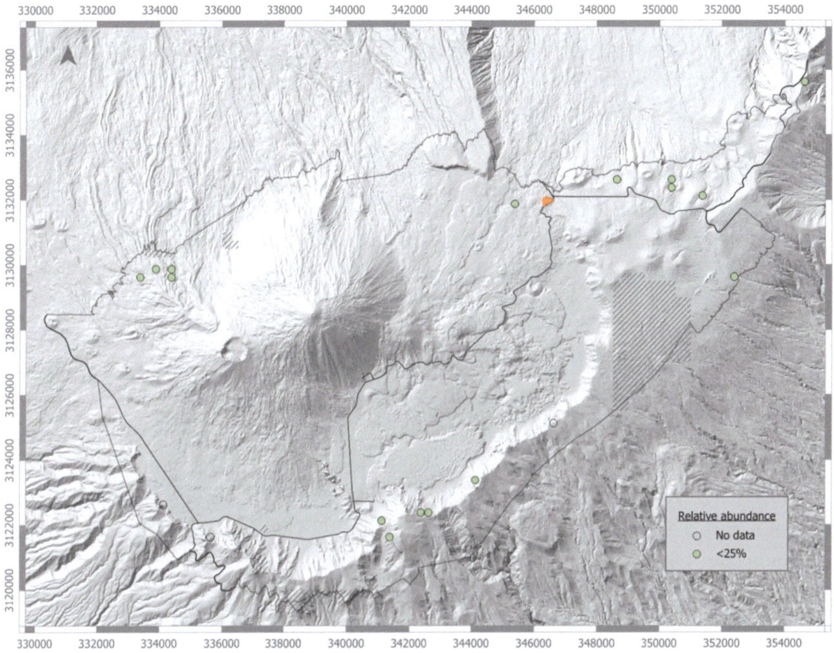

Fig. 3.48 *Asplenium adiantum-nigrum* distribution, N° Subgrids (250 × 250 m): **19**/3740

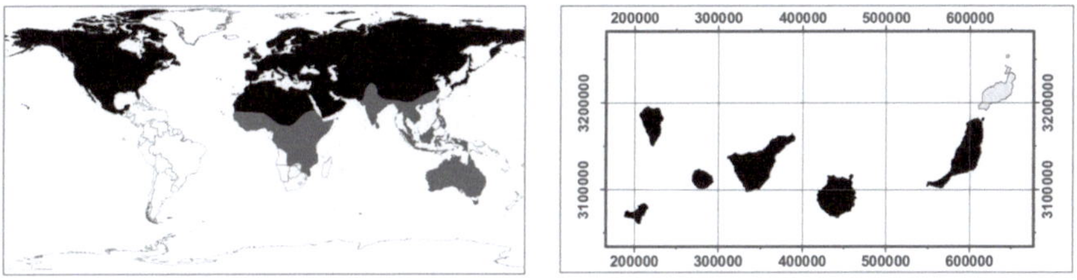

Fig. 3.49 Present on Canary Islands except Lanzarote Island

Habitat and Phytosociology

It appears preferably in shady and humid places, rock crevices, earthy slopes, and in siliceous or calcareous substrates, more or less deep and with a protosoil of considerable thickness.

Species included in the class ***Asplenietea trichomanis***, which groups the Holarctic chasmophytic communities of rocks and vertical walls that receive direct rainfall; formed by small vascular perennial rupicolous plants, pteridophytes and hemicryptophytic and chamaephytic spermatophytes, not nitrophilous, whose roots are housed in narrow fissures and grooves existing in rocks and walls. Within it, it is a characteristic species and bioindicator of ***Androsacetalia vandellii***, an order that represents the silicicolous and calcifuge Meso-cryotemperate subhumid-hyperhumid rupicolous chasmophytic communities of Eurosiberian distribution, as well as western Mediterranean, Supra-Orotemperate Submediterranean humid-hyperhumid and Meso-Cryomediterranean dry-humid.

Conservation Status

In almost all (18) of the subgrids in which this species was found, no damage of any kind was observed. In the rest (1), this variable was not specified at the time of sampling.

Exsiccata (Herbarium Speciemns)

Tenerife: Parque Rural de Anaga, El Pijaral, 12.02.1979, E. Beltrán (TFC 20943); Cumbres de la Orotava, 1000 m.s.n.m., 21.08.1979, W. Wildpret, I. La Serna, L. Negrín & A. Losada Lima (TFC 24668 + duplic.); Agua García, 06.1980, W. Wildpret & A. García Gallo (TFC 10300); Las Cañadas del Teide, Guajara, 2650 m.s.n.m.,15.05.1983, J.S. Socorro (TFC 18481); Las Cañadas del Teide, Degollada de Ucanca, 2350 m.s.n.m.,12.06.1983, J.S. Socorro (TFC 18486). Inventario de la Flora y Vegetación del parque Nacional del Teide (2001–2003) V.E. Martín Osorio, W. Wildpret & B. Hernández Bolaños: Montaña del Cedro, 07.05.2001, W. Wildpret de la Torre (TFC 44872); Guajara, nacimiento del Bco. del Río (42C 3b), 15.05.2003, F. Mesa (TFC 44225); Degollada de Guajara (42C 4), 15.05.2003, Y. de Rioja Santos (TFC 44210); Pico Viejo, orientación NE (35A), 11.06.2003, Y. de Rioja Santos (TFC 44207); P.N. del Teide, 05.06.2003, V.E. Martín Osorio & W. Wildpret de la Torre (TFC 44976); Guajara, cerca de El Capricho (42C 10 d), 17/06/2003, A. Montañés Rodríguez (TFC 44258); Roques de García, 13/11/2004, W. Wildpret de la Torre (TFC 44873).

Gran Canaria: Valleseco, Bco. Oscuro, 05.1980, C. Suárez (TFC 11109).

La Palma: Caldera de Taburiente, Paso de los Andenes, 2300 m.s.n.m., orientación norte, 10.2001, P.L. Pérez de Paz, M.C. León Arencibia & C. Hernández Padrón (TFC 43376).

3.5.16 Asplenium onopteris *L. var.* onopteris

Sp. Pl.: 1081 (1753)

Asplenium adiantum-nigrum L. var. *acutum.* (Willd.) Mnzs.
Asplenium adiantum-nigrum L. subsp. *onopteris* (L.) Heufler. *In Verh. Zool-Bot. Vereins Wien*, 6: 310 (1856)
Asplenium acutum Willd.
Asplenium-nigrum auct.
Asplenium productum Lowe.
Asplenium virgilii Bory. *Exp. Sci. Morée, Bot.*: 289 (1832)

*"**onopteris**": "donkey fern" Greek, -**onos**: donkey*

Not endemic, Certain native
Family: *Aspleniaceae*
Common name: Doradilla negra, Culantrillo negro de monte, Falzia negra, Adianto negro, Culantrillo mayor (ES); Western black spleenwort (EN)
Biotype: hemicryptophyte
Sporulation: all year round
Chromosome N°: 2n = 72; n = 36

Description

Rhizome creeping, sometimes branched, with palea narrowly lanceolate and filiform at the apex, dark brown to black, with narrow cells and thick walls. **Fronds** 15–45 cm, persistent; petiole 1/2–3/5 of the length of the blade, blackish brown to purplish black, shiny on the upper side (Fig. 3.50); blade (5.5) 9–25 (30) × 5.5–17 (20) cm, 3 (4) pinnate, triangular-lanceolate, long attenuate-caudate at the apex, dark green, leathery and very shiny on the upper side; pinnae 10–25 pairs, caudate; in the lower and middle ones, curved-antrarse; pinnules linear to ovate, with long, acute teeth. **Sori** inserted near the costule, oblong or linear. Spores (26) 28–32 (34) µm in diameter, ovoid, dark brown to black-brown, with echinulate-reticulate and warty perisory.

Fig. 3.50 Frond of *Asplenium onopteris*, April 2009. (Photo credit: Victoria Eugenia Martín Osorio)

Distribution

It is found in the Macaronesia region (except Cape Verde) and western and central Europe (disjunct in Ireland). Frequent in the Balearic Islands and in the coastal regions of the peninsula, scarcer towards the interior (0–1900 m a.s.l.). Present throughout the Canary Islands.

In the Teide National Park it is found in a very dispersed manner in the central part of the interior and the south-eastern Pared de Las Cañadas, and the NW slopes of Pico Viejo (Cerro or Montaña Las Chajoras); and in a more isolated and punctual manner in the Barranco de Boca Tauce and the northern slopes of Roque del Almendro; between 2100 and 2450 m a.s.l. (Figs. 3.51 and 3.52).

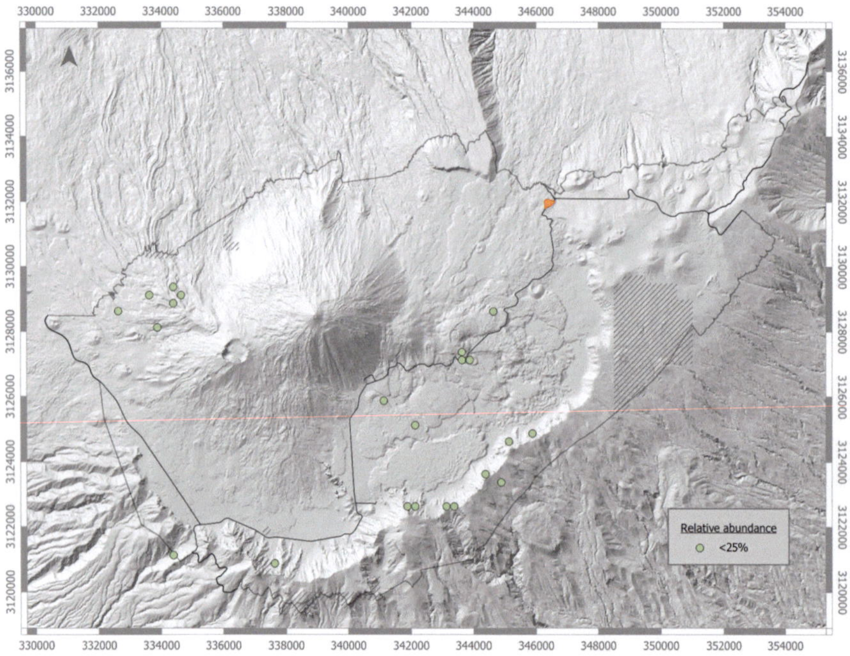

Fig. 3.51 *Asplenium onopteris* distribution, N° Subgrids (250 × 250 m): **22**/3740

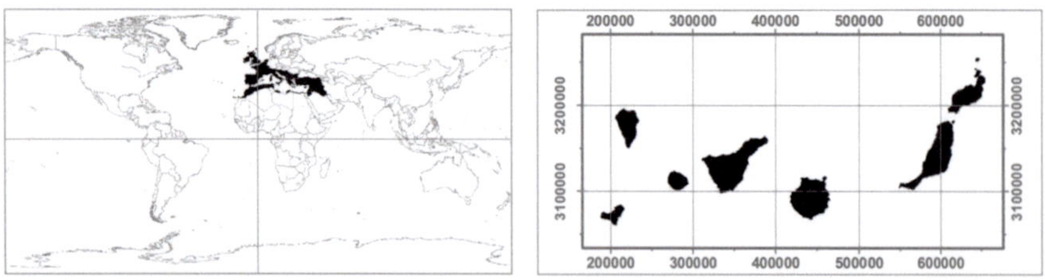

Fig. 3.52 Present on all the Canary Islands

Habitat and Phytosociology

It is found in somewhat humid and shady places, in cracks in rocks or walls and in preferably acidic or decalcified substrates. It is a characteristic species and bioindicator of ***Quercetalia ilicis***, an order that represents the communities of climatophilous and sometimes edaphoxerophilous forests of at least dry ombrotype (holm and cork oak forests, or gall oaks, wild olives or carob trees); where it intervenes in the associations:

In the Canary Islands, it is present in the communities of the ***Pruno hixae-Lauretea novocanariensis*** class, in mature forests of the "Monte Verde" both mesophytic and xerophytic; Infra-Meso-Mediterranean dry-humid hyperoceanic cloudy bioclimates.

Conservation Status

In all the subgrids in which this species was found (22), no damage of any kind was observed.

Exsiccata (Herbarium Specimens)

Tenerife: Volcán de Garachico, El Tanque, 09.06.1984, W. Wildpret (TFC 19881); Los Silos, Canal Las Carvas, 07.08.1991, V. Lucía Sauquillo & C. Acosta (TFC 38165); Los Silos, Mte. del Agua, parte central de la pista principal, 22.08.1994, ejusd. (TFC 38250 + duplic.).

El Hierro: Alrededor de Sabinosa, talud, Cruz de los Reyes, 21.07.1983, E. Beltrán, A. Losada Lima & C. Hernández Padrón (TFC 18316).

Gran Canaria: Bco. de Lina, Monte de Inagua, 08.1984, MN & M.C. Gil (TFC 23650).

3.5.17 **Athamanta montana** *(Webb* **ex** *Christ) Spalik & Wojew. & S.R.Downie*

First published *in* Edinburgh J. Bot. 58: 337 (2001)

Todaroa montana Webb *ex* Christ, H. (1887). Spicilegium canariense, *Engl. Bot. Jahrb.*, 9, 86–172.
Todaroa montana Webb. *in* Bourg. exs. st. II (1855) sine numero determ. Cosson.
Tinguarra montana (Webb *ex* Christ) A. Hans. et Kunk.

Endemism, Certain native
Family: *Apiaceae*
Common name: Cañaheja blanca (ES); Canary giant mountain carrot (EN)
Growth form: hemicryptophyte
Flowering: Mai–June

Fruiting: June–July
Pollination: insect pollination (entomogamy)
Dispersal: wind dispersal (anemochory)
Chromosome N°: 2n = 22

Description

Hemicryptophyte 55–100 cm tall, leafy and pubescent. Stems erect, crass, sparsely striated, rounded fistulous and branched.

Radical leaves densely peciolated, with petiole up to 44 cm long and 30 cm wide, broad laminae, arranged in three rows, lower segments medium sized and upper barely hirtate, ciliate, vaginate; vaginae amplexicaules with reddish striae; encircling leaves triangular, segments ovate-triangular, on large radical leaves numbering ca. 1200 or so per leaf; lamina 4–6 mm wide, bi- or tetrapinnate, lobes broadly ovate and densely toothed; upper leaves much smaller and pinnatifid

Involucre and involucellum very short, (Fig. 3.53) 5-split, laciniae setose. Umbels of about 20 rays, sparsely scabro-pubescent, 3 cm long, not very thick. Calyx dentate, sparsely acute. Flowers somewhat whitish (Fig. 3.54). Petals oblong, intact with long mucronate, inflexed apex.

Fig. 3.53 Umbels of *Athamanta montana* in Ayosa, 24 June 2024. (Photo credit: Victoria Eugenia Martín Osorio)

Fig. 3.54 *Athamanta montana* flowers Ayosa, 24 June 2024. (Photo credit: Victoria Eugenia Martín Osorio)

Fruits oblong with very scarcely attenuated apex, very hispid, striate-pubescent, crass and short. Ribs not very prominent, obtuse, two lateral ones broad, endosperm oblong, in transverse sections. Middle part of the fruit swollen, with non-contiguous commissures, vitae divided or in two to four; stylopodia convex and glabrous; styles erect; anthers rounded.

Distribution

Endemism of western Canary Islands and Gran Canaria. On Tenerife it is distributed between 400 and 1600 m a.s.l. On Teide National Park it only appears around La Crucita and areas near La Tarta, near the side of the road (Figs. 3.55 and 3.56).

Habitat and Phytosociology

It is usually found on shaded rocky outcrops and in nitrophilous environments besides roads. It takes part in some communities, some of which are included in the ***Chamaecytiso angustifolii-Pinetea canariensis*** class, which corresponds with canary pine forests and summit broom shrublands and herbaceous summit communities of Teide violet.

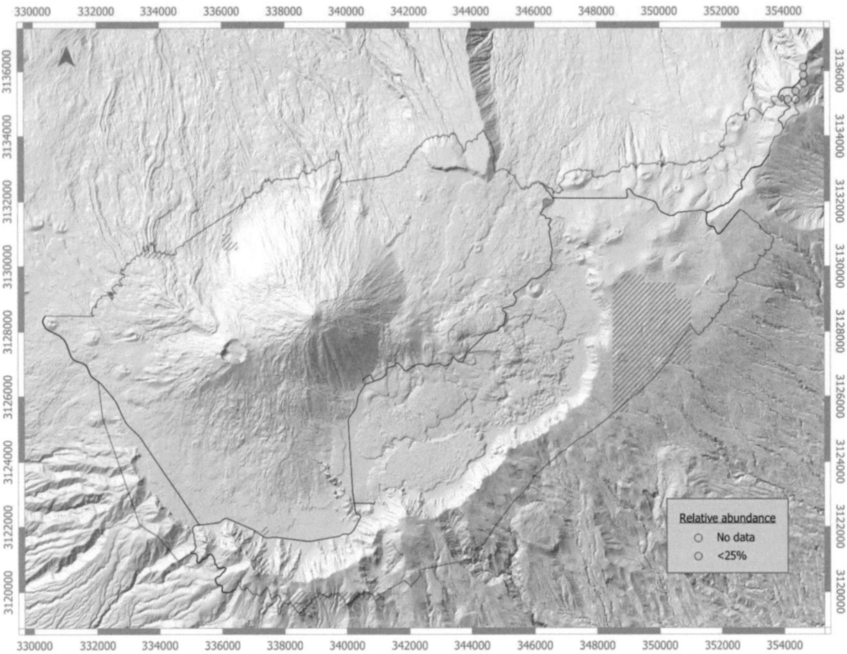

Fig. 3.55 *Athamanta montana* distribution, N° Subgrids (250 × 250 m): **6**/3741

Fig. 3.56 Endemism of western Canary Islands and Gran Canaria

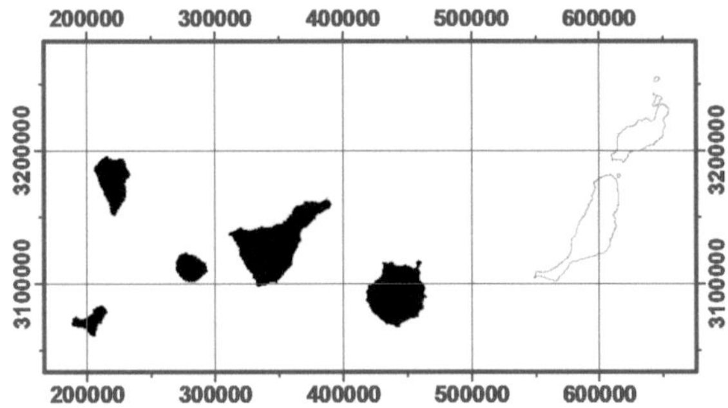

Conservation Status

On nearly all subgrids where this species is present, no damages were observed (5). In the rest (1) this variable was not considered.

Threat Status

R (Barreno et al. 1984)

Germplasm Banks

Jardín Botánico Canario Viera y Clavijo.

Exsiccata (Herbarium Specimens)

Tenerife: Pico de las Flores-Fuente de Joco, 13.06.1987, P.L. Pérez de Paz (TFC 25392 + Dupl.); Fuente de Joco-Cumbre de Pedro Gil, 15.05.1988, P.L. Pérez de Paz & C. Hernández Padrón (TFC 27460, 27461—Fitoquímica); Ibid., 15.05.1988, P.L. Pérez de Paz (TFC 667, 25702); Km. 26 Ctra. de Las Cañadas, 28.10.1994, J.R. Acebes (TFC 41138); Pico de Yeye-Teno, Diciembre 1994, S. Delgado & M. del Arco Aguilar (TFC 41137 + Dupl.).

Gran Canaria: Risco del Hurón, Mayo 1979, C. Suárez & P.L. Pérez de Paz (TFC 11002); San Mateo, 28.03.1988, P.L. Pérez de Paz (TFC 27459); Utiaca-San Mateo, 850 m.s.n.m., 28.03.1988, ejusd. (TFC 660, 25695); Parque Natural Bco. de Guayadeque-Cauce del Bco.-Cueva Labrada, 3.03.1990, F. Amor & P.L. Pérez de Paz (TFC 30824); Parque Natural Bco. de Guayadeque-Cmo. de Pajonales, 25.03.1990, ejusd. (TFC 30765); Parque Natural Bco. de Guayadeque, 17.06.1990, F. Amor (TFC 30516).

3.5.18 Bencomia exstipulata *Svent.*

Add. Fl. Canar.: 25 (1960).

Endemism, Certain native
Family: *Rosaceae*
Common name: Rosal del guanche/Bencomia de cumbre (ES); Guanches' rose/ Summit Bencomia
Growth form: nanophanerophyte
Flowering: April–Mai
Fruiting: June–July
Pollination: wind/insect pollination (anemo-/entomogamy)
Dispersal: anemochory/barochory/maybe zoochory

Description

Shrub 50–100 cm tall, monoecious or unisexual-female, depressed-compact habit, usually branching tortuous and intricate branching base; hard woody, ca. 10 cm diameter or broader from the base; bark glabrous, shiny-golden, drooping into large, hyaline, yellow-white, rough, rhizomatous-pergaminaceous blades on younger branches; branches short, smaller branches horizontally inserted and erect-patent.

Leaves 8–12, compound, with patent tomentum, subrosulate to the apex of the branches, transovate-apiculate, imparipinnate, (Fig. 3.57) 3–4 consecutive leaves opposite each other, the petiole of the median blade stipitate, to the base dilated and amplexicaules, (like the rachis) pubescent-glandular and white-hairy; leaves sessile, upper leaves larger, oblong or oblong-lanceolate, apiculate, margin crenulate, intensely green, glossy, somewhat sticky, with hairy midrib on obverse,

Fig. 3.57 (**a**) Teide, Wolfredo Wildpret Botanical Garden, June 2006 (**b**) Teide, Wolfredo Wildpret Botanical Garden, Nov. 2003. (Photos credit: Victoria Eugenia Martín Osorio)

white-flocculent, with prominent midrib on reverse; leaves withered, soon deciduous. Leaflets with toothed margins.

Inflorescences simple or racemose, erect, uni- or bisexual; male inflorescence almost always simple, usually in the bisexual ones amentiform (rarely in the unisexual ones), erect, 4–8 cm long, peduncle somewhat divided, tomentose, white-hairy, uni or bibracteate, bracts trifoliate (or upper simple), petiole equally elongated as terminal leaflet, base dilatated and stipulate, stipe triangular-subulate, tomentose-hairy, leaflets bracteate linear-lanceolate, cuneate, acute, apex crassly toothed; lower spike lax, dense to the apex, occasionally with some apiculate female flowers (sometimes basal); flowers sessile, tribracteate, bracts ovate, acute, very concave, pergaminaceous-hyaline, margin scarious-lacerate, externally hairy; calyx very small, subcarnose, discoidal with somewhat rounded divisions; petals ovate-elliptic, 6 mm long, 4 mm wide, apex mucronulate-tomentose, green, margin glabrous, hyaline, trinervate, inferiorly concave, deflexed-revolute towards the top, glabrous on the obverse, white-hairy on the reverse; filaments 30–40, exceeding the petals in length; female inflorescence less racemose (usually 3–5 bunches), rigid, erect-patent on the short, bracteate peduncle; spike 2/3 shorter than male, dense, ovate-elongate, ca. 15 mm diameter on the bracteate peduncle, bracts obovate-acute, membranous, hyaline, hairy; flowers sessile, sepals lanceolate, acute, 5–6 mm long, 2–3 mm wide, apex cartilaginous, white-penicillate, deflexed, ciridinous, white-hyaline-margined, 5—nerved, glabrous on obverse, hairy on reverse; styles 2, divergent, 1 mm long, white-penicillate, very deciduous; ovary globose, glabrous.

Fruits depressed-globose, apex concave-crateriform, ca. 5 mm in diameter, glabrous, dry, spongy, not roughened by desiccation, whitish; nuculanium suboctahedral, 4.5 mm long, 2 mm wide, elongate angulose-costulate, surface started-rugose, chestnut-coloured, 4-loculate, monospermous (rarely two seeds).

Remarks:

> "*Seed production can be quite high, especially, in bigger individuals, where female flowers dominate. Recent studies seem to confirm that this is an endemism originating due to geographic isolation during a recent evolutive radiation. It presents a tremendously slow population dynamic with barely any recruitment, which is compensated by the longevity of adult individuals.*" (Translated from: Marrero et al. 2004)

Distribution

This species belongs to a genus endemic of the Macaronesian region, concretely to Canary Islands and Madeira, although this species only appears on the summits of Tenerife and La Palma, always above 1800 m a.s.l. On Tenerife it is found in the southwestern section of Las Cañadas del Teide (Cañada de El Cedro), between 2100 and 2200 m a.s.l. It was also planted in the Wolfredo Wildpret Botanical Garden of the Teide National Park (Figs. 3.58 and 3.59).

Habitat and Phytosociology

It takes part in rocky shrub communities (cracks or ledges of vertical walls with little soil), probably on areas which have served as refuge from former herbivory by goats. Species characteristic of the ***Spartocytision supranubii*** alliance, which

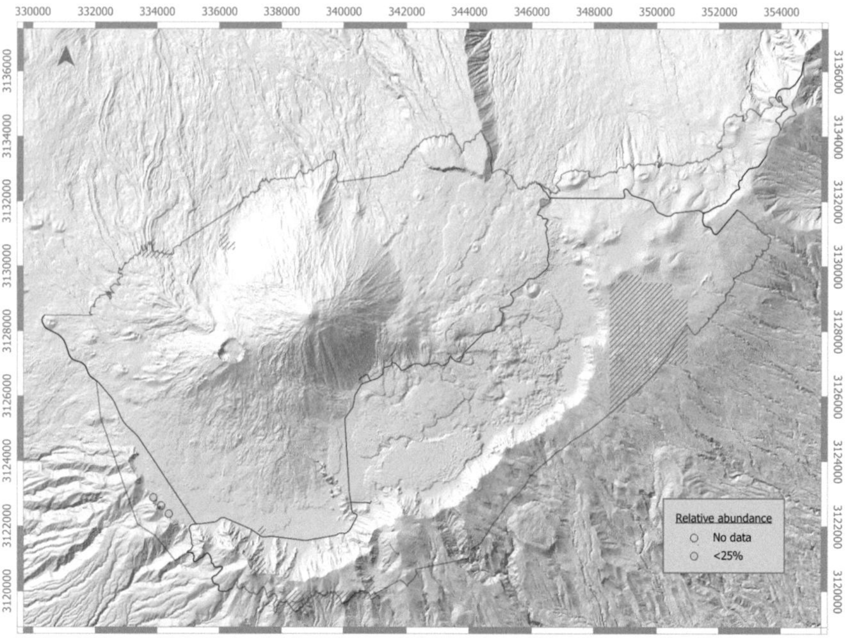

Fig. 3.58 *Bencomia exstipulata* distribution, N° Subgrids (250 × 250 m): **3**/3740

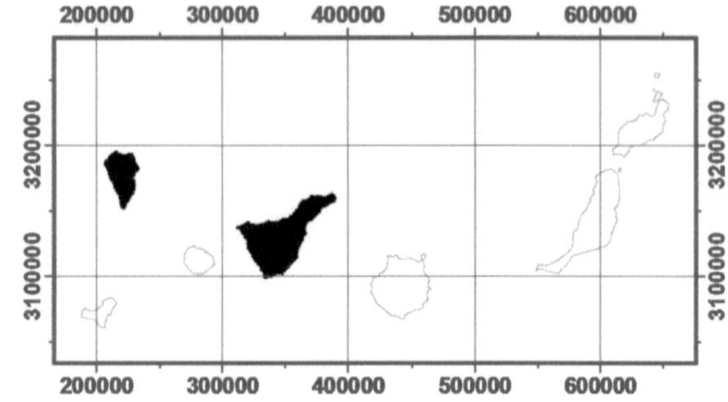

Fig. 3.59 Endemism of the islands of Tenerife and La Palma

groups the climatophilous or edaphoxerophilous high summit shrublands of brooms, flatpods and others, sometimes serial or as gradient of canary island pine; Meso-Oromediterranean dry-humid in Tenerife and La Palma.

Conservation Status

According to the sampling done on two (2) of the subgrids where it was found, conservation status was high, meaning no damages of any kind were identified. On the other known population, this variable was not taken into account.

> "*On high threat status due to the low number of individuals constituting the known populations, as well as due to indirect risks derived from the freeing of barbary sheep and moufflons in the National Parks of 'La Caldera de Taburiente' and of 'El Teide' respectively. Part of the populations from Tenerife are easily accessible to collectors and florists.*" (Translated from: Gómez Campo et al. 1996)

> "*Bencomia exstipulata is a canary endemism of which little more than 650 individuals survive, 580 of which are the results of a number of reinstatement campaigns by Teide and Taburiente National Parks. It is only found on two miniscule localities, and its disappearance from some sectors has been confirmed, therefore it could be included in the maximum threat category (CR o EN). Nonetheless, it cannot be currently considered that there is a negative trend in any of the parameters stipulated by the IUCN (range, occupation, habitat quality, number of localities or number of mature individuals), because the protection regime and actions which have been carried out guarantee its conservation, at least on the medium term and in recent years an increase of individuals has been observed. Without the existence of a continuous decrease or of severe fluctuations, the species can only be classified as Vulnerable (VU D1) because it has less than 1.000 mature individuals.*" (Translated from: Bañares et al. 2004)

> "*Species with only two natural populations of scarce number of individuals and on different islands. There are also reintroduced populations which have managed to increase the number of specimens up to 654. […] The main threat is derived from its lacklustre reproductive strategy and the scarce number of natural individuals in existence. Nonetheless it shows a degree of population stability. […] IUCN category for Spain and globally: VU D1. Legal protection figures: Berna, CNEA (E), Canarias (E), Orden Gobierno de Canarias 20/2/91 (Anexo I).*" (Translated from: Marrero et al. 2004)

Threat Status

E (Barreno et al. 1984)
Anexo I: Estrictamente protegida (Gobierno de Canarias 1991)
EN (Gómez Campo et al. 1996)
CR B1 + 2de (Bañares et al. 1998)
END (VVAA 2000)
En peligro de extinción (Gobierno de Canarias 2001)
VU–D1 (Bañares et al. 2004)
VU–D1 (Moreno 2008)
E**n peligro de extinción** (Gobierno de Canarias 2010)
E**n peligro de extinción** (Ministerio de Medio Ambiente, Medio Rural y Marino 2011)

Germplasm Banks

Banco de Germoplasma de E.T.S. Ingenieros Agrónomos de Madrid.
Jardín Botánico Canario Viera y Clavijo.

Exsiccata (Herbarium Specimens)

Tenerife: Las Cañadas del Teide, 17.05.1981, CRP (TFC 24498); Boca Tauce, 2050 m.s.n.m., 21.05.1983, S. Socorro (TFC 34260). Inventario de la Flora y Vegetación del Parque Nacional del Teide (2001–2003) V.E. Martín Osorio, W. Wildpret & B. Hernández Bolaños: Alrededores de la oficina del P.N. del Teide (30D 32 a), 19/11/2004, J.S. Martín (TFC 44882 + dupl.). **La Palma**: Casa Forestal Roque Faro "Cultivada", 22.03.2000, P.L. Pérez de Paz (TFC 42621 + Dupl.); Hoya de Lucía-Cumbre La Palma "Hojas", 22.03.2000 (TFC 42622).

3.5.19 Bethencourtia palmensis *(C.Sm.) Choisy*

in C.L.von Buch, Phys. Beschr. Canar. Ins.: 148 (1828)

Canariothamnus palmensis (C.Sm.) B.Nord. *in* Compositae Newslett. 44: 27 (2006), not validly publ.
Cineraria palmensis (C.Sm.) Nees *in* Horae Phys. Berol.: 115 (1820)
Senecio palmensis C.Sm. *in* Abh. Königl. Akad. Wiss. Berlin 1816–1817: 370 (1819)
Senecio palmensis (C. Sm. *in* Buch) Link DC., *Prodr.*, VI, 411; Link, *in* Buch, Phys. Beschr. Can. Ins., 148.

Endemism, Certain native
Family: *Asteraceae*
Common name: Cinco uñas o Turgaite (ES); La Palma groundsel (EN)
Growth form: nanophanerophyte

Flowering: July–October
Fruiting: August–November
Pollination: insect pollination (entomogamy)
Dispersal: wind dispersal (anemogamy)
Chromosome N°: n = 10

Description

Small decumbent shrub, suffruticose, glabrous and glaucous; stems pale-dark, newer ones shiny, older ones longitudinally cracked. Lignetum pale flavescent hard, crossed by a central pith, moderately branched. One year old branches densely foliated from the base to the apex, on the lower part the leaves are very drooping, on which white, sericeous axes are visible.

Leaves glabrous, glaucous, fleshy, compact, arranged in a spiral line, 2.5–6 cm long, lanceolate, cuneate, towards the base narrow, with cuneate petiole, thick incised apex and pl. 3–5, linear-lanceolate. Teeth erect, linear-lanceolate, very intact, acute, 1–10 mm long, 0.6–1.5 mm wide at the bottom, triangular lanceolate-linear, terminating in horned corpuscle. Apex of branches ending in a compound corymb, with linear leaf axis 6–20 mm long 0.6–2 mm wide. Lower branches 3–10 cm long, with a corymb-like development towards the centre, lower twigs with narrow linear leaf axis, 1–1.5 mm long, pubescent ones 1.5 mm long, or with simple corymb towards the lower central part. Head oblong-ovate, pedicellate. Pedicels slender, filiform, 4–8 mm long, of 2–4 narrow linear, pubescent. Involucre 1-serial, 5-phyllous, or at the bottom, one of the two leaflets enlarged, scales pedicellated. Leaflet 3 mm long, 0.6–1 mm wide, glabrous, pale flavo-virent, oblong-lanceolate-linear, obtuse, convex, margin hardly scarious, upper part shortly ciliate. Receptable small, 0.5 mm median diameter, convex, slightly alveolate.

Flower head 9-flowered, heterogamous, flowers yellow, (Fig. 3.60) 3 outer female, liguliform, about 4 mm long, glabrous. Tubes filiform, from the base spreading in very infundibuliform way, at the end subcorneous, broad; ligule 1 mm wide, oblong-elliptic, pierced by pl. 4 longitudinal striae, obscure, which in pairs may converge at apex tooth; apex with 3 very small triangularly obtuse ending teeth. Style of stems stems filiform, glabriuscule, apex subtruncate, subrevolute, exserted. Hermaphrodite flowers tubular, 3 mm long. The tube, also extending from the base, apex infundibuliform, ending in 5 obtuse oblong teeth. Reproductive apparatus subexternal. Anthers recurved, apex ending in an oblong-lanceolate appendage. Filament glabrous. Style narrowly recurved, reflexed, apex truncate-peniciform. Achene with pappus 4 mm and, very long if the pappus is smaller, cylindrical-linear, towards the base attenuate, pale flavo-dark, longitudinally striate, hairs obtuse, short, arrect, arranged in longitudinal series, bristle like. Little nectar. Pappus sub-1-seriate white, at base setae in dense and easily distinguishable rings, unequal, the outer ones smaller, denticulate, the small teeth of which are formed by a single cell, erect and short, the upper ones more elongated

Fig. 3.60 Flower detail, September, 2022. (Photo credit: Victoria Eugenia Martín Osorio)

Distribution

Endemism of the islands of Tenerife and La Palma.

It is distributed predominantly on the highest summits, being much rarer at lower altitudes (500–2400 m a.s.l.). On Tenerife it is especially found in the southernmost sectors of the Wall of Las Cañadas, where it can reach 2600 m a.s.l.

Within the National Park it is also found on La Fortaleza, Montaña de Diego Hernández, Cañada de Las Pilas, Topo de la Grieta, Barranco del Río—Montaña Guajara—El Riachuelo, Roques de García, El Sombrerito, Roque del Almendro, Boca Tauce and Cañada de El Cedro. Also within the Botanical Garden of Teide (Figs. 3.61 and 3.62).

Habitat and Phytosociology

Rock plant which can be found on cracks of cliffs and rocks, generally in the upper Mesomediterranean and Supramediterranean bioclimatic belts. It is a characteristic and bioindicator species of the communities of the ***Greenovio aureae-Aeonietea*** class, perennial vegetation which groups evergreen comophytic and casmo-comophytic communities.

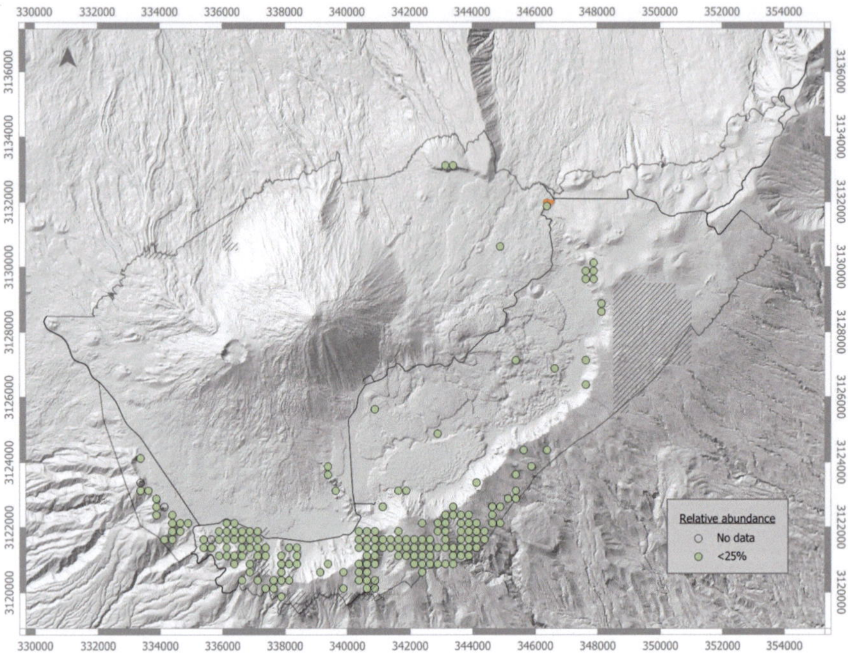

Fig. 3.61 N° Subgrids (250 × 250 m): **183**/3741

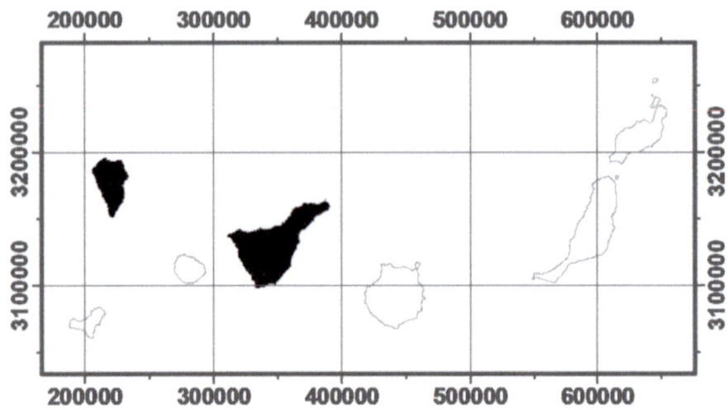

Fig. 3.62 Endemism of Tenerife and La Palma

Conservation Status

On 97% (178) of subgrids where this species was found, no damages of any kind were detected. Its conservation status was deemed as medium or low in only 2% (4) of sub-grids where this taxon was found. On the rest (1), this variable was not measured during sampling

> "*Rare endemism which can be locally frequent in some cases. It is predominately distributed on the highest summits of the islands, being much rarer at lower altitudes. It doesn't appear to suffer from any direct threats. Nonetheless, poaching or the geologic instability of the young volcanic substrates it colonises could be a source of damage for its populations.*" (Translated from: Gómez Campo et al. 1996)

Threat Status

R (Barreno et al. 1984)
Anexo II: Protegidas (Gobierno de Canarias 1991)
LR ca (Bañares et al. 1998)

Germplasm Banks

E.T.S. Ingenieros Agrónomos de Madrid.

Exsiccata (Herbarium Specimens)

Tenerife: 10.06.1969, Bramwell 1899 (TFC 480); Las Cañadas, 28.07.1971, F. Estévez & A. Acuña (TFC 8421); Degolladas de Ucanca y Riachuelo-Mña. La Corona, 10.09.1979, T. Bravo (TFC 18326); Bco. Pasajirón-Vilaflor, 2.03.1980, W. Wildpret, M. del Arco Aguilar, J.R. Acebes, C. León Arencibia, L. Negrín & AVR (TFC 25137); 12.10.1982, W. Wildpret, E. Beltrán Tejera, RLN & S. Socorro (TFC 13155 + Dupl.); Pasajirón-Vilaflor, 19.10.1982, J.R. Acebes & P.L. Pérez de Paz (TFC 16220 + Dupl.); Boca Tauce, aprox. 2050 m.s.n.m., 21.05.1983, S. Socorro (TFC 34264); Altos de Arico, 1750 m.s.n.m., 29.08.1984, P.L. Pérez de Paz & M. del Arco Aguilar (TFC 30337 + Dupl.); Ibid., 21.11.1994, P.L. Pérez de Paz, A. García Gallo & ADH (TFC 36683 + Dupl.); El Pinalete, 1150 m.s.n.m., 6.11.1995, C. León Arencibia & ARB (TFC 38384). Inventario de la Flora y Vegetación del Parque Nacional del Teide (2001–2003) V.E. Martín Osorio, W. Wildpret & B. Hernández Bolaños: Por debajo de El Sombrero, hacia el Sur, en Los Retamares (41D 42 a), 29/04/2003, A. Montañés Rodríguez (TFC 44848); Guajara (42C 4 d), 23/06/2003, Y. de Rioja Santos (TFC 44255); Jardín Botánico (P.N. Teide) (Ñ23–30D 11 b), 23/06/2004, B. Hernández Bolaños (TFC 45395); Roques de García, 13/11/2004, W. Wildpret de la Torre (TFC 45446).

La Palma: Caldera de Taburiente, 25.07.1974, M. del Arco Aguilar (TFC 4285, 27063 + Dupl.); Los Brecitos-La Caldera, 900 m.s.n.m., 20.07.1977, M. del Arco Aguilar, C. León Arencibia & I. La Serna (TFC 32914); Morro Negro, 25.08.1979, W. Wildpret, P.L. Pérez de Paz, M. del Arco Aguilar & I. La Serna (TFC 16002 + Dupl.).

Uses and/or Medicinal Properties

> "*Repellent, toxic. Useful part: extract, juice. Use case: insecticide.*" (Translated from: Pérez de Paz and Hernández Padrón 1999)

3.5.20 Bituminaria bituminosa *(L.) C.H. Stir*

Bothalia, 13: 318 (1981)

Aspalthium bituminosum (L.) Fourr. *in* Ann. Soc. Linn. Lyon, n.s., 16: 365 (1868)

Aspalthium frutescens Medik. *in* Vorles. Churpfälz. Phys.-Ökon. Ges. 2: 380 (1787), nom. superfl.
Bipontinia bituminosa (L.) Alef. *in* Pollichia 22–24: 123 (1866)
Lotodes bituminosum (L.) Kuntze *in* Revis. Gen. Pl. 1: 193 (1891)
Psoralea bituminosa L. *in* Sp. Pl.: 763 (1753)

Not endemic, Probable native
Family: *Fabaceae*
Common name: Tedera o Hediondo (ES); Pitch trefoil (EN)
Biotype: subshrub (chamaephyte)
Flowering: all year round
Fruiting: all year round
Pollination: insect pollination (entomophilous)
Dispersal: –
Chromosome N°: 2n = 20

Description

Herb (10) 20–150 cm, pluricaule. Stems little branched, more or less sericeous. Basal leaves larger than the cauline leaves, the latter 4–18 cm; stipules 4–15 × 1.3–3 mm, lanceolate, generally cuspidate, with 3–7 marked nerves; petiole 15–90 mm, with 4 (6) ribs, with a groove or channel between the 2 adaxial ribs, more or less sericeous; rachis 3–17 mm, similar to the petiole; Leaflets 122–90 × 3–30 mm, elliptic or lanceolate, mucronate or muticous, emarginate, truncate or rounded at the apex, with hairs more or less dispersed on the upper surface, sericeous on the nerves on the underside, with a 1–2 mm setose-sericeous petiole.

Inflorescences usually much longer than the axilla leaves, rarely of equal size, globose or ovoid, (Fig. 3.63) with 15–25 flowers; bracts 6–15 mm, with several nerves, frequently cuspidate-awned, sericeous on the underside, glabrous on the upper surface, the outer ones of the involucre bifid or trifid, the inner ones of the inflorescences generally entire and more or less deltoid; flowers with pedicel c. 1 mm, sericeous-hirsute. Calyx 12–18 mm, with numerous setaceous hairs concentrated mainly on the nerves; tube 5–7 mm, shorter than the teeth, upper teeth 5–8 × 1–2 mm narrowly triangular, flat, the lateral ones 5–9 × 1–1.5 mm, narrowly triangular, flat, the size of the upper ones, the lower one 7–11 × 1.5–3 mm, lanceolate, somewhat concave, often longer than the rest of the teeth. Standard 11–18 × 4–4 5 (7) mm, elliptic, emarginate at the apex, with a 4–6 5 mm claw; wings 10–18 × 1.7–2 (3) mm, acute at the apex, with a 4.5–5 5 (8) mm claw; keel 8.5–14 × 1.3–2 5 mm, with a 4.5–7 mm claw. Androecium with all stamens more or less of the same length; staminal tube very developed; anthers 0.2–0 4 (0.5) mm. Ovary densely sericeous.

Fruit 13–19 mm; body of shoot 4–5 5 × 2.5–3 mm, ovoid, setose-sericeaus, with a crown of 0.5–1 × 0.5 mm spicules, generally purple at the apex and 2 spicules 2–4 × 0.2–0.5 mm, white, arched, horn-shaped at the apex of the belly, with

Fig. 3.63 *Bituminaria bituminosa*, Oct. 2015. (Photo credit: Victoria Eugenia Martín Osorio)

numerous sessile and black glands over the entire surface; beak 9–16 × 1.5–2.5 mm, ensiform, flattened, straight or curved, scabrid. Seeds 2.5–3 × 4–5 mm.

Distribution

Distributed throughout the Mediterranean region, the Canary Islands and the islands of Madeira and Porto Santo. Between 0 and 1250 m above sea level

Bituminaria bituminosa is considered an introduced species within the Teide National Park (Martín Osorio et al. 2006). Within this park, according to the sampling carried out, it is widely distributed, occurring on the escarpment of La Fortaleza and in sectors of the Wall, from Cañada de El Cedro, passing through Boca Tauce and the interior slopes of Ucanca, to Roques de García and Montaña Guajara, from where it reaches over the edge and colonises the southern slope of said Wall (Barranco Erís, W-Montaña Guajara); 1950–2600 m above sea level (Figs. 3.64 and 3.65).

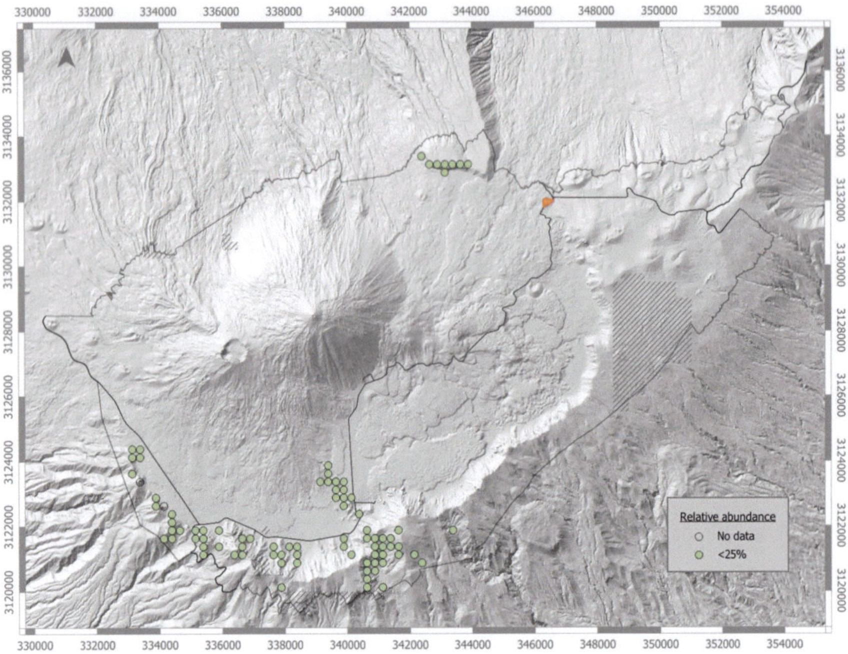

Fig. 3.64 *Bituminaria bituminosa* distribution, N° Subgrids (250 × 250 m): **85**/3740

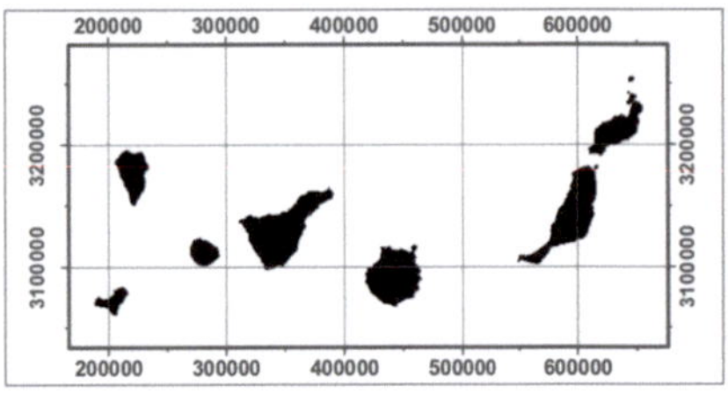

Fig. 3.65 Present on all the Canary Islands

Habitat and Phytosociology

The species is found in nitrified environments at the base of cliffs or on the edges of forests, on clayey, marly and limestone substrates, rarely on sand, schist or quartzite rocks.

It is a characteristic species and bioindicator of ***Lygeo sparti-Stipetea tenacissimae***, a type of vegetation that groups together communities of Mediterranean xerophytic tufted basophilous perennial grasslands, from open lawns to closed grasslands, in which bulbous plants, perennial herbaceous plants and microfruticose plants can also be found on grasslands which develop on soils with a loamy or clayey texture, rich in bases, calcareous, gypsum, dolomitic or ultramafic, which are not subject to salinization or temporary hydromorphism processes and which correspond primarily to serial stages of neutrophilous and basophilous forests and summit shrublands of ***Quercetea ilicis***. They have a wide Mediterranean distribution in the arid-subhumid Thermo-Supra-Mediterranean bioclimatic zones from balanced Euoceanic to moderate continental.

It also intervenes in ***Artemisietea vulgaris*** communities, a type of vegetation that groups nitrophilous, biennial or perennial ruderal communities, developed on soils enriched in nitrogenous or phosphorous nutrients, due to the activities of man and livestock.

Conservation Status

In practically all the subgrids in which this species was found (81), no damage of any kind was observed. Only in three (3) of them, damage was detected (browsing, high mortality without regeneration, etc.), but this did not affect more than 25% of the specimens. In the rest (1), this variable was not specified at the time of sampling.

Remarks:

Dickson et al. (1987) warn that in the Teide National Park, the introduced species have been dispersed mainly by the practice of grazing, stopped by the declaration of National Park in 1954; and by tourism, which in recent years has encouraged the entry of a relatively recent flow of invasive plants.

Germplasm Banks

E.T.S. Ingenieros Agrónomos de Madrid.

Exsiccata (Herbarium Specimens)

Tenerife: La Laguna, 16.02.1982, A. García Gallo (TFC 22266); Laderas del Risco de las Nieves, 10.01.1983, M. Arco Aguilar, P.L. Pérez de Paz, W. Wildpret (TFC 25376); La Laguna, 23.05.1985, W. Wildpret, A. García Gallo, E. Beltrán (TFC 22267); La Laguna, 23.05.1985, ejusd. (TFC 22268, 22269, 22270); Los Silos, Pina, 10.06.1992, V. Lucía Sauquillo, C. Acosta (TFC 38189); La Matanza, Casa de El Caletón, 16.01.1994, F. Ramallo Rodríguez (TFC 37261); Los Silos, junto a Charcos de Erjos, 23.07.1994, V. Lucía Sauquillo (TFC 38194); Casillas Ijuana, Ladera Norte, 600 m.s.n.m., 19.01.96, R. Racca (TFC 38957); Las Casillas, Bco. Ijuana, ladera Norte, 26.04.1996, R. Racca (TFC 39121); Barranco Ijuana, ladera Norte, 350 m.s.n.m., 18.06.1996, R. Racca (TFC 39152); Inventario de la Flora y Vegetación del Parque Nacional del Teide (2001–2003) V.E. Martín Osorio, W. Wildpret & B. Hernández Bolaños: Guajara, cerca de El Capricho (42C 10 c), 17/06/2003, A. Montañés Rodríguez (TFC 44269).

La Palma: Cumbre Vieja, 12.1987, FCR (TFC 601, 25604).

La Gomera: San Sebastián, Vegaipala, 30.04.1989 (TFC 29105 + duplic.); Pista del basurero de Agulo, 08.05.1989, PL. Pérez de Paz, M. Arco Aguilar, M. Salas (TFC 25299).

Gran Canaria: Monte Pajonales, 01.06.1985, M. Nogales, (TFC 20149); Monte Inagua, 02.1985, ejusd. (TFC 20150 + duplic.); Artenara-Cruz de Acusa, 1100 m.s.n.m., 15.03.1992, M. Salas Pascual (TFC 37030); San Nicolás de Tolentino, Bajando por Bco. de Inagua, 1.6.1996, W. Wildpret, E. Beltrán Tejera, V.E. Martín Osorio & M. Salas Pascual (Proyecto Dir. Flora.) (TFC duplic. 39197).

Fuerteventura: Jandía, La Pared, 26.02.1993, P.L. Pérez de Paz (TFC 35786).

Lanzarote: Alegranza, Ladera exterior sur de la Caldera (pocos ejemplares), 17.05.1989, C. Suárez (TFC 30423, 30424); San Bartolomé, Tomaren, 24.2.1995, J.A. Reyes-Bentancort & M.C. León Arencibia (TFC 37741).

Uses, Exploitation and/or Medicinal Properties

> "*Vulnerary, against bee stings, hypocholesterolemic; toxic: erythematous, vesicant; fodder; dye: yellow-greenish. Useful part: leaves, tops; whole plant. Instructions for use: external use: decoction, poultice; fodder: fresh (toxic)/dry (hay).*" (Translated from: Pérez de Paz and Hernández Padrón 1999)

3.5.21 Bromus madritensis L.

First published in Cent. Pl. I: 5 (1755)

Anisantha madritensis (L.) Nevski *in* Trudy Sredne-Aziatsk. Gosud. Univ., Ser. 8b, Bot. 17: 21 (1934)

Bromus diandrus var. *madritensis* (L.) DC. *in* Cat. Pl. Horti Monsp.: 11 (1813), nom. illeg.

Bromus madritensis subsp. *eumadritensis* Maire & Weiller *in* Fl. Afrique N. 3: 237 (1955), not validly publ.

Bromus muralis var. *madritensis* (L.) Huds. *in* Fl. Angl., éd. 2: 51 (1778), nom. illeg.

Bromus rubens subsp. *madritensis* (L.) Bonnier & Layens *in* Tabl. Syn. Pl. Vasc. France: 369 (1894)

Not endemic, Probable native

Family: *Poaceae*

Common name: Aceitillo, Balango o Espiguilla de Madrid (ES); Compact brome (EN)

Biotype: annual (therophyte)

Flowering: March–May

Fruiting: March–May

Pollination: self pollination (autogamy)

Dispersal: wind dispersal (anemochory)

Description

Annual plant with stems up to 60 cm, erect or ascending, glabrous or scabrid below the panicle (Fig. 3.66).

Leaves with 2–3 mm fimbriate ligule and blade up to 20 × 0.7 cm, flat, generally tomentose, sometimes villous near the ligule; the lower ones with a hairy sheath with short and patent hairs of c. 0.5 mm or with long and flexuous hairs of 1.5 mm; the upper ones with glabrous or hairy sheaths.

Panicle loose or dense, with generally semi-whorled branches; branches up to 4 cm, with 1–5 spikelets. Spikelets of 25–35 mm, glabrous or hairy, with 4–12 flowers. Unequal glumes, very acute with a median nerve reaching the apex; The lower one 9–10 mm, linear-subulate; the upper one 12–15 mm, linear-lanceolate. Lemma 15–19 mm, lanceolate with scarious margin and bidentate and scarious apex; awn 18–25 mm, longer than the lemma, generally curved when mature, inserted 2–3 mm below the apex of the lemma. Palea shorter than the lemma, linear-elliptic, ciliate in the upper half. Androecium with two stamens; anthers 0.6–1.2 mm.

Distribution

We can find it in the W and S of Europe, in the Mediterranean region and the Macaronesia Islands (except Cape Verde). Present on all the islands of the Canary Archipelago (Fig. 3.68).

Fig. 3.66 *Bromus madritensis*. (Photo credit: GG Casanova)

In the Teide National Park it is found in a very dispersed way in areas of the Ridge (La Crucita, near Montaña del Limón, and south of Izaña); very punctually, in some sectors of the Las Cañadas Wall (Cañada de La Grieta, Barranco del Río, Montaña Guajara, lower area of the Riachuelo, southern slopes of Roque del Almendro and La Sombrera, and Morro del Tiro del Guanche), and outside it, near the road that goes to Chío, on the lava flows of Las Narices del Teide; between 2000 and 2450 m a.s.l. It was also found in the Teide, Wolfredo Wildpret Botanical Garden (Figs. 3.67 and 3.68).

Habitat and Phytosociology

It normally appears in grasslands and roadsides. It is present in communities of ***Stellarietea mediae***, a type of annual anthropogenic nitrophilous vegetation, made up of therophytes or small geophytes of a weedy, ruderal, or scionitrophilous character, with a Holarctic distribution and cosmopolitan irradiation.

It is a characteristic species of ***Thero-Brometalia***, subnitrophilous therophytic vegetation of the western Mediterranean, Ibero-Atlantic and Tingitano, dry, subhumid, humid; which shows some Cantabrian-Atlantic and Canarian irradiations. It preferably settles in abandoned fields of crops and roadsides.

It can also be found within ***Sisymbrietalia officinalis***, a moderately nitrophilous ruderal and roadside vegetation, with a vernal or summer phenology, widely distributed throughout the Mediterranean region and other Holarctic territories with a Mediterranean bioclimate.

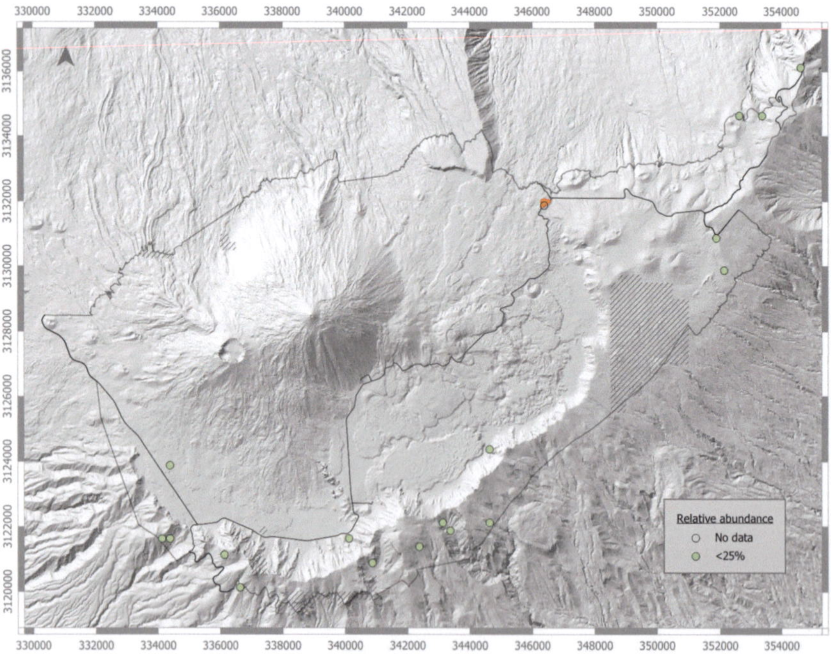

Fig. 3.67 *Bromus madritensis* distribution, N° Subgrids (250 × 250 m): **18**/3740

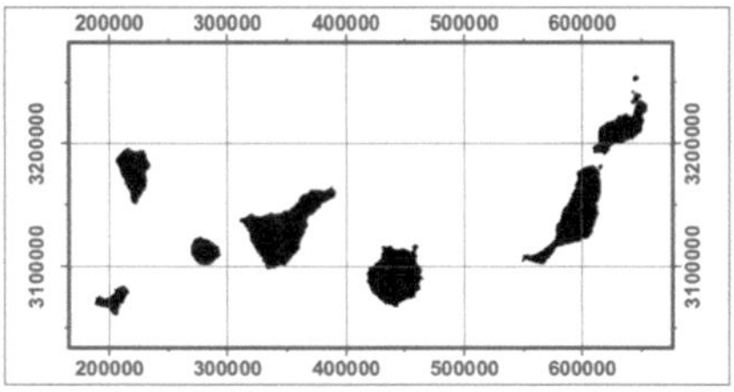

Fig. 3.68 Present on all the Canary Islands

Conservation Status

In most of the subgrids in which this species was found (16), no damage of any kind was observed. Damage was only found on more than 25% of the specimens in one (1) of them. In the rest (1), this variable was not specified at the time of sampling.

Exsiccata (Herbarium Specimens)

Tenerife: Inventario de la Flora y Vegetación del Parque Nacional del Teide (2001–2003) V.E. Martín Osorio, W. Wildpret & B. Hernández Bolaños: Al NE de la Cañada del Cedro (41A 20), 17.05.2001, L. Afonso, A. Hernández, Y. de Rioja & F. Tapia (TFC 44880); Degollada de Guajara (42C 5 c), 21.05.2003, F. Mesa (TFC 44230); Al SW de La Tarta (31A), 05.04.2001, L. Afonso Jiménez & A. Hernández Luis (TFC 44909); Bar del cruce entre La Orotava y La Esperanza, en El Portillo, 25.06.2003, F. Mesa, A. Montañés & Y. de Rioja (TFC 44293); Entrada al Jardín Botánico (30D 11 b) (T19), 17/06/2004, B. Hernández Bolaños (TFC 45400).

3.5.22 Bromus tectorum *L.*

First published in Sp. Pl.: 77 (1753)
Anisantha tectorum (L.) Nevski *in* Trudy Sredne-Aziatsk. Gosud. Univ., Ser. 8b, Bot. 17: 22 (1934)
Bromus lateripronus St.-Lag. *in* Ann. Soc. Bot. Lyon 7: 121 (1880), nom. superfl.
Bromus sterilis var. *tectorum* (L.) Kuntze *in* Trudy Imp. S.-Peterburgsk. Bot. Sada 10: 249 (1887)
Bromus tectorius Dulac *in* Fl. Hautes-Pyrénées: 96 (1867), nom. superfl.
Bromus tectorum var. *genuinus* Godr. *in* J.C.M.Grenier & G.Godron, Fl. France Corse 3: 583 (1856), not validly publ.

Not endemic, Probable native
Species introduced in Teide National Park
Family: *Poaceae*
Common name: Espiguilla colgante (ES); Cheat Grass (EN)
Biotype: Annual (therophyte)

Flowering: March–May
Fruiting: March–May
Pollination: –
Dispersal: –
Chromosome N°: 2n = 14

Description

Annual with stems up to 50 cm, erect or ascending, generally with retrorse indumentum below the inflorescence.

Leaves with 2.5–4 mm ligule, fimbriate, and blade up to 14 × 0.5 cm, with ciliate margin; the lower ones with a sheath of double, patent indumentum, with short hairs of c. 0.5 mm and other long hairs of c. 2.5 mm. Very flexuous panicle, with spikelets arranged unilaterally; whorled branches, much longer than the spikelets, flexuous, with 1–3 spikelets.

Spikelets 20–27 mm, glabrous, with 5–10 flowers, generally patent or pendulous (Fig. 3.69). Upper flowers of each spikelet generally sterile. Glumes with scarious margin and upper part and nerves that do not reach the apex, generally violet; the lower one 6.5–10 mm, triangular-lanceolate, acute; the upper one 10–14 × c. 2 mm, lanceolate-elliptic, generally emarginate. Lemma c. 17 × 2.7 mm, lanceolate-elliptic, emarginate, broadly scarious; awn 10–15 mm, straight, inserted c. 4 mm below the apex of the lemma. Palea c. 10 × 0.2 mm, linear, very slightly emarginate, with ciliate margin on the upper 2/3. Androecium with three stamens; anthers c. 0.8 mm.

Distribution

It can be found in Europe, temperate Asia, northern Africa and the Macaronesia region. In the Canary Islands it has only been found on the islands of La Palma, Tenerife and Gran Canaria.

Bromus tectorum is considered an introduced, invasive and transforming species within the Teide National Park (Martín Osorio et al. 2006). Within this, although it

Fig. 3.69 (**a**) Siete Cañadas, May 2004. (Photo credit: Beatriz Hernández) (**b**) Jardín Botánico June 2006. (Photo credit: Victoria Eugenia Martín Osorio)

seems to preferentially colonise the areas of the Pared, it is widely distributed throughout the entire territory, between 1600 and 2900 m a.s.l., except for the slopes and lava flows of the Teide-Pico Viejo and Montaña Blanca complex. However, and although still very sparsely, according to the sampling carried out, it is already beginning to colonize these latter sectors as well (Figs. 3.70 and 3.71).

Habitat and Phytosociology

The taxon is mainly ruderal-nitrophilous in highland grasslands. It is present in communities of the class ***Stellarietea mediae***, a type of anthropogenic annual nitrophilous vegetation, made up of therophytes or small geophytes of a weedy, ruderal, or scionitrophilous character, with a Holarctic distribution and cosmopolitan irradiation.

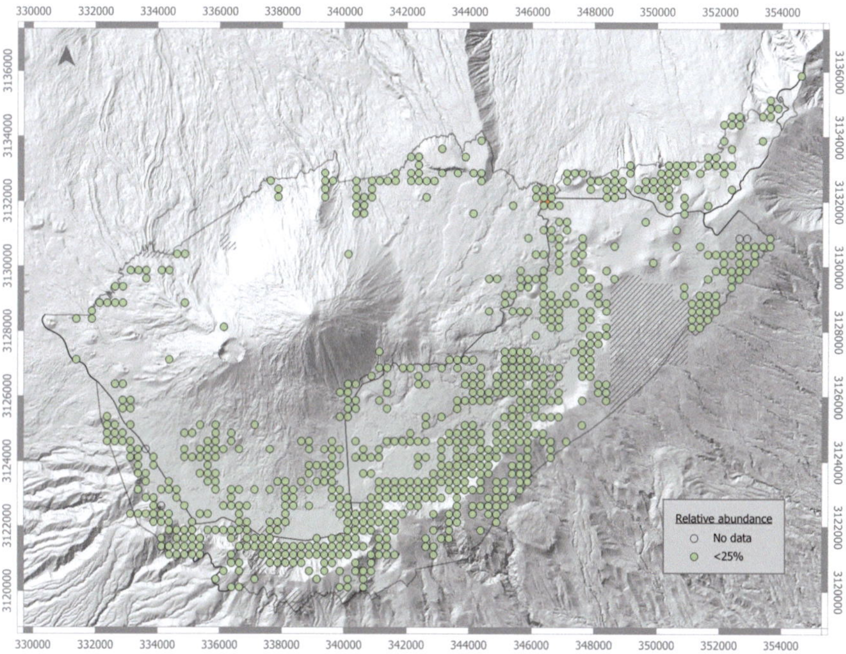

Fig. 3.70 *Bromus tectorum* distribution, N° Subgrids (250 × 250 m): **820**/3740

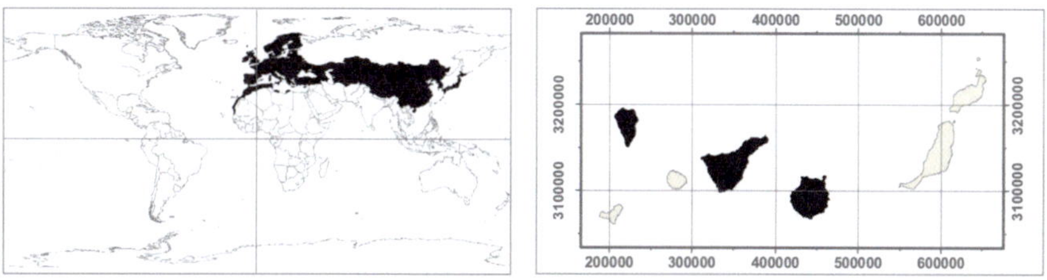

Fig. 3.71 Present on La Palma, Tenerife and Gran Canaria Islands

It is also found in the community described as Cerrillar-Crespar de las Cañadas, ***Arrhenathero calderae-Plantaginetum webbii***, colonizing the slopes of gelifract debris of the Pared.

Conservation Status

In practically all the subgrids in which this species was found (786), no damage of any kind was observed. Only in thirty (30) of them, its condition was considered to be medium or low.

Remarks:

Dickson et al. (1987) warn that, in the Teide National Park, the introduced species have been dispersed mainly by the practice of grazing, stopped by the declaration of National Park in 1954; and by tourism, which in recent years has encouraged the entry of a relatively recent flow of invasive plants. Of the 83 introduced species, 30 were considered well established, 15 of them were grasses, with ***Bromus tectorum*** and ***Festuca myuros*** standing out as the most abundant.

According to Martín Osorio et al. (2006), between 2003 and 2005, technical staff from the Teide National Park developed a "Programme for the eradication of introduced flora" with 67 catalogued taxa, on which a total of 34 are being acted upon with elimination programmers, with ***Avena barbata*, *Bromus tectorum*, *Chenopodium album*, *Chenopodium vulvaria*, *Dittrichia viscosa*, *Lactuca serriola*, *Marrubium vulgare*** and ***Petroselinum crispum*** standing out for their abundance and distribution. At present, work is still being carried out to eliminate exotic plants from the protected area.

Furthermore, it is made clear, using Dukes and Money (2004) as a reference, that the case of ***Bromus tectorum*** could be considered a clear example of the most pronounced effect that invasions have on the dynamics of terrestrial ecosystems due to the alteration of the nitrogen balance in the receiving ecosystem. The magnitude of this effect is sufficient to compromise the stability of the affected ecosystems. It is a species that competes for space with endangered species such as ***Helianthemum juliae***.

Exsiccata (Herbarium Specimens)

Tenerife: Bandas del Sur, Playa de las Américas, Médano, excursión, 4.2.1968, E. Sventenius & W. Wildpret (TFC 4162); Bco. de Tabares, 9.3.1970, E. Beltrán Tejera (TFC 35503); Las Chumberas, la Cuesta, 18.3.1971, ? (TFC 34541); Circo de la Caldera de Las Cañadas, 12.6.1976, E. Beltrán Tejera (TFC 25868); Las Cañadas, La Fortaleza, 13.6.1980, E. Beltrán Tejera (TFC 35259); Ayosa, borde carretera, 6.5.1981, M. del Arco Aguilar (TFC 24387 duplic. 4); Ayosa, 6.5.1981, ejusd. (TFC 24959); Cumbres de Arafo, Valles del Volcán de las Arenas, 9.5.1981, P.L. Pérez de Paz, W. Wildpret & M. Arco Aguilar (TFC 23727 + duplic.); Ibid, 9.5.1981, ejusd. (TFC 23751, duplic. 3); Ibid, 1700 m.s.n.m., 9.5.1981, ejusd. (TFC

23732); Las Cañadas del Teide, Bco. del Riachuelo, 2100 m.s.n.m., 7.5.1983, J.S. Socorro & W. Wildpret (TFC duplic. 18478); [Como *Bromus* cf. *tectorum* L.] Adeje, Mña. Colorada, 16.4.1987, P.L. Pérez de Paz & M. del Arco Aguilar (TFC 24509); Pinares de Adeje, 20.4.1987, P.L. Pérez de Paz, M. del Arco Aguilar & M. Chacana (TFC 24076). Inventario de la Flora y Vegetación del Parque Nacional del Teide (2001–2003) V.E. Martín Osorio, W. Wildpret & B. Hernández Bolaños: Al N de la Montaña del Limón (31A 2), 4.04.2001, L. Afonso Jiménez & A. Hernández Luis (TFC 44910); Cañada de Los Guancheros—La Fortaleza (30A 24), 9.07.2001, Y. de Rioja Santos (TFC 44265); Ladera Sur del Teide, hacia el teleférico (35D 28 d), 24.03.2003, Y. de Rioja Santos (TFC 44858); Parque Nacional del Teide, 13.11.2003, W. Wildpret (TFC 45416); Cañada de Diego Hernández, 12/05/2004, B. Hernández Bolaños, V.E. Martín Osorio & W. Wildpret de la Torre (TFC 45437); Área de Ayosa, carretera de La Esperanza, 08/06/2005, V.E. Martín Osorio, W. Wildpret & B. Hernández Bolaños (TFC 44981 + Duplic.).

Gran Canaria: Artenara, Mña. del Brezo, 15.3.1991, M. Salas Pascual (TFC 37129).

3.5.23 Bufonia paniculata *F. Dubois* ex *Delarbre*

Fl. Auvergne ed., 2, 300 (1800).

Bufonia macrosperma Gay *ex* gren. & Godron. *Fl. France*, 1: 248 (1847)
Bufonia paniculata Dub. ssp. *teneriffae* (Christ) Kunk.
Bufonia teneriffae (Christ) Kunk.

Not endemic, Certain native
Family: *Caryophyllaceae*
Common name: Escobilla estrellada (ES); Panicled Bufonia (EN)
Biotype: therophyte
Flowering: May
Fruiting: May
Pollination: –
Dispersal: –
Chromosome N°: 2n = 32

Description

Annual or biennial herb, apparently sometimes perennial, 10–30 (50) cm. **Stems** erect or ascending, branched at least in the upper part, with branches often divaricate, scabrid-puberulous in the lower part, glabrous in the rest.

Leaves up to 15 (30) mm, setaceous, mucronate. Inflorescences elongated, raceminform or spiciform. **Flowers** pedicelate or subsessile, the outer pedicels being c. 3.5 mm, the inner ones 4–4.3 mm-lanceolate, acute, acuminate, with 5 nerves-at least at the base and scarious margin generally more than 0.2 mm wide (Fig. 3.72).

Fig. 3.72 *Bufonia paniculata*. (Photo credit: G. García Casanova)

Petals small, shorter than the sepals, at least than the inner ones. Stamens generally 4. Styles up to 1/2 the length of the ovary. **Capsule** 2–2.3 mm, smooth, papery, yellowish-green, generally dispersed. Seeds (1.5) 1.6–2 mm, with dorsal tubercles generally short, obtuse.

Remarks:

According to the consulted bibliography, there are southern populations of the species ***B. perennis*** that have been named as ***B. paniculata var. tuberculata*** (Loscos) Rivas Goday & Borja *in* Anales Inst. Bot. Cavanilles, 19, 328 (1961).

Distribution

It is distributed throughout the Iberian Peninsula, central and southern France, Switzerland, Italy, Greece, Lebanon and Syria, between 600 and 1800 m a.s.l.

In the Canary Islands it is found in La Palma, La Gomera, Tenerife and Gran Canaria [ssp. *teneriffae* (Christ) Kunkel].

In the Teide National Park its distribution is closely related to the rugged areas of the Las Cañadas Wall and Roques de García (2000–2600 m a.s.l.). Only very occasionally in the interior of the Depression (Figs. 3.73 and 3.74).

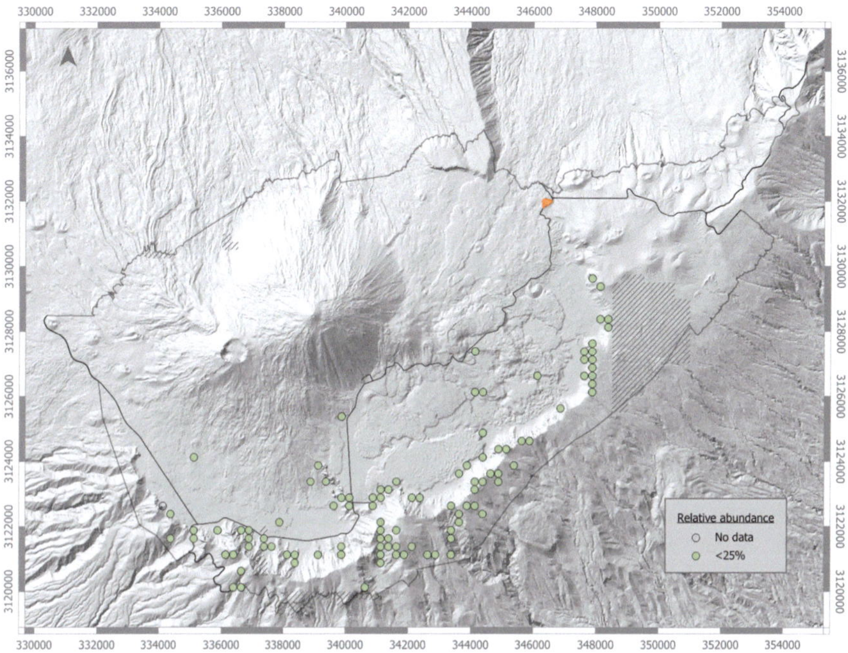

Fig. 3.73 *Bufonia paniculata* distribution, N° Subgrids (250 × 250 m): **102**/3740

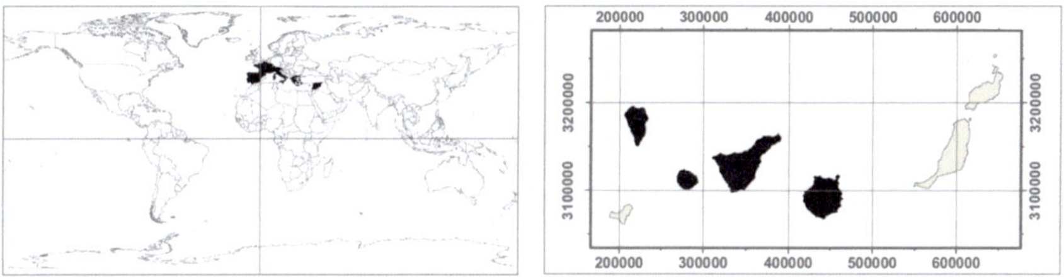

Fig. 3.74 Present on La Palma, La Gomera, Tenerife and Gran Canaria Islands

Habitat and Phytosociology

The taxon grows in uncultivated fields, wastelands and rocky terrain. It is a characteristic species and bioindicator of the order ***Chamaecytiso-Pinetea canariensis***, in pine forests and broom shrublands, a dry-humid Meso-Oro-Mediterranean bioclimate of the central-western Canary Islands. It is also found in the community described as Cerrillar-Crespar, which preferentially colonizes the slopes of gelifract debris of the Pared de las Cañadas: ***Arrhenathero calderae-Plantaginetum webbii***.

Conservation Status

In practically all the subgrids in which this species was found (98), no damage of any kind was observed.

Threat Category

R (Barreno et al. 1984)—Como *B. teneriffae* Christ.
VU (Bañares et al. 2004).
VU B2ac(iv) (Moreno 2008).

Exsiccata (Herbarium Specimens)

Tenerife: Las Cañadas, Cañada de Diego Hernández, 2.2.1973, P.L. Pérez de Paz (P TFC: 265; TFC 13148); Las Cañadas, acantilados, 12.6.1976, E. Beltrán Tejera (TFC 13081); Las Cañadas, 17.5.1981, CRP (TFC 24501); Roques de Chavao, cerca de Boca de Tauce, 18.6.1983, W. Wildpret & J.S. Socorro (P TFC: 290; TFC 13165); Cabecera de Bco. Pedro Díaz sobre Bailadero, 29.5.1986, W. Wildpret, P.L. Pérez de Paz, M. del Arco Aguilar & O. Rodríguez Delgado (TFC 28101); Inventario de la Flora y Vegetación del Parque Nacional del Teide (2001–2003), V.E. Martín Osorio, W. Wildpret & B. Hernández Bolaños: Por debajo de El Sombrero, hacia el Sur (41D 41 d), 29/04/2003, Y. de Rioja Santos (TFC 44856); Guajara, nacimiento del Bco. del Río (42C 14b), 22.5.2003, A. Montañés (TFC 44277); Cañada de Diego Hernández, 12/05/2004, B. Hernández Bolaños, V.E. Martín Osorio & W. Wildpret de la Torre (TFC 45439).

La Palma: Cumbres de la Caldera, Refugio de los Roques, 23.7.1973, P.L. Pérez de Paz (TFC 13055).

Gran Canaria: Bco. de Guayadeque, 19.12.1979, I. La Serna & B. Méndez (TFC 7435).

3.5.24 Bystropogon canariensis *(L.) L'Hér.*

Sertum Anglicum, 12 (1788)

Mentha canariensis L. *Species Plantarum*, 2: 578 (1753)
New appointment for Teide National Park

Endemism, Certain native
Family: *Lamiaceae*
Common name: Poleo de monte (ES); trans: Wild pennyroyal (EN)
Growth form: nanophanerophyte
Flowering: summer
Fruiting: end of summer
Pollination: insect pollination (entomogamy)
Dispersal: wind dispersal (anemochory)
Chromosome N°: 2n = 42, 48

Description

Nanophanerophyte with hairy branches in the annual shoots, always with a long indumentum, internodes 0.7–12 cm apart. **Leaves** variable in size, measuring from 1.5 to 10.25 × 1–5 cm, crenulate or crenulate serrate and slightly revolute, sometimes leathery, more or less hairy green on the upper side and green to green-tinged on the underside, ovate to lanceolate, more or less cordate or attenuate at the base, obtuse or subacute, rarely acute; petiole with long indumentum, measuring 0.5–3 cm, always shorter than the blade. Semireticulate simple veins prominent on the underside, indumentum always present on both leaf surfaces. Indumentary hairs of two types, some more or less long multicellular and articulated, composed of elongated cells, generally on the nerves and scattered on the upper and lower side in greater or lesser numbers, the other small uni- or bicellular hairs densely distributed on the lower side of the leaves and absent on the upper side. Essential glands hyaline, sometimes yellowish, always present on both sides of the leaves, more abundant on the underside.

Inflorescence in panicle with dichotomous branching from the peduncle (Fig. 3.75). Peduncle with long indumentum, axillary opposite, one to three in number, up to 5.2 cm long, occurring on both primary and secondary branches of annual shoots or younger branches. Bracts of the isomorphic dichotomous apexes, sometimes the first pair in the form of leaflets, from 3.5 (6) × 1 mm the first pair,

Fig. 3.75 Inflorescence of the wild pennyroyal, August 2019. (Photo credit: Victoria Eugenia Martín Osorio)

progressively smaller to 0.75–1 c 0.1–0.25 mm the floral ones, lanceolate-linear, acuminate, provided with defined central nerve, with long indumentum on the margins and on the external side, glabrous on the internal side

Flowers shortly pedicellate or subsessile arranged in more or less dense globose glomeruli—glomerulated dicasia. Flower solitary at the apex of all dichotomies, more or less pedicellate, pedicels of the first 0.5–3 mm. Calyx infundibuliform flared, (1.5) 2–2.5 mm long with long indumentum on the outside, 10 nerves of which the 5 main ones are more prominent than the secondary ones, teeth triangular, short, acuminate or slightly subulate at the aepx, of less length than the tube (1/3 to 1/2 than the tube length), with long indumentum on the margins. Weak carpostegium. Corolla white, outer one (3.5) 4–4.5 mm long, tube 1.5–2 (2.5) mm; upper lip chordate or better square forked (0.75) 1–1.25 (1.5) mm, lower lip trilobed, central lobe reed-shaped, entirely or weakly notched (1) 1.5–2 (2.25) mm long, larger than the lateral ones which range in length from (0.75) 1–1.25 (1.5) mm, entire. Androsterile individuals with reduced corollas varying in size from 2.25 to 2.5 mm are common. Stamens with filaments 0.5–1 mm long anteriorly, anthers 0.3–0.5 mm long, style 1–1.75 mm and large stigma lacinia 0.5–0.75 mm giving an overall length of (1.5) 2–2.5 mm. Nucules chestnur brown, (0.75) 1 × 0.5–0.6 mm, rounded at the apex, pericarp tuberculate to the naked eye and papillose-glandular to the scanning electron microscope.

Remarks:

Tipos: *Mentha canariensis* Linn.; *Bystropogon canariense* L'Hérit.!; 2, Teneriffe, 1778; Fr. Masson s.n. (BM!; lecto).

Distribution

Canary endemism, present on all western Canary Islands and Gran Canaria.

In the Teide National Park it has only been found around the area of La Crucita, near La Tarta and, on a one-off basis, on the northwestern slopes of Pico Viejo (Mancha Ruana—2225 m a.s.l.) and on Montaña del Cedro (Figs. 3.76 and 3.77).

Habitat and Phytosociology

Normally, it usually appears in the typical environments of the fayal-brezal, more or less humid, also in rockrose shrublands in fayal-brezal succession ecotones and in degraded humid pine forests as well as on walls and ledges of ravines. It is a characteristic species and bioindicative of the ***Myrico fayae-Ericion canariensis*** alliance, which groups summit shrublands and serial microforests of heathers of Canary Islands cloud forests in which Canary heather is predominant (*Erica canariensis*).

Conservation Status

On all subgrids where this taxon was found its conservation status was high, meaning no damages of any kind were detected.

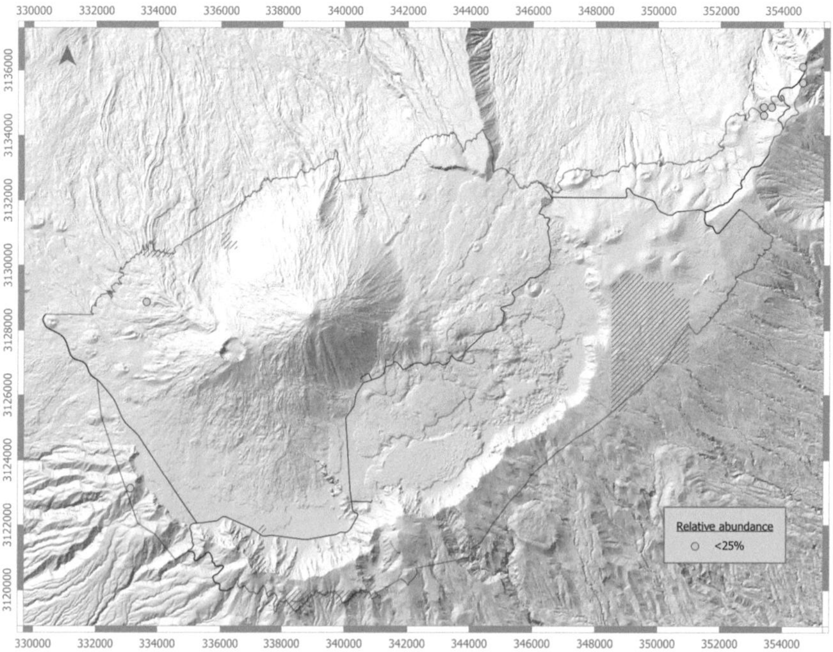

Fig. 3.76 *Bystropogon canariensis* distribution, N° Subgrids (250 × 250 m): **7**/3740

Fig. 3.77 Endemism of the western and central islands

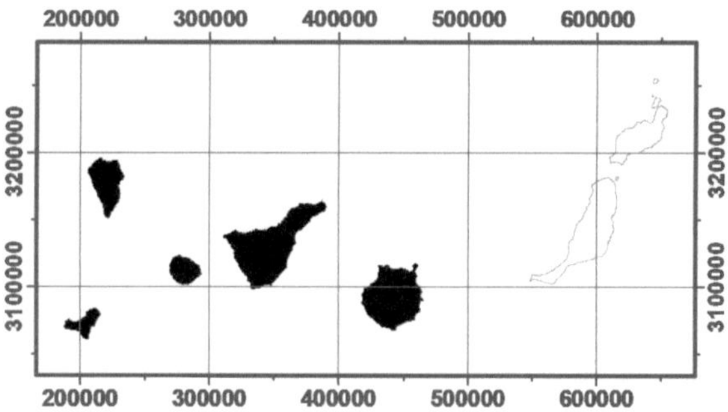

Threat Status

LR ca (Bañares et al. 1998)

Germplasm Banks

Banco de Germoplasma de E.T.S. Ingenieros Agrónomos de Madrid.
Jardín Botánico Canario Viera y Clavijo

Exsiccata (Herbarium Specimens)

Tenerife: In sylvis laurorum supra Güímar, 8.IV.1855, E. Bourgeau 1472 e.p (FI: *in* Herb. Webbianum s.n.!, W: 23220!).-Ibid., 8.IV.1855, ejusd. (K. MA: 162292!; W: 58722, 74436, s.n., 2455!).- Tenerife cult. *In* Hortus Milfortianum y Bco. Badajoz, s.d., Webb 17 y à specimen Herb. D.C. - lui inscrititan "Teneriffa Güímar sur la laves"-, *in* Herb Webbianum s.n. (FI!).- Barranco Badajoz "*in* faucibus" 4–500 m, 3.VI.1901, J. Bornmüller 2770 (K, W: 3978!, G!).- Güímar, Bco. del Río "*in* rupestribus umbrosis" 500 m, 9.VI.1901, ejusd. 2771 (W: 3979!).- Bco. Badajoz (Güímar), 6.VI.1894, R.P. Murray (K, BM).- Canary s.l., s.d., Despréaux s.n. (K).- Bco. E de la Orotava 1000–3000 pies, VIII.1862, C.B. Clarke 1948 (K).- Barranco near Güímar, 21-V.1890, R.P. Lowe 185 (K).- Tamadaya, VII.1855, Perraudiére (BM, W: 52030!).- "*In* lauretis supra Güímar", 8.IV.1855, ejusd. (Herb. H.F. hauce 11509 *in* BM; W: 52057!).- Güímar "planta rarísima", s.d., C. Bolle (W!).- Ibid., "locus natalis unicus", 1856, ejusd. (W: 282775!).- Tenerife s.l., s.d., Broussonet (MA: 106776!).- Güímar. Bco. del Río, 24.VIII. 1946, L. Ceballos & Ortuño (Ma: 106775!, Herb. Esc. Técn. Sup. de Ing. de Montes de Madrid s.n.!).- Sobre Aguamansa hacia Pedro Gil 2100 m, 10.X.1960, E. Sventenius (ORT: 8868, 13091, 13095).- Arafo 1800 m, 19.VI.1962, ejusd. (ORT: 13092).- Ibid. 1600 m, ejusd. (ORT: 13093).- Güímar: Bco de Badajoz, 25.III.1944, ejusd. (ORT: 13094).- Cumbre sobre Aguamansa, 29.IX.1943, ejusd. (ORT: 13064).- Arafo "Hacia el volcán" 900 m, 22.VIII.1944, ejusd. (ORT: 1687, 13060).- Aguamansa, barranco cerca del pueblo 1100 m 28° 21′ N-12° 48′ W, 30.IV.1977, C. Jarvís 487 (RNG!).- Ladera de Güímar, 2.V.1971, E. Barquín *ex* Herb. TFC s.n. (RNG!).- Ibid., V. 1976, Beltrán, E., (TFC: 6855, Duplic. K).- Ibid., 9.V.1971, Pérez de Paz, P.L., (TFC: 6856).- Ibid (canal) 475 m, I.XI.1975, W. Wildpret et al. (TFC: 6858, Duplic O, 6859, 6860, 6861, 6862).- Ibid, (canal del Río), 29.VII. 1976, Acebes, J.R. & P.L. Pérez de Paz (TFC: 6854, Duplic. BM).- Ibid (canal) 1000 m, VII.1977, Beltrán, E. et A. Bañares (TFC: 6850, Duplic. J.E., COI).- Ibid.: Cueva Negra, 5.V.1975, Mesa, R., (TFC: 6857).- Altos de Güímar (canal), 27.IV.1974, Acebes, J.R., (TFC: 2704).- Bco. de Badajoz (Gúímar) 6.VII.1975, Mesa, R., (TFC: 2742, 2743)-. Ibid., 6.V.1975, ejusd. (TFC: 2746).- Bco. al N del de Badajoz (Güímar), 2.XI.1974, Barquín, E., et al. (TFC: 2717).- Altos de Arafo (alrededores del canal), 27.IV.1974, Beltrán, E., (TFC: 2703).- Altos de Arafo, 29.VII. 1974, Acebes, J.R., (TFC: 2714).- Bco. de Los Eres (Arafo), VII.1974, Méndez, B., (TFC.2707).- Lomo de Mena, 8.III. 1975, La Serna, I. et al. (TFC: 6853).- Bco. de Añavingo (Arafo), 23.V. 1973, Wildpret, W. et al. (TFC: 6852).- Aguamansa, 22.V.1977, La Serna, I., (TFC: 6843, Duplic. S; 6844, 6845, 6846, 6847, Duplic. W; 6848, Duplic. FI).- Ibid., 31.IV.1977, ejusd. (TFC: 6849).- Ibid., 25.IV.1975, Afonso, R., (TFC: 2744, 2745).- Ibid., 29.V.1975, Delgado, A. et al. (TFC: 2747, 2748, 2749, 3399, 3400, 3401, 3402, 3403).- Bco. del Agua (Güímar) 600–650 m, 4.I.1978, La Serna, I., (TFC: 6842).- Ayosa (Las Cañadas), 5.VI.1977, Pérez de Paz, P.L., (TFC: 6852, Duplic. TFMC).- Ladera de Güímar, 31.X.1975, Mesa, R., cult. En Jard. Exp. del Dpto. de Bot. de la Univ. de La Laguna, coll.: 14.IV.1978 (TFC: 6863).- Ibid, 31.X. 1975, ejusd., Ibid., coll.: 10.V.1979 (TFC: 7091).- Aguamansa, 27.V.1976, La Serna, I., Ibid., coll.: 4. IV.1978

(TFC: 6864).- Altos de Arafo (barrancos húmedos), 13.III.1979, Barquín, E., (TFC: 6978).- Ctra. de la Cumbre a Arafo, 1350 m, 18.III.1979, Beltrán, E., (TFC: 6979, 6980, 6981).- El Ancón (Ladera de Güímar) 700 m, 20.IV.1979, Mesa, R., (TFC: 6985, 6986).- Bco. de Jafoña (Altos de Arafo) 925 m, 2.VI.1979, Beltrán, E., (TFC: 7099).- Ibid., 21.VI.1979, La Serna, I. et al. (TFC: 7101).- Barranco en los Altos de Arafo, 1400 m, 160° SE, 21.VI.1979, ejusd. (TFC: 7102, Duplic. C, B, MAF, ORT; 7103, Duplic. C, COI); Ldra. de Güímar, 20.III.1971, Barquín Díez, E., (TFC 21679); Bco. de Araca, 11.IV.1976, Metlesics, (TFC 6091); Ldra. de Güímar, 14.IV.81, del Arco Aguilar, M., W. Wildpret de la Torre, P.L. Pérez de Paz, (TFC 13707); Bco. de Tamay, Arafo, 9.V.81, Pérez de Paz, P.L., M. del Arco Aguilar, W. Wildpret de la Torre, (TFC 23720); Bco. de Tamay, Arafo, 1300 m, 9.V.81, Pérez de Paz, P.L., M. del Arco Aguilar, W. Wildpret de la Torre, (TFC Duplicc. 32162); Valle del Volcán de las Arenas, Cumbre de Arafo, 9.V.81, del Arco Aguilar, M., W. Wildpret de la Torre, P.L. Pérez de Paz, (TFC 23730, 23753 Duplicatas 5); Altos de Arafo, Volcán de las Arenas, 1500 m., 6.V.81, del Arco Aguilar, M., (TFC 24381 Duplicatas 3); Fasnia, 4.V.84, Charpin, A., O. Rodríguez, (TFC 20096); Icod, Los Marquez, 25.VIII.87, Ardévol González, F., (TFC 38310); Bco. de Araca, 550 m.s.n.m., 30.X.88, MMG., (TFC 29765); Bco del Agua, Güímar, 18.IV.93, Pérez de Paz, P.L., (TFC Duplicc. 35644); Bco. del Agua (Río) Güímar, 850 m.s.n.m., 1994, del Arco Aguilar, M., (TFC Duplicc. 36450); Las Cañadas, 6.X.94., del Arco Aguilar, M., J.R. Acebes, EMD., I. La Serna Ramos, (TFC 41208).

3.5.25 Bystropogon origanifolius *L'Hér. var.* origanifolius

Sertum Anglicum, 12 (1788)

Bystropogon origanifolius var. *odoratissimus* (C. Bolle) Ceb. & Ort. non *B. odoratissimus* C. Bolle, *in Estudio sobre la Vegetación y la Flora Forestal de las Canarias Occidentales*, ed. 1ª, 425 (1951); ejusd. *in* ibid. ed. 2ª, 381 (1976);
Bystropogon plumosus L'Hér. Pitard & Proust Les Iles Canaries. Flore de L'Archipel, 302–303 (1908) e.p.; Lindinger *in* Beiträge zur Kenntnis von Vegetation...
Bystropogon plumosus var. *origanifolius* Christ *Bot. Jahrb.*, 9: 136 (1888); Kuntze *in Revision Generum Plantarum Vascularium omnium*, 513 (1891)

Endemism, Certain native
Family: *Lamiaceae*
Common name: Poleo de monte, Poleo de cumbre o Poleo montuno (ES); trans: Wild or Summit Pennyroyal
Growth form: nanophanerophyte
Flowering: June–August
Fruiting: July–September
Pollination: insect pollination (entomogamy)
Dispersal: self-dispersal (atelochory)
Chromosome N°: 2n = 42

Description

B. origanifolius: Nanophanerophyte with finely pubescent young or annual shoot branches (short type indumentum), internodes 0.25–8.25 cm apart.

Leaves 0.7–4.75 × 0.3–2.5 cm, weakly serrate, more or less crenulate-serrate, serrate or entire, hairy, green on the upper side and white, cinereous or rarely green-cinerous on the underside, lanceolate, ovate, elliptic, ovate-lanceolate or lanceolate-elliptic, more or less rounded or cordate at the base, acute, subacute or obtuse, rounded or pointed at the apex, petiole with short (0.2–1.25 cm) indumentum, always shorter than the blade. Simple veins prominent on the underside. Indumentum always present on both leaf surfaces. Indumentary hairs of one type only, short uni- or bicellular, rarely tricellular, densely distributed on the underside and less dense on the upper side. Underside whitish due to the massive and compact presence of white hairs. Hyaline, golden or chestnut-coloured essential glands on both sides of the leaves, often less visible on the underside due to the density of the hairs. Menthol odour.

Inflorescences in panicle with dichotomous branching (2–8 successive divisions) from the peduncle (Fig. 3.78). Opposite axillary peduncles, pubescent in numbers of one or two, up to 0.2–4.4 cm long, on both primary and secondary branches of annual shoots or younger branches. Bracts of the dichotomous apexes more or less isomorphous, progressively decreasing in size, the first pait being larger (2.0–9.0 × 0.4–2.5 mm) stipitate or in the form of leaflets or reduced leaves, hairy (short indumentum) on margins and both sides, more on the outer than on the inner side, the rest lanceolate, lanceolate-linear, acute subulate or not, hairy on margins and outer side, glabrous or glabrous on the inner side; marginal and outer hairs long; florets (0.5) 0.75–2.0 × 0.2–0.5 mm; uninerved.

Fig. 3.78 Path to Pico Viejo, Aug. 2005. (Photo credit: Victoria Eugenia Martín Osorio)

Flowers in dichotomous cymes. Flower solitary at the apex of all dichotomies more or less pedicellate, pedicel of the first 0.5–3.0 (4.75) mm. Calyx tubular or tubular-campanulate (82.5) 3.0–4.0 (4.5) mm long, hairy externally, 10 nerves all conspicuous; teeth broadly triangular at base and long-finned, of equal or greater length than tube, occasionally somewhat smaller, hairy at the margins, long hairs giving feathery appearance. Corolla white, pinkish or white maculate, exserted about (4.25) 4.5–6.0 (6.25) mm long, tube (1.5) 2.0–4.0 mm; upper lid notched, 1.0–1.5 (2.0) mm; lower lip 3-lobed, central lobe more or less flabellate, scalloped or weakly emarginate, 1.5–2.0 (2.5) mm long, entire, scalloped. Andro-sterile individuals with reduced corolla varying in size from 3.0 to 3.5 (4.0) mm are common. Stamens with filaments, the largest (0.5) 0.75–1.0 mm long, anthers (0.3) (0.4)—0.5 mm long. Style 1.5–4.25 (5.0) mm and large ligules of the style 0.5–0.75 (1.0) mm for a total length of 2.0–4.75 (5.75) mm.

Nucules chestnut-brown, rounded at the apex, pericarp smooth to the naked eye and rough under scanning electron microscope; rougness quite different from that of *B. odoratissimus* C.Bolle and slightly different from that of *B. plumosus* (L.f) L'hér and *B. wildpretti* I. La Serna.

B. origanifolius* var. *origanifolius: **Leaves** lanceolate, ovate to elliptic; acute, to obtuse, rounded or pointed at the apex. **Calyces** (2.5) 3.0–4.0 (4.5) mm long. Androfertile **corollas** (4.25) 4.5–6.0 (6.25) mm long.

Distribution

The species is found on the islands of El Hierro, La Palma, La Gomera, Tenerife and Gran Canaria, but the var. *origanifolius* is only found on Tenerife and La Gomera.

In the Teide National Park it reaches up to 2400 m a.s.l. and its distribution area is mostly related with the Canary Pine forest fringe corresponding to the western flank of Las Cañadas (Figs. 3.79 and 3.80).

Habitat and Phytosociology

On Tenerife, *B. origanifolius* is found from 250 (NW) to 500 (S) m a.s.l. (domain of ***Aeonio-Euphorbion canariensis***), up to the summits (2400 m a.s.l.) where on occassion it forms ample, closed populations (Montaña de los Poleos, El Chinyero).

It is a characteristic species of mountain communities of the "pine tree—tree lucerne" class (***Chamaecytiso-Pinetea canariensis***) and it can be found as settler of badlands or accompanying the undersotry of pristine pine forest stands on lava flows, as well as on mixed pine forest stands. Locally very abundant in the climatic domains of pine forests of Guía de Isora and Chío where it occupies vast areas together with *Chamaecytisus proliferus* (L. fil.) Link, preferably installing itself on older lava flows with initial soil formation.

It is a bioindicative species of the ***Cisto symphytifolii-Pinion canariensis*** alliance, which groups closed climatophilous and edaphoxerophilous meso-macroforests of Canary Pine Tree, Meso-Supramediterranean dry-humid, rarely below the clouds of western canary distribution. It is also present in ***Adenocarpo***

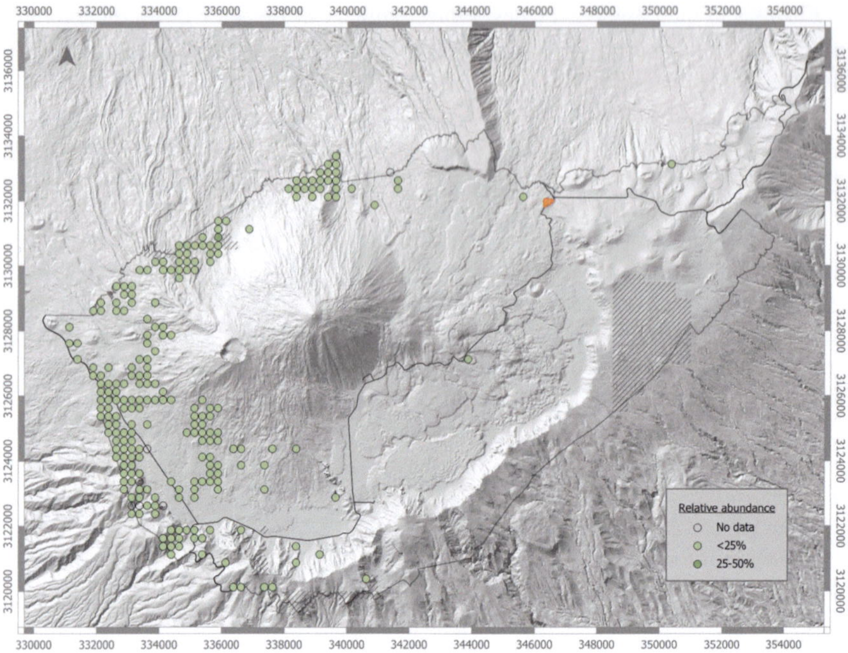

Fig. 3.79 *Bystropogon origanifolius* var. *origanifolius*, N° Subgrids (250 × 250 m): **233**/3740

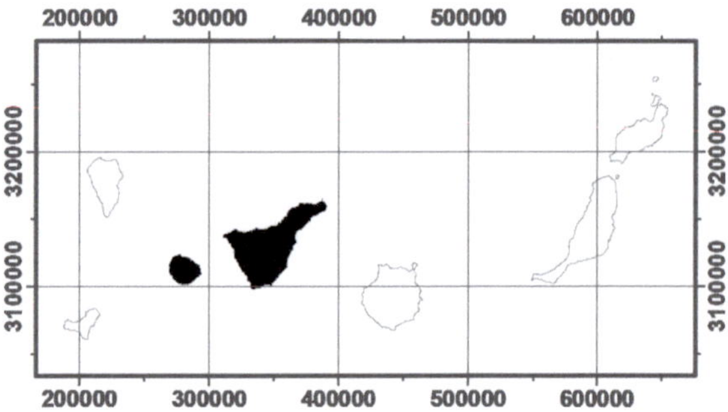

Fig. 3.80 Endemism of the island of Tenerife and La Gomera

viscosi-Chamaecytisetum angustifolii, flatpod-tree lucerne shrub, which constitutes the helophilous forest fringe of dry-humid Meso-Supramediterranean canary pine forests of Tenerife, in las Cañadas del Teide it is located like an arch nearly bordering the national park.

Conservation Status

On 94% (218) of subgrids where this taxon was found, no damages of any kind were detected. Only on 3% (8), damages were detected (herbivory, high mortality without regeneration, etc.), but never on more than 25% of individuals. Only in 1% (2) sub-grids damage was detected on more than 25% of individuals. Damages were observed. In the remaining, 2% (5) of sub-grids, assessment is missing.

Threat Status

Anexo III (Gobierno de Canarias 1991)
LR ca (Bañares et al. 1998)

Germplasm Banks

Banco de Germoplasma de E.T.S. Ingenieros Agrónomos de Madrid.
Jardín Botánico Canario Viera y Clavijo

Exsiccata (Herbarium Specimens)

Tenerife: "In vallibus et rupestribus Teneriffa legi propè Chasna", s.d., Webb 14 *in* Herb. Webbianum s.n. (FI!); "Plini Chasna region superior a la base des versans merid. des montagnes des Cañadas", s.d., ejusd. 14 *in* Herb. Webbianum s.n., e.p.: sched. 1, 2, 4 (FI!); "In siccis Teneriffa ad Vallem Acentexo", s.d., ejusd. s.n. *in* Herb. Webbianum s.n., e.p.: sched. 4 non 1, 2, 3, (FI!); Cult. *in* h. Milfordiani, s.d., *in* Herb. Webbianum s.n. (FI!); "In Teneriffa rupibus siccis legi in viâ inter Garachico et Montem Agua", s.d., Webb 14 *in* Herb. Webbianum s.n. (FI!); "Ad Portum Orotavensem in convallium puribus", 17.02.1845, Bourgeau 70 e.p.: sched. 2 non 1 (W: 106680!); "Traste de Dña. Beatriz, Chasna", July 1846, Bourgeau? s.n. *in* Herb. Webbianum s.n. (FI!); "In apricis pinetorum prope Chasna", 25.06.1855, H. de la Perraudière *in* Bourgeau: Pl. Canarienses 1471 (W: 2463, s.n., 74759!; MA: 162293!; FI *in* Herb. Webbianum s.n.!; K); Icod de los Vinos "in rupestribus", 200 m.s.n.m., 23.06.1901, J. Bornmüller 2773 (W: 3980!); "Prope Chasna", 1400 m.s.n.m., June 1904, O. Burchard 174 (B!); Ibid. s.alt., May 1906, ejusd. s.n. (ORT); Icod, July 1905, C. Sobrado (MA: 106781, 106783!); Barranco Guía, 9.06.1907, M. Mücke 2678 (B!); El Guincho-Garachico, 300 m.s.n.m., 25.05.1944, E. Sventenius (ORT: 1690); Valle Santiago hacia punta Lagarto, 1200 m.s.n.m., 18.04.1945, S. Sventenius (ORT: 13077, 13078, 13080); Cumbre du Portillo de la Villa, 16.11.1945, Bourgeau? 816 *in* Herb. Webbianum s.n. (FI!); Masca "el barranco", 1.04.1946, S. Sventenius (ORT: 13068); Masca, 19.07.1946, S. Sventenius (ORT: 13073); Cumbres de La Orotava "in rupestribus siccis", 16.08.1946, L. Ceballos et F. Ortuño (MA: 106780!); "En los claros del escobonal del monte Las Cumbres", 1750 m.s.n.m., 16.08.1946, ejusd. (Herb. Esc. Técn. Sup. de Ing. de Montes de Madrid s.n.!); Masca, 800 m.s.n.m., 14.04.1949, S. Sventenius (ORT: 13075); El Guincho-Garachico, 175 m.s.n.m., 15.06.1949, E. Sventenius (ORT: 9310, 13093); Ibid. s.alt., 29.07.1948, ejusd. (ORT: 13101); Icod de los Vinos, 23–29-31.07.1954, K. Lems 2080 (RNG!); Fuente del Cedro-Las Cañadas, 6.05.1956, E. Sventenius (ORT: 13091); Cordillera de Boca Tauce, 10.05.1956, ejusd. (ORT: 13098); Ibid., 25.06.1956, ejusd. (ORT: 13097); Masca, Sep.1965, D. Bramwell s.n. (BM); Tamaimo, 13.06.1968, S. Sventenius 30840 (ORT: 13063); Punta de Teno, 200 m.s.n.m., 20.06.1969, D. Bramwell 1736 (RNG!); Ibid., 150 m.s.n.m., 20.06.1969, ejusd. (TFC: 63!); El Retamar 2500 ft., 14.05.1970, V.F. Mulford 132 (RNG!); Vilaflor, 1300 m.s.n.m., 12.06.1971, E. Barquín *ex* Herb. TFC s.n. (RNG!); Vilaflor, 13.06.1971, P.L. Pérez de Paz (TFC: 6989); Bco. de la

Orchilla, 22.04.1973, E. Beltrán (TFC: 6990, 6991); Mña. de La Fortaleza, 19.05.1973, W. Wildpret, AA, E. Beltrán Tejera, P.L. Pérez de Paz, E. Barquín Díez, LG, CGR & H. Metlesics (TFC: 6992); Bco. del Cuervo-Trevejo, 25.07.1973, E. Beltrán (TFC: 6993); Mña. de Los Poleos, 1482 m.s.n.m., 13.04.1975, I. La Serna, M. del Arco Aguilar & J. Ariz. (TFC: 2724, 2725); Bco. La Vera-Vilaflor, 3.05.1975, I. La Serna & C.M. Álvarez (TFC: 2721, 2722, 2723); Mña. de Los Poleos, 24.05.1975, ejusd. (TFC: 2726, 2727, 2728, 2729, 2730, 2731, 2732, Duplic. ORT, JE, FI; 2733, 2734, 2735, 2736, 2737, 2738, Duplic. TFMC; 2739, Duplic. P, BM, MAF); Ibid., 8.06.1975, S. Alvarez (TFC: 2740, 2741); Bco. de Masca-cauce, 21.07.1975, I. La Serna, E. Beltrán Tejera, M. del Arco Aguilar, C. León Arencibia & JBD (TFC: 4161); Mte. del Agua-Los Silos, 690 m.s.n.m., pista forestal abandonada, 25.09.1975, M.C. León Arencibia, E. Beltrán Tejera & M. del Arco Aguilar (TFC: 4548, 4560); Teno, 25.11.1975, W. Wildpret, E. Beltrán Tejera & E. Barquín Díez (TFC: 6994); Bco. de Masca, 16.02.1976, E. Beltrán et A. Bañares (TFC: 7003); Ibid., 900 m.s.n.m., 1.05.1976, J.R. Acebes, ERH & M. del Arco Aguilar (TFC: 7002); Ibid., 850 m.s.n.m., ladera derecha, 1.05.1976, ejusd. (TFC: 7004, Duplic. O, JE, ORT); Ibid., 15.05.1976, I. La Serna, E. Beltrán Tejera, M. del Arco Aguilar & H. Metlesics (TFC: 7005, Duplic. BM; 7006; 7007); Carretera de Boca Tauce a Chío, 1280 m.s.n.m., 15.05.1976, I. La Serna, E. Beltrán Tejera & M. del Arco Aguilar (TFC: 6995); Mña. de Los Poleos, 15.05.1976, I. La Serna, M. del Arco Aguilar & E. Beltrán Tejera, cult. en el Jard. Exp. del Dpto. de Bot. de la Univ. de La Laguna, coll.: 12.05.1978 (TFC: 7032); Ibid., 15.05.1976, ejusd., ibid., coll.: 11.07.1978 (TFC: 7031); Altos de Icod, 15.03.1977, I. La Serna, ibid., coll.: 11.07.1978 (TFC: 7033); Masca, 800 m.s.n.m., 15.05.1976, I. La Serna, E. Beltrán Tejera & M. del Arco Aguilar, ibid., coll.: 11.07.1978 (TFC: 7034); Mña. de Los Poleos, 18.07.1976, S. Alvarez (TFC: 6996, Duplic. RNG, SEV, MAF, TFMC, COI; 6997; 6998; 6999; 7000; 7001, Duplic. MADM); Alrededores de Chío—Carretera Santiago del Teide—Chío, 700 m.s.n.m., 28° 14′ N to 13° 7′ W, 12.05.1977, C.E. Jarvis 664 (RNG!); Las Cañadas, 2.07.1977, I. La Serna, W. Wildpret de la Torre, C. León Arencibia, M. del Arco Aguilar & H. Metlesics (TFC: 7027, Duplic. BM, B); Caldera de Las Cañadas-Carretera de Boca Tauce a Chío, 2100 m.s.n.m., E-SE, 2.07.1977, ejusd. (TFC: 7024, Duplic. B, K; 7016, Duplic. S, W, MA; 7017; 7018; 7019, Duplic. SEV; 7020, Duplic. RNG, MAF, MADM; 7021, Duplic. COI; 7022, Duplic. G; 7023, Duplic. P, S); Ibid., 2050 m.s.n.m., 2.07.1977, ejusd. (TFC: 7008; 7009, Duplic. G; 7010); Carretera de Boca Tauce a Chío, 1200 m.s.n.m., zona de introgresión en el pinar de especies del K-E alto, 2.07.1977, ejusd. (TFC: 7013, 7014); Pinar de Chío, 1250 m.s.n.m., 2.07.1977, ejusd. (TFC: 7015, Duplic. P); Ibid., 1190 m.s.n.m., 2.07.1977, ejusd. (TFC: 7011, 7012); Ibid., 890 m.s.n.m., 2.07.1977. ejusd. (TFC: 7025, Duplic. FI); Tamaimo, 650 m.s.n.m., 2.07.1977, ejusd. (TFC: 7026); Bco. Mancha de Los Díaz, 3.07.1977, I. La Serna et al. (TFC: 7028, Duplic. K, W, MA; 7029); Bco. del Natero, 3.07.1977, ejusd. (TFC: 7030, Duplic. FI); Bco. del Rey-Arona, 900 m.s.n.m., 17.03.1979, M.C. León, E. Beltrán Tejera, M. del Arco Aguilar & J.R. Acebes (TFC: 7036); Ifonche-Arona-Altos del Bco. del Infierno, 24.03.1979, W. Wildpret, E. Beltrán Tejera, M. del Arco Aguilar, C. León Arencibia & J.R. Acebes (TFC:

7035); La Cuesta de la Villa, 360 m.s.n.m., 21.08.1979, I. La Serna, W. Wildpret, L. Negrín, A. Losada Lima & A. García Gallo (TFC: 7319); El Tanque, 400 m.s.n.m., 340° NW-N, 29.08.1979. ejusd. (TFC: 7337, 7338); Alrededores de San José de Los Llanos, 1000 m.s.n.m., 29.08.1979, ejusd. (TFC: 7339, 7340); Las Arenas, 1150 m.s.n.m., 22° N-NE, 29.08.1979, ejusd. (TFC: 7341, 7342, 7343, 7344); Sur del Cráter del Chinyero, 1430 m.s.n.m., 320° NW, 4.09.1979, ejusd. (TFC: 7345, 7346); Bco. de Pina-Los Silos, 28.08.1996, V. Lucía Sauquillo & C. Acosta (TFC 39294 + Dupl.); Bco. Los Carrizales-Teno, abril 2001, J. Ramos & I. La Serna, (TFC 43372). Inventario de la Flora y Vegetación del Parque Nacional del Teide (2001–2003), V.E. Martín Osorio, W. Wildpret & B. Hernández Bolaños: Pinar cara Norte del Teide, por el Lomo Mataperros (29B 8), 19/06/2003, Y. de Rioja Santos (TFC 44254).

Uses and/or Medicinal Properties

> "*Stomachic, expectorant, anticatarrhal, nasal decongestant; hepatotoxic in infants. Useful part: plant, leaves. User manual: infusion, salad.*" (Translated from: Pérez de Paz and Hernández Padrón 1999)

3.5.26 Carex demissa Hornem. subsp. demissa

First published in Fors. Oecon. Plantel., ed. 2: 826 (1806)

Carex tumidicarpa Andersson ssp. *cedercreutzii* Fagerstr. Act. Soc. Fauna et Flora Fenn., 79: 3 (1967)

Carex demissa subsp. cedercreutzii (Fagerstr.) Jac. Koopman, First published *in Carex Eur.* 1: 88 (2011) basionimo de *Carex tumidicarpa* subsp. *cedercreutzii* Fagerstr. (taxon endémico de Azores)

Not endemic, Probable native
Family: *Cyperaceae*
Common name: Cuchillera enana (ES); Commom Yellow-sedge
Biotype: hemicryptophyte
Flowering: April–June
Fruiting: April–June
Pollination: –
Dispersal: –
Chromosome N°: –

Description

Plant of variable height, 5–35 cm, green to yellowish green, forming small dense clumps.

Stems in great number, more or less curved, striated-furrowed, glabrous, with a subtriangular to rounded section, most of them leafy at the base, which can reach up to 1/3 of it.

Leaves equal to or exceeding the stems, sometimes smaller and even reduced to the sheath; leaf sheath yellowish-brown (5 cm), closed; ligule scarious-membranaceous-short and arched upwards; weakly patent, green, narrow (2.5 mm) blade, (Fig. 3.81) attenuating towards the apex, showing a keel and ending in a short mucron; edges, especially those at the apex, sometimes spiny, the same as the keel; mucron of triangular section.

Unisexual, monoecious **flowers**, grouped in spikes, 2–3(5), multiflorous female spikes, cylindrical-ovoid, more or less close to each other, pedunculated and arranged in the axil of a patent leaf with a closed sheath (5 mm); 2 sterile bracts at the base of the inflorescence with the same morphology as the fertile ones, in whose axils the utricle is found. Utricles yellowish-green, narrow (3–4 mm), flat-concave, long apiculate, ending in a weakly bifid and serrated face, veined, with 2 very pronounced lateral nerves like a keel, the internal ones (several) faint. The floral bracts are lanceolate, long attenuated towards the apex, green to halo-green at the margins, three-nerved, with very close nerves smaller than the utricles, reaching only the base of the face. Above the female inflorescence and finishing the stem, there is the male inflorescence, solitary, elongated and with little or no pedunculation, all the bracts being fertile and bearing in their axil three stamens with long anthers.

Distribution

It is widely distributed throughout the European continent, Russia and Western Asia. In the Canary Islands it only appears in Tenerife (Fig. 3.82).

In the Teide National Park (Fig. 3.83) it has been found occasionally on the western escarpment of the head of the Barranco del Río (Colorado de

Fig. 3.81 Artificial fountain in the Wolfredo Wildpret Botanical Garden of the Teide National Park, June 2004. (Photo credit: Beatriz Hernández)

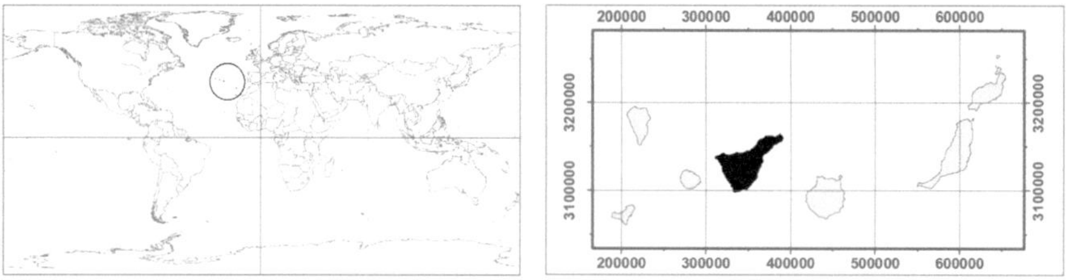

Fig. 3.82 Present on Tenerife

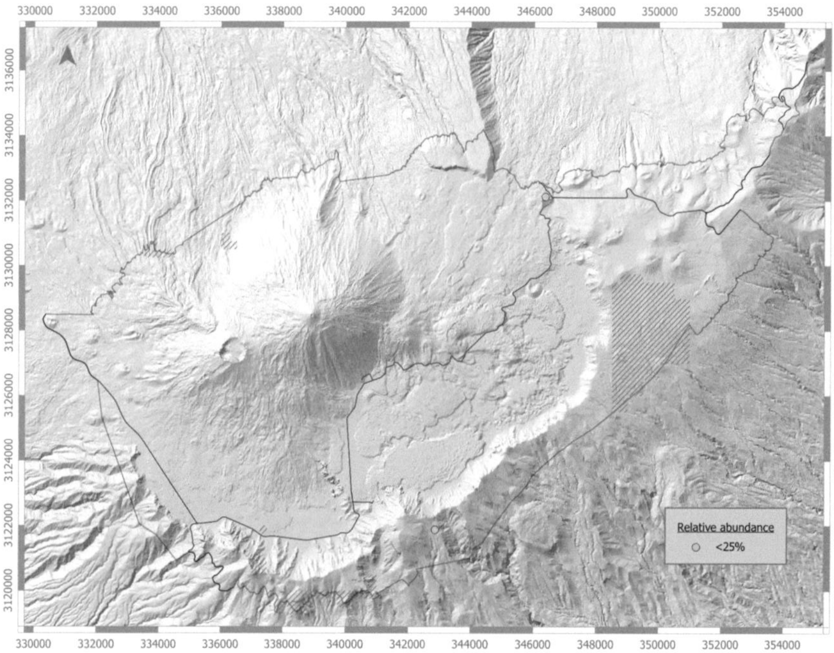

Fig. 3.83 *Carex demissa susp. demissa* distribution, N° Subgrids (250 × 250 m): **2**/3740

Guajara—2400 m a.s.l.) and in the artificial pond of the Teide, Wolfredo Wildpret Botanical Garden.

This taxon was cited as ***Carex tumidicarpa subsp. cedercreutzii*** Fagerstr. It is currently considered an endemic taxon of the Azores as ***Carex demissa subsp. cedercreutzii*** (Fagerstr.) Jac. Koopman, First published *in* Carex Eur. 1: 88 (2011). In the Canary Islands it needs taxonomic revision.

Habitat and Phytosociology

Within the Park, it is usually found in hygro-hydrophilous environments.

Present in the communities of ***Stellarietea mediae***, a type of anthropogenic annual nitrophilous vegetation, consisting of therophytes or small geophytes of a weedy, ruderal, or scionitrophilous character, with a Holarctic distribution and cosmopolitan irradiation.

Conservation Status

In the two (2) subgrids in which this species was found, no damage of any kind was observed.

Threat Category

DD (Bañares et al. 1998)

Exsiccata (Herbarium Specimens)

Tenerife: Las Cañadas del Teide, fuente de Guajara, 2400 m.s.n.m., 28.5.1983, J.S. Socorro & W. Wildpret (TFC 18433); Ibid, 13.7.1983, J.S. Socorro (TFC 18435); Bco. del Río, 13.7.1983, J.S. Socorro, W. Wildpret & M. León Arencibia (TFC 18434 + duplic.); Inventario de la Flora y Vegetación del Parque Nacional del Teide (2001–2003), V.E. Martín Osorio, W. Wildpret, B. Hernández Bolaños: Guajara (42C 13b), 19.5.2003, Y. de Rioja Santos (TFC 44233).

3.5.27 **Carex paniculata** *L. ssp.* **calderae** *(A. Hansen) Lewej. & Lobin*

Senckenbergiana biol., 67 (4/6), 437–447 (1986)

Carex calderae A. Hans. *Cuad. Bot. Can.*, 14/15: 699–702 (1972)
calderae: referring to the caldera or crater of a volcano.

Endemism, Certain native
Family: *Cyperaceae*
Common name: Cuchillera de Las Cañadas/Cerrillo de agua (ES); Caldera greater tussock-sedge (EN)
Growth form: hemicryptophyte
Flowering: April–June
Fruiting: June–Septembre
Pollination: wind-pollination (anemogamy)
Dispersal: water dispersal (hydrochory)
Chromosome N°: 2n = c.68

Description

Hemicryptophyte of a more or less intense green colour (Fig. 3.84). Stems of (0.4)–1–(1.5) m, erect or pendulous, triangular, the angles towards the inflorescence being acute and scabrous, sides striated

Fig. 3.84 El Riachuelo, June 2003. (Photo credit: Juvenal Soto Martín)

Leaves with very long blade, a little shorter than the stems, flat and folded at the base, long attenuate and pointed, striate, scabrous at margins and base; ligule slightly ovate to rounded, sometimes emarginate; lower sheaths sharp, blackish-brown, shiny, not tearing into fibres, upper sheaths green, striate, smotth and truncate at neck.

Panicle erect, with numerous spikelets, dense, oblong or narrowly lanceolate, (5)–12–(17) cm long, without foliaceous bracts. Spikelets 3–15 mm ovate, bracts at base elongate, strongly setose and toothed at margins. Glumes (3–4.7 mm) ovate to ovate-lanceolate, hyaline-green, acute, broadly hyaline-scarious at margins, uninerved, with slightly brownish base, smooth and glabrous, the male ones sharply lanceolate. Three stamens with linear, yellow-clear anthers (2.4–3 mm) with a hispid apiculum. Utricles concave-flat, concave or slightly convex (2.6–3 to 8 mm), generally shorter than the glumes, coriaceous, rounded to truncate at the base, somewhat stipitate with finely winged margins, scabrous on the anterior 2/3, ending in a long, clefted and bidentate beak, distinctly veined on the dorsum and not on the ventral side. Style narrowly thickened at the base with two long, exserted brown stigmas.

Nucules light-brown, ovoid-biconvex (1–1.75 mm), sections trigonous-rounded, slightly stipitate, attenuate at the apex and finely reticulate.

Distribution

Endemism of the island of Tenerife, present between 2000 and 2425 m a.s.l. In the Teide National Park it was found naturally in the Cañada de La Grieta, head of the Barranco del Río and western sector of Montaña Guajara; also in the Botanical Garden (Figs. 3.85 and 3.86).

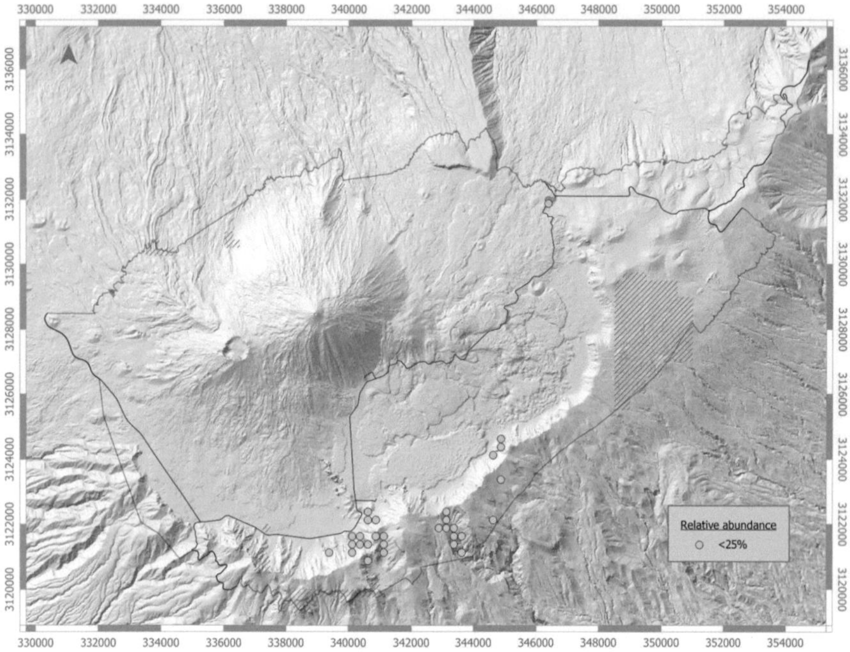

Fig. 3.85 *Carex paniculata* subsp. *calderae*, N° Subgrids (250 × 250 m): **30**/3740

Fig. 3.86 Endemism of the island of Tenerife

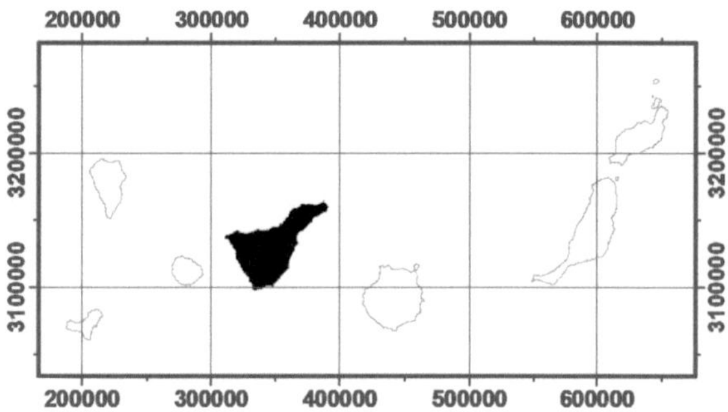

Habitat and Phytosociology

It forms dense populations on wet water spring fields and along the small watercourses of Las Cañadas.

Species characteristic of the ***Mentho-Caricetum calderae*** community installed on hyfro-hydrophilous Meso-Supramediterranean environments.

Conservation Status

On 97% (29) of subgrids where this taxon was found, no damages of any kind were detected. Only on 3% (1), damages were detected (herbivory, high mortality without regeneration, etc.), but never on more than 25% of individuals.

Threat Status

V (Barreno et al. 1984)
Anexo II: Protegidas (Gobierno de Canarias 1991).
VU D2 (Bañares et al. 1998).
VU D2 (VVAA 2000).
VU (Bañares et al. 2004).
VU D2 (Moreno 2008).

Germplasm Banks

Banco de Germoplasma de E.T.S. Ingenieros Agrónomos de Madrid.
Real Jardín Botánico de Madrid.
Jardín Botánico Canario Viera y Clavijo.

Exsiccata (Herbarium Specimens)

Tenerife: Las Cañadas, abril 1981, B. Cabezudo (TFC 20945); Cañada del Riachuelo, 13.06.1981, CRP (TFC 24495); Fte. de la C. de la Grieta, 2175 m.s.n.m., 2.10.1982, S. Socorro (TFC 18450); Bco. del Riachuelo-Las Cañadas del Teide, 2150 m.s.n.m., 12.10.1982, S. Socorro, W. Wildpret & E. Beltrán (TFC 18447); Cauce de Guajara-Las Cañadas del Teide, 2300 m.s.n.m., 6.03.1983, S. Socorro (TFC 18448 + Dupl.); Valle de Ucanca, 2300 m.s.n.m., 24.04.1983, S. Socorro & W. Wildpret (TFC 18425 + Dupl., 18426, 18427); Ibid., 24.04.1983, S. Socorro (TFC 18446 + Dupl.); Ibid., 7.05.1983, 2300 m.s.n.m., S. Socorro & W. Wildpret (TFC 13174—Palinoteca); Bco. del Río, 2100 m.s.n.m., 15.05.1983, S. Socorro (TFC 18445 + Dupl.); Ibid.- Afluente, 2200 m.s.n.m., 28.05.1983, S. Socorro & W. Wildpret (TFC 18443); Fte. de Veno, 2325 m.s.n.m., 31.05.1983, S. Socorro (TFC 18442); Bco. de los Riachuelos, 2200 m.s.n.m., 1.06.1983, S. Socorro (TFC 18449); Valle de Ucanca, 2300 m.s.n.m., 12.06.1983, S. Socorro (TFC 18428, 18429, 18430, 18441); Bco. del Riachuelo, 12.05.1984, A. Charpin & W. Wildpret (TFC 20118); Ctra. de Guajara-Las Cañadas, 25.04.1987, W. Wildpret, P.L. Pérez de Paz & Biondi (TFC 24122 + Dupl.) 24.04.1983, S. Socorro & W. Wildpret (TFC 18425 + Dupl., 18426, 18427); Ibid., 24.04.1983, S. Socorro (TFC 18446 + Dupl.); Ibid., 7.05.1983, 2300 m.s.n.m., S. Socorro & W. Wildpret (TFC 13174—Palinoteca); Bco. del Río, 2100 m.s.n.m., 15.05.1983, S. Socorro (TFC 18445 + Dupl.); Ibid.- Afluente, 2200 m.s.n.m., 28.05.1983, S. Socorro & W. Wildpret (TFC 18443); Fte. de Veno, 2325 m.s.n.m., 31.05.1983, S. Socorro (TFC 18442); Bco. de los Riachuelos, 2200 m.s.n.m., 1.06.1983, S. Socorro (TFC 18449); Valle de Ucanca, 2300 m.s.n.m., 12.06.1983, S. Socorro (TFC 18428, 18429, 18430, 18441); Bco. del Riachuelo, 12.05.1984, A. Charpin & W. Wildpret (TFC 20118); Ctra. de Guajara-Las Cañadas, 25.04.1987, W. Wildpret, P.L. Pérez de Paz & Biondi (TFC 24122 + Dupl.).

3.5.28 Carlina xeranthemoides *L.f.*

L.f., *Suppl.*, 349. 1781 [1782]

Carlina xeranthemoides Masson. Linn. fil., *Suppl.*, pág. 349; Lam., Enc., vol. 1, pág. 625; Willd., *Sp. pl.*, vol. 3, pág. 1697; Buch, *Canar.*, pág. 147 et 181; DC., *Prodr.*, vol. 6, pág. 548.
Carduus fruticosus Poir. *in* J.B.A.M.de Lamarck, Encycl., *Suppl.* 2: 199 (1811)

Endemism, Certain native
Family: *Asteraceae*
Common name: Malpica de cumbre/Alazor/Cardo de Cristo/Cardo salvaje (ES); Summit carline thistle (EN)
Growth form: Nanophanerophyte
Flowering: July–September
Fruiting: August–October
Pollination: insect pollination (entomogamy)
Dispersal: wind dispersal (anemochory)
Chromosome N°: 2n = 20

Description

Suffruticose shrub that can reach more than 50 cm in height, highly branched at the base, silver-tomentose, dense **leaves**, with simple corymb at the terminal part. Stems 25–45 cm, base 3–4 mm in diameter, rounded, lignetum hard, whitish, subflavescent, pith cylindrical; young plants with little bark, glabrous, softly yellow-greenish, crass, somewhat bruniaceous and rimose, with bark covered with a white tomentum, of weakly interwoven, dense and rough curly hairs that detaches easily.

Leaves cauline, lanceolate, very numerous, dense, prickly, very often with thorn-like, cottony-looking ends, distant from each other by almost 4 cm, 2.5 cm long, arranged in a spiral line.

Flowers arranged in multifloral, always erect, shiny flower heads (Fig. 3.87). Involucre hemispherical, 8–10 mm high, tomentose, hard, multiseriate, imbricate, with some foliar base, with leathery stems ending in a spine. Involucre gradually towards the spiny margin, sometimes exceeding the scales; ends of scales narrow, linear, apex spiny; pedicels of upper leaves similar, narrow and short; those below with broad base, lanceolate-linear, spiny; those below, oblong, apex with narrow appendage, margin lacerate-dentate, expanded, ending in recurved spines, oblong-linear, glabrous, tomentose above and with narrow linear, glabrous apex; margin ciliate-dentate; those within narrow-linear, rigid, subscarious, glabrous, 12 mm long, 1 mm wide, radial, white below, sub-integral, reflexed above, radial, sulphurous coloured, margin irregular, somewhat lacerate-dentate, apex ending in a spine. Receptable alveolate; margin of alveoli formed by 66 mm long paleae, provided with 3–5 filiform laciniae in the centre, unequal, pinnatipartite. Petals of the alveoli, white on the lower part, on the upper part subequal flavescent pappus, distinctly

Fig. 3.87 (**a**, **b**) Wolfredo Wildpret Botanical Garden, Teide National Park, Aug. 2008 (**c**) Siete Cañadas—Base of Guajara, Apr. 2009. (Photos credit: Beatriz Hernández)

divided, spiky setae. Flowers all tubular, hermaphrodite; receptacle paleae subequal, glabro-flavescent, 5-toothed, obtuse oblong-lanceolate teeth. Anthers with cylindrical tubes exceeding 2 mm, with long, linear horny appendage, flavescent terminally, caudate base plumose, prolonged on both sides; filament glabrous; style even in anthers, contiguous, ovate-oblong, papillose.

Distribution

It is found primarily in the central part of the island of Tenerife, with important populations in the summits of Arica and Vilaflor. In the National Park it reaches heights of up to 2600 m a.s.l. and its distribution is mostly related to the areas closest to the Wall of Las Cañadas, being quite rare in more northwestern areas and on the Teide-Pico Viejo edifice (Figs. 3.88 and 3.89).

Remarks:

As historical reference, we refer to Webb and Berthelot (1836–1850): "*this plant purely canarian, was found on rocky areas of Tenerife, above the spring Traste de Doña Beatriz, passed the village of Chasna, localised on the same spot in previous times by Masson, by using herbarium specimens of the Banksian Herbarium and revied in Buch; among rocks in the great circus (Wall of Las Cañadas) which formerly surrounded the crater of the island.*"

Currently, as with many other species, this plant has developed spectacularly in Las Cañadas and has even expanded downwards beyond the limits of this region.

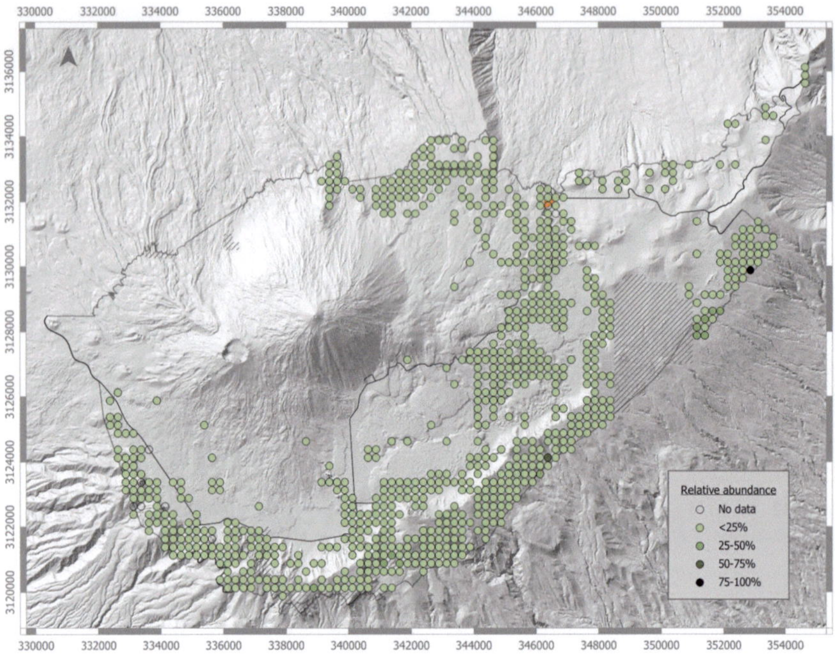

Fig. 3.88 *Carlina xeranthemoides* distribution, N° Subgrids (250 × 250 m): **851**/3740

Fig. 3.89 Endemism of the island of Tenerife

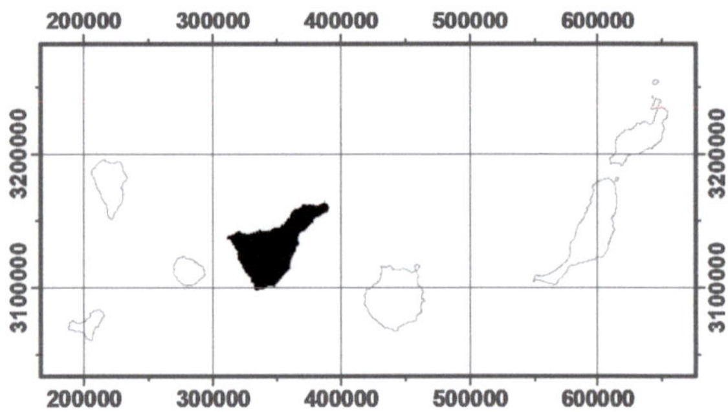

Habitat and Phytosociology

Although it reaches its optimum within the domains of the broom-flatpod summit shrubland, it is commonly found in ecotones of the pine forest by taking part in its understory. It develops well on pumitic, compacted soils, debris areas and cracks of basaltic rocks.

It intervened in the ***Chamaecytiso-Pinetea canariensis*** communities, vegetation class which groups coniferous forests and ***Chamaecytiso-Pinetalia canariensis***, the primecolonising and serial shrublands and dwarf shrublands of Tenerife and La Palma, as well as the shrublands of the volcanic cone of Teide: ***Spartocytisetalia supranubii***.

It also is characteristic of following vegetation units:

Cheirolopho teydis-Carlinetum xeranthemoides
Tolpido webbii-Micromerietum lachnophyllae

Conservation Status

On 88% (753) of subgrids where this taxon was found, no damages of any kind were detected. Only on 7% (57), damages were detected (herbivory, high mortality without regeneration, etc.), but never on more than 25% of individuals. Only in 4% (36) sub-grids damage was detected on more than 25% of individuals. Damages were observed. In the remaining (5) sub-grids, assessment is missing.

"This species is currently well represented in the central summit of the island of Tenerife, especially on its southern slope, where it forms vast stands. Nevertheless, all mid-century naturalists make reference to its extreme rarity, caused possible by past overgrazing. Currently this plant is highly valued by herbivores – rabbits and moufflon – especially by the latter one. Concretely, it appears as the second most consumed species by this ungulate, introduced to the island in 1971 and which represents a risk for many endemisms of the area." (Translated from: Gómez Campo et al. 1996)

Threat Status

R (Barreno et al. 1984)
Anexo II: Protegidas (Gobierno de Canarias 1991)
LR pm (Bañares et al. 1998)

Germplasm Banks

E.T.S. de Ingenieros Agrónomos de Madrid (*cantidad de muestra insuficiente*).
Jardín Botánico Canario Viera y Clavijo.

Exsiccata (Herbarium Specimens)

Tenerife: 18.06.1969, Bramwell 1793 (TFC 79); 1.08.1969, Bramwell 2131 (TFC 78); 28.07.1971, F. Estévez & A. Acuña (TFC 15995 + Dupl.); Cumbres de La Guancha, 17.09.1972, E. Beltrán et al. (TFC 33546); Mña. Guajara-Las Cañadas, 15.11.1972, J.R. Acebes & P.L. Pérez de Paz (TFC 8456, 23336); Aguamansa, 9.08.1977, H. Metlesics (TFC 3858); Cañadas, 20.08.1977, H. Metlesics (TFC 4435); Pinar de Chío-Entre Mña. Los Poleos y Boca Cangrejo, 4.09.1979, W. Wildpret, I. La Serna, L. Negrín, A. Losada Lima & A. García Gallo (TFC 24519 + Dupl.); Valle del Volcán de Las Arenas-Cumbres de Arafo, 9.05.1981, W. Wildpret, M. del Arco Aguilar & P.L. Pérez de Paz (TFC 23747 + Dupl.); Ibid., 1700 m.s.n.m., 1.09.1981, M. del Arco Aguilar & P.L. Pérez de Paz (TFC 13680 + Dupl.); Altos de Arafo, 13.09.1982, J.R. Acebes (TFC 16245 + Dupl.); Caldera de Pedro Gil-Altos de Arafo, 13.09.1982, J.R. Acebes & M. del Arco Aguilar (TFC 25452); Las Troviscas-Icod, 1600 m.s.n.m., 15.07.1986, F. Árdevol

González (TFC 19307); Altos de la Mesa-Araya-Candelaria, UTM 3613141, año 1993, W. Wildpret, MMG, O. Rodríguez Delgado & ECA (TFC 36130); Entre Escalona y Vilaflor, 1220 m.s.n.m., 28.08.1997, A. Santos (TFC 41564). Inventario de la Flora y Vegetación del Parque Nacional del Teide (2001–2003), V.E. Martín Osorio, W. Wildpret & B. Hernández Bolaños: Guajara, por Montaña Colorada (42C 16 a), 25/06/2003, A. Montañés Rodríguez (TFC 44283).

Uses and/or Medicinal Properties

> "*Antispasmodic; doffer plant. Useful part: tops, whole plant (fodder). User manual: infusion, branching.*" (Translated from: Pérez de Paz and Hernández Padrón 1999)

3.5.29 Cedrus atlantica *(Endl.) Manetti* ex *Carrière*

Traité Gén. Conif., 285 (1855)

Abies atlantica (Endl.) Lindl. & Gordon *in* J. Hort. Soc. London 5: 214 (1850)
Cedrus libani subsp. *atlantica* (Endl.) Batt. & Trab. *in* Fl. Algérie Tunisie: 397 (1905)

Not endemic, species introduced into gardens and new record in the Teide National Park
Family: *Pinaceae*
Common name: Cedro del Atlas, Cedro atlántico o Cedro plateado (ES); Atlas Cedar (EN)
Biotype: phanerophyte
Flowering: October–November
Fruiting: Two years after flowering
Pollination: anemogamy
Dispersal: anemo and zoochory

Description

Evergreen tree, with a thick trunk and irregularly whorled branching, up to 50 m tall, with an erect stem and non-curved branch tips (Fig. 3.90).

Leaves 1–3 cm, acicular, rigid, arranged mainly in fascicles at the apex of the brachyblasts.

Male cones solitary in the centre of the leaf fascicles. Strobili 5–8 cm, doliform, large erect, maturing bi- or triennially, with deciduous scales when mature, the tectrices tiny and included, the seminiferous subtriangular, with winged seeds.

Fig. 3.90 Gardens of the Bar of El Portillo Bajo, Aug. 2008. (Photo credit: Beatriz Hernández)

Distribution

Originally from the mountains of Algeria and Morocco, it was introduced to the Canary Islands as an ornamental garden plant. Apart from its presence on the island of Tenerife, we have no evidence that it has also been introduced into garden areas on other islands.

In the Teide National Park it is found in the gardens of El Portillo Bajo and Alto, although outside of these areas, there are still individual specimens in the interior of Las Cañadas that have not been eliminated: near Montaña Mostaza and Minas de San José (2000–2275 m a.s.l.) (Figs. 3.91 and 3.92).

Habitat and Phytosociology

This species usually colonizes the mountain slopes of Algeria and Morocco (Atlas), forming forests between 1200 and 2800 m above sea level, in places where the level of precipitation is between 500 and 1700 mm/year. Although it also forms large pure forests, it can be accompanied by firs, junipers, holm oaks, maples, cork oaks, oaks and gall oaks. Outside these areas, its presence is mainly due to cultivation in garden areas. According to Barbero and Quezel (1975), it is a characteristic species

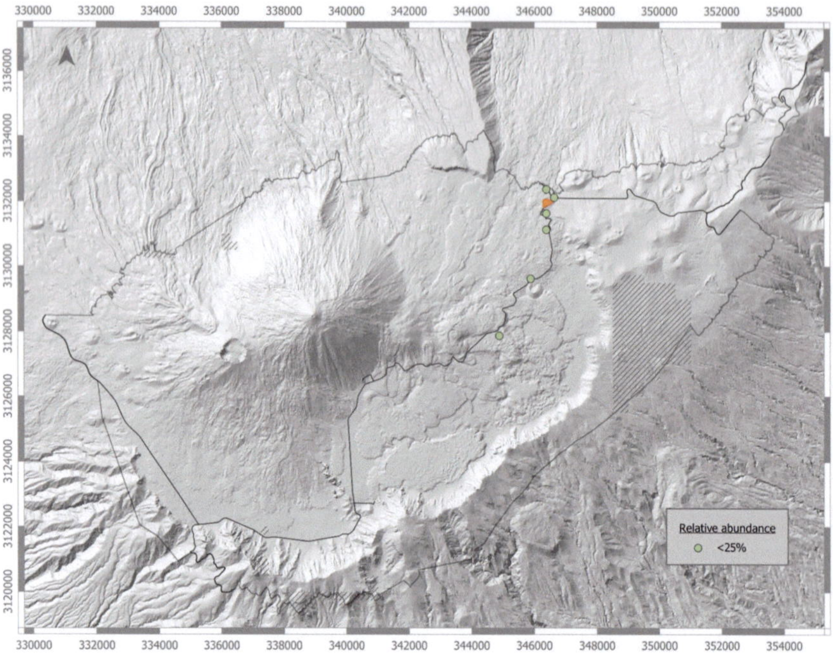

Fig. 3.91 *Cedrus atlantica* distribution, N° Subgrids (250 × 250 m): **6**/3740

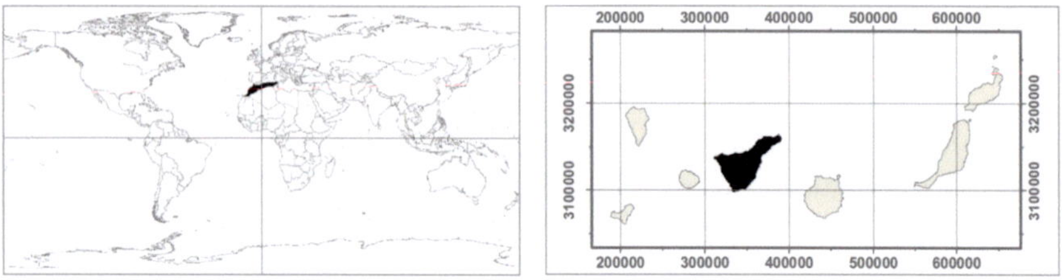

Fig. 3.92 Present on Tenerife Island

of ***Querco-Cedretalia atlanticae***, a type of vegetation that colonizes calcareous and metamorphic substrates at high altitudes in North Africa.

Conservation Status

In all the subgrids in which this species was found (6), no damage of any kind was observed.

Germplasm Banks

E.T.S. Ingenieros Agrónomos de Madrid.
Real Jardín Botánico de Madrid.

3.5.30 Cerastium sventenii *Jalas*

Ann. Bot. Fenn., 3: 133 (1966).

Cerastium arvense Linn. Webb *in* Barker-Webb & Berthelot, *Phyt. Canar.*, 152 (1836–1850).

Endemism, Certain native
Family: *Caryophyllaceae*
Common name: Rilla (ES); Sventenium's mouse-ear (EN)
Growth form: hemicryptophyte/chamaephyte
Flowering: Mai–June
Fruiting: June–July
Pollination: insect pollination (entomogamy)
Dispersal: self dispersal (atelochory/barochory)

Description

Perennial plant with 50 cm tall stems, (Fig. 3.93) knobby base, densely covered in patent glanduliferous hairs.

Leaf lanceolate, acute, 27 mm long and 4 mm wide, 4.5–7 times as long as wide, stiff. Stiff hairs pubescent (or sparsely glabrous), prominent rounded lower midrib; sclerenchymatous cell on midrib abaxially located.

Inflorescence multifloral with pedicellate peduncle, fruiting, erect. Sepals 5–6.5 (4–7) mm long, obtuse, glandular-pilose, scarious-margined. Petals, according to Webb, twice as long as the sepals. Capsule (5) 5.5–7 mm long, straight, teeth sub-straight, margins not revolute. Outer and inner epidermis of mature crass-walled capsule with protected cells. **Seeds** chondriospermous 0.9–1.3 mm in diameter, warty, slightly rounded.

Fig. 3.93 Montaña de Guajara, June 2003. (Photo credit: Victoria Eugenia Martín Osorio)

Distribution

Canary Island endemism found on Las Cañadas del Teide (Tenerife), between 2200 and 3200 m a.s.l. Also, in the islands of La Palma and El Hierro.

Within the Teide National Park, it is located in northeastern rocky outcrops of Montaña de La Angostura, El Roque de La Grieta, Montaña Guajara and sectors of the Wall to the south of Llano de Ucanca; also in areas near the crater of Pico Viejo and the sloped of Teide, between the Corredor de Mario and of La Isla (NE side) (Figs. 3.94 and 3.95).

Habitat and Phytosociology

It is a plant remarkably associated to fissure habitats. It usually appears on vertical cliffs and cavities, preferably in shadowy areas. Occasionally, it is associated with *Arabis caucasica.*

It is a characteristic species of the ***Sileno berthelotianae-Cerastietum sventenii***, community for rupicolous cracks and rocky outcrops of phonolytic and trachybasaltic composition oriented to N and NW of the dry Supramediterranean belt. It is situated at altitudes between 2200 and 3000 m a.s.l., but always on wind exposed situations in the area of *Spartocytisetea supranubii*, *Spartocytision* and *Violion cheiranthifoliae*. This community substitutes *Cheilantho guanchicae-Aeonietum smithii* above 2200 m a.s.l.

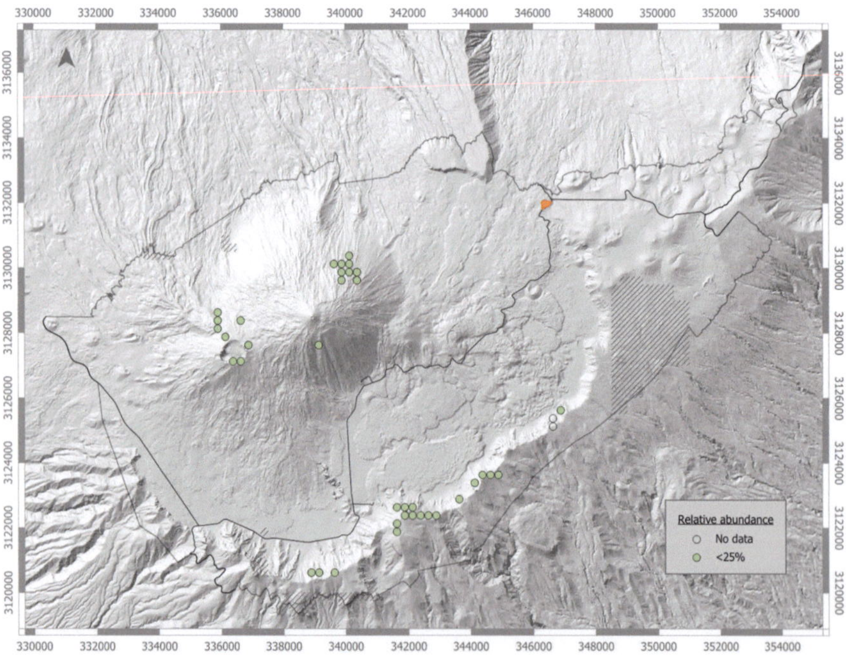

Fig. 3.94 *Cerastium sventenii* distribution, N° Subgrids (250 × 250 m): **39**/3740

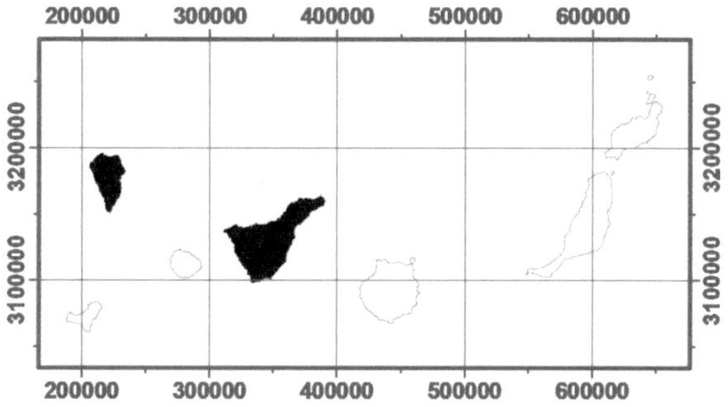

Fig. 3.95 Endemism of the islands of Tenerife, La Gomera and La Palma

Remarks:

"It appears as rarity on the Canary Islands. Collected on the high summit of the island of La Palma called El Lomo del Biscayno, at an altitude of 1000 hexafeet above sea level, where Etesian winds blow, together with Viola palmensi and Arabide albida, on fissured of rocks and as a thick lawn." (From Webb in Barker-Webb & Berthelot, loc. cit., p. 152)

According to Jalas *(*1966*), all specimens seen by him of Cerastium sventenii come from the island of La Palma. He even assumes that two individuals of C. arvense found in the herbarium of DeCandolle, given by Lowe and Möschl (1964) and originating from this island, belong to C. sventenii. At the same time, he mentions that numerous works where C. arvense is cited for La Palma, refer to the citation of Webb. Additionally, C. arvense has been cited for Tenerife by Christ (1885, pág. 488) and recapped by Schenck (1907, p. 397) and Möschl (loc. cit.).*

Conservation Status

According to the monitoring done, nearly all subgrids were this taxon was found (38) showed a very high conservation status without damages of any kind to individuals. Only on one sub-grids, the population of Montaña Guajara, damages, like herbivory, were observed but not on more than 25% of individuals.

"*Very scarce.*" (Sventenius 1946)

"*Canary Island endemism with a dozen populations which harbour somewhat under 2.000 individuals. Its principal threats are introduced herbivores which hinder the expansion of its populations. […] Traditionally, the impact of introduced herbivores (rabbits, mouflons, goats, barbary sheep, etc.) has been cited as the main threat factor for this species. Currently, although herbivours tend to be controlled, this aspect is still the main source of pressure on this taxon. […] All individuals are found within natural protected areas: National Park of Caldera de Taburiente (LIC), Natural protected area Cumbre Vieja (LIC), Rural Park of Frontera (LIC) and P. National Park of Teide (LIC) and practically all are found also within 'Hábitats de Interés Comunitario'. Accessions exist in Germplasm Banks (E.T.S. Ingenieros Agrónomos, Jardín Botánico Viera y* Clavijo*). […] Legal protection figures: Canarias (SAH), Orden Gobierno Canarias 20/2/91 (Anexo II).*" (Mesa Coello et al. 2004)

Threat Status

V (Barreno et al. 1984)
Anexo II: Protegidas (Gobierno de Canarias 1991)
EN (U.I.C.N., 1994)
EN B1 + 2ce (Bañares et al. 1998)
V (Walter and Gillett 1998)
Sensible a la alteración de su hábitat (Gobierno de Canarias 2001)
EN B1 + 2 ce (VVAA 2000)
EN B2 ab (iii, v) (Bañares et al. 2004)
EN B2 ab (iii, v) (Moreno 2008)
Especie de interés para los Ecosistemas canarios (Gobierno de Canarias 2010)

Germplasm Banks

E.T.S. de Ingenieros Agrónomos de Madrid.
Jardín Botánico Canario Viera y Clavijo.

Exsiccata (Herbarium Specimens)

Tenerife: Montaña de Guajara, Agosto, Amaro Luis (TFC 4400); Topo de la Grieta, 31.05.1973, W. Wildpret, H. Metlesics, M. Galván, C. Gil, P.L. Pérez de Paz & E. Beltrán (TFC 3896); Topo de la Grieta, 31.05.1973, H. Metlesics (TFC 3892, 3893, 3894); Las Cañadas, 2200 m.s.n.m., 2.06.1973, H. Metlesics (TFC 4390); Cañada de Diego Hernández, 2.06.1973, P.L. Pérez de Paz (TFC 13146); Topo de la Grieta, 2.06.1975, P.L. Pérez de Paz (TFC 4951); Las Cañadas, 2.07.1977, W. Wildpret, C. León, M. del Arco & I. La Serna (TFC 13053); Pared norte de Guajara-Las Cañadas, 18.07.1982, T. Bravo (TFC 13131); Oeste de la Pared de Diego Hernández-Las Cañadas, 18.06.1983, W. Wildpret & S. Socorro (TFC 13166). Inventario de la Flora y Vegetación del Parque Nacional del Teide (2001–2003) V.E. Martín Osorio, W. Wildpret & B. Hernández Bolaños*:* Roques al Sur del Llano Ucanca (41D 36 d), 21/04/2003, A. Montañés Rodríguez (TFC 44846); Roques al Sur del Llano Ucanca (41D 37 c), 21/04/2003, Y. de Rioja Santos (TFC 44859); Cerca del Corredor de La Corbata (35B 47 c), 12/05/2003, Y. de Rioja Santos (TFC 44214); Ladera Norte del Teide (35B 9 b), 28/05/2003, F. Mesa (TFC 44217)

La Palma: Los Andenes-Cumbre de La Palma, 16.06.1995, P.L. Pérez de Paz (TFC 38783)

3.5.31 Chamaecytisus proliferus *(L.f.) Link ssp.* angustifolius *(Kuntze) G. Kunkel*

Vieraea, 8(2), 354 (1980)

Cytisus proliferus L. fil. var. *nanus* Kuntze. *Rev. Gen. pl.*, 1: 178 (1891)

Endemism, Certain native
Family: *Fabaceae*
Common name: Escobón de monte o de pinar/Tagasaste (ES); Tree lucerne (EN)
Biotype: dwarf tree (nanophanerophyte)
Flowering: April–July
Fruiting: May–September
Pollination: animal dispersal (entomogazoochory)
Chromosome No: 2n = 48

Description

Nanophanerophyte 0.5–3 m tall, greyish to silvery in appearance. Leaflets linear-lanceolate, lanceolate, narrowly oblanceolate, blade frequently more or less closed in V, apex acute and obtuse, mucronulate, base acute, petiolulate; upper surface glabrous to sparsely sericeous; lower surface sericeous-applied hairs-, greyish, more hairy edges; central leaflet 17–40 (42) × 4–8 (14) mm, [L/A Fc = (3.6) 4–7 (8)].

Flowers 20–25 mm, white (Fig. 3.96). Standard 19–27 × 12–18 mm, blade oval to broadly oval, emarginate at the apex, upper surface almost entirely hairy, hairs more or less long applied; [L/A E = 1.1–1.7]; after anthesis the leaf blade of the standard is reflexed. Seeds with an average weight of 0.01440–0.01904 g.

Fig. 3.96 (**a**) El Sombrerito, next to the Vilaflor road, Apr. 2004 (b) Teide, Wolfredo Wildpret Botanical Garden, Aug. 2008. (Photo credits: Beatriz Hernández)

Distribution

It is found on the islands of La Gomera and Tenerife. Its presence is doubtful on the island of El Hierro (Acebes Ginovés et al. 2010) (Figs. 3.97 and 3.98).

On the northern slope of the island of Tenerife, its distribution begins from the upper limits of the Montaña Verde, approximately at 1500 m a.s.l.; on the southern slope, it is found from 400 m a.s.l., being more abundant in the pine forest area.

In the Teide National Park it is also closely related to the edge of the Canary Island pine forest, although it begins to enter inland bordering the Llano de Ucanca through the area of the Roques de García and reaching up the southern slopes of Pico Viejo; and from the Cañada de La Angostura towards the North, reaching areas close to the Minas de San José and Montaña Rajada.

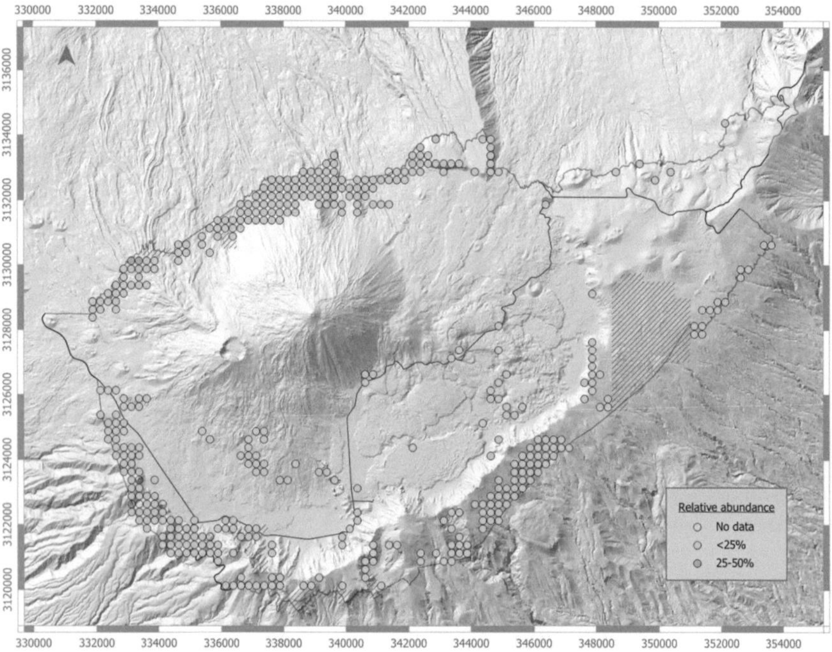

Fig. 3.97 *Chamaecytisus proliferus* subsp. *angustifolius* distribution, N° Subgrids (250 × 250 m): **425**/3740

Fig. 3.98 Endemism of the island of Tenerife and La Gomera

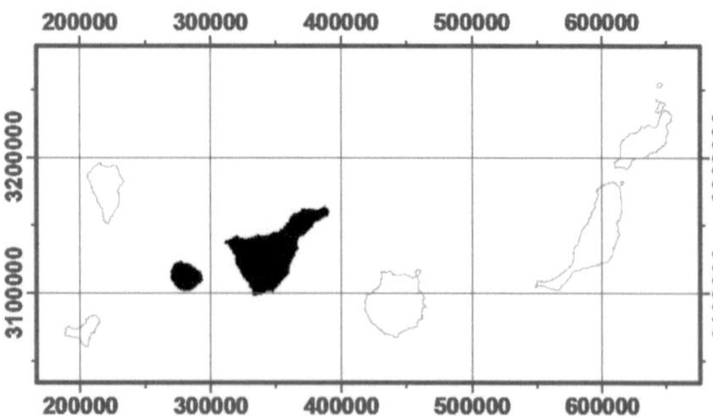

Habitat and Phytosociology

It is part of the characteristic communities of ***Chamaecytiso-Pinetea canariensis***, a vegetation class that includes the coniferous forests ***Chamaecytiso-Pinetalia canariensis***, the retamoid shrubland and the colonizing and serial dwarf smountain srublands of Tenerife and La Palma, as well as the Oromediterranean shrublands of the Teide volcanic cone ***Spartocytisetalia supranubii***; climatophilous, edaphoxerophilous, pyrophilous or serial, endemic to the midlands and peaks of the Central-Western Canary Islands, typical of dry-humid Meso-Oromediterranean bioclimates, of the Western Canary biogeographic subprovince.

In particular, it is presented as a characteristic species of the ***Cisto symphytifolii-Pinion canariensis*** alliance, which groups closed climatophilous and edaphoxerophilous meso-macroforests, well characterized by the Canary Island pine, dry-humid Meso-Supra-Mediterranean, rarely sub-cloudy, with a western Canary Island distribution.

It also characterizes or intervenes in the following associations:

Adenocarpo viscosi-Chamaecytisetum angustifolii, Codesar-Escobonal, which constitutes the heliophilous fringe of dry-humid Meso-Supra-Mediterranean pine forests on the island of Tenerife; in the Cañadas del Teide it is located in an arc almost bordering the protected area. ***Cheirolopho teydis-Carlinetum xeranthemoides***, a replacement scrub, sometimes acting as permanent vegetation, of the summit broom and Canary Island pine forests on soils made of old materials of the Teide Wall or Precaldera building, exposed to strong dry-subhumid Meso-Supra-Mediterranean sunlight. Its proliferation is frequent after some fires such as the one in the southern area of El Sombrerito.

Conservation Status

In 96% (409) of the subgrids in which this taxon has been found, its conservation status was high, as no damage of any kind was observed. Only in 2% (9), damage was detected (browsing, high mortality without regeneration, etc.), but this did not affect more than 25% of the specimens. Damage was only seen on more than 25% of the specimens in 1% (3) of the sub-grids where it appeared. In the rest (4), this variable was not specified.

Threat Status

nt (Barreno et al. 1984)
Anexo III (Gobierno de Canarias 1991)
LR pm (Bañares et al. 1998)

Germplasm Banks

Jardín Botánico Canario Viera y Clavijo.

Exsiccata (Herbarium Specimens)

Tenerife: Cumbres de Arafo, 1200 m.s.n.m., 3.04.1983, P.L. Pérez de Paz & C. Hernández Padrón (TFC 34267 + Dupl.); Pista ppal. Monte del Agua-Los Silos, 23.02.1997, V. Lucía Sauquillo & C. Acosta (TFC 40640 + Dupl.). Inventario de la Flora y Vegetación del Parque Nacional del Teide (2001–2003) V.E. Martín Osorio, W. Wildpret & B. Hernández Bolaños: Carretera de La Esperanza, P.N. del Teide, 19/11/2004, J.S. Martín (TFC 44884 + dupl.)

La Gomera: Fortaleza de Chipude, 7.04.1991, ADH, MMG & M. Salas Pascual (TFC 35256).

Uses, Exploitation and/or Medicinal Properties

> "*Fodder; nutritious; wood: firewood, forks; crafts: tools, spoons. Useful part: branches, seeds: toasted and ground. Instructions for use: fodder (of poorer quality than ssp. proliferus); browsing; crafts, tools.*" (Pérez de Paz and Hernández Padrón 1999)

3.5.32 Cheirolophus teydis *(C. Sm.* in *Buch) G. López*

Buch, L. von (1825). *Phys. Bes. der Canar. Inseln*, Berlín, pg. 147. Anales Jard. Bot. Madrid 38: 153 (1982)

Centaurea teydis C.Sm. in Abh. Königl. Akad. Wiss. Berlin 1816–1817:369 (1819)
Centaurea arguta Nees *in* Horae Phys. Berol.: 116 (1820)
Centaurea teidis Steud. *in* Nomencl. Bot., ed. 2, 1: 323 (1840), orth. var.
Cheirolophus argutus (Nees) Holub *in* Preslia 45: 143 (1973)
Ptosimopappus argutus (Nees) Boiss. *in* Diagn. Pl. Orient., ser. 2, 3: 60 (1856)

Endemism, Certain native
Family: *Asteraceae*
Common name: Cabezón de la cumbre o Centaurea del Teide (ES) Teide Centaury (EN)
Biotype: shrub (chamaephyte)
Flowering: July–September
Fruiting: September
Pollination: insect pollination (entomogamy)
Dispersal: wind dispersal (anemochory)
Chromosome N°: 2n = 30

Description

Shrub that can reach up to 1.5 m in height. Its lower part woody, with narrow, elongated branches, 30–60 cm high or more, with the tips covered in resin, glabrous, green, the lower part provided with many leaves, few branches, monocephalous.

Lower stems 2–4 mm in average diameter, round, furrowed-striated, glabrous, resinous tips subglabrous, green, or the lower ones purple, provided with leaves. Monocephalous or few branched. Hard whitish lignetum, crossed by a yellowish cylindrical medulla, with a thin tasteless or somewhat balsamic-tasting bark.

Leaves compact, numerous, the lower ones denser, larger, wider, straight, 2.5–7.5 cm long or more, 5/4–5 pl. 4–6 mm wide, linear-lanceolate, attenuate on both sides, with winged petiole attenuated towards the base; base equally dilated, acute, compact and scarcely serrated; the upper acute ones more distant, smaller, narrow, the upper ones subintact, sub-approximate to the stem, linear-lanceolate, 1.5–2.5 mm wide, 12.5–25 mm long, completely flat, coriaceous, with points on both sides, resinous-viscous. After loss of leaves, lower stem brown and scarred up to the petiole. Stems never simple, 30.5–45.5 cm, from the top of the base or towards the middle, branch into one or two branches of about 20–30 cm, erect, linear-lanceolate leaves, subintact, terminal subcorneal head, more distant from the capitulum always protected, gradually crossed by a cover of scales, fruiting drooping.

Capitulums always erect, multifloral (Fig. 3.99), narrowed towards the peduncles in the form of nails, exceeding 1.5 mm in average diameter. Involucre ovate-subhemispherical, narrow base of the apex, 2 cm high, 1.25 cm or more in average diameter; scales closely imbricated 8–9-serial, with swollen peduncular cyme, 6 mm long; upper leaves with a continuous cover of scales, 1 mm long, 1.5 mm wide below, oblong-lanceolate, obtuse, entire or with subdentate apex, the rest narrowly accumbent, cartilaginous, green, convex on the outside, concave on the inside, with resinous tips on both sides, subcarious and faintly brown towards the margin, slightly dilated on the upper side, irregularly lacerated, on the outer side ovate-oblong, 2–8 mm long, 2–3 mm wide, middle oblong-linear, 10–14 mm long, 3–3.5 mm wide, on the inner side narrow, thin and short, 12–14 mm long, 1–2 mm wide, entire, acute, subhyaline. Receptacles hairy-fimbriliferous, entire acute setae 2 mm long, white, upper part spotted.

Fig. 3.99 (**a**, **b**) Wolfredo Wildpret Botanical Garden of Teide National Park, Aug. 2008 (**c**) Guajara, Apr. 2009. (Photo credits: Beatriz Hernández)

Yellowish **flowers** extending beyond the involucre with subequal longitudinal divisions, about 2.5 cm long, glabrous, 5-toothed; teeth close together, narrow, linear. Peripheral flowers not very graceful, styles simple, puberulous with penicilliform pubescence, slightly protected. Small hermaphrodite flowers; anthers cylindrical, slightly excerpted, white, 6 mm long; anther appendages oblong-linear, obtuse, crowned, bisected base; filaments subpapillose; styles undivided, excerpted. Achene oblong-linear, shiny, cinereous, darkly spotted, at first glance glabrous; mature achene deciduous, on the other hand, the achene in the capitulum may be smaller, 3 mm long, linear, glabrous, cinereous, longitudinally striated, lateral basilar areola, pluriserial, imbricate protected pappus, ciliate-dentate complanate setae 1 mm long, both the outer and inner ones.

Remarks:

Like most species of this genus, it does not present problems for its reproduction by seeds.

Distribution

It is found on the islands of La Palma and Tenerife. In the Teide National Park, its distribution is closely related to the interior and exterior sectors of the Pared de Las Cañadas and Roques de García, between 1875 and 2625 m a.s.l., extending from the area of El Portillo to Boca Tauce (Figs. 3.100 and 3.101).

Habitat and Phytosociology

It is usually found in environments of debris, slag, cliff cracks and in compacted pumice soils. Occasionally, it comes into contact with pine forests. It intervenes preferably in the communities of the summit broom included in the alliance ***Spartocytision supranubii***, which groups the mountain, climatophilous or edaphoxerophilous mountain shrublands, sometimes catenal or serials of the Canary pine forests; It is a La Palma and Tenerife-type plant, dry-humid Meso-Oro-Mediterranean. It also characterises or intervenes in the following communities:

Arrhenathero calderae-Plantaginetum webbii,

Cheirolopho teydis-Carlinetum xeranthemoides, a replacement scrub, sometimes acting as permanent vegetation, of the summit broom and Canary Island pine forests on soils made of old materials from the Pared or pre-caldera building of Teide, denuded and exposed to strong sunlight in the dry-sub-humid Meso-Supra-Mediterranean. Its proliferation is frequent after some fires such as the one in the southern area of El Sombrerito.

Tolpido webbii-Micromerietum lachnophyllae, a substitution thyme that colonizes volcanic deposits and denuded soils of pine forests, cedar forests and broom shrublands, dry-subhumid Meso-Supramediterranean, endemic to the island of Tenerife between 1930 and 2400 m a.s.l.; it is frequent in the upper part of La Fortaleza and old stony areas of La Pared, or Edificio Precaldera de Siete Cañadas.

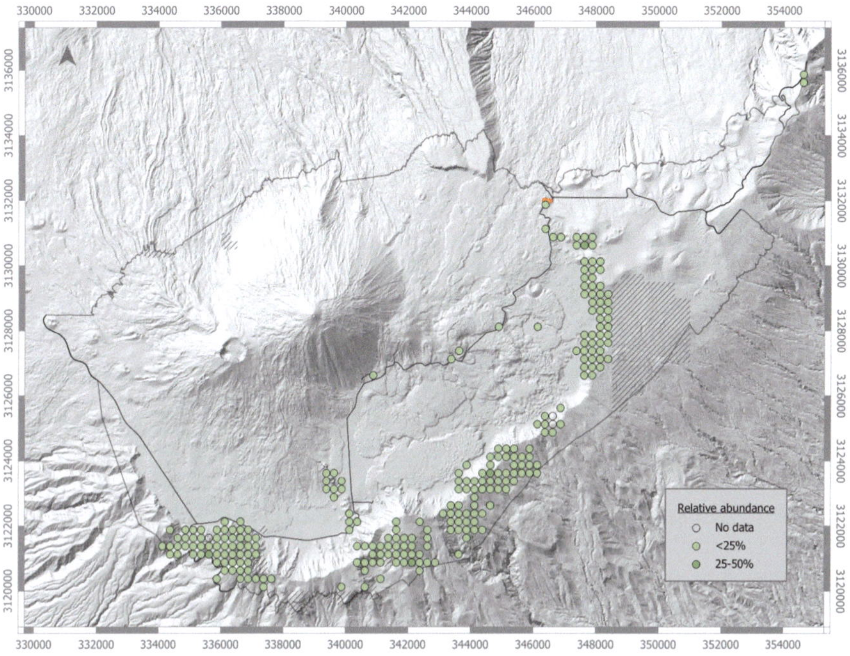

Fig. 3.100 *Cheirolophus teydis* distribution, N° Subgrids (250 × 250 m): **248**/3740

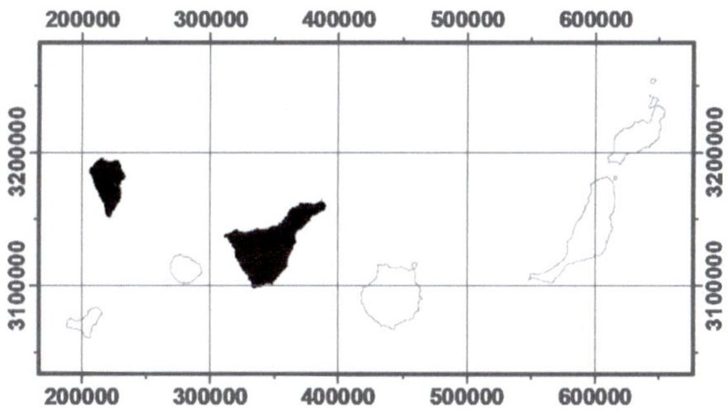

Fig. 3.101 Endemism of the island of Tenerife and La Palma

Conservation Status

In 97% (241) of the subgrids in which this taxon was found, no damage of any kind was observed. Only in 2.5% (6), damage was observed (browsing, high mortality without regeneration, etc.), but this did not affect more than 25% of the specimens. Only damage was detected on more than 25% of the specimens in 0.5% (1) of the sub-grids where it appeared.

> "*Most of the species that make up this genus in the Canary Islands have fairly small areas. However, this one is relatively well represented in the central part of the island of Tenerife, with several populations within the national park of 'El Teide' and in its peripheral protection zone. Among its natural predators we must highlight rabbits, which have a strong impact on the seedlings, and a certain species of dipteran whose larvae attack the seeds before they ripen. Visitors also tend to cut off their heads.*" (Gómez Campo et al. 1996)

Threat Status

R (Barreno et al. 1984)
Anexo II: Protegidas (Gobierno de Canarias 1991)
LR ca (Bañares et al. 1998)

Germplasm Banks

E.T.S. Ingenieros Agrónomos de Madrid.
Jardín Botánico Canario Viera y Clavijo.

Exsiccata (Herbarium Specimens)

Tenerife: Valle del Volcán de Las Arenas-Cumbres de Arafo, 9.05.1981, P.L. Pérez de Paz, W. Wildpret & M. del Arco Aguilar (TFC 23758 + 8 Dupl.); Ibid., 1.09.1981, W. Wildpret, M. del Arco Aguilar & P.L. Pérez de Paz (TFC 24082 + Dupl.); Los Roques de García-Las Cañadas, 10.10.1982, S. Socorro (TFC 13151 + Dupl.); Boca Tauce-Las Cañadas, 13.06.1987, P.L. Pérez de Paz (TFC 25395); Ibid., 30.07.1988, P.L. Pérez de Paz (TFC 30055 + Dupl.); Ibid., 30.07.1988, P.L. Pérez de Paz & J. Castañeda (TFC 30056—Testigo Análisis Fitoquímico. INPO); Arico-Las Cañadas del Teide, 2300 m.s.n.m., 11.08.1998, A. Santos (TFC 41706). Inventario de la Flora y Vegetación del Parque Nacional del Teide (2001–2003) V.E. Martín Osorio, W. Wildpret & B. Hernández Bolaños: Pared de Las Cañadas—Risco Verde, 29/08/2001, W. Wildpret de la Torre (TFC 44893 + dupl.).

Uses, Exploitation and/or Medicinal Properties

> "*Hypoglycemic, erythematous. Useful part: leaves, pre-flowering tops. Instructions for use: decoction, infusion.*" (Pérez de Paz and Hernández Padrón 1999)

3.5.33 Cistus osbeckiifolius *Webb* ex *H.Christ ssp.* osbeckiifolius

Spic. Can. in Engl. Bot. Jahrb., 9 (h.1): 96 (1888)

Rhodocistus osbeckiifolius Webb *ex* Grosser *in* H.G.A.Engler (ed.), Pflanzenr., IV, 193: 12 (1903), not validly publ.
Cistus osbeckiaefolius Webb. Bourg., I: 543.

Endemism, Certain native
Family: *Cistaceae*
Common name: Jara de Las Cañadas o Amagante de cumbre (ES) Teide rock-rose (EN)

Biotype: shrub (chamaephyte)
Flowering: May–July
Fruiting: July–August
Pollination: insect pollination (entomogamy)
Dispersal: wind dispersal (anemochory)
Chromosome N°: 2n = 18

Description

Chamaephyte that can reach 1 m in height, with a squat appearance and tortuous branches. **Leaves** narrow, lanceolate to elliptic, acute and obtuse, scarcely acuminate, attenuate base, three-veined, bracts ovate or rounded, sericeous, densely hairy, with silvery ciliated edges, rough surface, whitish tomentum; broad, orbiculate, spiny, linguliform, calyx-like leaves with a short apex.

Flowers up to 5 cm; pink-purple petals (Fig. 3.102). Stigma capitate; brown, ovate, densely hairy capsules, attenuate towards the apex; individual valve acuminate.

Lectotypus:

La Degollada del Cedro, 1846, Bourgeau 543 (holo-: P; iso-: K).

Fig. 3.102 (**a**, **b**) Wolfredo Wildpret Botanical Garden of Teide National Park, June 2004. (Photo credits: Beatriz Hernández) (**c**) La Fortaleza, June 2004. (Photo credit: Victoria Eugenia Martín Osorio)

Remarks:

In Demoly et al. (2006) it is stated that there also exists *C. osbeckiifolius* ssp. *tomentosus* Bañares & Demoly, subsp. nov. (Tágara, 1400–1650 m a.s.l.), which differs from ssp. *osbeckiifolius* by its height, which can reach 350 (400) cm (as against 120 or, more rarely, 155 cm), its frequently larger leaf blades (up to 80 (90) × 35 mm), always less than 3.5 times longer than wide, provided with simple hairs on the upper surface, and only a few simple hairs scattered on the nerves on the lower surface, its shorter leaf sheaths (6–8 mm long against 8–10 mm), its much longer inflorescences (from 20 (rarely 13) to 30 cm against 8–15 cm), with some simple hairs and some fasciculated on the peduncle, holding a greater number of flowers (18–42 (56) against 7–15), and its more pointed capsules (with long spaces of 3–4 mm at mid-height against 4–5 mm). Germination trials have shown that seed reproduction is an effective method of propagation.

Distribution

It is found only on the island of Tenerife. According to the sampling carried out, in the Teide National Park it is found naturally in La Fortaleza, Hoya del Montón de Trigo, western sectors of Sombrero de Chasna and Risco de La Magdalena (in the vicinity of Boca Tauce), between 2100 and 2400 m a.s.l. Also in the Teide, Wolfredo Wildpret Botanical Garden (Figs. 3.103 and 3.104).

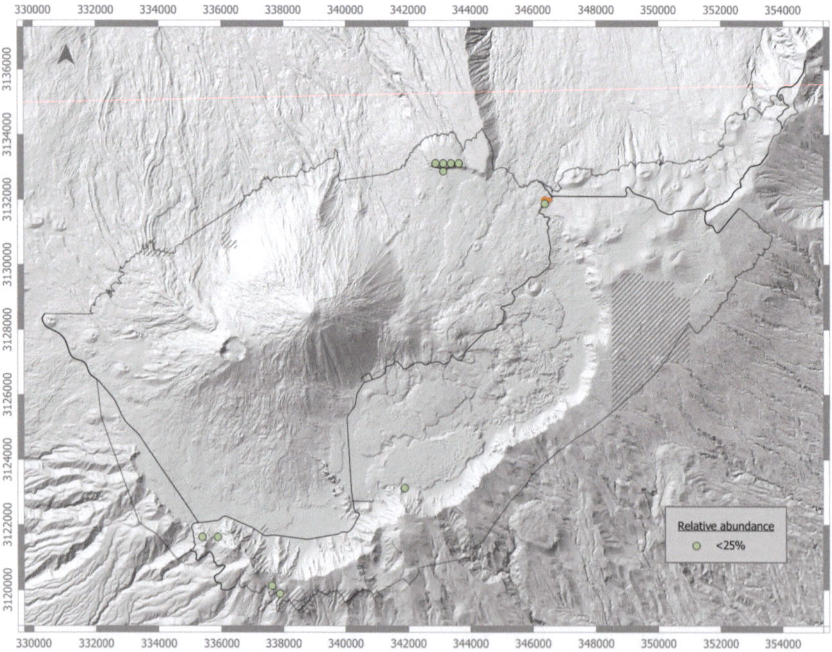

Fig. 3.103 *Cistus osbeckiifolius* subsp. *osbeckiifolius* distribution, N° Subgrids (250 × 250 m): **11**/3741

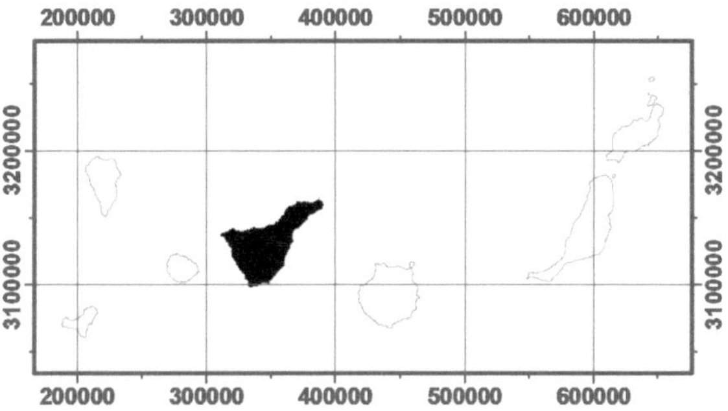

Fig. 3.104 Endemism of the island of Tenerife

In Demoly et al. (2006), using the quotes from Bañares et al. (1992, 2010) and Marrero Gómez et al. (1999), the following distribution areas are established:

La Fortaleza (Degollada del Cedro—locus classicus, UTM 28RC54333), at the foot of the escarpment of the edge of Las Cañadas: 300 individuals.

Risco de La Magdalena (2150 m a.s.l.): 210 individuals.

Llano de Las Mesas (2399 m a.s.l.): 190 individuals.

Fuente de la Junquera: 189 individuals.

Montaña de Guajara (UTM 28RCS417229—Program for the recovery of endangered species in the Teide National Park): 112 adults and 27 descendants in juvenile state.

Martín Osorio et al. (2010) speak of four natural populations and one introduced (Fuente de la Junquera, La Fortaleza, Guajara, Llano de las Mesas-Sombrero de Chasna and Riscos de la Magdalena), the one with the highest density being that of La Fortaleza, although this was reduced after the fire episode that occurred in 2007.

Habitat and Phytosociology

Although it is possible to observe this plant growing on cliffs, it develops best in environments of rubble and scree, especially in those places where volcanic ash has helped to form a more favourable substrate. Generally, on soils poor in material with acidic geological characteristics (pumice and phonolites).

Characteristic species and bioindicator of the ***Cisto symphytifolii-Pinion canariensis*** alliance, which groups closed climatophilous and edaphoxerophilous meso-macroforests, well characterised by the Canary Island pine, dry-humid Meso-Supra-Mediterranean, rarely sub-cloudy, of western Canary Island distribution.

It also gives character to the ***Sideritido solutae-Pinetum canariensis*** association; Tenerife Mediterranean climatophilous and edaphoxerophilous series, pluviseasonal oceanic Thermo-Supra-Mediterranean dry-subhumid semihyperoceanic of the forests of Pinus canariensis and ***Sideritis soluta*** with ***Chamaecytisus angustifolius*** and ***Argyranthemum adauctum* ssp. *dugourii***.

Conservation Status

According to the sampling carried out, in practically all the subgrids in which it was found, no damage of any kind was observed. Only in one of them, located in the surroundings of La Fortaleza, its condition was considered medium, that is, damage was observed (browsing, high mortality without regeneration, etc.), but this did not affect more than 25% of the specimens.

> "*This species has always been cited in scientific literature as being very rare and it continues to be so today, even though at first glance there are no adverse factors that would allow us to explain its extreme degree of rarity. However, it is possible to suppose that its populations have been decimated due to the intense grazing that supported the vegetation of the peaks of Tenerife for centuries, only being eradicated from these territories twenty years ago.*
>
> *At present, it can be observed that from old specimens that still survive on cliffs, stands of young plants have formed in the foothill areas, and there seems to be an increase in the number of individuals.*" (Translated from: Gómez Campo et al. 1996)

> "*It seems evident that the different current populations have been related to each other in the past, along the upper edge of the Pinar. The dozens of individuals that survive are still subject to pressure from goats and their seedbeds and from rabbits that have been present for 900 years. Batista et al. (2001) have shown that the fragmentation of the primitive area has led to […] a genetic erosion […] that could compromise the capacity of this species to adapt to other changes in its environment. The integrity of the current populations is therefore quite important for the survival of the species. Therefore, C. osbeckiifolius ssp. osbeckiifolius can be considered EN (but very close to CR). The previous classification in the VU category (*Marrero Gómez et al. 1999*) included the population of the Tágara valley, corresponding to ssp. tomentosus […].*" (Translated from: Demoly et al. 2006)

> "*The species is currently in danger for several reasons; the area of occupation, the presence of less than 5 natural populations, and the continuous decrease in the number of mature individuals. Only in one of them has the high degree of genetic diversity been verified. The cause of the decline and death of some specimens when they reach about 60 cm in height is unknown. There is also a risk of fire, such as the one that occurred in 2007 that destroyed an entire subpopulation.*"
>
> "*[…] It suffers predation by herbivores, landslides and fires. Even in the absence of animal pressure, it has a tendency towards so-called 'genetic erosion', that is, a reduction in polymorphisms that could compromise the capacity of this species to adapt to other changes in the environment. The integrity of current populations is therefore necessary for the survival of the species.*"
>
> "*[…] There are populations that, due to their poor state of conservation, need urgent and precise in situ measures. An example is the population of La Fortaleza, which saw its numbers decimated by the fire that occurred in 2007 and which also suffers predation by herbivores present in the Park, mainly the rabbit.*" (Translated from: Martín Osorio et al. 2010)

Threat Status

E (Barreno et al. 1984).
Anexo I: Estrictamente protegidas (Gobierno de Canarias 1991).
VU D2 (Bañares et al. 1998).
V (Walter and Gillett 1998).

VU D2 (VVAA 2000).
VU (Bañares et al. 2004).
EN B2ab(v) (Moreno 2008).
EN B2ab(v) (Bañares et al. 2010).

Germplasm Banks

Conservatoire Botanique National de Brest—France (Demoly et al. 2006).
E.T.S. Ingenieros Agrónomos de Madrid.
Jardín Botánico Canario Viera y Clavijo.

Exsiccata (Herbarium Specimens)

Tenerife: 18.07.1969, Bramwell 2101 (TFC 110); La Fortaleza, 19.05.1973, P.L. Pérez de Paz (TFC 23338 + Dupl., 23340); Ibid., 13.06.1980, E. Beltrán Tejera (TFC 20906 + Dupl.); Ibid., 13.06.1984, S. Socorro, (TFMC 1608 & 1609); Ibid., Junio 1987, A. Bañares (TFC 25068); Inventario de la Flora y Vegetación del Parque Nacional del Teide (2001–2003) V.E. Martín Osorio, W. Wildpret & B. Hernández Bolaños: Al lado de la casa del Parque, 31.05.2001, L. Afonso, A. Hernández, Y. de Rioja & F. Tapia (TFC 44920); Jardín Botánico del Parque Nacional del Teide, 13.11.2003, V.E. Martín Osorio & W. Wildpret de la Torre (TFC 44927); Por debajo de El Sombrero, hacia el Sur, en Los Retamares (41D 44 c), 29.04.2003, F. Mesa & Y. de Rioja Santos (TFC 44262); La Fortaleza, 05.06.2003, V.E. Martín Osorio & W. Wildpret de la Torre (TFC 44948).

3.5.34 Cistus symphytifolius *Lam. var.* symphytifolius

Dict. Encycl. Méth. Bot., II, 15 (1786)

Cistus berthelotianus var. symphytifolius (Lam.) Pit. *in* J.-C.M.Pitard & L.Proust, Iles Canaries: 110 (1909), nom. superfl.
Cistus vaginatus. Ait., *in* Hort. Kew., ed. I, vol. 2, pág. 232 (1789).
Cistus vaginatus Dryand. in W.T.Aiton, Hortus Kew. 3: 304 (1811)
Rhodocistus berthelotianus Spach *in* Ann. Sci. Nat., Bot., sér. 2, 6: 367 (1836)

Endemism, Certain native
Family: *Cistaceae*
Common name: Amagante de Pinar/Jara/Jarón (ES); Pine forest Rockrose (EN)
Biotype: shrub (chamaephyte)
Flowering: spring–summer
Fruiting: summer
Pollination: insect pollination (entomogamy)
Dispersal: wind dispersal (anemochory)
Chromosome N°: 2n = 18; n = 9

Description

Erect, very variable **shrub**, 50–70 or even 100 cm tall, open, very branched, more or less covered with stellate and simple hairs. Brown branches.

Leaves petiolate, fused at the base forming a long sheath, ovate-oblong or oblong-elliptic, acute, subacute or subobtuse, flat or somewhat wavy, 35–100 × 15–60 mm, deep green, light green or whitish, membranous or hardened, flexible, upper surface viscid, warty or subsericeous, subglabrous, densely tomentose or hirsute, lower surface thin or densely tomentose, reticulate, with 3 very marked nerves; winged petiole, more or less hirsute. Inflorescences paniculate-cymose, terminal, with 2–9 flowers. **Flowers** about 5 cm in diameter. Sepals 5, ovate or subcordate, asymmetrical, acuminate, 10–18 × 6–10 mm, the outer ones smaller than the inner ones, foliaceous, green or whitish, somewhat pubescent. Calyx normally deciduous when in fruit. Petals reddish pink, with a small yellow spot at the base, ovate-triangular, obcordate, somewhat curled, with the apex denticulate and the base cuneate, 15–30 × 12–25 mm. Pistil 15–25 mm long. Style geniculate (Fig. 3.105), not straight. Capsules globose or ellipsoidal, 8–12 mm long, brown, somewhat pubescent or almost glabrous, with 5 valves.

Fig. 3.105 Detail of the amagante flower, March 2012. (Photo credit: Victoria Eugenia Martín Osorio)

Remarks:

> *"This species is unique in that it is represented by two infraspecific taxa: one more northerly (nominal taxon), and another adapted to the warmer and drier conditions of the southern slopes, C. symphytifolius var. villosus. This differs from var. symphytifolius in that its leaf blades are densely covered with simple hairs on the upper surface and with glandular hairs, less numerous, on the lower surface."* (Translated from: Demoly et al. 2006)

Distribution

This species of *Cistus* is the most widely distributed and is found on the islands of La Palma and Tenerife. In the former, it is present in the northern half surrounding the Caldera de Taburiente (700–2000 m above sea level); in Tenerife, it also extends along the northern slope (600–2000 m above sea level), except in the Anaga massif (Demoly et al. 2006). In the Teide National Park it was found in bordering areas in the north (northwestern flank of the slopes of Teide: from Lomo La Perdiz to Los Frontones) (Figs. 3.106 and 3.107).

It is also found in southern border areas around the Las Cañadas wall: in the Cañada del Capricho and, on its southern slope, between the Roque del Almendro and the Sombrero de Chasna and in areas close to the Barranco del Río and Altos de Arico. It can reach altitudes of up to 2400 m above sea level.

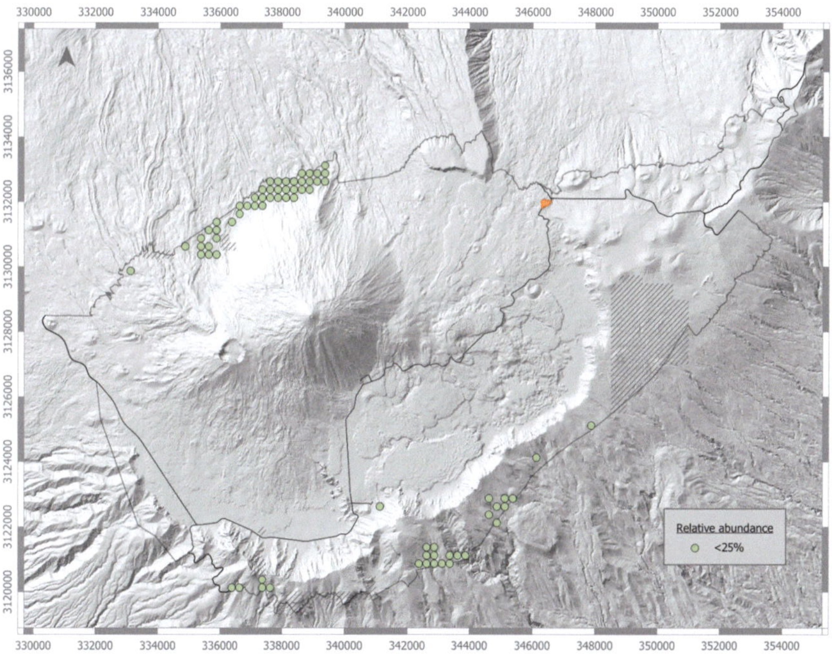

Fig. 3.106 *Cistus symphytifolius* Lam. var. *symphytifolius*, N° Subgrids (250 × 250 m): 71/3741

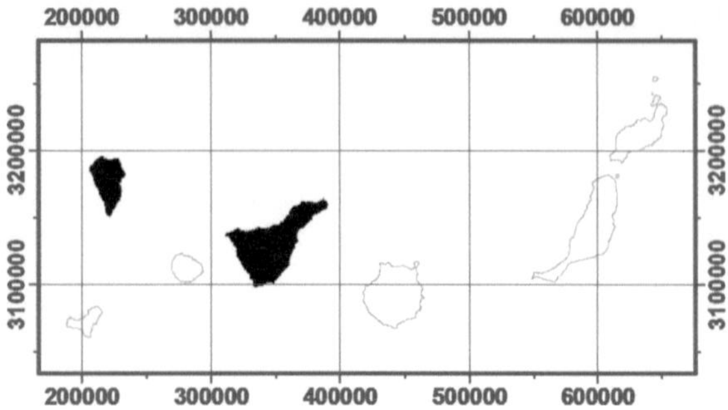

Fig. 3.107 Endemism of the island of Tenerife and La Palma

Habitat and Phytosociology

It grows preferably on soils developed from acidic or basic geological materials corresponding to Volcanic Series II and III; under climatic conditions, depending on the location, with temperatures between 9 and 18 °C and rainfall ranging between 400 and 1000 mm on average per year (bioclimatic type: Meso to Thermo-Mediterranean, dry to sub-humid). This taxon shows a special primocolonizing potential in different edaphic situations, which facilitates its wide chorological diffusion and the existence of numerically rich populations (Demoly et al. 2006).

It is a characteristic species and bioindicator of ***Cisto symphytifolii-Pinion canariensis***, an alliance that groups closed climatophilous and edaphoxerophilous meso-macroforests, well characterized by the Canary Island pine, dry-humid Meso-Supra-Mediterranean, rarely sub-cloudy, of western Canary Island distribution. While its upper limit marks the beginning of the summit broom, at its lower limit, it can mix with the Monteverde formations.

It also occurs in ***Adenocarpo viscosi-Chamaecytisetum angustifolii***, Codesar-Escobonal, which forms the heliophilous fringe of dry-humid Meso-Supra-Mediterranean pine forests on the island of Tenerife; in the Cañadas del Teide it is located in an arc almost bordering the protected area.

Conservation Status

In almost all the subgrids in which it has been found, no damage of any kind was observed. Only in one of them, located in the vicinity of the Barranco del Dornajito, south of the Roque del Almendro, was damage observed (browsing, high mortality without regeneration, etc.), but this did not affect more than 25% of the specimens.

Threat Status

nt (Barreno et al. 1984).
LR pm (Bañares et al. 1998).

Germplasm Banks

E.T.S. Ingenieros Agrónomos de Madrid.

Exsiccata (Herbarium Specimens)

Tenerife: (Enviadas), X. 1972, Bramwell, Leg. K. Lems, (TFC 4359); El Realejo, Pinar, 16.IV.72, Beltrán Tejera, E., (TFC 23412); La Esperanza, 26.V.73, La Serna Ramos, I., (TFC 23357); Bco. Tamay, Güímar, 17.V.75, Benítez, AD., (TFC 42882); Altos de la Corujera, VI.79, Bañares, A., (TFC 21191); Madre del Agua. Agua García, 24.4.80, Wildpret de la Torre, W., A. García Gallo, (TFC 10351, Duplicata); Bco. de Tamay, Arafo, 1200 m, 9.V.1981, Pérez de Paz, P.L., M. del Arco Aguilar, W. Wildpret de la Torre, (TFC 13694); Bco. de Erques, Canal y pista, 1000 m.s.m., 16.VIII.88, Rodríguez Delgado, O., (TFC 27545); Candelaria, altos de la mesa, (UTM 362–3140), 1993, MMG., O. Rodríguez Delgado, W. Wildpret de la Torre, ECA., (TFC Duplic. 36134); Bco. del Agua, (Río) Güímar, 850 m.s.n.m., 1994, del Arco Aguilar, M., (TFC 36449); La Quinta, Santa Úrsula, 25.V.94, Pérez de Paz, P.L., W. Wildpret de la Torre, Enrique., Montse., F. Ramallo Rodríguez, (TFC 37505); Charcos de Erjos, Los Silos, NO Charcas, 9.VII.95, Lucía Sauquillo, V., C. Acosta, (TFC 38153).

Uses, Exploitation and/or Medicinal Properties

> "*Dental, nutritional, bad forage. Useful part: root, seeds, whole plant. Instructions for use: infusion (mouthwash), ground seeds: gofio de guanches.*" (Pérez de Paz and Hernández Padrón 1999)

3.5.35 Cystopteris diaphana (Bory) Blasdell subsp. diaphana

Cystopteris viridula (Desv.) Desv. *In Mém. Soc. Linn. Paris*, 6, 264 (1827).
Cystopteris fragilis subsp. *diaphana* (Bory) Litard. *in* Bull. Géogr. Bot. 21: 20 (1911)
Asplenium diaphanum (Bory) Lojac. *in* Fl. Sicul. 3: 402 (1909)
Cystopteris diaphana (Bory) Blasdell *in* Mem. Torrey Bot. Club 21(4): 47 (1963)
Cystopteris fragilis var. *diaphana* (Bory) Rosenst. *in* Contr. Fl. Bolivia 1: 9 (1910)
Polypodium diaphanum Bory *in* Voy. îles Afrique 1: 328 (1804)

Not endemic, Certain native
Family: *Aspleniaceae*
Common name: Helecho de manantial (ES) greenish-bladder fern (EN)
Biotype: hemicryptophyte
Sporulation: February–November
Reproduction: water pollination (hydrogamy)
Dispersal: wind dispersal (anemochory)
Chromosome N°: 2n = 84; 168, 252

Description

Rhizome short and creeping, with approximate internodes, loosely covered by lanceolate, brown palea.

Fronds up to 40 cm; petiole shorter than the blade, generally straw-coloured, with some palea at the base; blade 2-pinnate, with the last segments 1-pinnatipartite, ovate-lanceolate; rachis sometimes with multicellular, glandular hairs; pinnules oblong, cuneate at the base, with the last order segments toothed; teeth generally emarginate, the secondary nerves ending within the emarginations.

Sori discrete or, less frequently, confluent at maturity; indusium ovate to suborbicular, glandular or rarely glabrous (Fig. 3.108). Sporangia subsessile. Spores 36–50 µm, with lacunar spines.

Distribution

It is found in Central and Southern America, Northern and Eastern Africa, Southwestern Europe and Western Spain, Navarre and Granada. In the Canary Islands it is only present on the islands of El Hierro, La Palma, La Gomera, Tenerife and Gran Canaria (Fig. 3.109). In the Teide National Park (Fig. 3.110) it appears very sparsely and punctually in areas of La Pared (Cañadas de la Angostura, Western Guajara Mountain and near Roque del Almendro), and in the lava flows of Cerro Las Chajoras (NW-Pico Viejo), between 2200 and 2500 m above sea level.

Fig. 3.108 *Cystopteris diaphana*. (Photo credit: G. G. Casanova)

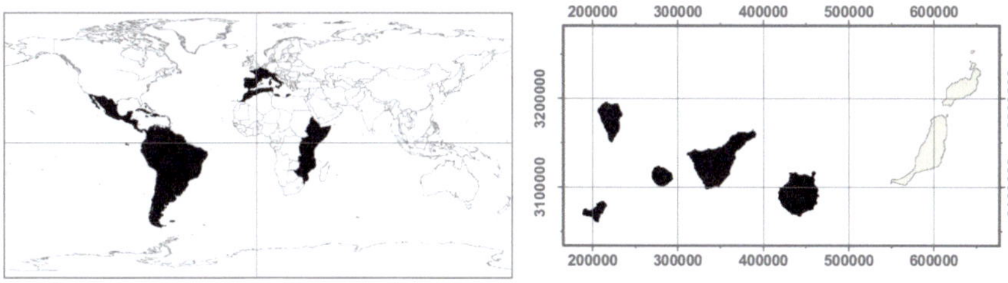

Fig. 3.109 Present on present on the islands of El Hierro, La Palma, La Gomera, Tenerife and Gran Canaria

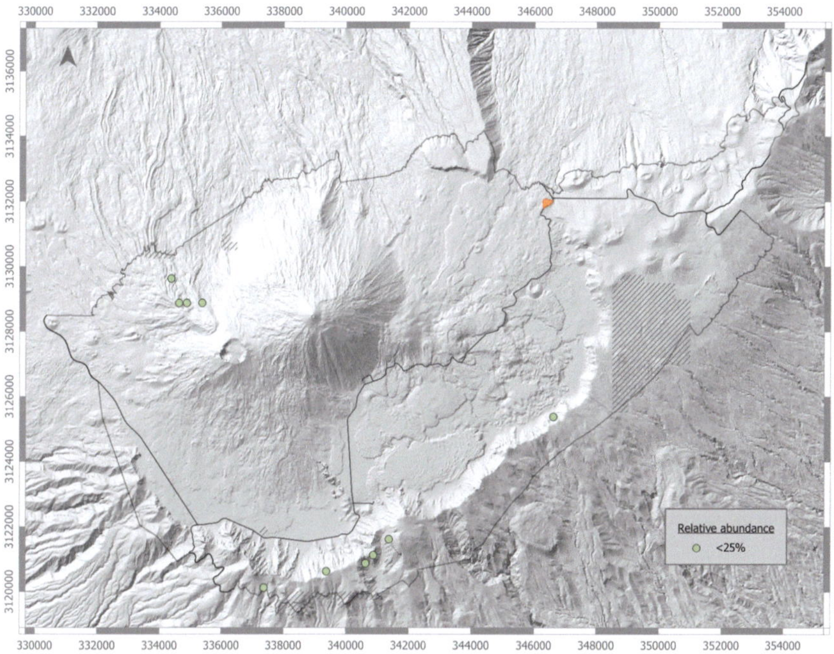

Fig. 3.110 *Cystopteris diaphana* distribution, N° Subgrids (250 × 250 m): **10**/3740

Habitat and Phytosociology

It lives in rock crevices and slopes, in humid and shaded places, near water channels or in oceanic environments. It is a characteristic species and bioindicator of ***Selaginello denticulatae-Anogrammion leptophyllae***, an alliance that groups the associations of earthy, shady and cool slopes, Thermo-Meso-Mediterranean subhumid-humid hyperoceanic-oceanic, with a wide Mediterranean distribution.

Conservation Status

In none of the subgrids in which this species was found (10), damage of any kind was observed.

Exsiccata (Herbarium Specimens)

Tenerife: Pico del Inglés (Sánchez Velázquez 1998). Inventario de la Flora y Vegetación del Parque Nacional del Teide (2001–2003) V.E. Martín Osorio, W. Wildpret & B. Hernández Bolaños: Montaña de La Angostura (36D 36 a), 28.08.2001, L. Afonso Jiménez (TFC 44263); Ladera NW de Pico Viejo (35A 23b), 10.6.2003, F. Mesa (TFC 44215); Ladera NW de Pico Viejo (35A 23a), 10.6.2003, Y. Rioja Santos (TFC 44223); Ladera Sur del Teide (35D 18 a), 23/04/2003, A. Montañés Rodríguez (TFC 44855).

La Gomera: Agua de Los Llanos, 940 m.s.n.m., 24.6.2002, E. Ojeda-Land (TFC 44044); Agulo, Bco. de Meriga, 26.6.2002, E. Ojeda-Land (TFC 44041).

3.5.36 Cystopteris fragilis *(L.) Bernh.*

In Schrader, H.A., *Neus J. Bot.*, 1(2), 27 (1805)

Polypodium fragile L. *Sp. Pl.*:1091 (1753) ["*F. fragile*", basión.].
Polypodium filix-fragile L. *Fl. Suec. ed.*, 2: 374 (1755).

Not endemic to the Canary Islands, Certain native
Family: *Aspleniaceae*
Common name: Quebradizo, Culantrillo blanco, Helecho vejigoso o Helecho de manantial (ES); Brittle Bladder Fern (EN)
Biotype: hemicryptophyte
Sporulation: April–September
Reproduction: water pollination (hydrogamy)
Dispersion: wind dispersal (anemochory)
Chromosome N°: –

Description

Rhizome short and creeping, with approximate internodes, covered by remains of petioles, and at the apex, by lanceolate, brown palea.

Fronds fasciculate, (Fig. 3.111) sometimes with various types of glandular hairs; petiole generally shorter than the blade, straw-coloured to dark brown, with some palea at the base; blade 2(3)-pinnate, lanceolate to ovate-lanceolate; pinnae subopposite to alternate, the basal pair frequently distant and shorter than the previous one; pinnules entire to 1-parted, with generally toothed margin; segments of the last order with obtuse, acute or emarginate teeth. Indusium lanceolate to ovate-lanceolate, generally glabrous.

Spores 35–55 (60) μm, echinate with conical spines with a smooth surface or slightly striated at the base.

Remarks:

According to Salvo Tierra (1990), sporulation occurs from June to November.

Fig. 3.111 *Cystopteris fragilis* near La Tarta, Teide, National Park, June 6, 2016. (Photo credit: Victoria Eugenia Martín Osorio)

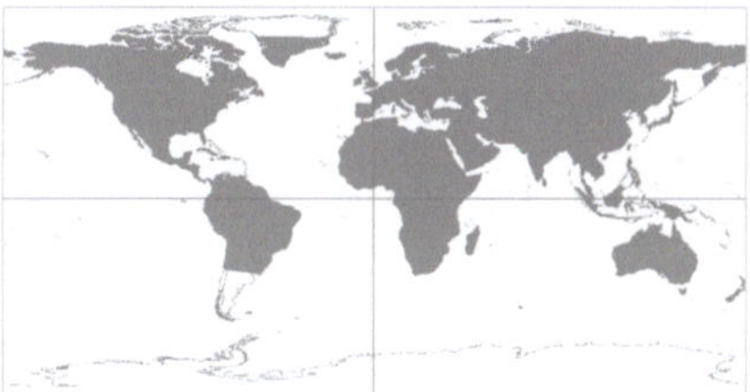

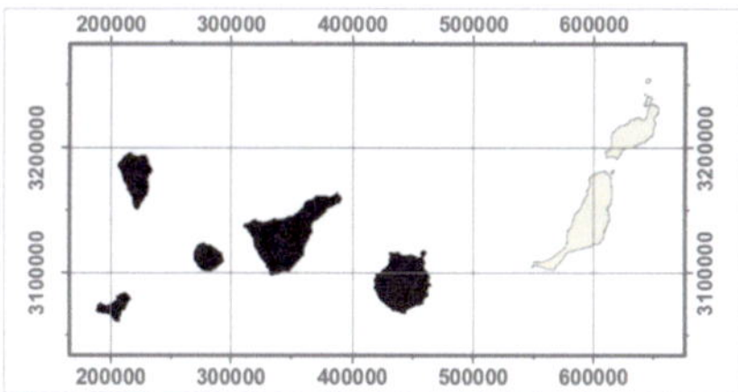

Fig. 3.112 Throughout the Canary Islands, except Lanzarote and Fuerteventura

Distribution

Subcosmopolitan. It reaches up to 83°N in the Northern Hemisphere (Salvo Tierra 1990). Distributed throughout the Macaronesian region; temperate oceanic Thermo and Meso bioclimate, as well as Thermo-Meso-Supra-Mediterranean. Throughout the Canary Islands, except Lanzarote and Fuerteventura (Fig. 3.112).

In the Teide National Park (Fig. 3.113) it was found mainly in areas of the Dorsal, in the heights of the La Orotava Valley (La Crucita, La Tarta, Montaña del Limón, northern slopes of Izaña and Montaña Guamasa); much more dispersed throughout the interior of Las Cañadas, areas of the Pared and also occasionally in La Fortaleza (1950–2400 m a.s.l.).

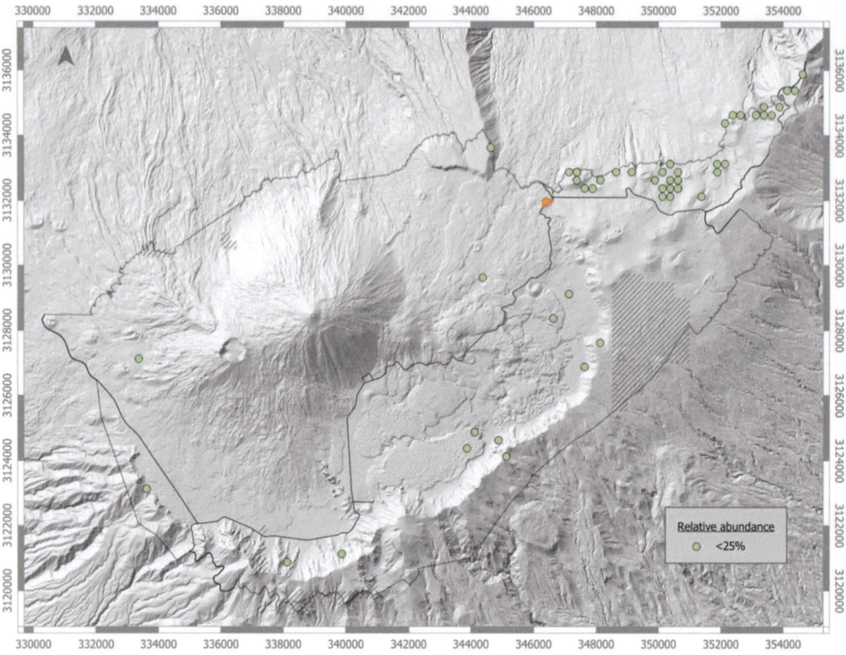

Fig. 3.113 *Cystopteris fragilis* distribution, N° Subgrids (250 × 250 m): **48**/3740

Habitat and Phytosociology

It is found in damp and shady places, springs and seeping sites, coexisting with bryophytes; in rock fissures, walls or earthen slopes with low sunlight.

It is a characteristic species and bioindicator of ***Asplenietea trichomanis***, a type of vegetation that groups the Holarctic chasmophytic communities of rocks and vertical walls that receive direct rain. They are formed by small vascular perennial rock plants, pteridophytes and hemicryptophytic and chamaephytic spermatophytes, not nitrophilous, whose root system is housed inside narrow fissures and grooves existing in rocks and walls that are moistened during the rains.

It can also be present in the following units:

Anomodonto viticulosi-Polypodietea cambrici
Greenovio aureae-Aeonietea

Conservation Status

In practically all the subgrids in which this species was found (47), no damage of any kind was observed. Only in one (1) of them, its conservation status was determined to be medium.

Exsiccata (Herbarium Specimens)

Tenerife: Como *C. grex fragilis*: El Pijaral, Vaguada húmeda próxima al camino de bajada a Almáciga, junto con varios ejemplares de *P. incompleta* (UTM 28RCS3833159), 1.6.2001, Ricardo González (TFC 43338); Las Mercedes, 800 m, Las Vueltas 600–800 m (Pitard and Proust 1908); Mercedes forest north of Las Mercedes, 920 m (LID 1968).

Como *C. fragilis*: Las Mercedes, 600 m, 21.2.1944 (ORT 1875), Vueltas de Taganana, 29.10.1949 (ORT 17928) (Santos Guerra & Fernández Galván *in* I.S.H.A.P.A., 1977); Las Mercedes, 15.2.1970, M.C. Gil (TFC 33680); Inventario de la Flora y Vegetación del Parque Nacional del Teide (2001–2003) V.E. Martín Osorio, W. Wildpret & B. Hernández Bolaños: Pared de los Helechos, en la Carretera General, pasando La Crucita, 12/06/2003, V.E. Martín Osorio & W. Wildpret de la Torre (TFC 44950).

3.5.37 Dactylis metlesicsii *Schönfelder & Ludwig*

Willdenowia, 26, 217–223 (1996)

Endemism, Certain native
Family: *Poaceae*
Common name: Jopillo de cumbre (ES); Summit cock's foot (EN)
Biotype: hemicryptophyte
Flowering: May–July
Fruiting: June–August
Pollination: wind pollination (anemogamy)
Dispersal: short distance dispersal (atelochory)
Chromosome No: 2n = 14

Description

Perennial herb, densely cespitose, (Fig. 3.114), 70–100 cm tall. Leaf rigid erect, mostly at the base, blade flat, 25 cm long, 3–6 mm on each side, glaucoviridi, also rigid and erect when dry, ligule 3–7 mm long, acute. Culminates in 2 or 3 rigid erect leaves, panicle (4–) 8–16 (–20) cm long, branches increasingly less erect, lower branches sessile or 0.5–4 cm pedunculated.

Spicule 4–5 mm long, 2–4 flowers. Lower glume 3–4 mm long, gradually awned, without prominent vein, the upper glume as carine and toothed as ciliate, without upper part of the lemma, rarely with short tip, membranous stems, acute, near the lemma. Anthers (1.2–) 1.7–2.4 mm long, yellowish.

Fig. 3.114 (**a**) Wolfredo Wildpret Botanical Garden of Teide National Park, Aug. 2008. (Photo credit: Beatriz Hernández) (**b**) La Fortaleza 24 June 2015. (Photo credit: Victoria Eugenia Martín Osorio)

Distribution

It is found only on the island of Tenerife. Within the Teide National Park it was located at the top of Lomo de la Burra (Tigaiga), western flank of Montaña El Cerrillar, southeastern slopes of Montaña de La Angostura, Roque de La Grieta and Montaña Guajara, and at the foot of the slopes that flow towards Llano Ucanca; between 2000 and 2325 m a.s.l. (Figs. 3.115 and 3.116).

Habitat and Phytosociology

A ruderal-nitrophilous species, usually found on the edges of roads or huts. It can also appear as a companion in the hydrophytic vegetation of Las Cañadas del Teide or in areas of rocky ground, preferably of a phonolitic nature, within the domain of the summit broom. As a secondary habitat, it can also be found in open pine forest areas.

This species appears within the domain of ***Spartocytisetum supranubii***; climatophilous and edaphophilous Tenerife Mediterranean pluviseasonal oceanic Mesomediterranean upper-Oromediterranean dry-subhumid euoceanic series of the summit shrublands of ***Spartocytisus supranubius*** and ***Adenocarpus viscosus*** with ***Descurainia bourgeauana*** and ***Nepeta teydea var. teydea***.

It also participates in the community ***Sideritido solutae-Pinetum canariensis; climatophilous*** and edaphophilous Tenerife Mediterranean pluviseasonal oceanic Thermo-Supramediterranean dry-subhumid semihyperoceanic series of the forests

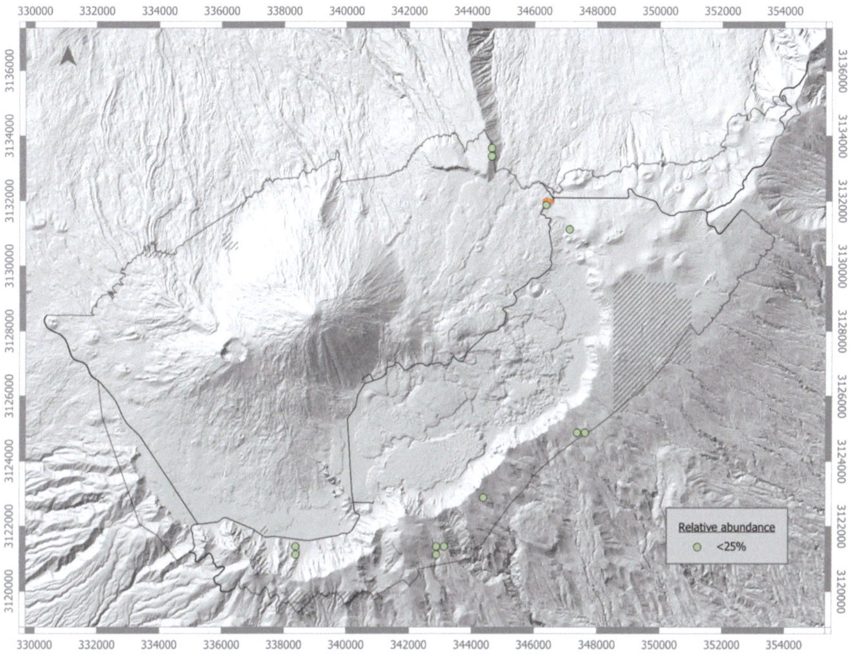

Fig. 3.115 *Dactylis metlesicsii* distribution, N° Subgrids (250 × 250 m): **12**/3741

Fig. 3.116 Endemism of the island of Tenerife

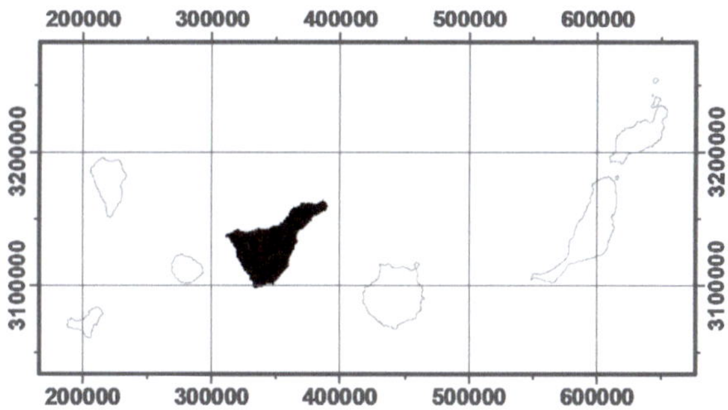

of ***Pinus canariensis*** and ***Sideritis soluta*** with ***Chamaecytisus angustifolius*** and ***Argyranthemum dugourii***.

Conservation Status

Of the five populations studied of this Tenerife endemic species, in two of them no specimens have been found in the sampling phase (Barranco del Riachuelo and Cañada de Diego Hernández), the rest present an acceptable state of conservation. The populations are located within the Natural Park of La Corona Forestal (LIC) and in the Reserve Zone of the Teide National Park (LIC). Furthermore, its habitat is registered in Directive 92/43/CEE. Among the threat factors are: the transformation of water courses, collecting and harvesting, predation and potential accidents (fires, landslides, droughts, etc.). Proposed conservation measures: Monitoring and storage of seeds in germplasm banks. Legal protection figures: Canary Islands (SAH) (Martín Osorio et al. 2004).

In almost all the sub-grids where this taxon was found, no damage of any kind was observed. Only in one of them, located in Lomo de la Burra (Tigaiga), damage was observed (browsing, high mortality without regeneration, etc.), but this did not affect more than 25% of the specimens.

Threat Status

Sensible a la alteración de su hábitat (Gobierno de Canarias 2001).
EN B2ab (iii, iv, v) (Bañares et al. 2004).
EN B2ab (iii, iv, v) (Moreno 2008).
Vulnerable (Gobierno de Canarias 2010).

Exsiccata (Herbarium Specimens)

Tenerife: Topo de la Grieta-Las Cañadas, 2.06.1973, P.L. Pérez de Paz (TFC 24068); Cañada de Diego Hernández, 2.06.1973, ejusd. (TFC 24147); El Cabezón, 4.06.1973, H. Metlesics (TFC 4345); Bco. del Riachuelo, Mayo 1981, CRP (TFC 33571 + Dupl.); Bco. del Río, 2200 m.s.n.m., 28.05.1983, W. Wildpret & S. Socorro (TFC 18476); Inventario de la Flora y Vegetación del Parque Nacional del Teide (2001–2003) V.E. Martín Osorio, W. Wildpret & B. Hernández Bolaños: Guajara, al Norte del Paisaje Lunar (42C 21 d), 02.06.2003, Y. de Rioja Santos (TFC 44218); La Fortaleza, 05.06.2003, V.E. Martín Osorio & W. Wildpret de la Torre (TFC 44947); Ruta de Siete Cañadas (Alto de ladera—30D 23—UTM: 0347108–3131175), 15.06.2005, V.E. Martín Osorio & W. Wildpret (TFC 44983 + Dupl. 1–5); Vertiente Sur de Mña. Guajara (42C 21 b y d), 02/06/2003, L. Afonso, A. Hernández, Y. de Rioja & F. Tapia (TFC 45425).

3.5.38 **Descurainia bourgeauana** ***(E. Fourn.) Webb*** **ex** ***O.E. Schulz***

Engler Pflanzenr., IV, 105(Heft 86): 345 (1924)

Sisymbrium bourgaeanum Webb *ex* Fourn. *Recherch. Crucif.*, 55 (1865)
Hesperis bourgaeana (E.Fourn.) Kuntze *in* Revis. Gen. Pl. 2: 934 (1891)
Descurainia bourgaeana Webb. Bourg., *Pl. Can. Exsics.* 1268 (1855) Nomen nudum.
Descurainia bourgaeana Webb *ex* Christ. *Bot. Jahrb.*, 9: 89 (1888) *in* synon.

Endemism, Certain native
Family: *Brassicaceae*

Common name: Hierba pajonera o amarilla, Pajonera de cumbre (ES); Flixweed (EN)
Biotype: shrub (chamaephyte)
Flowering: April–June
Fruiting: May–July
Pollination: insect pollination (entomogamy)
Dispersal: wind dispersal (anemochory)
Chromosome No: n = 7; 2n = 14

Description

Branched shrub, woody from the base, 25–65 cm tall; young stems erect, greyish, more or less glabrous above, densely leafy below; mature bark wrinkled and fissured.

Leaves in bi-pinnatisect axillary fascicles, ovate, sessile, up to 4 cm long, greyish-green, tomentose with branched hairs interspersed with glandular trichomes, primary segments uncommon, tri-dentate to the apex; secondary segments sparsely acute. Uppermost leaves petiolate, pinnate; segments linear-lanceolate, often tri-dentate.

Inflorescences usually simple, erect, pedicels 6–10 mm long, sparsely pubescent with branched and glandular hairs. Sepals 2.0–2.5 mm with spreading glands and occasionally long, sparsely branched hairs. Petals 3–4 × 1.2–1.6 mm, blade oblong, obtuse, gradually narrowing into a claw about 1/3 petal length. Stamens 2.5–3.0 mm. Silique pedicels patent-ascending (Fig. 3.117), 7–11 mm long; silique 11–17 mm × 0.75 mm, about 16-seeded; septum mononerved, more or less halfway up. Seeds 1.2 × 0.5 mm, brown, uniseriate.

Fig. 3.117 Siete Cañadas, May 2004. (Photo credit: Beatriz Hernández)

Remarks:

Morphologically very similar to D. lemsii but can be distinguished by its unusual primary leaf segments with 3 pointed apices and its patent fruiting pedicels with ascending silique. Type: Tenerife, Filo de las Cañadas, 7.7.1855 Bourgeau 1368 (FI, BM Isotipo).

Distribution

It can be found on the island of La Palma and Tenerife. In the latter it is widely distributed along the summit between 1700 and 3400 m above sea level, from La Crucita (beginning of the National Park boundaries on the La Esperanza Road) to Montaña Abejera, the eastern slopes of Teide, Montaña Blanca and up to the alignment of the Roques de García, both in areas within Las Cañadas and on the Wall and outer slopes. Much more dispersed and less abundant towards the western half of the National Park (Figs. 3.118 and 3.119).

Habitat and Phytosociology

Although it seems to be distributed indiscriminately on different types of substrates, it extends with greater preference on volcanic lapillis and among the scree of phonolitic rocks and debris slopes; in association with ***Spartocytisus supranubius***, ***Argyranthemum tenerifae***, ***Tolpis webbii*** and other local endemic species.

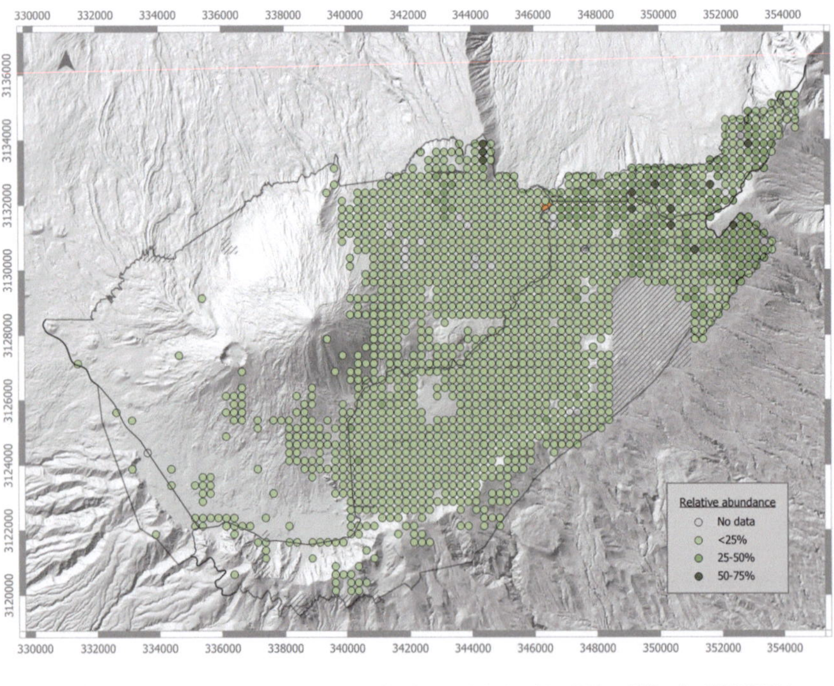

Fig. 3.118 *Descurainia bourgeauana* distribution, N° Subgrids (250 × 250 m): **1742**/3741

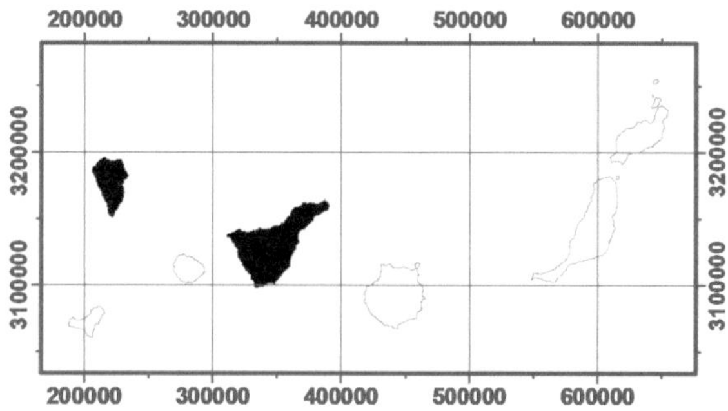

Fig. 3.119 Endemism of the island of Tenerife

This species falls within the domain of ***Spartocytision supranubii***, an alliance that groups the mountain summit shrublands of brooms, flatpods and summit-daisies, climatophilous or edaphoxerophilous, sometimes catenal or serial of the Canary pine forests; of distribution in La Palma and Tenerife, dry-humid Meso-Oromediterranean. Within this, it is characteristic or integral to the following associations:

Descurainio bourgeauanae-Spartocytisetum supranubii is distributed from the La Crucita-Izaña area, almost to the Roques de García, colonizing the pumice substrates and recent materials after the Las Cañadas landslide. This represents the mature stage of the summit brooms, developing preferably on volcanic materials prior to the landslide, belonging to the Upper and Lower Cañadas Building.

Within the ***Pterocephalenion lasiospermi*** suballiance, it also occurs in the following units:

Arrhenathero calderae-Plantaginetum webbii
Tolpido webbii-Micromerietum lachnophyllae
Arrhenathero calderae-Adenocarpetum viscosi

Conservation Status

In 75% (1309) of the subgrids in which this species was found, no damage of any kind was observed. In 1% (17), damage was detected (browsing, high mortality without regeneration, etc.), but this did not affect more than 25% of the specimens. Damage was observed on more than 25% of the specimens in 23% (398) of the subgrids. In the remaining 1% (18) of the sub-grids, this variable was not specified during sampling.

Threat Category

nt (Barreno et al. 1984).
LR pm (Bañares et al. 1998).

Germplasm Banks

E.T.S. Ingenieros Agrónomos de Madrid.
Jardín Botánico Canario Viera y Clavijo.
Real Jardín Botánico de Madrid (2003–2004).

Exsiccata (Herbarium Specimens)

Tenerife: Filo de Las Cañadas, 7.01.1855, Bourgeau (1268); Anaga, 800 m.s.n.m., 24.04.1855, Perraudiére 11204 (all BM); Volcán Teide, 2000 m.s.n.m., 3.07.1855, Perraudiére (s.n.); Las Cañadas, 3.04.1890, Murray (s.n.); Las Cañadas, 3.06.1890, Murray s.n. (all K); El Portillo-Las Cañadas, 20.04.1933, Asplund (864); El Portillo-Las Cañadas-1 km a lo largo de la carretera a Pio (Chio?), 22.06.1956, Lems (2690); Las Cañadas, Hierba Pajonera, Septiembre 1965, Bramwell (45); Mña. Arenas Negras-Las Cañadas del Teide, 2000 m.s.n.m., 22.10.1968, Bramwell (262); Entre el Portillo y El Cabezón, 2000 m.s.n.m., 15.04.1969, Bramwell 1312 (TFC 156); Las Cañadas-desde la base del Volcán Teide, 2300 m.s.n.m., 8.05.1969, Okpon (); 18.06.1969, Bramwell 2102 (TFC 155); La Fortaleza-Las Cañadas, 2100 m.s.n.m., 16.07.1969, Bramwell (2102); Cráter del Teide, 7000 ft, 15.05.1970, Mulford 149, 152 (all RNG); El Portillo-Las Cañadas, 200 (2000?) m.s.n.m., 10.04.1971, Bramwell & Humphries 3383 (RNG, BM); Las Cañadas, 26.05.1972, I. La Serna (TFC 35745); Mirador de Los Roques-Las Cañadas del Teide, 2000 m.s.n.m., 26.07.1972, Bramwell (3517); Las Cañadas del Teide, 2.05.1975, A. Delgado Benítez (TFC 42901); Aguamansa, 5.03.1981, R.Mª. Arnay (TFC 27189); 2.06.1979, E. Beltrán (TFC 7832, 7833, 7834, 7835, 7836, 7837, 7838, 7839, 7840, 7841); Güímar, 1700 m.s.n.m., 31.05.1993, M. del Arco Aguilar (TFC 36003); Izaña-Güímar, 2000 m.s.n.m., 31.05.1993, ejusd. (TFC 36005). Inventario de la Flora y Vegetación del Parque Nacional del Teide (2001–2003) V.E. Martín Osorio, W. Wildpret & B. Hernández Bolaños: Las Cañadas—¿Montaña Colorada?, 2069 m.s.n.m., 02.04.2001, L. Afonso, A. Hernández, Y. de Rioja & F. Tapia (TFC 44968); Cumbre Montaña de la Negrita, 2227 m.s.n.m., 03.04.2001, L. Afonso, A. Hernández, Y. de Rioja & F. Tapia (TFC 44969); Parque Nacional del Teide (Km. 31), 03/04/2001, Y. de Rioja Santos (TFC 44970); A unos 500 m. antes de la curva de "La Tarta" (Las Cañadas), 04.04.2001, L. Afonso, A. Hernández, Y. de Rioja & F. Tapia (TFC 44971); Final Pista 1, cerca Monumento Amnistía Internacional, 04.04.2001, F. Tapia (TFC 44973); Las Cañadas, Km. 34, 04.04.2001, L. Afonso, A. Hernández, Y. de Rioja & F. Tapia (TFC 44974); Lomo Tieso, por debajo del camino que va al Refugio de Altavista (35B 40), 14/05/2003, A. Montañés Rodríguez (TFC 44294); Siete Cañadas, zona de Diego Hernández, 12/05/2004, B. Hernández Bolaños, V.E. Martín Osorio & W. Wildpret de la Torre (TFC 45434); Cráter de la Montaña de Arenas Negras, Siete Cañadas (30D 33), 15/06/2005, V.E. Martín Osorio & W. Wildpret de la Torre (TFC 44990).

3.5.39 Descurainia gonzalezii *Svent.*

Bol. Inst. Nac. Inv. Agron., XIII (28) 17–19 (1953)

Sisymbrium gonzalezi Svent. Flora Selecta Canariensis, Cent. III núm. 218, 1950.

Endemism, Certain native
Family: *Brassicaceae*
Common name: Hierba amarilla o Pajonera de Pinar (ES); Pine forest Flixweed
Biotype: subshrub (chamaephyte)
Flowering: April–May
Fruiting: June–July
Pollination: insect pollination (entomogamy)
Dispersal: wind dispersal (anemochory)
Chromosome N°: 2n = 21

Description

Subshrub, 100–150 cm tall, very branched and woody. Erect, rigid branches, bark of young branches yellowy-silver.

Stem **leaves** dense, somewhat covered with withered leaves; bark of adult branches yellowy-grey. Flowering stems very long, very rigid and with the upper part half-naked. Leaves of the flowering stem linear-filiform, very or slightly pinnate, sharp, 3–5 cm long, 1 mm wide, white vein; in the adult stem, dense fascicles and protected axil.

Inflorescence barely branched, erect branch (Fig. 3.120); flowers densely clustered from the apex, yellow; elliptical bud not exceeded by the flowers; pedicel filiform, greyish-tomentose, three times longer than the sepal; Sepal 3 mm long and 1 mm wide, linear-lanceolate, obtuse, externally powdery-tomentose, erect-patent; petals barely twice as long as the sepal, limb oblong, obtuse, slightly undulated, uniformly sericeous; stamens 2.5–3.0 mm long, ovary approximately 2 mm long, slightly ovulate (6–12); style 1 mm.

Pedicels and **siliques** applied to the fruiting branches; siliques glabrous, linear-cylindrical, erect-arched, approximately 2 cm long and 1 mm in diameter, gradually attenuated, golden; seeds linear-oblong, compressed, slightly furrowed by a small granular striation, reddish in colour.

Distribution

It is found exclusively on the island of Tenerife. In the Teide National Park, we can find it widely on the southeastern outer slopes of the Pared de Las Cañadas; more concentrated in the Cañada de Diego Hernández, Risco Verde, near the Parador Nacional, northern sector of the Roques de García, surroundings of Boca Tauce and Cañada del Cedro. It extends from 2100 to 2500 m a.s.l. (Figs. 3.121 and 3.122).

Fig. 3.120 Siete Cañadas, western slope of Risco Verde, May 2004. (Photo credits: Beatriz Hernández)

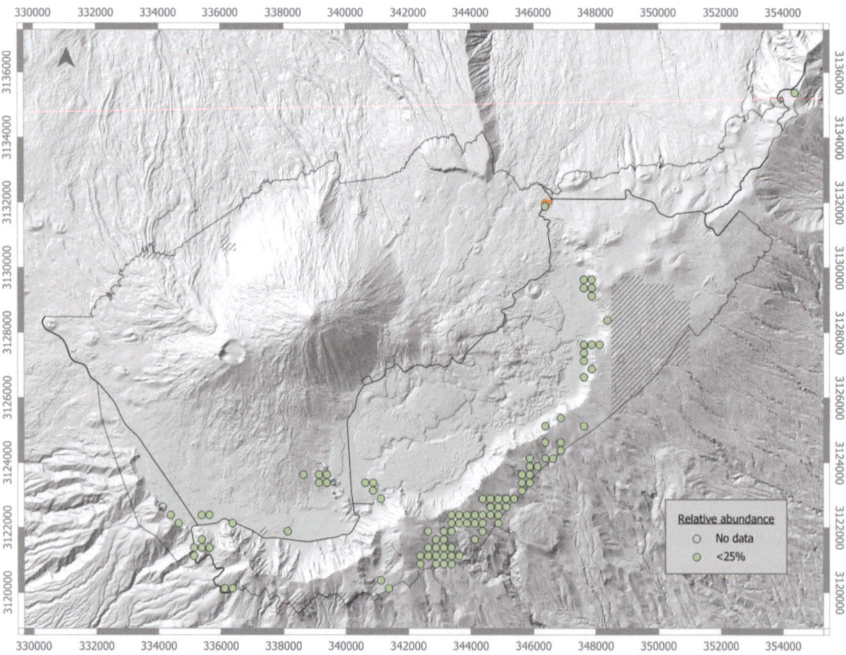

Fig. 3.121 *Descurainia gonzalezii* distribution, N° Subgrids (250 × 250 m): **98**/3741

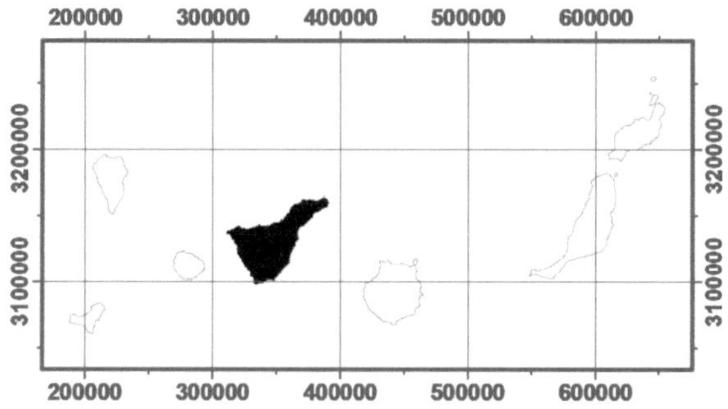

Fig. 3.122 Endemism of the island of Tenerife

Habitat and Phytosociology

It usually colonizes monolithic pyroclastic substrates, mostly welded in the form of flows (ignimbrites). It also grows among volcanic slags and dry stony ground.

This species appears within the ***Chamaecytiso-Pinetalia canariensis*** domain, which groups the pine and broom shrublands of the western Canary Islands. Within this order, it characterizes the dry upper Mesomediterranean subassociation of the retamares of the precaldera wall or building on its southern slope: ***Spartocytisetum supranubii-descurainietosum gonzalezii***.

It is also found in the community described as Cerrillar-Crespar, which preferably colonizes the slopes of gelifract debris of the Las Cañadas wall: ***Arrhenathero calderae-Plantaginetum webbii***.

Conservation Status

In 91% (89) of the subgrids in which this species was found, no damage of any kind was observed. Only in 5% (5), damage was detected (browsing, high mortality without regeneration, etc.), but this did not affect more than 25% of the specimens. Damage was only observed on more than 25% of the specimens in 2% (2) of the sub-grids. In the rest (2), this variable was not specified at the time of sampling.

Threat Category

"Vulnerable" (García Gallo et al. 1993).
V (Barreno et al. 1984).
Anexo II: Protegidas (Gobierno de Canarias 1991).
LR ca (Bañares et al. 1998).
V (Walter and Gillett 1998).

Germplasm Banks

E.T.S. Ingenieros Agrónomos de Madrid.

Exsiccata (Herbarium Specimens)

Tenerife: 14.04.1969, Bramwell 1280 (TFC 158); 30.04.1977, P.L. Pérez de Paz & C. Hernández (TFC 6443, 6499 + Dupl.); Inmediaciones del Campamento Madre del Agua-Granadilla—Vilaflor, 30.04.1979, P.L. Pérez & C. Hernández (TFC 6443, 6499); Pista forestal del Lomo Blanco-Vilaflor, 1600 m.s.n.m., 13.05.1979, ejusd. (TFC 8028, 8029, 8030, 8031, 8032); Pared Cañada de Diego Hernández-Arenas Negras-Ladera de escombros, 6.06.1982, T. Bravo (TFC 23764 + Dupl.); Los Picachos-Pinares de Arico, 9.03.1990, P.L. Pérez de Paz & M. del Arco Aguilar (TFC 37864 + Dupl.); Pista Vilaflor-El Río cerca campamento, 1600 m.s.n.m., 25.04.1990, M. Salas Pascual & M. del Arco Aguilar (TFC 37884). Inventario de la Flora y Vegetación del Parque Nacional del Teide (2001–2003) V.E. Martín Osorio, W. Wildpret & B. Hernández Bolaños: Al NE de la Tarta (25C 5 b), 2204 m.s.m., 03.04.2001, L. Afonso, A. Hernández, Y. de Rioja & F. Tapia (TFC 44966); Degollada de Guajara (42C 5 b), 22.05.2003, A. Montañés Rodríguez (TFC 44275); Por debajo de El Sombrero, hacia el Sur (41D 41 c), 29.04.2003, Y. de Rioja Santos (TFC 44857); Orientación Norte, 03/04/2001, L. Afonso, A. Hernández, Y. de Rioja & F. Tapia (TFC 45408); Risco Verde, Siete Cañadas, 12/05/2004, B. Hernández Bolaños, V.E. Martín Osorio & W. Wildpret de la Torre (TFC 45433); [Como D. cf. gonzalezii] Al Oeste de la Montaña del Roquillo o Volcán de Fasnia (37A 3 a), 09/05/2001, L. Afonso, A. Hernández, Y. de Rioja & F. Tapia (TFC 44902).

3.5.40 Descurainia lemsii *Bramwell*

Cuad. Bot. Canar., XVII, 19–26 (1973)

Endemism, Certain native
Family: *Brassicaceae*
Common name: Hierba amarilla o pajonera—Pajonera de Ayosa (ES); Ayosa's Flixweed (EN)
Biotype: subshrub (chamaephyte)
Flowering: March–July
Fruiting: May–September
Pollination: insect pollination (entomogamy)
Dispersal: wind dispersal (anemochory)
Chromosome N°: n = 7; 2n = 14

Description

A loose-growing **subshrub**, 50–75 cm tall, with erect branches, brown bark, with a short silvery pubescence on younger branches.

Leaves bipinnatisect, suberect, lanceolate, 2–4 (−6) × 0.8–1.0 cm, sessile or shortly petiolate, densely pubescent with branched hairs only; primary lobes linear or linear-lanceolate, 7.0 × 1.0–1.5 mm, shortly petiolate; secondary lobes linear to oval-lanceolate up to 2 mm, those near the apex slightly larger, acute. Upper leaves pinnate and sessile; linear-lanceolate lobes, those near the base filiform.

Inflorescences simple or sparsely branched, erect (Fig. 3.123). Pedicels up to 15 mm long, provided with branched and glandular hairs. Erect patent sepals of 3–4 mm, pubescent with glandular and branched hairs. Petals broadly ovate 5–6 × 3–4 mm, cuneate with a nail equal to the blade. Stamens 4–5 mm, stigma short and capitate, pedicels of the silique erect.

Siliques 20–35 × 8 mm, torulose, often irregularly curved, with 28–32 seeds. Seeds ovate, 1.5 mm, brown, uniseriate.

Remarks:

In the original description of the taxon, the number of seeds per silique is 14–16, while in the revision of ***Descurainia*** published in *Botanica Macaronesica*, 4, 47 (1977), the number of seeds reported is 28–32.

Type: Tenerife, Lomo de Pedro Gil, 1800 m. 14.4.1969 Bramwell 1299 (RNG).

Distribution

It is found only on the island of Tenerife, on high slopes of the dorsal mountain range and areas at the beginning of the Teide National Park on the La Esperanza road, between 1600 and 2375 m above sea level. Within the protected area it is found naturally in the area of La Crucita and near the Ladera de la Mesa or Montaña Marqués, north of Izaña. Also in the area of El Portillo Bajo and the Teide Botanical Garden (Figs. 3.124 and 3.125).

Habitat and Phytosociology

Heliophilous species, with certain nitrophilous appetites, which preferentially colonizes substrates with accumulations of loose volcanic material of the caliber of the lapillis or recent cones of filling of the formation of the La Esperanza Ridge. It is usually found in dispersed populations in open places at the upper limit of the Pinar,

Fig. 3.123 Surroundings of La Crucita, May 2005. (Photo credit: Victoria Eugenia Martín Osorio)

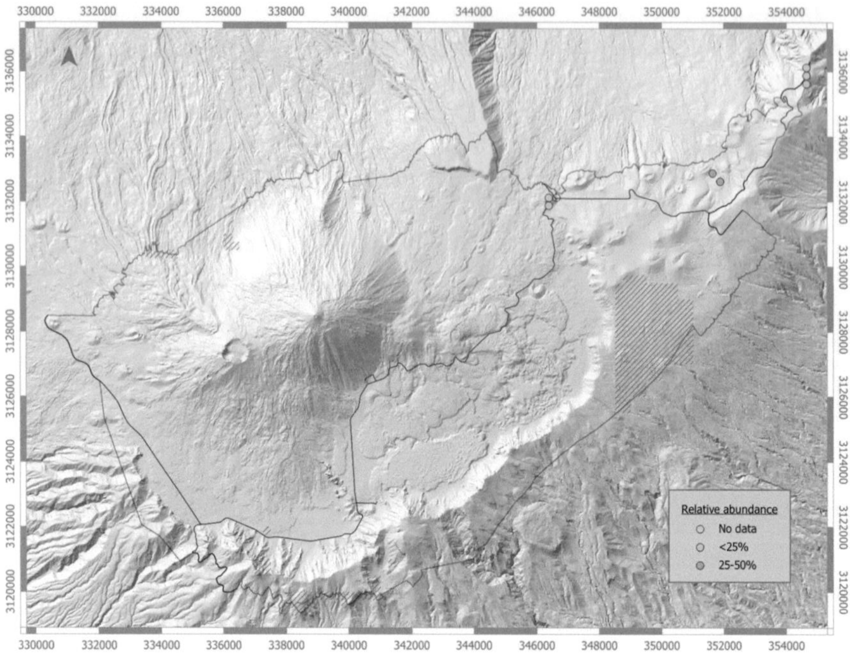

Fig. 3.124 *Descurainia lemsii* distribution, N° Subgrids (250 × 250 m): **8**/3741

Fig. 3.125 Endemism of the island of Tenerife

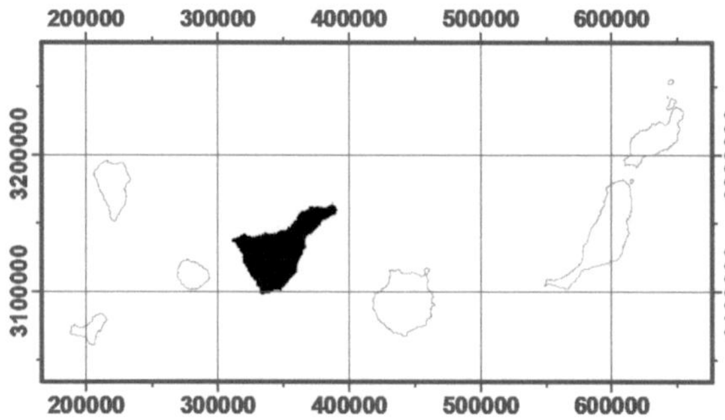

accompanying the summit brooms (dry mesophytic Mesomediterranean bioclimatic belt).

It is a characteristic species and bioindicator of ***Cisto symphytifolii-Pinion canariensis***, an alliance that groups closed climatophilous and edaphoxerophilous meso-macroforests, well characterized by the Canary Island pine, dry-humid Meso-Supra-Mediterranean, rarely sub-cloudy, with a western Canary Island distribution.

It is also characteristic of the ***Spartocytisetum supranubii-descurainietosum lemsii*** subassociation, which behaves as an introgression of the summit brooms in the dry upper Mesomediterranean belt in the eastern limits of the National Park (Martín Osorio et al. 2007).

Conservation Status

In 75% (6) of the subgrids in which this species was found, no damage of any kind was observed. In 12.5% (1), damage was detected (browsing, high mortality without regeneration, etc.), but this did not affect more than 25% of the specimens. In the remaining 12.5% (1), this variable was not specified at the time of sampling.

> "*Species with a not very extensive distribution area, but frequent locally and, for the moment, without significant risks for its populations.*" (Translated from: Gómez Campo et al. 1996)

Threat Category

nt (Barreno et al. 1984).
VU D2 (Bañares et al. 1998).
VU D2 (VVAA 2000).
VU (Bañares et al. 2004).
VU D2 (Moreno 2008).

Germplasm Banks

E.T.S. Ingenieros Agrónomos de Madrid.
Real Jardín Botánico de Madrid (2003–2004).
Jardín Botánico Canario Viera y Clavijo.

Exsiccata (Herbarium Specimens)

Tenerife: Borde superior del bosque de pinos sobre Aguamansa-Montaña Ayosa, 1950 m.s.n.m., 12.11.1965, Lems: 6269 (RNG); Ibid., 2.02.1966, Lems: 6905 (NRG); Lomo de Pedro Gil, 1800 m.s.n.m., 14.04.1969, Bramwell: 1299 (RNG); Borde superior del bosque de pinos sobre Aguamansa, 1800–2000 m.s.n.m., 14.04.1969, Bramwell: 1280 (RNG); Ibid., Nr. Izaña, 2000 m.s.n.m., 30.05.1969, Bramwell: 2284 (RNG); Ibid., Fuente de Joco, 1900 m.s.n.m., 2.04.1971, Bramwell: 3184 (RNG, BM); Aguamansa, 15.06.1973, H. Metlesics (TFC 4380); Ibid., 22.05.1977, I. La Serna (TFC 16888); Los Órganos-La Orotava, 26.06.1979, J.R. Acebes (TFC 20683 + Dupl.); Valle del Volcán de las Arenas-Cumbres de Arafo, 9.05.1981, P.L. Pérez de Paz, W. Wildpret & M. del Arco Aguilar (TFC 23734 + Dupl., 23755 + Dupl.); Cumbre de Pedro Gil-Ayosa, 20.06.1985, P.L. Pérez de Paz, I. La Serna & IMM (TFC 24750 + Dupl.); Izaña-Güímar, 1700 m.s.n.m., 31.05.1993, M. del Arco Aguilar (TFC 36004). Inventario de la Flora y Vegetación del Parque Nacional del Teide (2001–2003) V.E. Martín Osorio, W. Wildpret & B. Hernández Bolaños: Corona Forestal, Pinar. La Esperanza—Cañadas, 11.04.2001, V.E. Martín Osorio & W. Wildpret de la Torre (TFC 44908 + 5 Dupl.); La Crucita, 03.04.2001, Y. de Rioja (TFC 44967); Cerca de La Tarta, 02.07.2003, F. Mesa, A. Montañéz & Y. de Rioja (TFC 44866); Subida de La Tarta, 12/05/2004, B. Hernández Bolaños, V.E. Martín Osorio & W. Wildpret de la Torre (TFC 45436);

Cerca de carretera de zona de La Crucita, 12/05/2004, B. Hernández Bolaños, V.E. Martín Osorio & W. Wildpret de la Torre (TFC 45438); Carretera de La Esperanza—Pinar (fuera del Parque), 10/05/2005, B. Hernández Bolaños, V.E. Martín Osorio & W. Wildpret de la Torre (TFC 45452); Carretera de La Esperanza—Pinar (fuera del Parque), 1890 m.s.n.m., 10/05/2005, B. Hernández Bolaños, V.E. Martín Osorio & W. Wildpret de la Torre (TFC 45453 + Dupl.).

3.5.41 Dittrichia viscosa *(L.) Greuter*

Exsicc. Genav., 4, 71 (1973)

Conyza viscosa (L.) Steud. *in* Nomencl. Bot., ed. 2, 1: 415 (1840)
Cupularia viscosa (L.) Godr. & Gren. *in* Fl. France Corse 2: 181 (1850)
Erigeron viscosus L. *in* Sp. Pl.: 863 (1753)
Helenium viscosum (L.) Kuntze *in* Revis. Gen. Pl. 1: 343 (1891)
Inula viscosa (L.) Aiton *in* Hort. Kew. 3: 223. (1789)
Paniopsis viscosa (L.) Raf. *in* Fl. Tellur. 2: 49 (1837)
Pulicaria viscosa (L.) W.D.J.Koch *in* Syn. Fl. Germ. Helv.: 361 (1837)
Solidago viscosa (L.) Lam. *in* Fl. Franç. 2: 144 (1779)

Not endemic, Probable native
Family: *Asteraceae*
Common name: Altabaca, Mosquera, Olivarca, Olivarda, Tárrago, Altamaca (ES); Yellow fleabane (EN)
Biotype: subshrub (chamaephyte)
Flowering: August–November
Fruiting: August–November
Pollination: insect pollination (entomogamy)
Dispersal: wind dispersal (anemochory)
Chromosome N°: 2n = 18, 24

Description

Sub-shrub, with stems up to 150 cm, erect or ascending, very branched at the base.

Leaves (10–) 20–80 × (−100) × 3 (3–) 6–27 mm, oblong-lanceolate, sub-entire to markedly closed; the cauline leaves semi-amplexicaule.

Involucre (5.5–) 6.5–10 mm, densely pubescent-glandular, sticky. Hemiligulate flowers (7–) 8–11.6 mm, yellow sometimes with a slightly purple-tinged back (Fig. 3.126), extending beyond the involucre. Florets 5–7 mm, yellow rarely tinged with purple. Achenes of the flowers are flossulate, 2–2.6 (3) mm, with a pappus of 5–6.2 (−6.5) mm.

Fig. 3.126 *Dittrichia viscosa*. (Photo credit: GG Casanova)

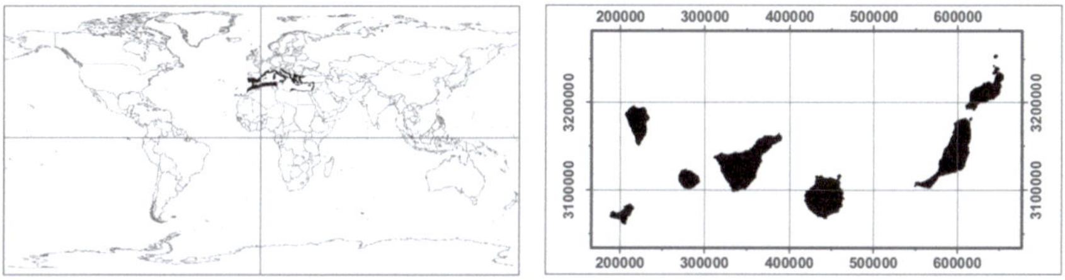

Fig. 3.127 Present on all the Canary Islands

Distribution

Widely distributed throughout the Mediterranean region, the Azores, Madeira and the whole of the Canary Islands (Fig. 3.127). In the Teide National Park (Fig. 3.128) two very localised but quite large populations were found: one in the Roques de García and another at the head of the Barranco del Río. Occasionally in the Cañadas de La Angostura and Diego Hernández (at the foot of the slopes of the Pared) and, outside of Las Cañadas, in Montaña Limón (highlands of the La Orotava Valley); between 2050 and 2400 m a.s.l.

It can be considered an introduced species to the Park, the areas most affected by this species in the National Park are usually those most frequented by visitors and around urbanised areas, roads and viewpoints; also the humid areas such as the Headwaters of the Barranco del Río and some natural springs, since they are frequented by animals that contribute to the dispersion of these species and especially,

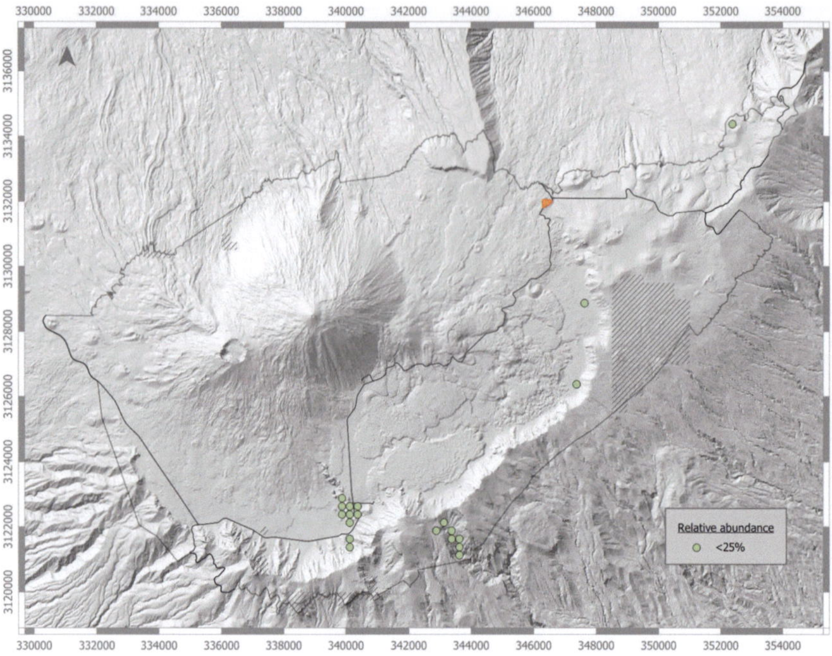

Fig. 3.128 *Dittrichia viscosa* distribution, N° Subgrids (250 × 250 m): **20**/3740

of herbs (Martín Osorio et al. 2006). The distribution of *Dittrichia viscosa* represents this phenomenon very well.

Habitat and Phytosociology

Ruderal plant, of humid environments, in grasslands, abandoned farms, roadsides and ditches, disturbed, stony and uncultivated lands; it prefers a certain soil humidity.

It is a characteristic species and bioindicator of ***Anisantho madritensis-Piptatherion miliacei***, an alliance that groups the associations of gramini-perennial herbaceous plants, structured by mijeras (large-sized perennial tussock grasses of the genus *Piptatherum*), which develop on deep soils in ruderal-nitrophilous environments and abandoned crops; of circum-Mediterranean distribution with Cantabrian-Atlantic and Canary-Madeiran Infra-Meso-Mediterranean semi-arid-sub-humid disjunctions.

Conservation Status

In practically all the subgrids in which this species was found (17), no damage of any kind was observed. Damage was found on more than 25% of the specimens in only three (3) of them.

Remarks:

According to Martín Osorio et al. (2006), in the years 2003–2005, technical personnel from the Teide National Park developed a "Program for the eradication of introduced flora" with 67 catalogued taxa on which a total of 34 are being acted upon with elimination programs, with ***Avena barbata***, ***Bromus tectorum***, ***Chenopodium album***, ***Chenopodium vulvaria***, ***Dittrichia viscosa***, ***Lactuca serriola***, ***Marrubium vulgare*** and ***Petroselinum crispum*** standing out for their abundance and distribution.

Germplasm Banks

Real Jardín Botánico de Madrid.

Exsiccata (Herbarium Specimens)

Tenerife: Valle de Tahodio, 12.8.1972, W. Wildpret & E. Beltrán Tejera (TFC 27173; 27174); Bco. de Añavingo, 26.5.1973, E. Beltrán Tejera (TFC duplic. 27165); Los Silos, Mte. del Agua, 17.8.1973, I. La Serna (TFC 23564 + duplic.); La Laguna (parasitada), 20.5.1974, J.D. Rodrigo (TFC 16916); El Escobonal, 9.10.1976, O. Rodríguez Delgado (TFC 27680); La Laguna, 14.10.1981, A. García Gallo (TFC duplic. 22088; 22089); Las Cañadas, Bco. de los Azulejos, 2050 m.s.n.m., 3.10.1982, J.S. Socorro (TFC 18455); Anocheza (huertas abandonadas), 4.5.1984, O. Rodríguez Delgado (TFC duplic. 27669); La Laguna, 4.5.1985, P.L. Pérez de Paz (P-TFC:496; TFC 13330 + duplic.); La Laguna, 27.8.1985, A. García Gallo & J.M. González (TFC 22722 + duplic. 22723); Ibid, 27.8.1985, ejusd. (TFC 22724); Los Silos, la Costa, 27.7.1995, V. Lucía Sauquillo & C. Acosta (TFC 38037 + duplic.); Bco. de Tahodio, Riscos en el Sur de Jardina, 11.4.1999, S. Becher & T. Himstedt (TFC 42267); La Laguna, Mña del Aire, 27.9.2001, J.R. Acebes (TFC 43296 + duplic.); AHIM Cent VI (2001) La Laguna, Mña. del Aire, 650 m.s.n.m., 27.9.2001, J.R. Acebes (TFC 43495); Bco. Hoya de las Limeras, 500 m.s.n.m., UTM 28RCS5949, 17.11.2002, M. Rodríguez & M.L. Rodríguez Navarro (TFC 44732).

La Palma: Velhoco, 11.1987, FCR (P-TFC: 538; TFC 25587).

Gran Canaria: Telde, Ctra. Cumbre, 15.6.1985, IMM (TFC 24578); Parque Natural de Guayadeque, cauce del Bco. a la altura de Cueva Labrada, 700 m.s.n.m., 13.10.1990, F. Amor & P.L. Pérez de Paz (TFC 30823).

Fuerteventura: Isla de Lobos, 26.9.1994, J.A. Reyes-Betancort (TFC duplic. 37941); Valle de Tetir, 27.9.1994, ejusd. (TFC 37942); San Bartolomé, detrás de Tamía, UTM 28RFT63332108, 21.12.1994, J.A. Reyes-Betancort (TFC 37451).

(¿?), 12.1968, A. Santos (TFC 251).

Uses, Exploitation and/or Medicinal Properties

> "*Vulnerary, diuretic, febrifuge, analgesic, astringent, stomachic, antirheumatic; toxic: insecticide. Useful part: flowering tops. Instructions for use: herbal tea, rubbing, infusion: foot baths, eye washes.*" (Translated from: Pérez de Paz and Hernández Padrón 1999)

3.5.42 Echium auberianum *Webb & Berthel.*

Phyt. Canar., 3(2;3): 42 (1844)

Echium bourgaeanum Webb. sensu Christ, *Bot. Jahrb.*, 9: 126 (1888), in syn., non Webb *ex* Coincy (1903).
Echium mordax Solander, *in* sched.

Endemism, Certain native
Family: *Boraginaceae*
Common name: Taginaste picante o Taginaste azul (ES); Spicy or Blue bugloss (EN)
Biotype: hemicryptophyte
Flowering: May–June
Fruiting: July–August
Pollination: insect pollination (entomogamy)
Dispersal: animal dispersal (zoochory)
Chromosome No: –

Description

Simple, **erect stem**, 1.2–1.5 m tall, about 7 cm in diameter at the bottom, thick, herbaceous, hairy, fistulous, leafy.

Basal stem leaves very close together, densely imbricated, rosette-shaped; erect, 16–20 × 0.7–1.5 cm, petiolate; divided on sides, linear-lanceolate, apex acute, with narrow petiole, semi amplexicaule, whitish hairy on both sides; protruding base densely covered, the lower vein prominent petiolate, densely provided on the underside with long, erect hairs, stem leaves, sometimes canaliculate, with apex that gradually reduces to the stem; leaves of the flowering branches more glabrous.

Flowers densely arranged in a large thyrsoid panicle (Fig. 3.129), the density of the flowers in an orderly manner, in alternate simple spikes or else bifid scorpioid. Elegantly revolute before flowering, successive from the bottom to the top; flowers supported by a short peduncle 1.25 cm long, each supported by short pointed lanceolate-linear divided bracts. Flowers in alternate spikes, distichous, continuous, dense, subsessile, tribracteolate, linear acute bracts equaling the calyx or else surpassing it. Calyx campanulate, 8–10 mm, 5-parted, in unequal parts, the two largest parts narrow and the smallest linear-lanceolate, acute, the lower middle part equaling the corolla, densely hairy on both sides, flavescent dried hairs. Corolla bluish 10–15 mm, infundibuliform, wide mouth, 5-lobed limb, rounded subequal lobes, pubescent on the outside, half an inch long; tube from base with annularly arranged hairs. Stamens included or occasionally slightly exceeding corolla; lower part of tube inserted on corolla; filaments unequal subulate, lower part inflated, whitish; anthers ovate, bilocular, locules parallel equal, rimula dehiscent longitudinally. Arched styles exceed stamens, hairy, upper part glabrous, highly bifid, whose parts are subequal and subulate. Ovary quadrilobular, lobes hemispherical, hypogynous fleshy disk mobile. The nutlet is often divided into three parts, ovate-trigonal, acuminate, lateral parts denticulate, on the back the nerve is prominent subcarinate, scrobiculate, cinereous. Fruits are brown, 2–2.5 mm, rough and conical.

Fig. 3.129 Surroundings of the volcano Siete Fuentes, May 2005. (Photo credit: Beatriz Hernández)

Remarks:

During the field sampling carried out for the project, a hybrid was found between this species and ***Echium wildpretii ssp. wildpretii*** recently described as ***Echium × hermanni*** (*Echium wildpretii × auberianum*) (Martín Osorio et al. 2015).

Distribution

Endemic species of the peaks of the island of Tenerife, mainly distributed in sectors of the northeastern half of the Cañadas del Teide, between 1950 and 2800 m a.s.l.: more continuously, from the area of Izaña, extending towards the interior of the Depression, to the vicinity of Risco Verde, Minas de San José and Montaña Blanca; more discontinuously, in a population that extends from Morro de La Angostura towards Topo de la Grieta at the top of La Pared; and more punctually to the west of Montaña Abejera Grande and to the north of Roques de García. Also, within the Teide Botanical Garden.

In recent years, a certain expansion towards the interior of the Las Cañadas caldera and towards the upper part of Montaña Blanca has been observed, although it was already cited by Lindinger (1926) in this place, as well as in La Fortaleza (where it was not found during sampling) and in the Güímar Valley at 2200 m a.s.l. outside the limits of the National Park (Figs. 3.130 and 3.131).

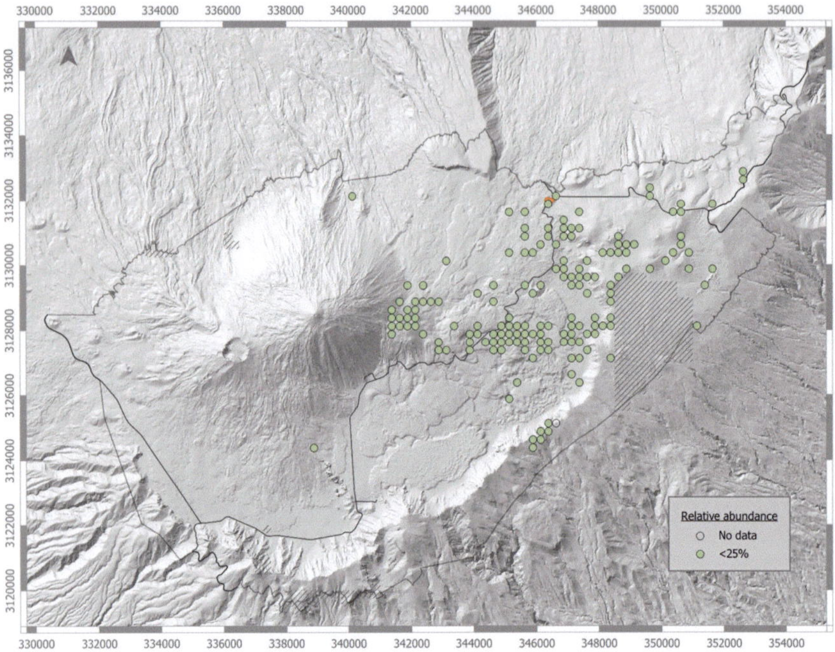

Fig. 3.130 *Echium auberianum* distribution, N° Subgrids (250 × 250 m): **160**/3741

Fig. 3.131 Endemism of the island of Tenerife and doubtful record in La Gomera

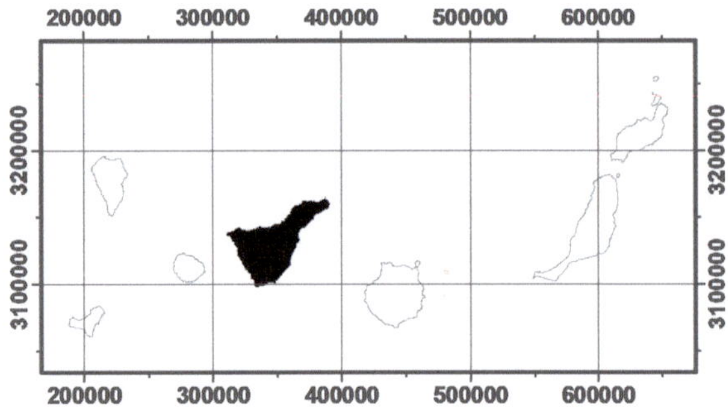

Habitat and Phytosociology

It is usually found in cinder cones and pumice slopes. It develops very well in areas of pumice soil with very sparse vegetation. Sometimes it can be found growing on land formed by alluvial deposits and colonizing some volcanic cones still devoid of plant cover. It is part of the communities of the ***Spartocytision supranubii*** alliance, Teide brooms or lower dry-humid Supra-Oromediterranean reeds of Tenerife.

It is a characteristic species of the ***Pterocephalenion lasiospermi*** suballiance, which groups permanent and serial dwarf shrublands associations, which colonize volcanic deposits and bare soils of pine forests and brooms, dry-subhumid Meso-Supramediterranean, endemic to the island of Tenerife. Within this, it is a taxon that is part of the ***Echietum auberiani*** association.

Furthermore, it is also a characteristic species and bioindicator of ***Juniperion cedri***, an alliance that groups the open edaphoxerophilous microforests of Canary cedars, which thrive in the leptosols of the escarpments of Las Cañadas del Teide and Caldera de Taburiente, from the upper horizon of the Mesomediterranean belt to the upper dry-humid Supramediterranean belt, outside the usual reach of fog.

Conservation Status

In 81% (130) of the subgrids in which this species was found, no damage of any kind was observed. In 14% (22), damage was detected (browsing, high mortality without regeneration, etc.), but this did not affect more than 25% of the specimens. Damage has only been observed on more than 25% of the specimens in 4% (7) of the sub-grids where it appeared. In the rest (8), this variable was not specified at the time of sampling.

> "*The fact that a large part of its populations are found within the Teide National Park represents a guarantee for the protection of the habitat of this rare endemic species of Tenerife. However, we must highlight that one of the most notable populations is crossed by a forest track with easy access both for unscrupulous visitors who cut its showy inflorescences, and for private seed collectors who then sell them to tourists. In this respect, we hope that the measures adopted by the National Park Management to close this road to vehicle traffic and increase surveillance will stop these actions.*" (Translated from: Gómez Campo et al. 1996)

Threat Category

E (Barreno et al. 1984)
Annex I: Strictly protected (Gobierno de Canarias 1991)
VU D2 (Bañares et al. 1998)
E (Walter and Gillett 1998)
VU D2 (VVAA 2000)
Sensitive to habitat alteration (Gobierno de Canarias 2001)
VU (Bañares et al. 2004)
VU D2 (Moreno 2008)

Germplasm Banks

E.T.S. Ingenieros Agrónomos de Madrid.
Jardín Botánico Canario Viera y Clavijo.

Exsiccata (Herbarium Specimens)

Tenerife: Montaña Blanca-entre escorias y pumitas, s.d., Auber (FI, P); Junto a montañas arenosas, 1778, Masson (BM); Los Azulejos, 2250 m.s.n.m., 22.06.1954, Lindquist (BM); Cañada de las Arenas Negras, Agosto 1965, Bramwell 65 (LIVU); Montaña de las Arenas Negras, 2200 m.s.n.m., 22.10.1968, Bramwell 266 (RNG); Ibid., 30.05.1969, Bramwell 1713 (LTR, RNG, SEV); Cañada de Diego Hernández, 31.07.1971, F. Estévez & A. Acuña (TFC 21940); Topo de la Grieta, 20.06.1982,

E. Beltrán Tejera, W. Wildpret, C. León Arencibia et al. (TFC 19909 + Dupl.); Mña. Mostaza-Las Cañadas, 27.06.1982, RL (TFC 25450).

3.5.43 Echium virescens *DC.*

Cat. Pl. Hortii. Monsp., 107 (1813)

Echium strictum Lam. in Tabl. Encycl. 1: 414 (1792), nom. illeg.
Echium candicans sensu DC. *Prodr.*, 10:17 (1846), pro parte.
Echium virescens var. angustissimum Bolle *ex* Christ *in* Bot. Jahrb. Syst. 9: 127 (1887)
Echium virescens var. *candollei* Coincy. *Bull. Herb. Boiss.*, sér. 2, 3: 269 (1903).
Echium bicolor Webb *in* Bourg. *Pl. Canar. Exsicc.*, 456 (1845), in sched.
Nueva cita para el Parque Nacional del Teide

Endemism, Certain native
Family: *Boraginaceae*
Common name: Taginaste azul de Tenerife (ES); Tenerife Blue bugloss (EN)
Biotype: chamaephyte—nanophanerophyte
Flowering: spring
Fruiting: spring
Pollination: insect pollination (entomogamy)
Dispersal: self or wind dispersal (autochory, anemochory)
Chromosome N°: 2n = 16, 24

Description

Shrub 1–2 m with densely hispid stems and papery bark. Leaves lanceolate, 8–12(−20) × 0.3–1.5 cm, attenuate, shortly petiolate; acute apex; prominent veins on the lower surface; dense indumentum, consisting of simple or rarely tuberculate setae, greyish to silvery-white.

Inflorescences narrowly spike-shaped, 10–30 cm; lateral cymes bifid; peduncle 5–10 mm, elongating in the fruit. Flowers subsessile. Calyx 5 mm, hispid, linear-lanceolate segments, acrescent in the fruit. Corolla pink or bluish (Fig. 3.132), narrowly infundibuliform, 6–10 mm, rounded lobes, obtuse, 2 mm diameter, the central vein hairy; Fleshy annulus, sparsely hairy with wavy margin. Stamens exserted; reddish filaments. Style exserted; apex shortly bifid-pubescent towards the base, 2–3 mm angular, rough nutlets.

Remarks:

var. *angustissimum* Bolle *ex* Christ, Bot. Jahrb. 9: 127 (1888): Leaves linear-lanceolate, up to 0.5 cm wide, the upper surface with pustular trichomes. Corolla shorter (6 mm). Endemic to Tenerife, present from 600 to 2000 m a.s.l.

Fig. 3.132 *Echium virescens*, April 2012. (Photo credit: Victoria Eugenia Martín Osorio)

Distribution

It is found only on the island of Tenerife. According to the bibliography consulted, it is distributed from 500 m a.s.l.: upper part of the Güímar Slope, in the Barranco de Badajoz, next to the gullies; In Agua Mansa, Los Órganos, Anaga, Barranco del Infierno, Valle de la Orotava, Guía de Isora, Monte de Los Silos and Masca. Lindinger (1926) mentions it in Los Realejos, Paso Alto de Santa Cruz, Adeje, Chasna (Bco. de Cuervo) and in the Igueste Valley (Figs. 3.133 and 3.134).

In the Teide National Park it reaches altitudes of 2200 m and was found in Lomo de Las Colmenas (SE-Topo de la Grieta), Barranco del Río and the northern sector of Roques de García (Las Escaleritas). It is a good indicator of the rise of some species to higher altitudes in response to the trend of rising temperatures.

Habitat and Phytosociology

Frequent in forest areas and the lower parts of the southern slopes, generally on cliffs and ravine sides.

It participates in the domains of the ***Cisto-Pinion canariensis*** alliance, which groups the closed climatophilous and edaphoxerophilous meso-macroforests, well

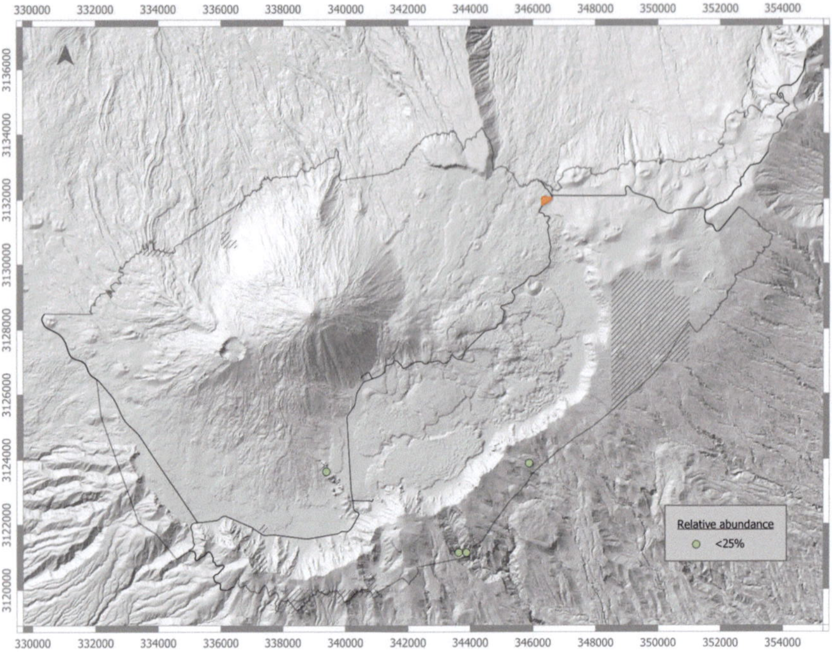

Fig. 3.133 *Echium virescens* distribution, N° Subgrids (250 × 250 m): 4/3741

Fig. 3.134 Endemism of the island of Tenerife and doubtful record in La Gomera

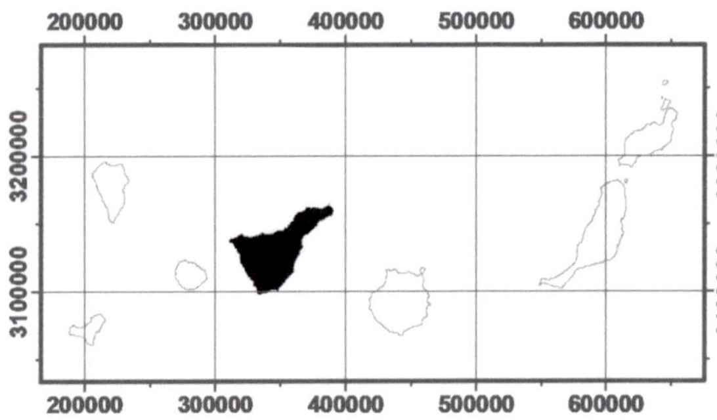

characterized by the Canary Island pine, dry-humid Meso-Supra-Mediterranean, rarely sub-cloudy, of western Canary distribution.

In general, it is found in communities of the ***Mayteno-Juniperion canariensis*** alliance, which groups the climatophilous sclerophyllous and scale-like microforests (wild olive groves and juniper groves), semi-arid-dry Infra-Thermo-Mediterranean, endemic to the Canary Islands, on well-structured but shallow soils.

It is also a member of the ***Kleinio-Euphorbietea canariensis*** class communities, which includes succulent shrublands: cardonales, tabaibales, or climatophilous and edaphoxerophilous toldares, as well as the "tabaibales amargos" and permanent or serial "baleras" in wastelands or alluvial deposits, typical of the arid basal areas of the Canary Islands, with a hyperarid-semi-arid Infra-Thermo-Mediterranean bioclimate.

Conservation Status

No damage of any kind was observed in any of the subgrids in which this species was found.

Germplasm Banks

E.T.S. Ingenieros Agrónomos de Madrid.
Jardín Botánico Canario Viera y Clavijo.

Exsiccata (Herbarium Specimens)

Tenerife: Aguamansa, 21.4.1935, Cuatrocasas (TFC 16509); Canal del Tesoro, Lomo de Mejiniar, 8.4.1972, P.L. Pérez de Paz (TFC 34032); Ctra. El Bailadero, 8.4.1972, W. Wildpret, A. Acuña, P.L. Pérez de Paz, C. Gil, J.R. Acebes & E. Beltrán Tejera (TFC 4463); Anaga, bajo Roque de los Pasos, 25.1.1975, W. Wildpret, J.R. Acebes, C. León Arencibia, E. Beltrán Tejera & I. La Serna (TFC 23476); Ladera de Güímar, 13.5.1975, ADB (TFC 42912); Valle de Masca, E. Beltrán Tejera & A. Bañares (TFC 4928); Bco. de Erques, 3.4.1976, M. del Arco Aguilar, P.L. Pérez de Paz, J.R. Acebes & C. León Arencibia (TFC 4406; 4280); Cumbres de Arafo, 18.4.1979, E. Beltrán Tejera (TFC 7970); Arico, Bco. Albarderos, 1300 m.s.n.m., 1.4.1980, W. Wildpret, J.R. Acebes, M. Arco Aguilar, C. León Arencibia, L. Negrín & AVR (TFC 25140 + 5 duplic.); Ladera de Agache, 6.3.1981, O. Rodríguez Delgado (TFC 12797); Cumbres de Arafo, Valle del Volcán de las Arenas, 9.5.1981, P.L. Pérez de Paz, W. Wildpret & M. Arco Aguilar (TFC 23754 + 10 duplic.); Ibid, 1700 m.s.n.m., 9.5.1981, ejusd. (TFC 13696); Bco. de Erques, 2.4.1984, O. Rodríguez Delgado (TFC duplic. 27611); ?, 12.5.1984, W. Wildpret & A. Charpin (TFC 20086); Arafo, Volcán de las Arenas, Caldera de Pedro Gil, 6.5.1985, M. Arco. Aguilar (TFC 24386 + 3 duplic.); La Ladera, 6.3.1988, O. Rodríguez & W. Wildpret (TFC 27610); Los Silos, Mte. del Agua, pista G. Cochinos, 25.4.1988, V. Lucía Sauquillo & C. Acosta (TFC 38125); Los Silos, Pista de los Madroños, 20.5.1989, V. Lucía Sauquillo & C. Acosta (TFC 38280); Ibid, 2.2.1992, ejusd. (TFC 39257); Bco. de Tahodio, cerca del Roque Lombarda, 27.2.1999, S. Becker & T. Himstedt (TFC 42045); ?, 6.2.1971, W. Wildpret & A. Santos (TFC 178); ?, 11.5.1971, M.O. Pino Tejera (TFC 698); ?, 28.3.1972, P.L. Pérez de Paz (TFC 800); ?, 1978, Bramwell (TFC 177).

3.5.44 Echium wildpretii *Pearson* ex *Hook. f. ssp.* wildpretii

Curtis' Bot. Mag. 128: t., 7847 (1902)

Echium bourgaeanum Webb. *in* Bourgeau, *Pl. Canar. Exsicc*., 895 (1845), in sched.
Echium bourgaeanum Webb *ex* Christ. *Bot. Jahrb*., 9: 126 (1888), in syn.
Echium bourgaeanum Webb *ex* Coincy. *Bull. Herb. Boiss*., sér. 2, 4: 275 (1903)

Echium wildpretii Pearson *ex* Hook fil.

Endemism, Certain native
Family: *Boraginaceae*
Common name: Taginaste del Teide—Taginaste rojo (ES); Red bugloss, Tower of jewels (EN)
Biotype: hemicryptophyte
Flowering: May–June
Fruiting: May–August
Pollination: insect pollination (entomogamy)
Dispersal: animal dispersal (zoochory)
Chromosome N°: 2n = 16

Description

Monocarpic shrub. Stem unbranched, up to 25 cm in vegetative state, 2–3 cm in diameter; dark brown, hispid. **Leaves** densely gathered at the stem apex, up to 50 cm, linear to oblanceolate, petiolate, acute to acuminate, both surfaces densely covered with long tuberculate trichomes; veins very inconspicuous on the underside.

Inflorescence 1.5–3.0 m, dense, thyrsoid (Fig. 3.135), almost conical, the central axis up to 3 cm in diameter; numerous dense lateral cymes, with many flowers. Flowers subsessile. Calyx segments 9–10 mm, ovate or lanceolate, slightly or densely hispid, obtuse to subacute, strongly accrescent in the fruit. Corolla red (turning blue when dry), 10–14 mm, broadly infundibuliform; lobes more or less equal, approximately 3 mm long, rounded, subglabrous to densely hispid; slightly annular, sparsely hairy, unequal, fleshy ring. Stamens exserted, glabrous, reddish. Style exserted, hispid, bifid. Fruit 2 mm, subconical, whitish, rough.

ssp. wildpretii: The inflorescence is wide in the centre; the calyx and corolla are densely hispid; the style deeply bifid.

Remarks:

Once the basal rosette of leaves has developed, flowering occurs 2 or 3 years later. When the inflorescence dries, its structure remains for some time devoid of flowers and yellowish-straw-coloured leaves with a dry conical appearance. In habit, *E. wildpretii* resembles the growth form of Afro-alpine Dendro-Senecio and Lobelia, which, according to Hedberg (1964), is an adaptation to the diurnal temperature range, the dense rosette of leaves closing at night to protect against frost.

A naturally occurring hybrid between ***Echium wildpretii ssp. wildpretii*** and ***E. auberianum*** has been described as ***Echium × hermanni*** (*Echium wildpretii × auberianum*) (Martín Osorio et al. 2015).

Fig. 3.135 *Echium wildpretii* on the slope of the Valle de Ucanca, 28 May 2011. (Photo credit: Victoria Eugenia Martín Osorio)

Distribution

Endemism of the peaks of the island of Tenerife (1900–2500 m above sea level), distributed in a dispersed manner throughout the NE-SW half of the interior of Las Cañadas del Teide, except on the northwestern flank (Edificio Teide-Pico Viejo). In a more concentrated and continuous manner, along the inner and outer slopes of the Wall; also, in La Fortaleza and southern slopes of Izaña (Fasnia Volcanoes). Some spectacular specimens can be found in the Teide Botanical Garden itself (Figs. 3.136 and 3.137).

Habitat and Phytosociology

It colonises rock fields, landslide cones and areas of pumice slopes, fissures and stony ground, and especially in sunny places where it forms large colonies. It is a member of the communities of the mountain legume scrub included in the ***Spartocytision supranubii*** alliance, which groups the high-growing mountain broom and rock-cut shrubs, climatophilous or edaphoxerophilous, sometimes catenal or serial from the Canary pine forests; with a distribution in La Palma and Tenerife, dry-humid Meso-Oro-Mediterranean. Within this, it is a characteristic species of the ***Spartocytisetum supranubii-echietosum wildpretii*** sub-association.

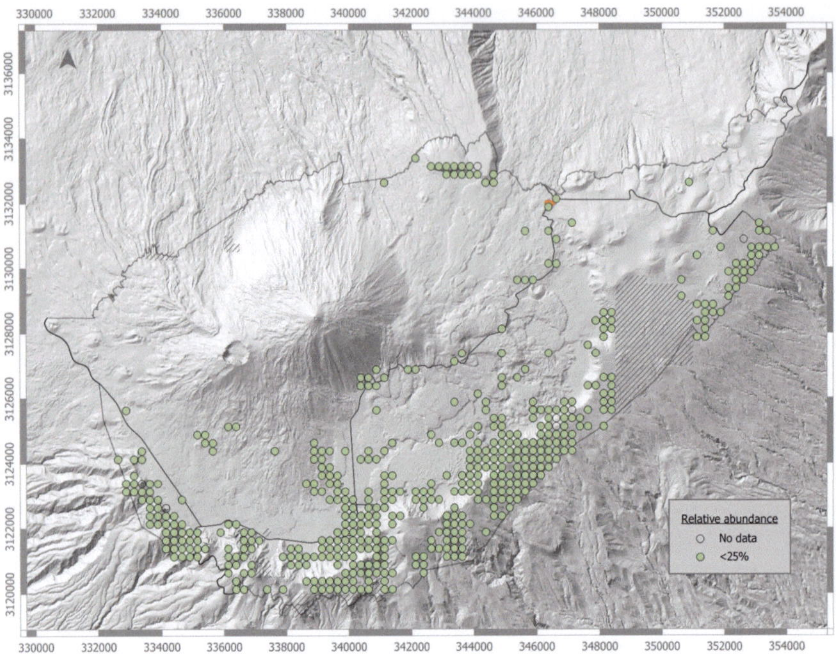

Fig. 3.136 *Echium wildpretii* subsp. *wildpretii* distribution, N° Subgrids (250 × 250 m): 493/3.741

Fig. 3.137 Endemism of the island of Tenerife

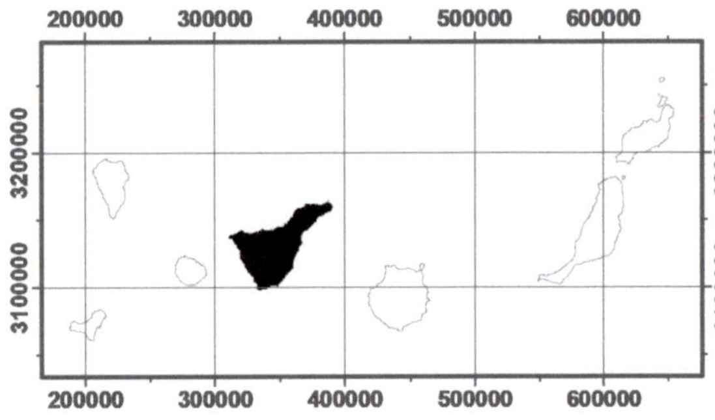

It is also a characteristic species and bioindicator of the ***Juniperion cedri*** alliance, which groups the open edaphoxerophilous microforests of Canary cedars, which thrive in the leptosols of the escarpments of Las Cañadas del Teide and Caldera de Taburiente, from the upper horizon of the Mesomediterranean belt to the upper dry-humid Supramediterranean belt, outside the usual reach of fog.

In addition, it can intervene in the following associations:

The community described as "Cerrillar-Crespar", a phytosociological association of the National Park that preferably colonizes the slopes of gelifract debris of the Pared:

Arrhenathero calderae-Plantaginetum webbii
Cheirolopho teydis-Carlinetum xeranthemoides
Arrhenathero calderae-Adenocarpetum viscosi

State of Conservation

In 97% (478) of the subgrids in which this taxon was found, no damage of any kind was observed. Only in 1.2% (6), damage was detected (browsing, high mortality without regeneration, etc.), but this did not affect more than 25% of the specimens. Only in another 1.2% (6), damage was observed on more than 25% of the specimens. In the rest (4), this variable was not specified at the time of sampling.

> "*Relatively frequent in some localities despite the pressure to which it is subjected by the large number of visitors to the summits of the island. Affected by the introduction of mouflons from the nineteen seventies onwards. The species can be classified as 'R' (IUCN Category: Species with small global populations that cannot be said to be currently 'endangered' or 'vulnerable' but are subject to risk. These species are generally found within limited geographic areas or habitats, or are distributed within a wider geographic area, but in small numbers).*" (Translated from: Gómez Campo et al. 1996)

Threat Category

nt (Barreno et al. 1984)
Anexo II: Protegidas (Gobierno de Canarias 1991)
LR ca (Bañares et al. 1998)
R (Walter and Gillett 1998)

Germplasm Banks

Jardín Botánico Canario Viera y Clavijo.

Exsiccata (Herbarium Specimens)

Tenerife: Ad rup. La Fortaleza del Teide, 5.04.1845, Bourgeau 895 (FI, P); En lo alto de Güímar, 4.04.1855, Perraudière (K); Agua Agria, Chasna, 2000 m.s.n.m., 27.06.1855, Perraudière (P); In rupibus praeruptis Cañadas del Teide, 4.07.1855, Bourgeau 1436 (BM, C, FI, K, P, Z); Filo de Las Cañadas, 2000 m.s.n.m., Junio 1911, Burchard 2 (CGE, Z); Cultivada en la Orotava, 1912, Pérez (K, 10 láminas; FI); Valle de Ucanca 2000 m.s.n.m., 22.05.1957, Larsen (C); Filo de Las Cañadas, 1.06.1963, Lansbohjsk (C); La Fortaleza, 2200 m.s.n.m., Agosto 1965, Bramwell 21 (LIVU); Topo de La Grieta, 2100 m.s.n.m., 2.11.1968, Bramwell 329 (RNG); Pinar de Vilaflor, 1900 m.s.n.m., 7.05.1969, Bramwell 2260 (RNG); Los Azulejos, 23.10.1969, Hansen (C); Cañada Blanca, Roques de García, Los Azulejos, La Fortaleza, 29.07.1971, 5.08.1971, F. Estevez & A. Acuña (TFC 21723 + Dupl.); Inventario de la Flora y Vegetación del Parque Nacional del Teide (2001–2003) V.E. Martín Osorio, W. Wildpret & B. Hernández Bolaños: Guajara, por la zona de El Capricho (42C 1 b), 15/05/2003, A. Montañés Rodríguez (TFC 44853).

3.5.45 Echium × hermanni *Martín Osorio, V.E. & Wildpret de la Torre, W.* (=Echium wildpretii × auberianum)

Makaronesia. 17, pp. 102–119. 2015
Nueva cita para el Parque Nacional del Teide

Endemism, Certain native
Family: *Boraginaceae*
Common name: Taginaste de Hermann
Biotype: phanerophyte
Flowering: May–June
Pollination: entomogamy

Description

A stem-like plant that can reach up to 3 m in height (Fig. 3.138). Inflorescence very similar to ***E. wildpretii* ssp. *wildpretii*** but somewhat looser and with bluish flowers. Basal leaves like ***E. wildpretii***, up to 50 cm linear-lanceolate, densely covered with long tuberculate trichomes, inflorescence up to 3 m, not very dense, thyrsoid, almost conical (Fig. 3.139)

Fig. 3.138 Pista de El Filo (**a**) Sept. 2018. (Photo credit: Javier Martín Carbajal) (**b**) June 2004. (Photo credit: Victoria Eugenia Martín Osorio)

Fig. 3.139 Wolfredo Wildpret Botanical Garden of Teide National Park, June 2003 (**a**, **b**) (Photo credit: Juvenal Soto Martín)

Remarks:

It is similar to the taxon ***E. auberianum*** in the flowers with the blue corolla and in the stamens that are not very prominent from the corolla. In general, as such a hybrid, it presents variable characteristics of the parents but generally, the morphology is of a red bugloss with blue flowers. There is a photographic record by Sergio Socorro in the town of Topo de la Grieta on June 25, 1983.

Type:

TFC 44926, Martín Osorio et al. (2003), Portillo Alto, Teide National Park, Tenerife, 2058 m a.s.l., 28°18′7″N-16°33′16″W.

Distribution

It is only found on the highest peaks of the island of Tenerife, between 2050 and 2350 m a.s.l. During the sampling it was located very punctually around El Portillo Alto, Cañada de Diego Hernández, Minas de San José, Lomo de la Gatera, Barranco del Río (more abundant), SW of Montaña Guajara, Llano de las Mesas and southern slopes of Pico Viejo. It also spontaneously occurs in the Wolfredo Wildpret Botanical Garden of the Teide National Park (Figs. 3.140 and 3.141).

Habitat and Phytosociology

It appears in the vicinity of the places where its parents are also found, ***Echium wildpretii ssp. wildpretii*** and ***Echium auberianum***. It preferably colonizes the substrates of loose volcanic material of a pumice nature and scree of considerable slope.

Characteristic of the ***Chamaecytiso-Pinetea canariensis*** communities, a vegetation class that includes coniferous forests, broom and serial colonizing

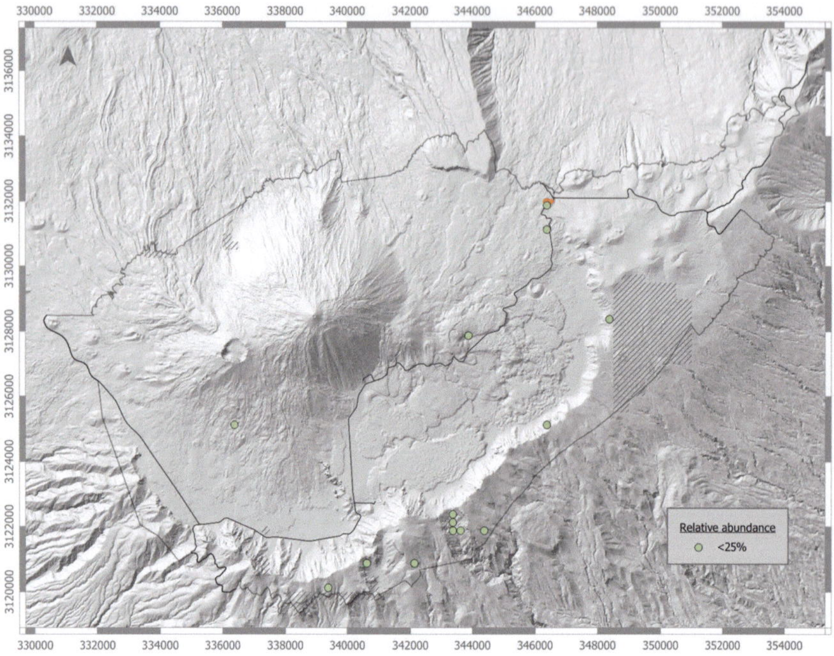

Fig. 3.140 *Echium × hermanni* distribution, N° Subgrids (250 × 250 m): **14**/3741

Fig. 3.141 Endemism of the island of Tenerife

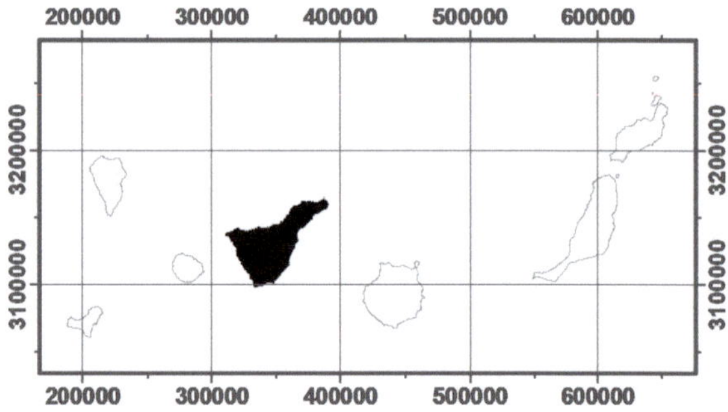

micro-brooms of Tenerife and La Palma, as well as oro-mediterranean dwarf shrublands of the Teide volcanic cone; climatophilous, edaphoxerophilous, pyrophilous or serial, endemic to the midlands and summits of the Central-Western Canary Islands, typical of dry-humid Meso-Oro-Mediterranean bioclimates, of the Western Canary Islands biogeographic subprovince.

It forms part of the summit legume thicket included in the ***Spartocytision supranubii*** alliance, which includes the climatophilous or edaphoxerophilous summit shrublands of broom and flatpod, sometimes catenal or serial, of the Canary pine forests; of distribution in La Palma and Tenerife, dry-humid Meso-Oro-Mediterranean.

Conservation Status

In practically all the subgrids in which this taxon was found, no damage of any kind was observed.

Exsiccata (Herbarium Specimens)

Tenerife: Inventario de la Flora y Vegetación del Parque Nacional del Teide (2001–2003) V.E. Martín Osorio, W. Wildpret & B. Hernández Bolaños: Next to the fountain in the Botanical Garden Visitor Centre, 13/11/2003, V.E. Martín Osorio, W. Wildpret de la Torre & B. Hernández Bolaños (TFC 44926).

3.5.46 Ephedra nebrodensis *Tineo*

First published *in* G.Gussone, Fl. Sicul. Syn. 2: 638 (1844)

Ephedra equisetiformis subsp. *nebrodensis* (Tineo) Bonnier & Layens *in* Tabl. Syn. Pl. Vasc. France: 374 (1894)
Ephedra major var. *nebrodensis* (Tineo) Hayek *in* Repert. Spec. Nov. Regni Veg. Beih. 30(1): 44 (1924)
Chaetocladus monostachys J.Nelson *in* Pinaceae: 162 (1866)
Ephedra atlantica Andr. *in* Bot. Jahrb. Syst. 64: 265 (1931)
Ephedra equisetiformis Webb & Berthel. *in* Hist. Nat. Iles Canaries 3(2; 3): 275 (1847)
Ephedra graeca C.A.Mey. *in* Bull. Cl. Phys.-Math. Acad. Imp. Sci. Saint-Pétersbourg 5: 35 (1845)
Ephedra major var. *atlantica* (Andr.) Maire *in* É.Jahandiez & al., Cat. Pl. Maroc 4: 921 (1941)

Not endemic, Certain native
Family: *Ephedraceae*
Common name: Efedra fina o Tepopote fino (ES); Narrow jointpine (EN)
Biotype: chamaephyte—nanophanerophyte
Flowering: spring–summer
Fruiting: summer
Pollination: wind pollination (anemogamy)
Dispersal: animal dispersal (endozoochory)
Chromosome N°: 2n = 14; n = 7

Description

Shrub up to 1 m, very branched, more or less fastigiate, twigs 0.4–0.7 mm in diameter, rough, not disjointed (Fig. 3.142).

Fig. 3.142 Minas de San José, June 2005. (Photo credit: Beatriz Hernández)

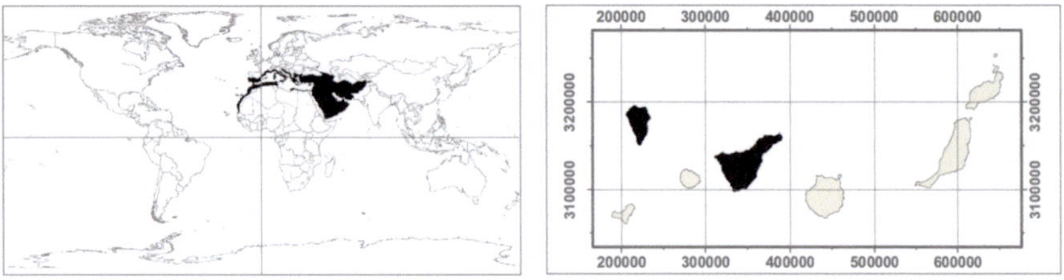

Fig. 3.143 Present on the islands of La Palma and Tenerife

Leaves up to 3 mm, completely membranous, the older ones dark brown. Male cones 2–4 pairs, in globose groups, sessile, 2–4 mm in diameter. Female cones with 1 seminal rudiment, shortly pedunculated, generally with two pairs of bracts, styliform tubule c. 3 mm. Syncarp 3–7 × c. 3 mm, ovoid, reddish or yellowish.

Remarks:

In Castroviejo (1986) it is recorded as ***Ephedra nebrodensis*** Tineo *ex* Guss. ***ssp. nebrodensis***, ***E. major*** being a synonym.

Distribution

Indigenous to Mediterranean and Macaronesian regions (Fig. 3.143). It is also distributed in western Asia, the eastern half of the Iberian Peninsula and the base of the Picos de Europa. In the Canary Islands it is present only on the islands of La Palma and Tenerife.

In the Teide National Park (Fig. 3.144) it is distributed in a very dispersed way throughout the interior of Las Cañadas: La Fortaleza, Montaña Negra, Las Mostazas, Montaña Rajada, Base de Risco Verde, Cañada de La Grieta, Barranco del Río, Roques de García and the western edges of Llano Ucanca. Specifically, also in Montaña Marqués, on the northern slopes of Izaña (2025–2325 m above sea level).

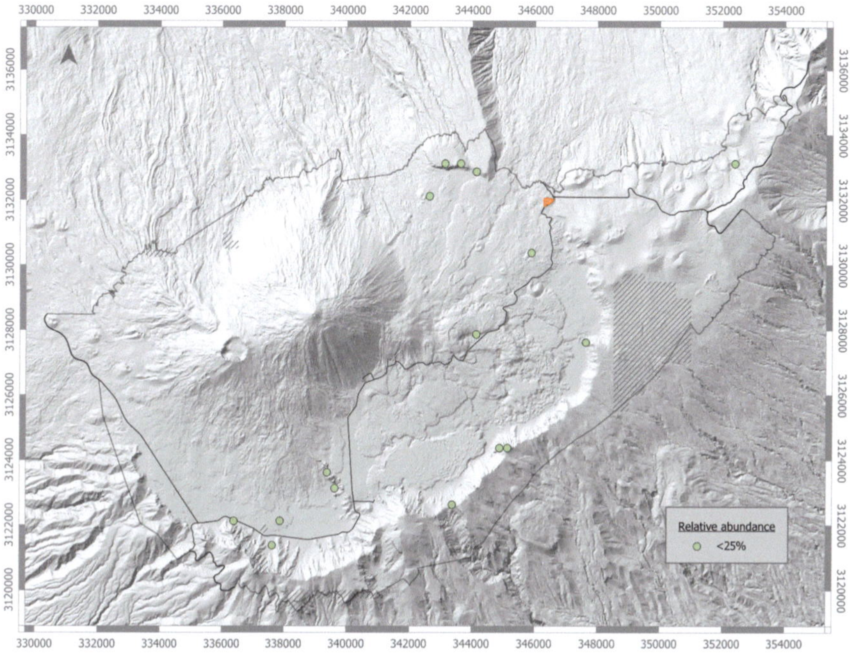

Fig. 3.144 *Ephedra nebrodensis* distribution, N° Subgrids (250 × 250 m): **16**/3740

Habitat and Phytosociology

It is native to rocky environments and xerophytic scrublands. In the Iberian Peninsula, it is a characteristic species and bioindicator of ***Rhamno lycioidis-Quercion cocciferae***, an alliance that groups the associations corresponding to the micro-forests and highlands of ***Pinus halepensis***, ***Quercus coccifera*** and ***Juniperus phoenicea***, semi-arid-subhumid Meso-Supra-Mediterranean and relict sub-Mediterranean Meso-Supra-Mediterranean, with a central-eastern Iberian Mediterranean, Betic and pre-Pyrenean distribution. In the Canary Islands, ***Ephedra nebrodensis*** is distributed in the potential vegetation of the cedar groves of the high Canarian summit, ***Juniperion cedri***. We had the opportunity to sample the species with Prof. Rivas Martínez in June 2017 (Fig. 3.145).

Conservation Status

According to the sampling carried out, in all the subgrids in which this species was found, no damage of any kind was observed.

> "*Very rare in Las Cañadas.*" (Sventenius 1946)

Threat Category

Anexo II: Protegidas (Gobierno de Canarias 1991).

Fig. 3.145 Prof. Salvador Rivas Martínez and Prof. Wolfredo Wildpret sampling at the San José Mines, Teide, June 17, 2017. (Photo credit: Victoria Eugenia Martín Osorio)

Exsiccata (Herbarium Specimens)

Tenerife: Inventario de la Flora y Vegetación del Parque Nacional del Teide (2001–2003) V.E. Martín Osorio, W. Wildpret & B. Hernández Bolaños: Minas de S. José, 15.05.2003, Y. de Rioja (TFC 44949); Mña. del Arroyo o la Zarza (por debajo de Izaña) (UTM 352429–3132944) (31A 19), 2060 m.s.m., alto pared de disyunción, 17/05/2005, V.E. Martín Osorio, W. Wildpret & B. Hernández Bolaños (TFC 46098); Minas de San José (UTM: 344283–3127679) (30A 47 a), 8/06/2005, 2335 m.s.m., entre rocas, V.E. Martín Osorio, W. Wildpret & B. Hernández Bolaños (TFC 45449).

Uses, Exploitation and/or Medicinal Properties

> "*Toxic; antirheumatic, sudorific, stimulant; vasoconstrictive, hypertensive, mydriatic. Useful part: young autumn stems. Instructions for use: decoction.*" (Pérez de Paz and Hernández Padrón 1999)

3.5.47 Erica canariensis *Rivas-Martínez, Martín Osorio & Wildpret*

Itinera Geobotanica, 18(2), 484 (2011)

New record for the Teide National Park

Endemism, Certain native
Family: *Ericaceae*
Common name: Brezo (ES); Canary tree heather (EN)
Biotype: nano—phanerophyte
Flowering: winter–spring
Fruiting: spring
Pollination: wind—insect pollination (anemogamy—entomogamy)
Dispersal: wind dispersal (anemochory)
Chromosome No: 2n = 24; n = 12

Description

Shrub or small tree of 1–4 (7) m young stems with inconspicuous ribs, yellowish brown bark, with dense, whitish indumentum, with patent hairs, some denticulate-equinate and others smooth, the latter more or less predominant in the shoots of the year (Fig. 3.146).

Fig. 3.146 *Erica canariensis* en La Fortaleza, June 2004. (Photo credit: Victoria Eugenia Martín Osorio)

Leaves 3–9 × (0.3) 0.5–0.7 mm, in whorls of 3–4, linear, erect-patent, so revolute that they do not allow the underside to be seen, glabrous or hairy, the young ones with very short glandular cilia.

Inflorescences very numerous, at the apex of short lateral branches, umbellate, with 1–3 flowers, with an involucre of bracteoles at the base, which form a paniculiform group. Pedicels (1.5) 2–5 (7) mm, relatively thick, glabrous, yellowish-brown; bracteoles 3, (0.6) 0.8–1.3 mm, inserted in the lower half of the pedicel, whorled or the two upper ones sometimes opposite-sometimes all at the base. Sepals 1.2–1.5, (2) mm, fused at the base, ovate-lanceolate, whitish, keeled in the upper half, cuculate, more or less membranous, glabrous, generally ciliolate. Corolla 2–3.5 (4) mm, bell-shaped or tubular-bell-shaped, white; lobes of c. 0.5 (1) mm, erect. Anthers 0.7–1 mm included, dorsifixed, with appendages 0.2–0.4 (0.6) mm, ovate-lanceolate to lanceolate-linear, rarely linear, denticulate; theca almost parallel. Ovary glabrous, style c. 2.5 mm, exserted, relatively thick; stigma discoidal, dilated at anthesis, c. 0.5 mm in diameter. Capsule c. 2 mm, obovoid or globose, glabrous. Seeds 0.4–0.6 mm, ellipsoidal, finely striate.

Erica canariensis Rivas-Martínez et al. 2011 differs from ***E. arborea*** L.: fastigiate tree 6–9 m, and trunk up to 2 m in diameter. Young stems with short hairs, those of older branches dense and simple, with a few glochid or compound branched trichomes. Leaves subflat, few revolute. Inflorescences racemose-panicle, with a long pyramidal shape. Corolla up to 5 mm long, briefly campanulate (Fig. 3.147).

Fig. 3.147 Detail of the corolla of *Erica canariensis* on La Fortaleza, June 2004. (Photo credit: Victoria Eugenia Martín Osorio)

Holotypus:

TFC 494208. La Hoya del Cedro, North face of Teide (Tenerife), 2163 m, 11/06/2003, in pine forest with broom. Cuadrícula 35A 4 a. UTM: 334000–3129750. Leg.: V.E. Martín Osorio.

Distribution

It is present in all the Canary Islands, generally between 700 and 1700 m above sea level, although it occasionally grows both above and below those limits.

In the Teide National Park, it has already reached, albeit occasionally, 2225 m above sea level (northwestern slopes of Pico Viejo). During the sampling it was also found occasionally, to the NW of Roque del Almendro, at the foot of the recent lava flows of Cañada de Diego Hernández and the northeastern slopes of Teide (Montaña Abejera and Echisere); and more extensively, in La Fortaleza and its surroundings (Figs. 3.148 and 3.149).

Habitat and Phytosociology

It preferentially colonizes the areas of ravines and cool sunny situations. It is found mainly in areas of the so-called Monteverde (fayal-brezal) and also in the undergrowth of the pine forest.

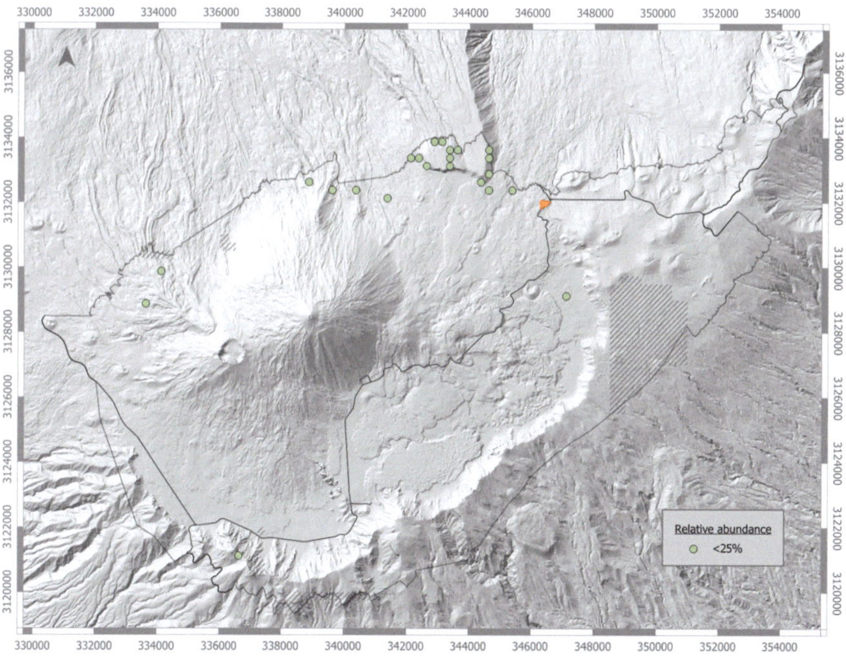

Fig. 3.148 *Erica canariensis* distribution, N° Sub-grids (250 × 250 m): **23**/3741

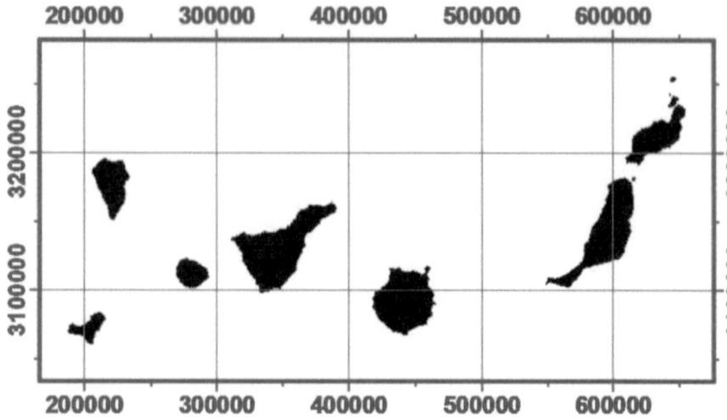

Fig. 3.149 Endemism of the Canary Island

It is a characteristic species and bioindicator of ***Pruno hixae-Lauretea novocanariensis***, a vegetation class that includes, first of all, the Canary-Madeiran climatophilous cloud forests and heminebulous lauroid forests. Within this, it is a member of ***Myrico fayae-Ericion canariensis***, an alliance that includes the high-altitude and serial ericaceous microforests of the Monteverde nebuloso of the Canary Islands, where the Canary heathers are predominant.

It can also be found in the community defined by ***Myrico fayae-Pinetum canariensis***; a climatophilous and edaphoxerophilous series of Tenerife, sub-cloudy, salic, Mediterranean, pluviseasonal, oceanic, Thermo-Meso-Mediterranean, subhumid, sub-hyperoceanic, of the ***Pinus canariensis*** and ***Myrica faya*** forests with ***Erica canariensis*** and Ilex canariensis. In addition, it can also appear as a subnebulous Thermo-Meso-Mediterranean faciation within the ***Cisto symphytifolii-Pinion canariensis*** alliance, which groups closed climatophilous and edaphoxerophilous meso-macroforests, well characterized by the Canary Island pine, dry-humid Meso-Supra-Mediterranean, rarely subnebulous, of western Canary Island distribution (***Sideritido solutae-Pinetum canariensis***).

Conservation Status

In all the subgrids in which this taxon was found, no damage of any kind was observed.

Threat Category

Anexo III (Gobierno de Canarias 1991)

Germplasm Banks

Jardín Botánico Canario Viera y Clavijo.

Exsiccata (Herbarium Specimens)

Tenerife: Orotava, 22.IV.1935, Cuatrocasas et al. (TFC 16539; La Esperanza, 21.III.59, Bellot & Casaseca, (TFC 10555, Duplic.); Las Mercedes, 2.V.73, La Serna Ramos, I., (TFC 23591); Anocheza (Monte), El Escobonal, II.77, Rodríguez

Delgado, O., (TFC 28022); La Laguna, Llano de los Viejos, III.79, MAT., (TFC 8647); La Laguna, S. Diego, 2.IV.79, MLRP., A. García Gallo, (TFC 20570); Bco. de Badajoz, Güímar, 1.III.1980, Beltrán Tejera, E., (TFC 8837); Madre del Agua, Agua García, 28.II.1980, García Gallo, A., C. Suárez, (TFC 10357); Madre del Agua. Agua García, 20.III.80, García Gallo, A., (TFC 10358); Canal de Anocheza, El Escobonal, Fasnia, 31.III.80, Wildpret de la Torre, W., J.R. Acebes, M. del Arco Aguilar, C. León Arencibia, L. Negrín, AVR, (TFC 25127 Dupl.); El Moquinal, 23.III.81, Méndez, B., M. del Arco Aguilar, J.R. Acebes, (TFC 20514); Las Mercedes, La Laguna, 3.II.82, García Gallo, A. (TFC 13089); Icod, Penichet, 7.IX.83, Ardévol González, F., (TFC 37984); Icod, Bco. de las Ánimas, 250 m, 26.VII.84, Ardévol González, F., (TFC 19312); Las Mercedes, 5.III.85, Méndez, B., (TFC 13275, P-TFC 440)); Finca Anocheza, detrás de la ladera (o Lomo Marrera), 25.III.85, Rodríguez Delgado, O., (TFC Dulplic. N. 28021); Bco. del Risco Azul, 500 m, *ex* Norte, 17.XII.88, MM., A. Díaz, (TFC 29858); Güímar, 600 m, Exp. N. 2.I.89, MMG. (TFC 29905); Monte de Las Mercedes, 15.II.92, Pérez Hernández, D., alumnos de 5 °C., P.L. Pérez de Paz, (TFC 35121); Anaga, Afur, Sabinar, 14.III.94, Rodríguez Delgado, O., M. del Arco Aguilar, (TFC Duplic. N. 33424); Anaga, El Moquinal, 16.V. 1995, Acebes, J.R., (TFC 43374 y 8 duplicatas); Las Casillas, Bco. de Ijuana, 21.III.96, Racca, R., (TFC 38997); Las Mercedes, 28.III.97. Beltrán Tejera, E., (TFC 40770); Los Silos, La Juncia, 12.I.97, VLA. (TFC 40002 y Duplicc.); Cuenca del Valle Vegas, Bco. de Tahodio, 28.III.99, Becker, S., T. Himstedt, (TFC 42039). Inventario de la Flora y Vegetación del Parque Nacional del Teide (2001–2003), V.E. Martín Osorio, W. Wildpret de la Torre & B. Hernández Bolaños: Al Este de La Hoya del Cedro (35A 4 a), 11/06/2003, Y. de Rioja Santos (TFC 44208); Pinar cara Norte del Teide, por encima de Cuchillos Marcados (29B 7 a), 18/06/2003, F. Mesa (TFC 44237).

Uses, Exploitation and/or Medicinal Properties

> "*Diuretic, antiseptic: treatment of urinary tract and insect bites; hypocholesterolemic; fodder; forestry interest: wood, firewood, charcoal, forks; crafts. Useful parts: flowers, leaves, branches, flowering tops; whole plant. Instructions for use: infusion, poultice; wood, firewood, charcoal.*" (Pérez de Paz and Hernández Padrón 1999)

3.5.48 Erigeron calderae *A. Hansen*

First published in Bol. Mus. Munic. Funchal 44: 175 (1992)

Erigeron cabreae Dittrich. *Candollea*, 36, 243–249 (1981)
Erigeron canariensis Šída f. *in* Preslia 70: 265 (1998)

Endemism, Certain native
Family: *Asteraceae*

Common name: Pompón de Las Cañadas (ES); Cañadas fleabane (EN)
Biotype: hemicryptophyte
Flowering: May–August
Fruiting: July–September
Pollination: –
Dispersal: wind dispersal (anemochorous)
Chromosome N°: 2n = 18

Description

Perennial herb up to 25 cm tall, stems ending in capitulum, the upper quarter 8 cm long; in the lower quarter leaves up to 10 cm long, oblong-spatulate with ciliate margin, alternate. Involucre oblique 3–4 seriate, flat-infundibuliform. Oblong involucre leaf externally covered with hairs. Receptacle flat, glabrous.

All **capitulums** of the typical flowers: between marginal female flowers ligulate, never elongated, papule and tubular hermaphrodite **flowers** yellow and flowers with filiform ligules sublilacine or of various colors. Oblong elliptic achene up to 2.5 mm long, whitish, adaxial and abaxial fascicle pale brown, appendage pedate whitish. Papule white spotted, uniseriate (Fig. 3.150). Between unequal rays (ca. 5.5 mm) scabrous up to 10 deciduous setae about 0.5 mm long.

Distribution

Endemism of the island of Tenerife that occupies small areas of the central cañadas (Montaña Blanca and surroundings of Montaña Rajada) and, more continuously, the eastern part of Las Cañadas del Teide (from the area of El Portillo, always

Fig. 3.150 Wolfredo Wildpret Botanical Garden of Teide National Park, (**a**) Aug. 2008, (**b**) Siete Cañadas, May 2005. (Photo credit: Beatriz Hernández)

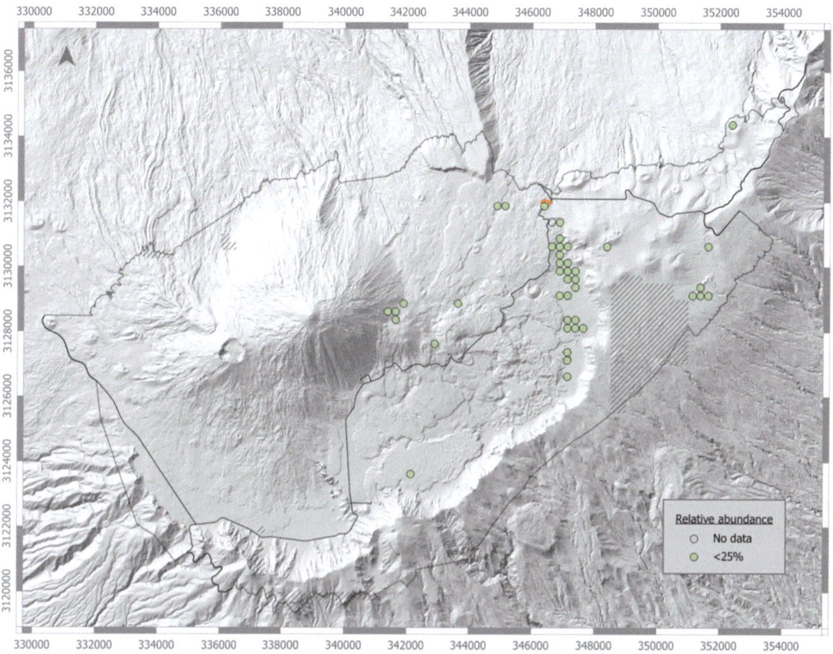

Fig. 3.151 *Erigeron calderae* distribution, N° Subgrids (250 × 250 m): **42**/3741

Fig. 3.152 Endemic of the Teide National Park on the island of Tenerife

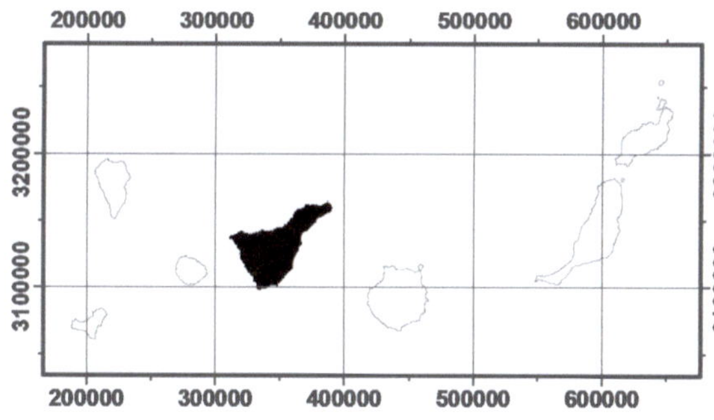

in locations close to La Pared, to Cañada de Las Pilas); about 2050–2725 m of altitude. Some isolated populations have been detected in the vicinity of the Sanatorium and in Lomo Hurtado (near La Fortaleza); outside Las Cañadas, in the surroundings of Izaña, Altos de Arico and Montaña Limón (highs of the valley of La Orotava). Also, in the Wolfredo Wildpret Botanical Garden of the Teide National Park (Figs. 3.151 and 3.152).

Habitat and Phytosociology

It lives in endorheic areas originated by contributions of alluvium and volcanic ash, where small artificial lakes are formed after thawing, also on pumice soils. The population located in Altos de Arico and around Izaña, lives on recent lapillis.

It is part of communities of the ***Spartocytision supranubii*** alliance, which groups the mountain summit shrublands of retamas and codesos, climatophilous or edaphoxerophilous, sometimes catenal or serial of the Canary pine forests; of distribution in La Palma and Tenerife, dry-humid Meso-Oromediterranean. It is also a characteristic species of the subassociation defined for the ***Descurainio bourgeauanae-erigerontetosum calderae*** National Park, which extends between the clayey retraction cracks originating in the temporarily flooded endorheic deposits of Siete Cañadas, Llanos de Maja, the dividing line between Montaña Blanca and Montaña Rajada, the base of the Altavista Trail and the base of the Fasnia Volcano (Martín Osorio et al. 2007).

Conservation Status

In 86% (36) of the subgrids in which this species was found, no damage of any kind was observed. Only in 12% (5), damage was detected (browsing, high mortality without regeneration, etc.), but this did not affect more than 25% of the specimens. In the rest (1), this variable was not specified at the time of sampling.

> *"Although it occupies a small area, its populations are clearly recovering, probably due to the cessation of grazing in these territories. However, we have observed some specimens attacked by rabbits, which is the most disturbing element of their populations at present. Human action due to trampling is also felt in some enclaves close to tracks and paths. The mouflon, not yet eradicated from its area, may also have a negative impact."* (Translated from: Gómez Campo et al. 1996)

Threat Category

VU D2 (Bañares et al. 1998).
VU D2 (VVAA 2000).
Sensible a la alteración de su hábitat (Gobierno de Canarias 2001).
VU (Bañares et al. 2004).
VU D2 (Moreno 2008).
Especie de interés para los Ecosistemas canarios (Gobierno de Canarias 2010).

Germplasm Banks

Jardín Botánico Canario Viera y Clavijo.

Exsiccata (Herbarium Specimens)

Tenerife: Cañada de Diego Hernández, 2.06.1973, P.L. Pérez de Paz (TFC 23342 + Dupl.); Alrededores Volcán de Arenas Negras-Las Cañadas, 2.07.1977, W. Wildpret, M. del Arco Aguilar, C. León Arencibia & I. La Serna (TFC

25072 + Dupl.); Cañada de la Grieta, Junio 1981, CRP (TFC 24491); Bajo Izaña-Altos de Güímar, 24.07.1981, W. Wildpret & M. del Arco Aguilar (TFC 24392 + Dupl.); Siete Cañadas-Cañada de Diego Hernández, 2.110 m.s.n.m., 11.07.1982, T. Bravo (TFC 23766). Inventario de la Flora y Vegetación del Parque Nacional del Teide (2001–2003) V.E. Martín Osorio, W. Wildpret & B. Hernández Bolaños*:* Ruta de Siete Cañadas, frente Volcán Arenas Negras (30D 32 d), lagunazo eventual al Pie de colada con *D. bourgeauana* (UTM: 347060–31306222) 15/06/2005, V.E. Martín & W. Wildpret (TFC 44984 + Dupl.); Llano de Maja, al NW del P.N. del Teide (30D 35 c), 31/05/2005, V.E. Martín, W. Wildpret & B. Hernández (TFC 44986).

3.5.49 Erysimum scoparium *(Brouss.* ex *Willd.) Wettst.*

Österr. Bot. Ztschr., 39, 283 (1889)

Dichroanthus scoparius Webb & Berth. *Phyt. Canar.*, 3(2/1): 68–69 (1836–1840).
Cheiranthus scoparius Brouss! Willd. enum. hort. Ver. pág. 681. non DC.
Cheiranthus cumbrae Link! *in* Buch Beschreib. Can. Ins. pág. 152.
Cheiranthus scaber Nob. ad seminum fasciculos.

Endemism, Certain native
Family: *Brassicaceae*
Common name: Alhelí de cumbre, de Las Cañadas o del Teide (ES); Teide wall-flower (EN)
Biotype: shrub (chamaephyte)
Flowering: February–June
Fruiting: May–August
Pollination: insect pollination (entomogamy)
Dispersal: wind dispersal (anemochory)
Chromosome No: 2n = 28; n = 14

Description

Chamaephyte with fruit-bearing stems, erect or ascending, knotty, rigid. Short, succulent, rigid, fastigiate branches, with very scarred, thickly rough leaves.

Leaves 3.75 cm long, 2–4 mm wide, up to the apex in compact branches, narrow, lanceolate, acuminate, concave, entire, cinereous-whitish. Long, white, acute, narrow, stellate hairs, 3–5 rays and, under a microscope, rough scabrous.

Purple **flowers** (Fig. 3.153), alternate or scattered, straight, rigid sub tetragonal or cylindrical peduncle, costate, striate below, scabrous, narrow, stellate hairs, 2 elongated rays.

Calyx erect, leaflet concave, the outer leaves cymbiform, apex obtuse, recurved, midrib carinate. Filaments filiform, anthers with sagittate base. Ovary subcylindrical, slightly compressed, densely hairy with very short hairs. Style cylindrical, 1 mm long, with fruit 2 mm long. Stigma capitate, round, finally sub bilobed.

Fig. 3.153 Teide National Park, May 2007. (Photo credit: Victoria Eugenia Martín Osorio)

Silique erect, narrow, tetragonous, cinereous-scabrous, hairs longitudinally stellate 2-rayed, valves succulent, thin reticulate dissepiment. Seed fusco-flavescent, clear, striate-pointed, wing dilated, yellowish. Cotyledons radicle elongated. Cotyledons ovate, apex subtruncate, externally pointed.

Distribution

It is found on the islands of La Palma and Tenerife. In Tenerife it ranges from 1500 to 3400 m above sea level (slopes of Teide). It is widely distributed throughout the National Park except for higher areas of the NE Ridge, slopes of the Teide-Pico Viejo stratovolcano (except on lava flows of Cerro de las Chajoras where it is quite abundant) and the interior of the western Cañadas (Llanos de las Chajoras, Boca Tauce and Llano de Ucanca) (Figs. 3.154 and 3.155).

Habitat and Phytosociology

Characteristic species and bioindicator of the legume thickets of the high mountains of Tenerife, ***Chamaecytiso-Pinetea canariensis***, a vegetation class that includes coniferous forests, the rhetamoid shrublands and the colonizing and serial dwarf mountain shrublands of Tenerife and La Palma, as well as the Oromediterranean dwarf shrublands of the Teide volcanic cone; climatophilous, edaphoxerophilous, pyrophilous or serial, endemic to the midlands and summits of the Central-Western Canary Islands, typical of dry-humid Meso-Oromediterranean bioclimates, of the Western Canary biogeographic subprovince.

It is part of the communities included in the ***Pterocephalenion lasiospermi*** suballiance and, in particular, of the ***Erysimo scoparii-Pterocephaletum lasiospermi*** association, dry-subhumid Upper-Meso-Supramediterranean, typical of lithosols and eroded soils with little soil.

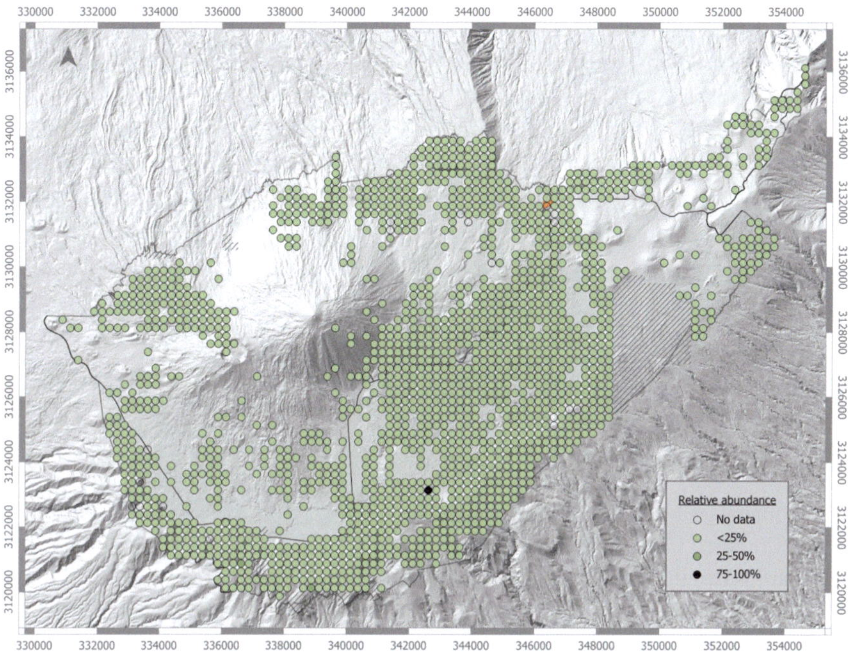

Fig. 3.154 *Erysimum scoparium* distribution, N° Subgrids (250 × 250 m): 1725/3741

Fig. 3.155 Endemism of the islands of Tenerife and La Palma

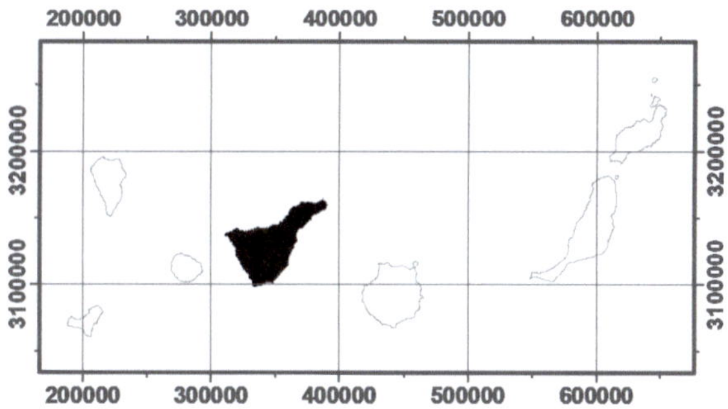

It also occurs in the following units:

Adenocarpo viscosi-Chamaecytisetum angustifolii
Arrhenathero calderae-Plantaginetum webbii
Cheirolopho teydis-Carlinetum xeranthemoides
Tolpido webbii-Micromerietum lachnophyllae
Arrhenathero calderae-Adenocarpetum viscosi

State of Conservation

In 90% (1555) of the subgrids in which this taxon was found, no damage of any kind was observed. In 8% (134), damage was detected (browsing, high mortality without regeneration, etc.), but this did not affect more than 25% of the specimens. Damage was only observed on more than 25% of the specimens in 1% (23) of the sub-grids where it appeared. In the rest (13), this variable was not specified at the time of sampling.

Threat Category

nt (Barreno et al. 1984)
LR pm (Bañares et al. 1998)

Germplasm Banks

E.T.S. Ingenieros Agrónomos de Madrid.
Jardín Botánico Canario Viera y Clavijo.
Real Jardín Botánico de Madrid.

Exsiccata (Herbarium Specimens)

Tenerife: Excursión Valle Santiago, 3–4.02.1968, W. Wildpret de la Torre (TFC 4173); Ibid., 3–4.02.1968, W. Wildpret & Alumnos (TFC 4180 + Dupl.); Las Cañadas, 26.05.1973, I. La Serna (TFC 35733); Pista de Lomo Blanco, 13.05.1979, P.L. Pérez de Paz & C. Hernández Padrón (TFC 20448); Bco. de los Albarderos—Arico, 1300 m.s.n.m., 1.04.1980, W. Wildpret, J.R. Acebes, M. del Arco Aguilar, C. León Arencibia, L. Negrín & AVR (TFC 25143 + Dupl.); Caldera de Pedro Gil, 6.05.1981, M. del Arco Aguilar (TFC 24384 + Dupl.); Valle del Volcán de Las Arenas-Cumbres de Arafo, 1700 m.s.n.m., 9.05.1981, W. Wildpret, M. del Arco Aguilar & P.L. Pérez de Paz (TFC 13692 + Dupl., 23735 + Dupl.); Las Cañadas, 27.06.1982, RLN (TFC 13119 + Dupl.); Ctra. Vilaflor, 15.05.1984, P.L. Pérez de Paz, I. La Serna & L. Negrín (TFC 34277 + Dupl.); Cumbre Mña. Arguazo, 5.11.1984, O. Rodríguez Delgado (TFC 27560); El Portillo-Las Cañadas, 15.07.1989, P. Valdés (TFC 29215); Agua Agria-Vilaflor, 28.05.1993, M. del Arco Aguilar (TFC 36058). Inventario de la Flora y Vegetación del Parque Nacional del Teide (2001–2003), V.E. Martín Osorio, W. Wildpret de la Torre & B. Hernández Bolaños: Al SW del Llano de La Rosa (31A 13 c); UTM: 3133762–35352019, 09/04/2001, A. Hernández Luis & L. Afonso Giménez (TFC 44896); Roques al Sur del Llano Ucanca (41D 35 d), 14/05/2003, A. Montañés Rodríguez (TFC 44299)

La Palma: Caldera de Taburiente, 26.07.1974, M. del Arco Aguilar (TFC 4287, 27093); Pico de la Cruz, 1.04.1977, C. León Arencibia, M. del Arco Aguilar, J.R. Acebes & JL (TFC 6519).

Gran Canaria: Cruz de Tejeda, 23.03.1973, E. Beltrán Tejera (TFC 8461); Los Pechos, 24.06.1983, W. Wildpret, P.L. Pérez de Paz & L. Negrín (TFC 32067).

Uses, Benefits and/or Medicinal Properties

> "*Cardiotonic, hypoglycemic. Useful part: whole plant. Instructions for use: infusion.*" (Translated from: Pérez de Paz and Hernández Padrón 1999)

3.5.50 Ferula linkii *Webb*

Phyt. Canar., 3(2/2), 160–161 (1836–1850)

Ferula aurea Link. *in* Buch., *Phys. Beschr. Can.*, Ins.: 152, excl. syn.
Peucedanum aureum DC. *Bot. reg.*, table 559, non Soland; DC., *Prodr.*, vol. 4, pág. 178, pro parte.
Ferula communis subsp. *linkii* (Webb) Reduron & Dobignard *in* Index Syn. Fl. Afrique N. 2: 95 (2011)

Endemism, Certain native
Family: *Apiaceae*
Common name: Cañaheja—Cañalé—Cañaleja—Julán (ES); Link's Giant Fennel (EN)
Biotype: hemicryptophyte
Flowering: January–June
Fruiting: March–July
Pollination: insect pollination (entomogamy)
Dispersal: wind dispersal (anemochory)
Chromosome No: n = 11; 2n = 22

Description

Stems erect, up to 3 m tall, rounded, striated, never sulcate, glabrous, with extensive, vaguely visible pith, branched above, alternate branches, the upper ones opposite and subcorymbose.

Leaves very divided, with a wide circle, laciniae linear-filiform, rigid, acute, canaliculate, short, subelongate, petiolate, petioles of the radical leaves long, dilated base; petioles of the leaves of the stems in wide evaginate, striated, semiamplifying stems dilated.

Central **umbels** (Fig. 3.156), sparsely pedunculated, peduncles 6–20 mm long, 16–24-rayed, rays subequal, striated, glabrous. There is no involucre. Umbelles 20–30-rayed. Involucel submonophyle, linear leaflet, acute, shorter than the umbellate. Flowers yellow, sparse. Petals ovate, concave, glabrous, with inflexed, incurved, sulcate acumen, entire. Anthers didymous, filaments recurved. Styles divergent, reflexed, apex capitate.

Fruits ovate, 9 mm long, 6 mm wide, mature, surrounded by a black band, with a flat, compressed back, dilated margin, complanate band. Mericarp in 5 pairs, 3

Fig. 3.156 Siete Cañadas, at the W foot of Risco Verde, May 2004. (Photo credit: Beatriz Hernández)

filiform on the dorsal part, prominent and at equal distance, 2 lateral with dilated margin, secondary pairs obsolete. Vita in 2–3 valleculae, superficial. Wide commissure, sub-6-vitata, vitae distant. Carpophore filiform, bipartite. Stylopodia conical. Seeds are flat and convex.

Distribution

Endemism of the Canary Islands, which is present on all the islands except Fuerteventura (Figs. 3.157 and 3.158).

In Tenerife it has a notable altitudinal distribution: from 200 m above sea level to 2400 m above sea level. Within the Teide National Park it was found in: La Fortaleza, between Montaña Arenas Negras and Montaña Chusqueros (beginning of Siete Cañadas), Risco Verde, eastern sectors of Montaña de Pasajirón, Barranco del Río, E-Llano de las Mesas, Riscos de Magdalena and Roques del Cedro. Also in the Botanical Garden of the Teide National Park.

Habitat and Phytosociology

It is usually present in different communities of the different bioclimatic zones of the island. In Las Cañadas its preferred habitat is usually stony and/or rocky environments.

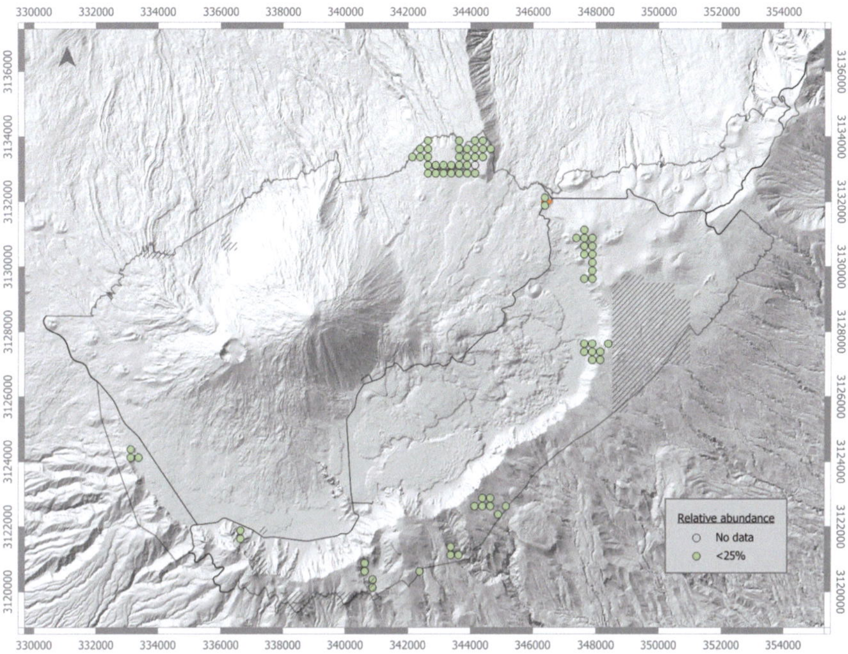

Fig. 3.157 *Ferula linkii* distribution, N° Subgrids (250 × 250 m): **74**/3741

Fig. 3.158 Present on all islands except Fuerteventura

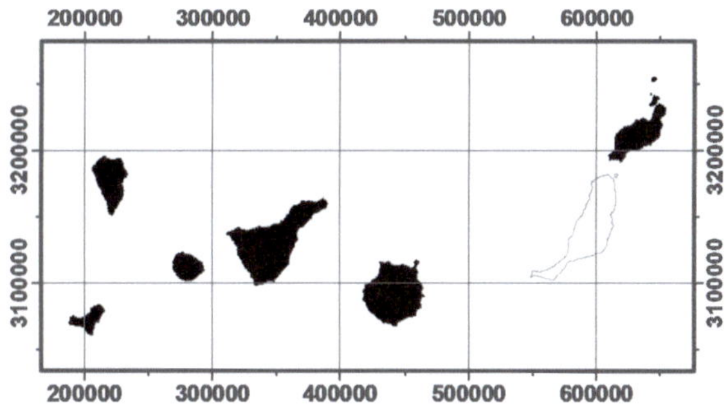

It is part of the communities included in the alliance ***Spartocytision supranubii***, dry-humid lower Supra-Oromediterranean Teidean brooms of La Palma-Tenerife. It is also found in the community described as Cerrillar-Crespar, a phytosociological association of the National Park that preferentially colonizes the slopes of gelifract debris of the Pared de Las Cañadas: ***Arrhenathero calderae-Plantaginetum webbii***.

Conservation Status

In 93% (69) of the subgrids in which this species was found, no damage of any kind was observed. Only in 1% (1), damage was detected (browsing, high mortality without regeneration, etc.), but this did not affect more than 25% of the specimens.

Damage has only been observed on more than 25% of the specimens in 4% (3) of the subgrids where it appeared. In the remainder (1), this variable was not specified at the time of sampling.

Threat Category

nt (Barreno et al. 1984).
LR pm (Bañares et al. 1998).

Germplasm Banks

E.T.S. Ingenieros Agrónomos de Madrid.
Jardín Botánico Canario Viera y Clavijo.
Real Jardín Botánico de Madrid.

Exsiccata (Herbarium Specimens)

Tenerife: W. Wildpret, A. Santos, E. Beltrán & P.L. Pérez de Paz (TFC 826); La Ladera-Canal, 1000 m.s.n.m., 4.05.1984, A. Charpín & O. Rodríguez Delgado (TFC 27958); Tierra del Trigo-Los Silos, 9.04.1989, P.L. Pérez de Paz (TFC 28882 + Dupl.); Ibid., 9.04.1989, P.L. Pérez de Paz & D. Pérez (TFC 33808 + Dupl.); Valle del Palmar, 9.04.1989, ejusd. (TFC 33813 + Dupl.); Icod, 5.01.1993, W. Wildpret & P.L. Pérez de Paz (TFC 35401); Bco. Ijuana-Ladera S-La Pardilla, 450 m.s.n.m., 18.03.1996, R. Racca (TFC 39185); Bco. Cuevas Negras, 6.04.1996, V. Lucía Sauquillo & C. Acosta (TFC 38086); Tierra del Trigo-Los Silos, 16.03.1997, V. Lucía Sauquillo & C. Acosta (TFC 40057); Pista entre Jardina y el fondo del Bco. de Tahodio, 9.04.1999, S. Beckerr. & Th. Himstedt (TFC 42046).

Gran Canaria: Bco. Oscuro, marzo 1980, C. Suárez (TFC 11023); Ctra. de Teror-Valleseco, 14.06.1986, CSR (TFC 31276); Utiaca-San Mateo, 900 m.s.n.m., 28.03.1988, P.L. Pérez de Paz (TFC 25696, 27469 + Dupl.).

Uses, Exploitation and/or Medicinal Properties

> "*Galactogenic (in goats); anti-inflammatory, antipyretic, antinociceptive; colorant: yellowish green. Useful part: leaves, stems and roots. Instructions for use: poultice, fodder; extract: experimental.*" (Translated from: Pérez de Paz and Hernández Padrón 1999)

3.5.51 Festuca agustinii *Linding.*

In Abh. Auslandsk., Reihe C, Naturwiss. 8: 298 (1926)

Festuca teneriffae Hack. in Oesterr. Bot. Z. 53: 35 (1903), nom. illeg.
Festuca bornmuelleri Hack.

Festuca linkiana Litard. in Candollea 10: 116 (1945), nom. superfl.
Festuca filiformis Chr. Sm. *ex* Link *in* Buch. Phys. Beschr. Canar. Ins.: 139 (1828), nom. illeg.
Nueva cita para el Parque Nacional del Teide

Endemism, Certain native
Family: *Poaceae*
Common name: Cerrillo de risco (ES); Rock fescue (EN)
Biotype: hemicryptophyte
Flowering: spring
Fruiting: summer
Pollination: wind pollination (anemogamy)
Dispersal: wind—insect dispersal (anemochory–entomochory)
Chromosome No: 2n = 14

Description

Perennial plant with robust, densely cespitose shoots (Fig. 3.159), up to 50 cm. Rolled leaves. Membranous, lacerated ligule. Erect, branched **inflorescence**, more or less erect, compact or extended branches. Unequal glumes, awned. Lemmas convex, shortly awned. Very short awns.

Fig. 3.159 Wolfredo Wildpret Botanical Garden of Teide National Park, Nov. 2003. (Photo credit: Victoria Eugenia Martín Osorio)

Distribution

Endemism of the western and central Canary Islands (Figs. 3.160 and 3.161).

In Tenerife, according to the consulted bibliography, it is found in Las Mercedes, Aguamansa and Cumbres de Pedro Gil. Within the Teide National Park, it was found in La Fortaleza and interior slopes of the Cañada de Las Pilas and La Angostura, between 2100 and 2300 m a.s.l. Very punctually in the Altos de Arico.

Habitat and Phytosociology

It is usually distributed in areas of pine forest and green mountain. It is found in associations of the ***Greenovio-Aeonietea*** class, perennial vegetation, which groups comophytic and chasmo-comophytic permanent perennial communities, endemic to the Canary-Madeiran biogeographic subregion.

Conservation Status

In 73% (8) of the subgrids in which this species was found, no damage of any kind was observed. In 18% (2), damage was detected (browsing, high mortality without regeneration, etc.), but this did not affect more than 25% of the specimens. In the remainder (1), this variable was not specified at the time of sampling.

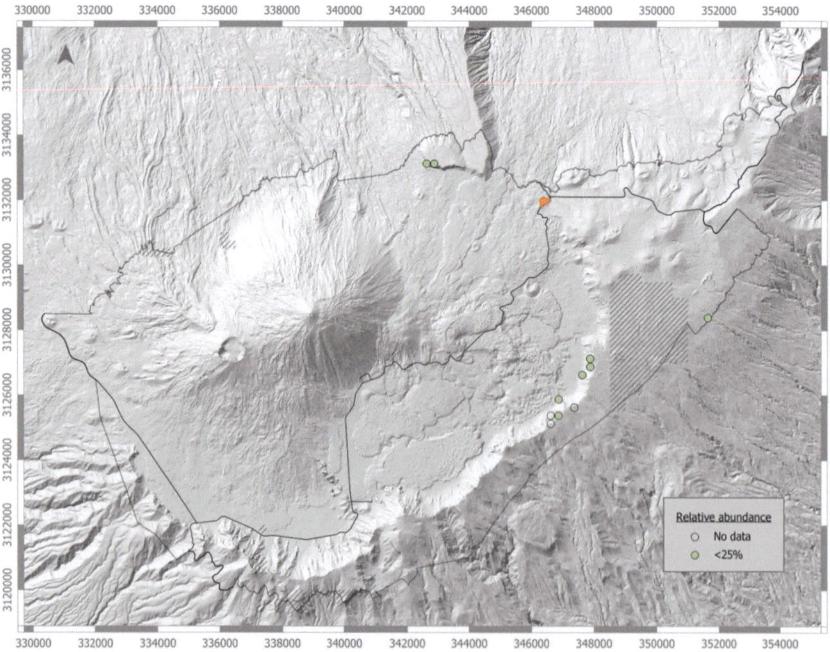

Fig. 3.160 *Festuca agustinii* distribution, N° Subgrids (250 × 250 m): **11**/3741

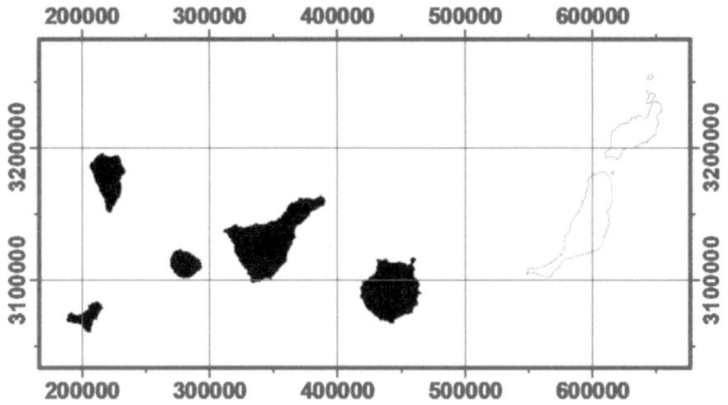

Fig. 3.161 Endemism of the central and occidental islands

Threat Category

R (Barreno et al. 1984)
LR pm (Bañares et al. 1998)

Germplasm Banks

E.T.S. Ingenieros Agrónomos de Madrid.

Exsiccata (Herbarium Specimens)

Tenerife: Hoya Fría, Altos de Arafo, Ladera Oeste, 27.6.1971, E. Barquín Díez (TFC duplic. 34571); Bco. del Río, 27.6.1971, E. Barquín Díez (TFC 34572); [Como *Festuca bornmuelleri* Hackel]: Arafo, Añavingo, 25.5.1973, H. Metlesics (TFC 4355); Ayosa, Cumbre de Pedro Gil, P.L. Pérez de Paz, I. La Serna & IMM (TFC 24739, 3 duplic.); Bco. de Añavingo, 26.6.1973–2.1992, P.L. Pérez de Paz & M.C. León Arencibia (TFC 35158); Valle del Volcán de Las Arenas, Arafo, 9.5.1981, W. Wildpret, M. Arco Aguilar & P.L. Pérez de Paz (TFC 23708 duplic.); Ibid, Arafo, 9.5.1981, ejusd. (TFC 23745, 5 duplic.); La Orotava, cerca de los Órganos de Aguamansa, Roque de La Escalera, 19.5.1981, M. Arco Aguilar & R. Aaron (TFC 24937, duplic.); Güímar-Agache, La Ladera, Canal 750 msm., 30.12.1987, W. Wildpret & O. Rodríguez Delgado (TFC duplic. 28104); Bco. de Los Dolores, 30.4.1992, P.L. Pérez de Paz (TFC 35220); Las Cañadas, 1992–1.1994, J.R. Acebes, P.L. Pérez de Paz, C. León Arencibia et al. (TFC duplic. 36414); Roque de Enmedio, 16.5.2002, J.P. Oval, A. Acevedo & J. Matos (TFC 43714). Inventario de la Flora y Vegetación del Parque Nacional del Teide (2001–2003), V.E. Martín Osorio, W. Wildpret de la Torre & B. Hernández Bolaños: La Fortaleza, 2100 m.s.n.m., 06/06/2004, V.E. Martín Osorio & W. Wildpret de la Torre (TFC 45440 + 2 Dupl.); Pared de disyunción (31A 19), 17/05/2005, V.E. Martín, W. Wildpret & B. Hernández (TFC 44977 + Dupl.); Refugio de Ayosa, carretera de La Esperanza, 08/06/2005, V.E. Martín Osorio, W. Wildpret de la Torre & B. Hernández Bolaños (TFC 44980 + Dupl.).

El Hierro: [Como *Festuca* cf. *bornmuelleri* Hackel]: Ctra. Frontera, El Golfo, 7.7.1985, P.L. Pérez de Paz (TFC 24601 + duplic.).

La Palma: [Como *Festuca* cf. *agustinii* Lindinger]: Los Andenes, Cumbres de La Palma, 16.6.1995, P.L. Pérez de Paz (TFC 38725).

La Gomera: [Como *Festuca* cf. *agustinii* Lindinger]: Agando, 20.7.1988, P.L. Pérez de Paz, J.R. Acebes & M. Arco Aguilar (TFC 28769).

3.5.52 Festuca myuros *L.*

First published *in* Sp. Pl.: 74 (1753)

Avena muralis Salisb. *in* Prodr. Stirp. Chap. Allerton: 122 (1796), nom. superfl.
Distomomischus myuros (L.) Dulac *in* Fl. Hautes-Pyrénées: 91 (1867)
Mygalurus caudatus Link *in* Enum. Hort. Berol. Alt. 1: 92 (1821), nom. superfl.
Vulpia murorum Gray *in* Nat. Arr. Brit. Pl. 2: 123 (1821 publ. 1822), nom. superfl.
Vulpia myuros (L.) C.C.Gmel. *in* Fl. Bad. 1: 8 (1805)
Vulpia vaginata St.-Lag. *in* A.Cariot, Étude Fl., éd. 8. 2: 938 (1889), nom. superfl.

Not endemic, Probable native
Family: *Poaceae*
Common name: Greñita
Biotype: therophyte
Flowering: March–June
Fruiting: March–June
Pollination: wind pollination (anemogamy)
Dispersal: wind dispersal (anemochory)
Chromosome N°: 2n = 21, 42

Description

Annual with stems up to 60 cm, erect, striate, glabrous (Fig. 3.162).

Leaves with 0.1–0.5 mm truncate ligule, blade up to 18 cm long and 0.2–1.5 (−2.5) mm wide, setaceous and convolute with striate upper surface, slightly pubescent and glabrous or more or less scabrid lower surface.

Panicle (1–) 3–30 cm, oblongoid or narrowly oblongoid, more or less dense, sometimes subraceniform, long-pedunculated or with base covered by the last foliate sheath, with erect-patent and scabrid branches. Spikelets 6–17 mm, with 3–10 flowers; the 1–3 upper ones male or sterile. Glumes shorter than the flowers or sometimes the one longer than their length; The lower one (0.3–) 1–6 (−10) mm, linear subulate, uninnerved, glabrous; the upper one (2.7–) 3.5–9 (−13) mm, lanceolate-subulate, sometimes with an awn up to 2 mm, with narrow scabrous margin glabrous or scabrid. Lemma (4–) 5–9 (−11) mm, narrowly lanceolate, with 5 slightly marked nerves scabrid at least in the upper half, with an awn of 8–30 mm; callus 0.1–0.2 mm, orbicular. Palea 5–7 mm, with antrorse-scabrid keels. Androecium with 1(3) stamens. Anthers 0.2–1.1 mm.

Fig. 3.162 Cañada de Diego Hdez., May 2004. (Photo credit: Victoria Eugenia Martín Osorio)

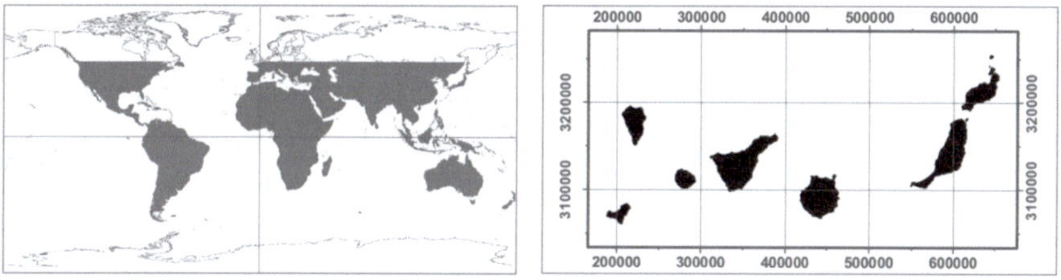

Fig. 3.163 Present in all the Canary Islands

Distribution

Subcosmopolitan; can be found in Macaronesia, Europe, N Africa and Asia. Present in all the Canary Islands (Fig. 3.163). Within this area it is widely distributed in inland areas of Las Cañadas and the northern limit of the protected area; much more dispersed, less abundant or non-existent in areas of La Dorsal, La Pared and the Pico Viejo-Teide-Montaña Blanca complex and lava flows; between 1600 and 3700 m above sea level, also present in the Pico area (Fig. 3.164).

Remarks:

Sventenius (1946) cited it as "*found at 3650 m*".

Habitat and Phytosociology

It is usually found in areas of grasslands, herbaceous lands and abandoned farms. In communities of the order ***Tuberarietalia guttatae***, of xerophilous therophytic grasslands, with superficial roots that colonize oligotrophic skeletal soils, with a gravelly, silty or sandy texture, but with a certain cohesion; Thermo and Mesomediterranean environments.

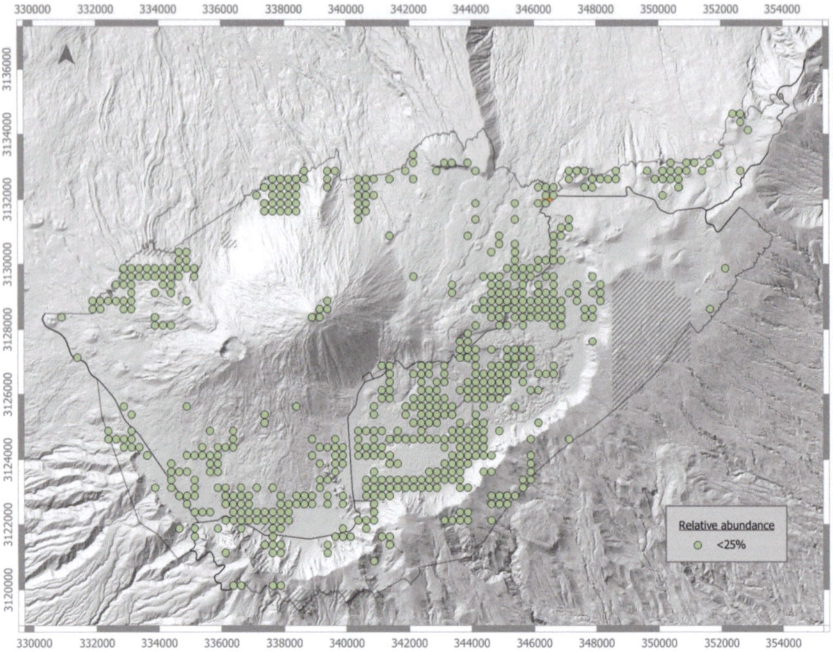

Fig. 3.164 *Festuca myuros* distribution, N° Subgrids (250 × 250 m): **614**/3740

Within the Park, we can find it intervening in the associations ***Arrhenathero calderae-Plantaginetum webbii***, which colonizes the slopes of gelifract debris of the Pared de Las Cañadas; and ***Festuco myuri-Laphangietum teydei*** (=***Vulpio myuri-Laphangietum teydei***), established in fumaroles of the Rambleta, base of the cone (3550 m a.s.l.) Supramediterranean environments.

Conservation Status

In most of the subgrids in which this species was found (587), no damage of any kind was observed. Only in two (2) of them, damage was detected (browsing, high mortality without regeneration, etc.), but this did not affect more than 25% of the specimens. And damage was only found on more than 25% of the specimens in twenty-five (25) of the subgrids where this taxon appeared.

In 1987, Dickson, Rodríguez and Machado published a catalogue of invasive plants of the high peaks of Tenerife (above 1800 m a.s.l.) and especially those of the Teide National Park. They warn that the introduced species have been dispersed mainly by the practice of grazing, which disappeared with the declaration of a National Park in 1954, and by tourism, promoting the entry of a relatively recent flow of invasive plants. Of the 83 introduced species in the catalogue, 30 were considered well established, 15 of them were grasses, with ***Bromus tectorum*** and ***Vulpia myuros*** being the most abundant. Martín Osorio et al. (2006) cite ***Festuca*** (=***Vulpia***)***myuros*** as a well-established species.

Exsiccata (Herbarium Specimens)

Tenerife: Campus Universitario, 2.3.1970, E. Beltrán Tejera (TFC 35507); Valle de Guerra, 5.5.1973, I. La Serna (TFC 35559); Las Mercedes, El Batán, 14.4.1975–8.1999, P.L. Pérez de Paz, RAL, W. Wildpret & J.A. Reyes-Betancort (TFC 42221); Las Cañadas, Montaña Rajada, 20.6.1982, E. Beltrán Tejera, T.B. & M.C. León Arencibia (TFC 32713); Las Cañadas del Teide, Bco. del Riachuelo, 2100 m.s.n.m., 7.5.1983, J.S. Socorro & W. Wildpret (TFC 18480); Bco. del Riachuelo, 12.5.1984, W. Wildpret (TFC 20059); La Laguna, 10.2.1985, A. García Gallo (TFC 22587; 22588 + duplic. 22589); Güímar, Agache, Lomo de Marrera, 17.2.1986, O. Rodríguez Delgado & W. Wildpret (TFC 28110); Güímar, Agache, Arrastradero de Arriba, Lomo de Mena, 21.2.1986, O. Rodríguez Delgado (TFC 28111); Los Silos, 23.3.1988, E. Beltrán Tejera, A. Losada Lima, J.L.R. & M.C. León Arencibia (TFC duplic. 32775); Ibid, 23.3.1988, ejusd. (TFC 32792); Bco. Hondo, Pista de Las Raíces, 23.3.1990, M. del Arco Aguilar & M. Salas Pascual (TFC 38420); Pinar sobre Agache, barranco, 1300 m.s.n.m., 20.4.1990, M. Salas Pascual (TFC 37868); Las Yedras, 24.4.1992, W. Wildpret & M.C. León Arencibia (TFC 35392); El Portezuelo, 30.1.1993, W. Wildpret & T. Jiménez Felipe (TFC duplic. 36328); Tegueste, Pico Español, 11.5.1993, W. Wildpret, A. García Gallo & M.T. Jiménez Felipe (TFC duplic. 36359); Guía de Isora, Acantilados de Isorana, 24.2.1996, P.L. Pérez de Paz, C. Hernández Padrón & D. Hernández Pérez (TFC duplic. 38476); Los Silos, Erjos, 16.3.1977, V. Lucía Sauquillo, J.A. Reyes-Betancort, M. del Arco Aguilar & A. Díaz Hernández (TFC 40095); Ibid, 26.3.1997, ejusd. (TFC 40094 + duplic.); Los Silos, Las Moradas, 6.4.1997, V. Lucía Sauquillo, A.D. Hernández, J.A. Reyes-Betancort & Estrella (TFC 40041); Los Silos, pista de Los Madroños, 18.5.1997, V. Lucía Sauquillo & J.A. Reyes-Betancort (TFC 40638 + duplic.); Las Lagunetas, UTM 28RCS 6245, 2.6.2001, M. López Rodríguez & M.L. Rodríguez Navarro (TFC 44768 + duplic.); Las Lagunetas, 30.5.2003, 362/3144, M. López Rodríguez & M.L. Rodríguez Navarro (TFC 44533). Inventario de la Flora y Vegetación del Parque Nacional del Teide (2001–2003) V.E. Martín Osorio, W. Wildpret & B. Hernández Bolaños: Al NW de la Montaña del Limón (31A 1 d), 08.04.2001, A. Hernández Luis & L. Afonso Jiménez (TFC 44895); Al W de la Montaña del Arroyo o de la Zarza (31A 18 b), 23.04.2001, L. Afonso Jiménez & F. Tapia (TFC 44868); Zapatito de la Reina, 14.04.2003, V.E. Martín Osorio & W. Wildpret de la Torre (TFC 44878); El Bastión, base NO del cono del Teide (35B 27 d y 35B 27 b), 3000 m.s.n.m., 08.05.2003, F. Mesa (TFC 44852 + dupl.); Degollada de Guajara (42C 5 c), 21.05.2003, F. Mesa (TFC 44227); Entrada al Jardín Botánico (30D 11 b) (T19), 17/06/2004, B. Hernández Bolaños (TFC 45397); Cañada de Diego Hernández, 12/05/2004, V.E. Martín Osorio, W. Wildpret de la Torre & B. Hernández Bolaños (TFC 45423); Jardín Botánico (P.N. Teide) (Ñ25), 17/06/2004, B. Hernández Bolaños (TFC 45411).

[Como *Vulpia* cf. *myurus*]: Bco. de Tahodio, Ladera derecha, 11.3.1978, W. Wildpret, M. del Arco Aguilar & I. La Serna (TFC 32499); Las Cañadas, La Fortaleza, 13.6.1980–12.1992, E. Beltrán tejera & M.C. León Arencibia (TFC 35258).

El Hierro: Malpaso, 10.4.1979, W. Wildpret, C. Hernández Padrón, P.L. Pérez de Paz & M. del Arco Aguilar (TFC 8276); El Cres, El Turrón, 10.4.1979, ejusd. (TFC 8248 + duplic.); Tiñor, 4.1979, ejusd. (TFC 8050); Tiñor, 4.1979, ejusd. (TFC 8035); Sobre Frontera, 23.3.1998, P.L. Pérez de Paz & GCT (TFC 42476; 42475) (Ver Observs. PLP Termófilos).

La Gomera: Parque Nacional de Garajonay, 8.5.1984, A. Charpin & P. Ranero (TFC 20009); Mirador de Tajaque, 3.5.1994, P.L. Pérez de Paz & D. Pérez Hernández (TFC 36833).

La Palma: Tagoja, 30.4.1992, P.L. Pérez de Paz (TFC duplic. 35211); Los Andenes, 2200 m.s.n.m., frecuente en ciertos lugares nitrofilizados de La Cumbres de La Palma, 8.8.1998, P.L. Pérez de Paz (TFC 41629 + duplic.).

Gran Canaria: Moya, Casa del Cura, 5.1980, P.L. Pérez de Paz & C. Suárez (TFC 11138); San Mateo, 21.4.1987, C. Suárez (TFC 31667); Cruz de Tirma, 16.7.1993, M. Salas Pascual (TFC 37132).

Lanzarote: San Bartolomé, Tomaren, 24.2.1995, J.A. Reyes-Betancort & M:C. León Arencibia (TFC 37807). ?, Montaña Breña, 13.4.1971, Anón. (TFC 26950).

3.5.53 Filago pyramidata *L.*

Sp. Pl., 1199 (1753)

Filago germanica subsp. *pyramidata* (L.) Arcang. *in* Comp. Fl. Ital.: 378 (1882), nom. illeg.
Filago germanica var. *pyramidata* (L.) Gaudin *in* Fl. Helv. 5: 253 (1829)
Gifola pyramidata (L.) Dumort. *in* Fl. Belg.: 69 (1827)
Gnaphalium germanicum var. *pyramidatum* (L.) Lej. *in* Rev. Fl. Spa: 173 (1825)
Gnaphalium pyramidatum (L.) Lam. *in* Encycl. 2: 760 (1792)
Impia pyramidata (L.) Bluff & Fingerh. *in* Comp. Fl. German. 2: 343 (1825)

Not endemic, Probable native
Family: *Asteraceae*
Common name: Yesquerilla tiesa (ES); Broadleaf cottonrose (EN)
Biotype: therophyte
Flowering: April–August
Fruiting: April–August
Pollination: insect pollination (entomogamy)
Dispersal: self dispersal (ballistochory)
Chromosome N°: 2n = 28

Description

Plants are greyish in colour (Fig. 3.165). Stems up to 44 cm, erect, ascending or decumbent, simple or very branched, especially in the upper part. **Leaves** 3–25 × 1–5.5 mm, oblong or spatulate.

Glomerules with (3–) 5–16 capitula, not or slightly exceeded by the involute leaves. Involucres 4.3–6.5 × 2–4 mm, with 5 angles, very marked and with 4–6 rows of bracts. External bracts with rigid and divergent awn. Central part of capitula with 4–8 (−10) female flowers of (1.7–) 2–2.4 mm and with (3–) 4–8 hermaphrodite flowers of (1.7–) 2–2.4 mm. Achenes 0.5–0.8 × 0.15–0.3 mm. Pappus (1.8–) 2–2.5 mm, with numerous persistent hairs.

Distribution

It is found in southern areas of S and W Europe, NW Africa, S Asia and throughout the Macaronesian region. Present in all the Canary Islands (Fig. 3.166).

In the Teide National Park (Fig. 3.167) it extends in a more or less dispersed way, between 1950 and 2200 m a.s.l.: Montaña Limón and surroundings, Montaña Guamasa, Asiento de Las Gallinas (E-El Portillo Alto), Coladas de relleno de las Cañadas de Diego Hernández, La Grieta, Montón de Trigo and Cañada Blanca; Llano de Ucanca and Natero de Chasogo (N-Cañada del Cedro).

Fig. 3.165 *Filago pyramidata*. (Photo credit: GG Casanova)

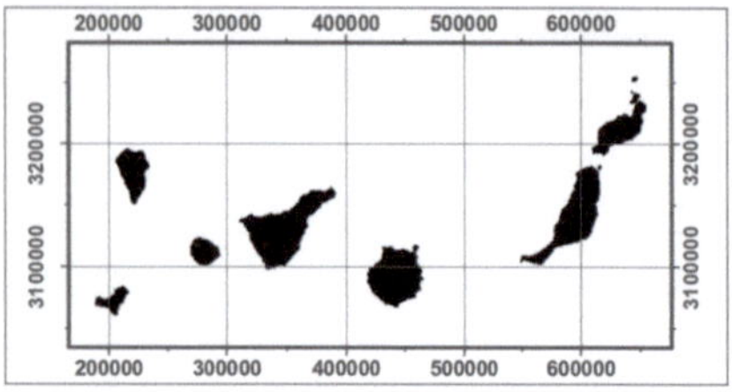

Fig. 3.166 Present on all the Canary Islands

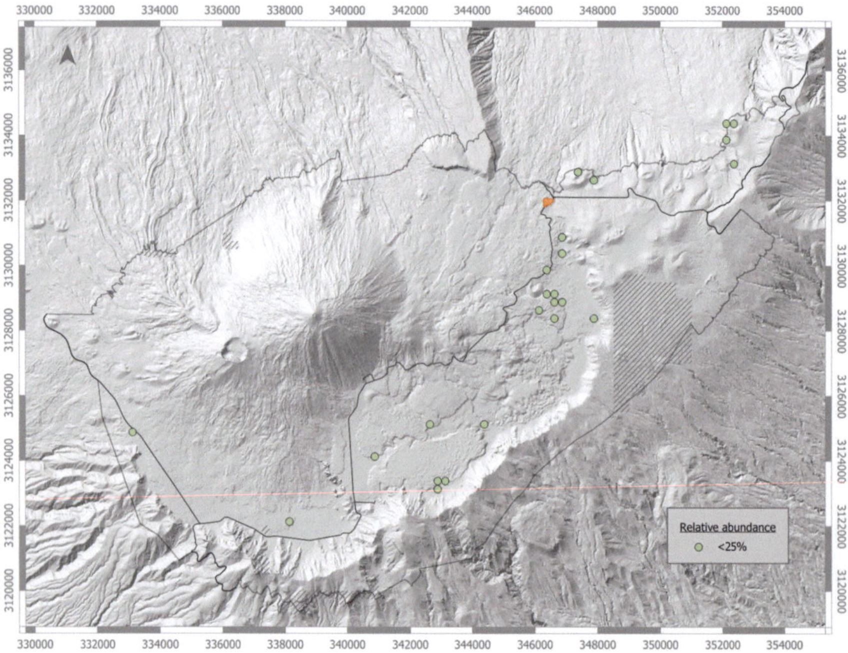

Fig. 3.167 *Filago pyramidata* distribution, N° Subgrids (250 × 250 m): **24**/3740

Habitat and Phytosociology

It is a characteristic species of the class ***Stellarietea mediae***, a type of anthropogenic annual nitrophilous vegetation, made up of therophytes or small geophytes of a weedy, ruderal, or nitrophilous character, with a Holarctic distribution and cosmopolitan irradiation. Specifically, it intervenes within the ruderal, nitrophilous and semi-nitrophilous herbaceous communities, order ***Thero-Brometalia***, subnitrophilous therophytic vegetation of the western Mediterranean, Ibero-Atlantic and Tingitano, dry, subhumid, humid, which shows some Cantabrian-Atlantic and Canarian irradiations. It preferably settles in abandoned crop fields and roadsides.

It can also appear in elements of the order ***Tuberarietalia guttatae*** xerophilous therophytic grasslands; Thermo and Mesomediterranean environments; oligotrophic skeletal soils, with a gravelly, silty or sandy texture, but with a certain cohesion.

Conservation Status

In practically all the subgrids in which this species was found (23), no damage of any kind was observed. Only in one (1) of them, its conservation status was determined as low.

Exsiccata (Herbarium Specimens)

Tenerife: Puertito de Güímar, 2.5.1971, E. Barquín Díez (TFC 21688); Cardonal de Güímar, 2.1979, W. Wildpret, P.L. Pérez de Paz & C. Hernández Padrón (TFC 20771); Santa María del Mar, 1.3.1979, W. Wildpret & M. del Arco Aguilar (TFC 20673); Taganana, 31.3.1979, P.L. Pérez de Paz (TFC 16644); Bco. de San Andrés, 24.3.1979, P.L. Pérez de Paz (TFC duplic. 33963); Santa Cruz, Presa de los Campitos, W. Wildpret, C. León Arencibia, M. del Arco Aguilar & J.R. Acebes (TFC 8860); El Escobonal, Bco. de Erques, 6.3.1981, O. Rodríguez Delgado & P.G. Cabrera (TFC 12878); El Escobonal, Laja de Erques, 23.3.1981, O. Rodríguez Delgado (TFC 12879); Lomo entre Bco. Afoche y La Ladera—Agache, 24.4.1981, E. Beltrán Tejera & O. Rodríguez Delgado (TFC 12886); Valle del Volcán de las Arenas, Cumbres de Arafo, 9.5.1981, W. Wildpret, M. del Arco Aguilar & P.L. Pérez de Paz (TFC 23712 + duplic.); Lomo entre Bco. Espigón y Bco. Pasada-Agache, 19.5.1981, E. Beltrán Tejera & O. Rodríguez Delgado (TFC 12881); Borde de Autovía TF-21, Túneles de Güímar, 19.11.1981, O. Rodríguez Delgado (TFC 12882); El Escobonal, Bco. de Erques, 4.4.1982, O. Rodríguez Delgado (TFC 12952); Ctra. entre Santiago del Teide y Masca, 24.4.1982, P.L. Pérez de Paz, J. R. Acebes, C. Hernández Padrón E. Beltrán Tejera & L. Negrín (TFC duplic. 33936); La Ladera, Camino Real, Agache, 19.3.1984, O. Rodríguez Delgado (TFC 27676); Bco. de Las Ánimas, 250 m.s.n.m., 26.7.1984, F. Ardévol González (TFC 38349); Lomo de Marrera, 17.2.1986, O. Rodríguez Delgado & W. Wildpret (TFC 27725); Arrastradero de Arriba, Lomo de Mena, 21.2.1986, O. Rodríguez Delgado (TFC 27727); Pájara de Abajo, 24.2.1986, ejusd. (TFC 27728); La Ladera, 6.3.1986, O. Rodríguez Delgado (TFC 27726); Lomo de La Cresta, Ladera bajo el Mirador, 17.3.1986, O. Rodríguez Delgado (TFC 27724); Pinar sobre el Bailadero, 29.5.1986, W. Wildpret, P.L. Pérez de Paz, M. del Arco Aguilar & O. Rodríguez Delgado (TFC 27723); La Laguna, Ma. Guerra, 11.4.1988, W. Wildpret, A. García Gallo & ECA (TFC 26504; 26505; 26506; duplic. 26507); Sobre Agache, 1200 m.s.n.m., 20.4.1990, M. Salas Pascual & M. del Arco Aguilar (TFC 37879); Fasnia, 30.3.1996, SICAH, GCMT (TFC 40252).

El Hierro: Los Lajiales, 22.3.1978, W. Wildpret et al. (TFC 8144); Sabinar, 23.3.1978, W. Wildpret, M. del Arco Aguilar, P.L. Pérez de Paz & J.R. Acebes (TFC 8258).

Gran Canaria: Lomo del Marco, 5.1980, P.L. Pérez de Paz & C. Suárez (TFC 11214); Los Pechos, 6.5.1995, W. Wildpret & V.E. Martín Osorio (TFC 37915).

Como *Filago* cf. *pyramidata* L.: Alrededores de Andén Verde, 400 m.s.n.m., 31.4.2003, S. Rivas-Martínez, M. del Arco Aguilar, J. R. Acebes & M.L. López (TFC 44356).

Lanzarote: Trasera del Colegio "Adolfo Thopman", 4.4.1993, J.A. Reyes-Betancort (TFC 37351); Haría, Altos del Bco. Madre del Agua, 23.2.1994, J.A. Reyes-Betancort, W. Wildpret & P.L. Pérez de Paz (TFC 39648 + duplic.); Bco. de Teneguime, 30.3.1994, J.A. Reyes-Betancort (TFC 38599); Teguise, Vega de San José, 12.3.1996, J.A. Reyes-Betancort, A. Losada Lima & C. León Arencibia (TFC 39745); Teguise, Lomo Blanco, 12.3.1996, ejusd. (TFC 39753); Haría, La Montaña, 25.4.1996, J.A. Reyes-Betancort (TFC 39652; 39654).

Fuerteventura: Bco. de Vinamar (Invent. 538 PLP), 26.2.1993, P.L. Pérez de Paz (TFC 35795).

3.5.54 Galium aparine *L.*

Sp. Pl., 108 (1753)

Galium barrelieri Salzm. *in* Flora 4: 107 (1821)
Galium ellipticum Willd. *ex* Hornem. *in* Hort. Bot. Hafn., Suppl.: 17 (1819)

Not endemic, Probable native
Family: *Rubiaceae*
Common name: Amor de jardinero o de hortelano, Cuajaleche, Raspilla, Raspilla cuajaleches, Raspasalla o Iguaje (ES): Cleavers, catchweed, goosegrass (EN)
Biotype: therophyte
Flowering: March–June
Fruiting: March–June
Pollination: autogamy
Dispersal: epizoochory
Chromosome N°: 2n = 22

Description

Stems up to 150 cm, climbing, very branched, markedly scabrous, with retrorse thorns at the angles and setose hairs above the nodes.

Leaves (10–) 20–60 × (1.5–) 3–8 mm, in whorls of 6–8 pieces, oblong-lanceolate or spatulate, abruptly mucronate, with retrorse thorns at least on the margin and above the vein.

Inflorescence narrowly cylindrical or subovoid (Fig. 3.168), with partially axillary cymes of 1–7 flowers. Pedicellate flowers. Whitish corolla; lobes 0.8–1.4 mm, ovate or triangular. **Anthers** 0.1–0.2 mm, ovoid. Ovary hairy. Mericarps 2.5–4 mm, densely covered with uncinate setose hairs with a markedly tuberculate base.

Fig. 3.168 Detail of *Galium aparine*. (Photo credit: G.G. Casanova)

Distribution

It is distributed throughout Europe, Asia, widely in the Macaronesia region and introduced in various parts of the globe. It is found throughout the Canary Islands (Fig. 3.169).

In the Teide National Park (Fig. 3.170) it appears occasionally at the head of the Barranco del Río and is more widespread in areas of the Dorsal: around La Tarta and Montaña Limón, northern slopes of Izaña and slopes of Caramujo (1900–2200 m a.s.l.).

Habitat and Phytosociology

It mainly inhabits crop areas, grasslands and undergrowth thickets. It is a characteristic species and bioindicator of ***Galio aparines-Urticetea maioris***, a class of Holarctic vegetation that groups anthropogenic, meso-hygrophytic perennial escio-nitrophilous communities, but with many biennial and annual species, in which large hemicryptophytes and geophytes with exuberant foliage, sometimes umbelliferous, as well as scandent plants, are abundant. They are most represented in temperate bioclimate territories; they are also frequent in rivers, irrigated lands and humid Mediterranean forests, and in thermoboreal forests. It can be present in communities of the ***Artemisietea vulgaris*** class, which groups nitrophilous, biennial or perennial ruderal communities, developed on soils enriched in nitrogenous or phosphorous nutrients, due to human activities and livestock. They are widely spread throughout the Mediterranean rain-seasonal and temperate Afro-Eurasian territories.

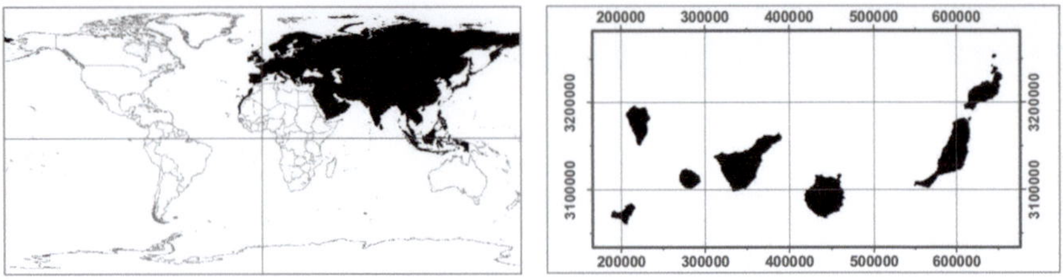

Fig. 3.169 Present on all the Canary Islands

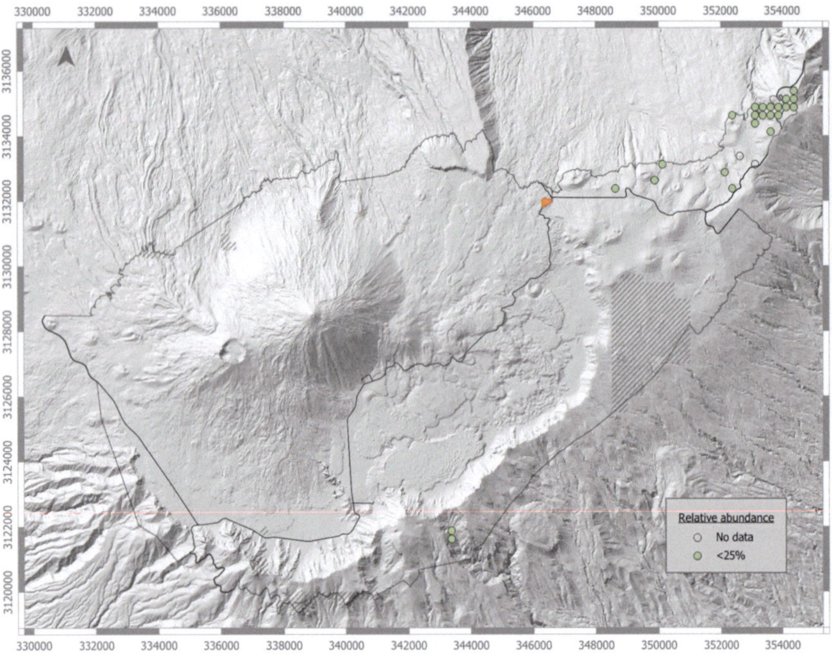

Fig. 3.170 *Galium aparine* distribution, N° Subgrids (250 × 250 m): **25**/3740

Conservation Status

In almost all the subgrids in which this species was found (23), no damage of any kind was observed. In the two (2) remaining, this variable was not specified at the time of sampling.

Exsiccata (Herbarium Specimens)

Tenerife: Taganana, E. Rodríguez Henríquez (TFC 4420); Bco. Tabares, 12.3.1970, E. Beltrán Tejera (TFC 35483); Canal del Tesoro, El Bailadero, 8.4.1972, E. Beltrán Tejera (TFC 16038); La Laguna. Mesa Mota, 3.4.1973, I. La Serna (TFC 23573); Agua García, Madre del Agua, 24.4.1980, A. García Gallo (TFC 10463 + duplic.);

La Laguna, Avenida de la Trinidad, 17.3.1980, AVR (TFC 8507); Ladera de Güímar, 16.4.1981, W. Wildpret, P.L. Pérez de Paz & M. del Arco Aguilar (TFC 13752); La Laguna, 21.1.1982, A. García Gallo (TFC 22393); Ibid, 16.2.1982, ejusd. (TFC duplic. 22394); La Laguna, 3.3.1982, A. García Gallo (TFC 22395 + duplic. 22396, 22397); Bco. del Río, 15.5.1983, J.S. Socorro (TFC 18459); Bco. de Badajoz, 12.3.1984, O. Rodríguez Delgado (TFC 29164); La Ladera, Camino Real, 19.3.1984, O. Rodríguez Delgado (TFC 27915); Arrastradero de Arriba, Lomo de Mena, 21.2.1986, O. Rodríguez Delgado (TFC 29165 + duplic. 29166); Arrastradero de Arriba, Lomo de Mena, ejusd. (TFC duplic. 29167); Mña. Gonzalianez, La Laguna, 16.4.1988, W. Wildpret, A. García Gallo & E. Carqué (TFC 26634; 26635 + duplic. 26637); Los Silos, sobre Las Canteras, UTM 28RCS 2038, 23.6.1989, V. Lucía Sauquillo & C. Acosta (TFC 38270); Arico, Madre del Agua (exterior), 29.4.1990, C.D. Hernández (TFC 33378); Arico, entrada Madre del Agua, 4.1990, ejusd., pliego citado en Tesis de Licenciatura, sin pliego en el Herbario (TFC 33379); Tegueste, La Padilla, 20.7.1993, T. Jiménez Felipe & EGG (TFC duplic. 36394); El Sauzal, El Puertito, 4.3.1994, F. Ramallo Rodríguez (TFC duplic. 37571); Los Silos, Tierra del Trigo, 10.5.1996, V. Lucía Sauquillo & M. del Arco Aguilar (TFC 39245 + duplic.); Los Silos, Vinatea, 9.3.1997, V. Lucía Sauquillo & C. Acosta (TFC 40058); Anaga, Sendero entre Jardina y la casa Forestal abandonada, Bco. de Tahodio, 5.4.1999, S. Becker & T. Himstedt (TFC 42054); Bco. Hoya de las Limeras, 300 m.s.n.m., UTM 28RCS 5950, 14.4.2002, M. Rodríguez López & M.L. Rodríguez López (TFC 44668). Inventario de la Flora y Vegetación del Parque Nacional del Teide (2001–2003) V.E. Martín Osorio, W. Wildpret & B. Hernández Bolaños: Al SW de La Tarta (31A 4), 05/04/2001, F. Tapia & Y. de Rioja Santos (TFC 44905); Guajara, Nacimiento del Barranco del Río (42C 14 c), 22/05/2003, A. Montañés Rodríguez (TFC 44238).

El Hierro: El Cres, 4.1979, W. Wildpret, M. del Arco Aguilar, P.L Pérez de Paz & C. Hernández Padrón (TFC 7900); Entre Sabinosa y Los Llanitos, 24.3.1998, P.L. Pérez de Paz & GCT (TFC 42420) (Termófilos 206–201/97) (Ver Inventario n.941 en la libr. n.18 de PLP).

Gran Canaria: Monte de los Tilos de Moya, 16.4.1973, E. Beltrán Tejera (TFC 33777); Moya, Bco. Oscuro, 18.2.1980, C. Suárez & A. García Gallo (TFC 11066); Bco del Cernícalo, Primavera de 1989, S. Ortega González & E. Fraile Bonafonte (TFC 30479).

Lanzarote: Altos del Bco. Chafarís, 23.2.1994, W. Wildpret, J.A. Reyes-Betancort & P.L. Pérez de Paz (TFC 39621); Yaiza, Atalaya de Femés, 18.2.1996, J.A. Reyes-Betancort (TFC duplic. 39875); Yaiza, Ladera SW de Hacha Grande, UTM 28RFS 185963, 21.2.1996, J.A. Reyes-Betancort (TFC 39876); San Bartolomé, El Islote, 13.3.1996, J.A. Reyes-Betancort, C. León Arencibia & A. Losada Lima (TFC 39713); Guíñate, La Cerca, 12.3.1996, ejusd. (TFC 39722).

Fuerteventura: La Oliva, 3.5.1980, W. Wildpret, A. García Gallo, C. León Arencibia, I. La Serna & M. del Arco Aguilar (TFC duplic. 21000).

Marzo 1970, J.R. Acebes Ginovés & J. Sánchez (TFC 214); 8.4.1972, W. Wildpret, E. Beltrán & P.L. Pérez Paz (TFC 932).

Uses, Benefits and/or Medicinal Properties

> "*Diuretic, aperitif, dermatological, antiulcer, antiscorbutic, vulnerary, anti-inflammatory; curdles milk. Useful part: flowering tops. Instructions for use: infusion, juice, poultice, ointment.*" (Translated from: Pérez de Paz and Hernández Padrón 1999)

3.5.55 Gastridium phleoides *(Nees & Meyen) C.E. Hubb.*

Kew Bull., 9, 375 (1954)

Agrostis phleoides (Nees & Meyen) É.Desv. *in* C.Gay, Fl. Chil. 6: 320 (1854)
Calamagrostis phleoides (Nees & Meyen) Steud. *in* Nomencl. Bot., ed. 2, 1: 251 (1840)
Gastridium ventricosum subsp. *phleoides* (Nees & Meyen) Tzvelev *in* Bot. Zhurn. (Moscow & Leningrad) 51: 1104 (1966)
Lachnagrostis phleoides Nees & Meyen *in* C.G.D.Nees von Esenbeck, Gramineae: 14 (1841)

New record for the Teide National Park and the island of Tenerife

Not endemic, Probable native
Family: *Poaceae*
Common name: Barriguilla (ES); nit grass (EN)
Biotype: therophyte
Flowering: May–June
Fruiting: May–June
Pollination: wind pollination (anemogamy)
Dispersal: wind dispersal (anemochory)
Chromosome N°: –

Description

Annual plant with 12–66 cm stems, erect or geniculate-ascending, glabrous.

Leaves glabrous, with 4–6 mm ligule and blade up to 12 × 5 mm. Panicle 3–15 cm, contracted, narrowly ellipsoid, with short, scabrid branches and peduncles.

Spikelets 5.5–7.5 mm (Fig. 3.171). Lower glume 5.5–7.5 mm, upper 4–6 mm. Lemma 2–2.5 mm, with 4 apical setae; the central ones c. 0.2 mm, the lateral ones c. 0.5 mm, densely hirsute, with a 6–9 mm awn, geniculate and loosely twisted at the bottom, hirsute towards the middle of the lemma. Anthers c. 1 mm. Caryopsis of c. 1 × 0.5 mm

Distribution

Present in E Africa, SW Asia and introduced to other parts of the globe.

Fig. 3.171 Detail of *Gastridium phleoides*. (Photo credit: G.G. Casanova)

In the Canary Islands it is found on the islands of El Hierro, La Palma, Tenerife and Fuerteventura (Fig. 3.172).

In the Teide National Park (Fig. 3.173), it was found forming important populations in western areas of La Fortaleza, near Montaña Mostaza and lava fillings of Cañadas de Diego Hernández and La Angostura, slopes of Chafira (western flank of Llano Ucanca) and northern surroundings of Punta de Los Roques. It was found occasionally along the heights of the La Orotava Valley (Montaña Limón, Montaña Guamasa and El Portillo), Cañada de La Grieta and Barranco del Río. Between 1900 and 2250 m a.s.l.

Habitat and Phytosociology

It appears preferably colonizing rocky environments. It is a characteristic species of the ***Thero-Brometalia*** order, subnitrophilous therophytic vegetation of optimal western Mediterranean, Ibero-Atlantic and Tingitano, dry, subhumid, humid; which shows some Cantabrian-Atlantic and Canarian irradiations. It settles preferably in abandoned fields of crops and roadsides.

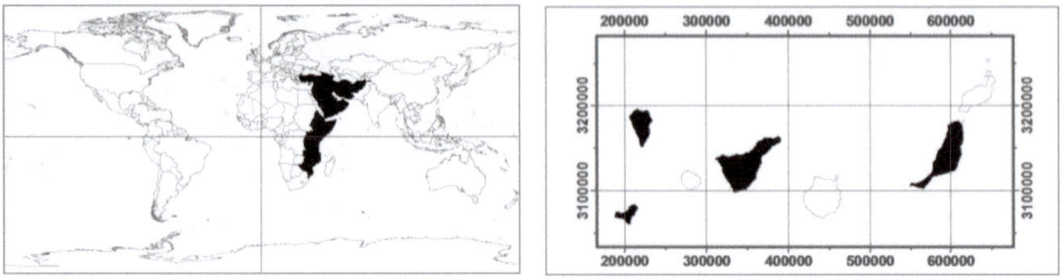

Fig. 3.172 Present on the islands of El Hierro, La Palma, Tenerife and Fuerteventura

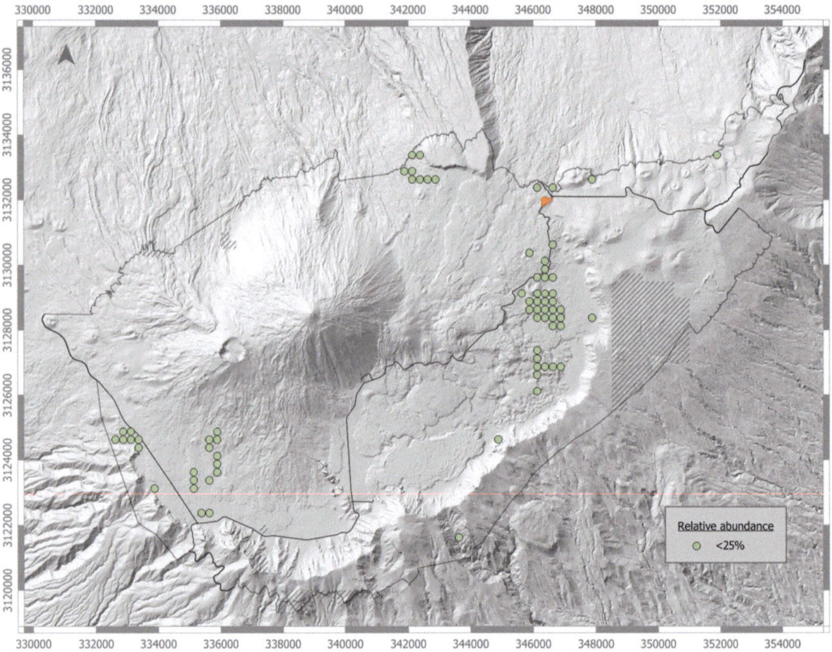

Fig. 3.173 *Gastridium phleoides* distribution, N° Subgrids (250 × 250 m): **70**/3740

Conservation Status

In practically all the subgrids in which this species was found (67), no damage of any kind was observed. In the three (3) remaining, this variable was not specified at the time of sampling.

Exsiccata (Herbarium Specimens)

Tenerife: Cerca del Roque Lombarda, Bco. de Tahodio, 3.5.1999, J.A. Reyes-Betancort (TFC 42148). Inventario de la Flora y Vegetación del Parque Nacional del Teide (2001–2003) V.E. Martín Osorio, W. Wildpret de la Torre & B. Hernández

Bolaños: Guajara, por Montaña Colorada (42C 15 c), F. Mesa, 09.06.2003, (TFC 44219).

Fuerteventura: Como *Gastridium* cf. *phleoides*: Al Este de Morro Velosa, 600 m.s.n.m., Macizo de Betancuria, 6.2002, S. Scholz (TFC 43874 + duplic.); Bco. de Vinamar, Ladera N, aprox. 400 m.s.n.m., Jandía, 5.5.2002, S. Scholz (TFC 43931).

3.5.56 Helianthemum juliae *Wildpret*

Vieraea 16:361–364 (1986)

Endemism, Certain native
Family: *Cistaceae*
Common name: Jarilla de Las Cañadas (ES); Cañadas Sunrose/Rockrose (EN)
Biotype: chamaephyte
Flowering: May–June
Fruiting: May–June
Pollination: entomogamy
Dispersal: –
Chromosome N°: 2n = 20

Description

Chamaephyte 30–60 cm tall, erect, with a woody trunk, twisted and dichotomously branched from the base; brown bark, sometimes reddish, easily detachable. Ascending, elongated branches; the basal ones perennial, somewhat woody, densely pubescent, slightly articulated, with short internodes and very marked leaf scars; middle and upper branches annual, glabrous, with the leaf whorls widely separated from each other. Hairs generally stellate, more rarely simple.

Dimorphic, sessile, revolute **leaves**; lower leaves densely hairy, canescent, linear-subspatulate; middle leaves less hairy, linear, acute, 1/4 larger, with hairs inserted preferably on the midvein; upper branches equal but glabrous, with the apex generally provided with 1–3 cilia, more rarely glabrous; Stipules subulate, equal per leaf, 2/3–1/2 shorter than these, ciliate like the leaves.

Inflorescence cymose, lax, elongated, with 5–12 unilateral flowers, bracteate. Floral pedicels always erect, 1 1/2 longer than the sepals, hairy, mainly at the apex. Bracts subulate, glabrous, 1/3 shorter than the pedicel. Outer sepals lanceolate, subglabrous, without prominent nerves, 1/2 smaller than the inner sepals. These are ovate-lanceolate, acute, approx. 6–8 mm, with four marked ribs, pale green and covered with scattered hairs, persistent, patent after fruiting. Yellow corolla (Fig. 3.174), 25–30 mm in diameter, regulate in appearance. Stamens with yellow filaments and auriculariform anthers. Style inflexed at the base, subclaviform towards the apex, surpassing the stamens.

Fig. 3.174 Teide National Park, June 2004 (**a**, **b**). (Photo credit: Victoria Eugenia Martín Osorio)

Capsule trigonous, obtuse, tomentose. Seeds numerous, irregular, 1 × 1 1/2 mm, polyhedral, black, finely tuberculated with white.

Remarks:

Although its germination potential is high, it has been observed that many seedlings die in the first stages of development. Germination tests carried out in an environment similar to that of nature have given satisfactory results.

Holotype: in Herb. TFC n° 23760 conservatus.

Distribution

Endemism of Las Cañadas del Teide, where only two populations were known, very close to each other, on the eastern edge of Las Cañadas (Cañada de Las Pilas, 2250 m a.s.l.). Until 1986 it had not been located outside its classic location, despite having been tracked in the Las Cañadas cirque with some intensity. During the 2001–2003 Inventory, two more populations were found, making a total of four known populations to date (Portillo Alto-Jardines de la Cruz Roja, Risco de Diego Hernández, Risco Verde and Risco de Las Pilas) and a small group of individuals planted in the Botanical Garden (Figs. 3.175 and 3.176).

Habitat and Phytosociology

Its populations are found in areas of the dry mesophytic Supra-Mediterranean bioclimatic zone where it is associated with a genuine legume thicket, made up of the following species: ***Spartocytisus supranubius***, ***Adenocarpus viscosus ssp. viscosus***, accompanied by ***Pterocephalus lasiospermus***, ***Cheirolophus teydis***, ***Pimpinella***

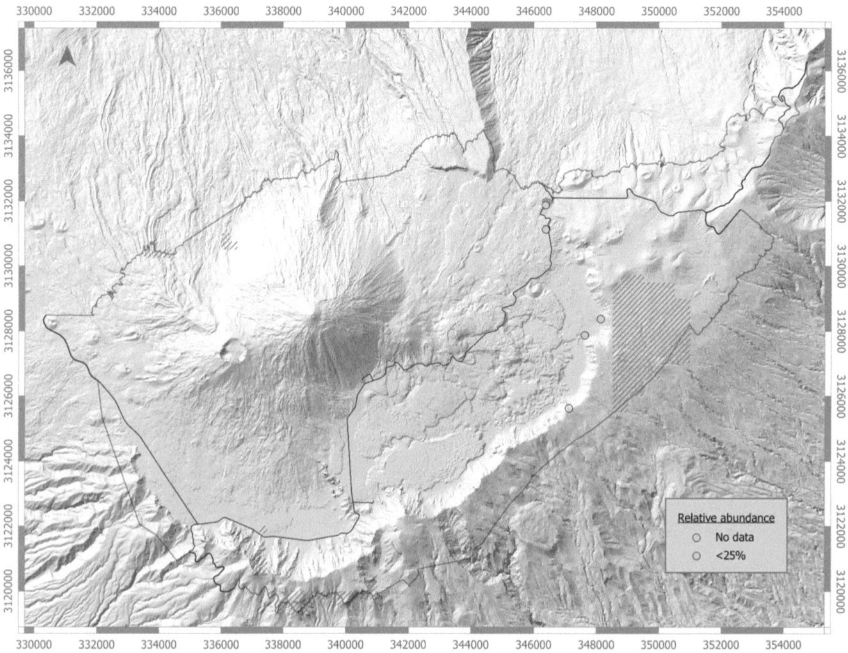

Fig. 3.175 *Helianthemum juliae* distribution, N° Subgrids (250 × 250 m): **5**/3741

Fig. 3.176 Endemism of the Teide National Park on the island of Tenerife

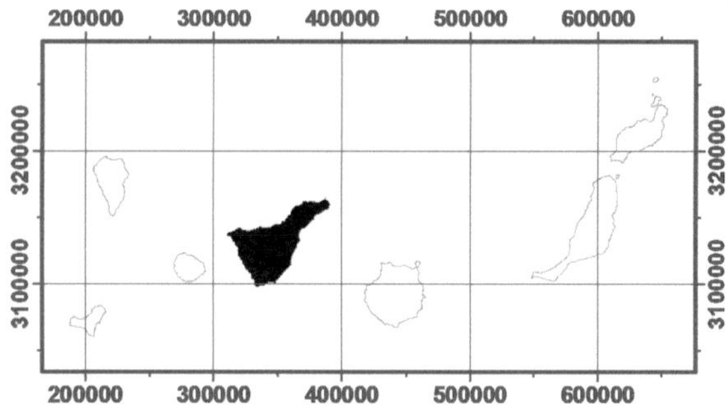

cumbrae*, *Nepeta teydea*, *Scrophularia glabrata and ***Tolpis webbii***, characteristics of the ***Spartocytision supranubii*** alliance.

> "*Species native to the high mountains of the Island of Tenerife, where it grows on ledges in cliff areas and occasionally on foothills, preferably in more or less shaded places facing northwest.*" (Translated from: Marrero Gómez et al. 2004)

Conservation Status

In most of the subgrids in which this species was found during the 2001–2003 sampling, no damage of any kind was observed. Only in the Risco de Diego Hernández was some damage detected (browsing, high mortality without regeneration, etc.). In the population of Risco de Las Pilas, this variable was not determined at the time of sampling.

Extremely rare taxon, of which only 166 wild individuals persist with regressive population trends. The following are noteworthy as threat factors: presence of introduced herbivores, scarce ecological plasticity of the species, a certain requirement for humidity in the substrate, infrequent in high mountains, with the current trend of increasing temperatures and decreasing seasonal rainfall and possible landslides. It is also affected by the presence of invasive exotic grass species that compete for the same ecological niche.

In the Teide National Park, a recovery programme is being developed which includes sending seeds to germplasm banks (E.T.S. of Agricultural Engineers, Jardín Canario Viera y Clavijo), intensive surveillance, control of introduced herbivores, study of its population biology, population reinforcements (of the 166 existing specimens, 63 are the result of this type of action). But in addition, the creation of new populations in suitable places for their self-maintenance is recommended, a study of their genetic variability and their inclusion in the Habitat Directive as a priority species (Marrero Gómez et al. 2004).

> "*Its recent description as a new species for science can serve as an indicator to highlight how rare and scarce this plant is, especially if we take into account that the peaks of Tenerife have been visited by botanists for more than two centuries. The small size of its populations makes this plant one of the rarest endemic species in the 'El Teide' national park. Although the only two known populations are within a protected area, this does not mean that they are out of danger, due to the presence of herbivores (rabbit and mouflon) in these territories.*" (Translated from: Gómez Campo et al. 1996)

Threat Category

Anexo I: Strictly protected (Gobierno de Canarias 1991)
EN (U.I.C.N., 1994)
CR B1 + 2ce (Bañares et al. 1998)
E (Walter and Gillett 1998)
EN D (VVAA 2000)
En peligro de Extinción (Gobierno de Canarias 2001)
CR B2 ab (v) c; C2 a (i); E (Bañares et al. 2004)
CR B2 ab (v) c (v); C2 a (i); E (Moreno 2008)
En peligro de Extinción (Gobierno de Canarias 2010)
En peligro de Extinción (Ministerio de Medio Ambiente, Medio Rural y Marino 2011)

Germplasm Banks

E.T.S. Ingenieros Agrónomos de Madrid.
Jardín Botánico Canario Viera y Clavijo.

Exsiccata (Herbarium Specimens)

Tenerife: Las Cañadas del Teide, 2.06.1973, W. Wildpret, H. Metlesics, E. Beltrán, CGR, P.L. Pérez de Paz & M. Fernández (TFC 23760 + Dupl.: Holotipo); Topo de la Grieta, 2.06.1973, P.L. Pérez de Paz (TFC 23760 + Dupl.); Cañada de Diego Hernández, 2.06.1973, CGR (TFC 24083 + Dupl.); Ibid., 18.06.1983, W. Wildpret & S. Socorro (TFC 13164, 32149); Risco Verde, 2250 m.s.n.m., 18.06.1983, ejusd. (TFC 34263). Inventario de la Flora y Vegetación del Parque Nacional del Teide (2001–2003) V.E. Martín Osorio, W. Wildpret de la Torre & B. Hernández Bolaños: Alrededores del Centro de Incendios del Parque Nacional del Teide, 25/06/2003, A. Montañés Rodríguez (TFC 44240); Jardín de la Casa de la Cruz Roja del Parque Nacional del Teide, 13/11/2003, V.E. Martín Osorio, W. Wildpret de la Torre & B. Hernández Bolaños (TFC 44929).

3.5.57 Hemionitis guanchica (Bolle) Christenh.

First published in Global Fl. 4: 15 (2018)

Cheilanthes guanchica Bolle *In Bonplandia*, 7, 107 (1859)
Allosorus guanchicus (Bolle) Christenh. *in* Willdenowia 42: 284 (2012)
Cheilanthes pteridioides subsp. *guanchica* (Bolle) O.Bolòs, Vigo, Masalles & Ninot *in* Fl. Man. Països Catalans: 1213 (1990)
Oeosporangium guanchicum (Bolle) Fraser-Jenk. & Pariyar *in* Annot. Checkl. Ind. Pterid. 1: 265 (2016)

Not endemic, Certain native
Family: *Sinopteridaceae*
Common name: –
Biotype: hemicryptophyte
Sporulation: April–July
Dispersal: wind dispersal (anemochory)
Chromosome N°: 2n = 120; n = 60 (allotetraploid)

Description

Rhizome short, covered with linear-lanceolate rusty paleas.

Fronds up to 24 cm; petiole brown, with abundant linear palea towards the base, rare in the upper part; lamina subequal or slightly smaller than the petiole, 2–3 pinnate, ovate-oblong to ovate-lanceolate (Fig. 3.177); rachis paleaceous; pinnules glabrous, ovate-lanceolate, lobed to 14-pinnatisect with lanceolate lobes. Pseudoindusium wide (0.5–1 mm), subcontinuous, sometimes lobed and then with subrectangular lobes, at the end rusty, with entire edge, sometimes slightly denticulate. **Spores** 43–65 µm, with crested-reticulate perisporium.

Fig. 3.177 *Hemionitis guanchica* February 2012. (Photo credit: Victoria Eugenia Martín Osorio)

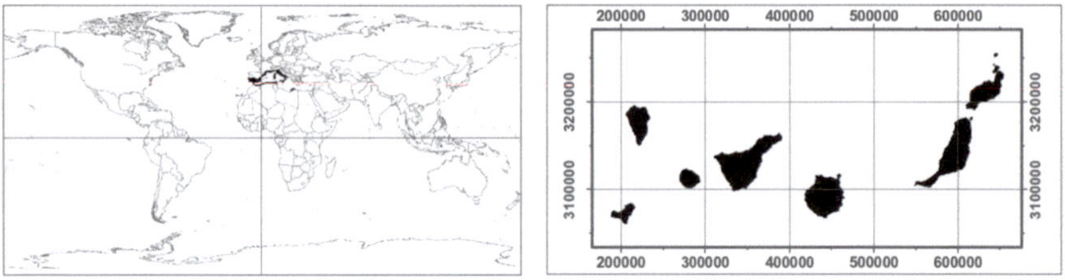

Fig. 3.178 Present on all the Canary Islands

Remarks:

Allotetraploid derived from the crossing of ***Ch. maderensis*** and ***Ch. pulchella***, it usually forms hybrids with ***Ch. acrostica***, which are difficult to identify.

Distribution

Its distribution is Mediterranean-Iberian and Canarian (Infra-, Thermo- and Mesomediterranean belts). It appears throughout the Canary Islands and in Madeira (Fig. 3.178).

In the Teide National Park (Fig. 3.179) it extends mainly along the interior of eastern Las Cañadas, La Fortaleza and El Portillo and, in a more dispersed form,

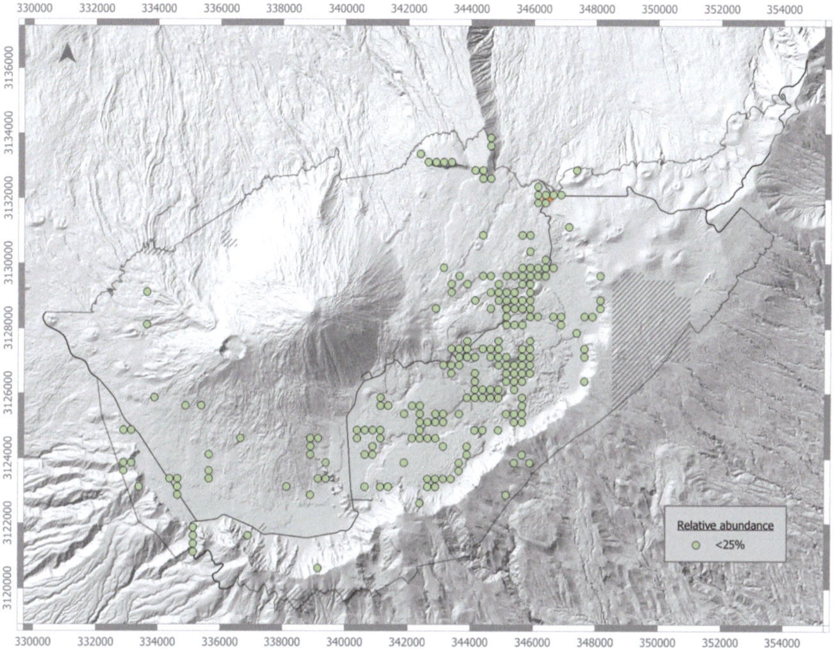

Fig. 3.179 *Hemionitis guanchica* distribution, N° Subgrids (250 × 250 m): **220**/3740

through western Las Cañadas (1900–2400 m a.s.l.). It can easily be found on a seeping wall at the entrance to the National Park near La Crucita.

Habitat and Phytosociology

The taxon is typical of rock crevices, mainly in those of basic silicates. It is well adapted to dry environments, such as south-facing earthen walls.

It is a characteristic species and bioindicator of ***Notholaeno marantae-Cheilanthetalia maderensis***, an order that represents ultramafic rock communities, mainly pteridophytic, Infra-Supra-Mediterranean arid-humid, with a Mediterranean distribution.

It can also intervene in ***Sileno berthelotianae-Cerastietum sventenii***, a rock association of phonolitic cracks and spurs and trachybasalts oriented to N and NW of the dry Supra-Mediterranean belt. It is found at altitudes between 2200 and 3000 m a.s.l., but always in windward situations within the scope of the communities of ***Spartocytisetea supranubii***, ***Spartocytision*** and ***Violion cheiranthifoliae***. This community replaces ***Cheilantho guanchicae-Aeonietum smithii*** at altitudes above 2200 m a.s.l.

Conservation Status

In all the subgrids in which this species was found (220), no damage of any kind was observed.

"*The plant is very rare in Las Cañadas and almost always very scarce.*" (Sventenius 1946)

Exsiccata (Herbarium Specimens)

Tenerife: Mña. Mostaza, 10.5.1969, A. Santos (TFC 23619); Ibid, 5.1969, ejusd. (TFC 33718); La Fortaleza, 1.1970, C. Gil Rodríguez (TFC 33717); La Fortaleza, 8.1971, A. A. González (TFC 33719); Guajara, 15.2.1972, M.C. Gil (TFC 33817); Bco. de Erques, 20.1.1973, C. Gil Rodríguez & C. León Arencibia (TFC duplic. 32776); Boca de Tauce, 3.1973 y 10.1990, ejusd. (TFC 32777); Arafo, 1970 y 1990, ejusd. (TFC 32778); Bco. de Araya, Igueste de Candelaria, 28.3.1972 y 10.1990, P.L. Pérez de Paz & C. León Arencibia (TFC 32781); ?, 4.1973, W. Wildpret, A. Acuña, H. Metlesics, E. Beltrán Tejera, L. Gallo, M.C. Gil & Ginés López (TFC 33792 + duplic.); Las Cañadas, Los Roques, 21.12.1979, A. Santos (TFC 23283); La Fortaleza, El Sombrero y El Cabezón, 28.7.1971 y 5.8.1971, F. Estévez & A. Acuña (TFC 21751); Los Realejos, 4.1974, C. L. García Palmero (TFC 28579); Barranquillo sobre El Bailadero, 29.5.1986, W. Wildpret, O. Rodríguez Delgado, P.L. Pérez de Paz & M. del Arco Aguilar (TFC 27505); Los Silos, Cuevas Negras, 20.5.1989, V. Lucía Sauquillo & C. Acosta (TFC 40613); Pinar de Chío, 23.2.1992, P.L. Pérez de Paz (TFC 35230); Guía de Isora, Malpaís de Chío, 13.3.1999, M.C. León Arencibia & M.M. Pérez (TFC 42017); Ibid, 13.3.1999, ejusd. (TFC 42023); Inventario de la Flora y Vegetación del Parque Nacional del Teide (2001–2003) V.E. Martín Osorio, W. Wildpret & B. Hernández Bolaños: Cañada del Montón de Trigo, junto al monte de la laguna, 01.01.2003, W. Wildpret de la Torre (TFC 44939); Roques al Sur del Llano Ucanca (41D 37 c), 21.04.2003, Y. de Rioja Santos (TFC 44232); Jardín Botánico (P.N. Teide) (P23) (30D 11 b), 17.06.2004, B. Hernández Bolaños (TFC 45401).

Como *Cheilantes sventenii* Benl.: Las Cañadas, La Fortaleza, 19.5.1973, P.L. Pérez de Paz (TFC 23337).

El Hierro: Taibique, 4.1979, P.L. Pérez de Paz & C. Hernández (TFC 8226); Malpaso, 10.4.1979, W. Wildpret, M. del Arco Aguilar, P.L. Pérez de Paz & C.Hernández (TFC 8270); La Chiquita, 700 m.s.n.m., 11.4.1979, ejusd. (TFC 7858).

Como *Cheilantes* cf. *guanchica* Bolle: El Julán, 800 m.s.n.m., 27.3.1988, M. del Arco Aguilar, P.L. Pérez de Paz, V. Lucía Sauquillo & M. Salas Pascual (TFC 30054).

La Palma: Como *Cheilantes sventenii* Benl.: Mazo, Poleal, 3.1968, A. Santos (TFC 23637).

Como *Cheilantes* cf. *guanchica* Bolle: Roque del Huso, 29.4.2000, E. Beltrán Tejera (TFC 42663).

La Gomera: El Cedro, 22.2.1973, C. Gil Rodríguez & C. León Arencibia (TFC duplic. 32776).

Como *Cheilantes* cf. *guanchica* Bolle: Roque de la Verruga, Ctra. Jerduñe, 700 m.s.n.m., 7.3.2002, R. González, González, M.V. Cabrera (TFC 43545).

Gran Canaria: Mte. de Inagua, Andenes de Tasarte, 12.1984, MN & M.C. Gil (TFC 23654); Artenara, El Pinar, 15.3.1992, M. Salas Pascual (TFC 37134).

Lanzarote: Tinajo, Lavas de Mña. Colada, 11.5.1996, A. Marrero & M. Glez. Martín (TFC 40108).

Fuerteventura: Como *Cheilantes* cf. *guanchica* Bolle: Tuineje, Mña. Carbón, 600 m.s.n.m., 2.1.2000, S. Scholz.

3.5.58 Hypericum grandifolium *Choisy*

Prodr. Monogr. Hypéric., 38 (1821)

Androsaemum grandifolium (Choisy) Rchb. *in* Icon. Fl. Germ. Helv. 16: 70 (1854)
Androsaemum webbianum Spach *in* Hist. Nat. Vég. 5: 418 (1836)
Hypericum canariense Webb & Berthel. *in* Hist. Nat. Iles Canaries 3(2; 1): 49 (1836), pro syn.
Hypericum erectum Sol. *ex* R.Br. *in* Abh. Preuss. Akad. Wiss., Phys.-Math. Kl. 1816–1817: 75 (1819), not validly publ.

Not endemic, Certain native
New record for the Teide National Park
Family: *Hypericaceae*
Common name: Corazoncillo, Malfurada o Malfurada grande,
Leña de Brujas, Flor de la cruz, Hierba de San Juan (ES); Large-leaved St. Johns-Wort (EN)
Biotype: tall shrub (chamaephyte—nanophanerophyte)
Flowering: early winter–summer
Fruiting: summer–autumn
Pollination: insect pollination (entomogamy)
Dispersal: wind dispersal (anemochory)
Chromosome N°: 2n = 40

Description

Erect **shrub**, branched from the base, 0.5–1 or up to 1.8 m tall, glabrous, without black glands. Young twigs reddish-brown, with 4 prominent longitudinal lines. Deciduous or semi-evergreen **leaves**, opposite, sessile, subamplexicaul, broadly triangular-ovate or oblong-ovate, with obtuse or rounded apex, cordate or rounded base, flat, 40–90 × 25–45 mm, papery, upper and lower surface dark green, lower surface not glaucous, with translucent glands.

Inflorescences in loose panicles, with 1–13 flowers. Star-shaped flowers, 3–4.5 cm in diameter. Sepals 5, 6–8 mm long, persistent in the fruit. Petals 5, golden yellow, not tinged with red, narrowly oblong-lanceolate, 18–22 mm long, deciduous after anthesis. Numerous stamens in five fascicles (Fig. 3.180). Styles 3. Coriaceous capsules, completely dehiscent, ellipsoidal, 8–13 mm long, dark brown when ripe. Seeds small and very abundant.

Fig. 3.180 *Hypericum grandifolium*, April 2009 (**a**) and May 2004 (**b**). (Photo credit: Victoria Eugenia Martín Osorio)

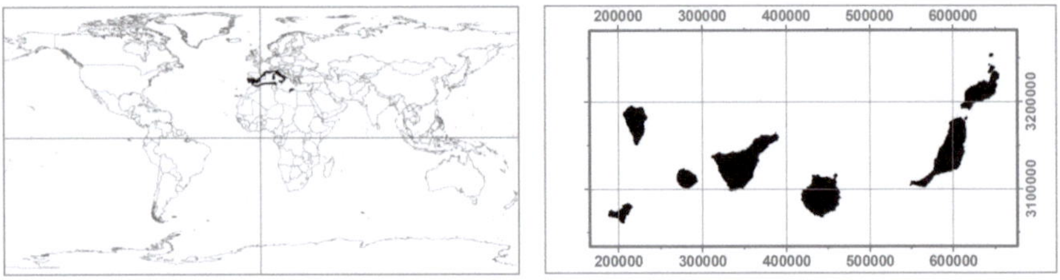

Fig. 3.181 Present on all the Canary Islands

Distribution

Macaronesian endemism. In the Canary Islands (Fig. 3.181) it is found on all the islands and widely distributed from sea to summit.

In the Teide National Park (Fig. 3.182) it reaches 2250 m a.s.l., and has been located between La Crucita, La Tarta and Cueva Los Lajones, Colorado Grande (highs of the La Orotava Valley), southern slopes of Izaña, Vista La Chapa (E-La Fortaleza) and Barranco del Río.

Habitat and Phytosociology

It reaches its optimum in the degraded scrublands of cloud forest communities, and is common in Monte Verde (laurel forest) and pine forest areas, occasionally found in higher areas (Thermo-Mesomediterranean, subhumid-dry).

It is a characteristic species and bioindicator of ***Pruno hixae-Lauretea novocanariensis***, a vegetation class that groups the Canarian-Madeiran climatophilous cloud forests and heminebulous lauroid forests, the serial ericoid or retamoid microforests and mountain shrublands, and the broadleaf microforests and rivular deciduous brambles, which together are commonly called "Monte Verde". In the Canary Islands these formations are typical of dry-humid hyperoceanic cloudy Infra-Mesomediterranean bioclimates. Within it, it intervenes in the order ***Andryalo***

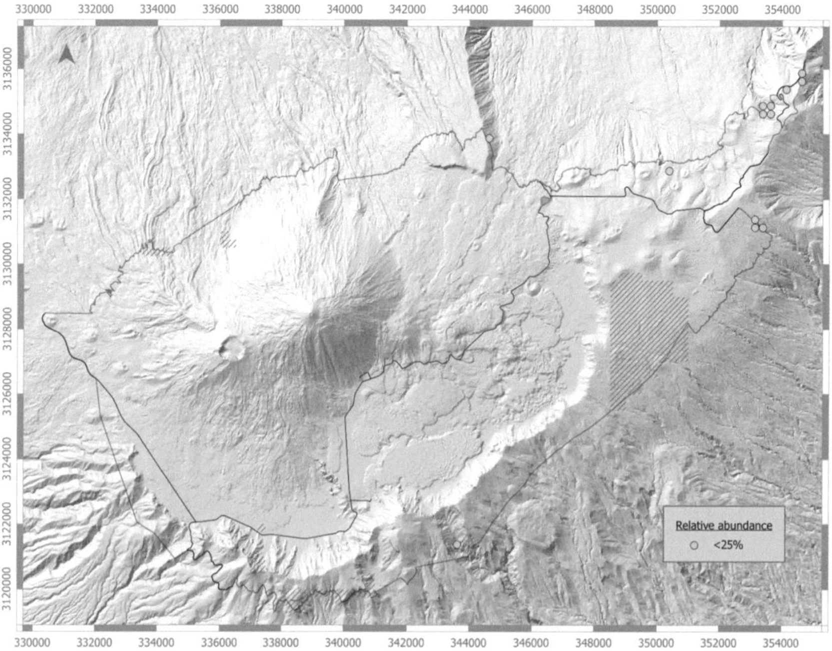

Fig. 3.182 *Hypericum grandifolium* distribution, N° Subgrids (250 × 250 m): **13**/3740

pinnatifidae-Ericetalia canariensis, which represents the ericoid or retamoid mountain shrublands and microforests, serial and fringe of the climatophilous lauroid cloud forests, of Canary-Madeiran distribution.

Conservation Status

In almost 70% (9) of the subgrids in which this species was found, no damage of any kind was observed. In the remaining 30% (4), damage was detected (browsing, high mortality without regeneration, etc.), but this did not affect more than 25% of the specimens.

Threat Category

LR pm (Bañares et al. 1998)

Germplasm Banks

E.T.S. Ingenieros Agrónomos de Madrid.
Jardín Botánico Canario Viera y Clavijo.

Exsiccata (Herbarium Specimens)

Tenerife: Las Mercedes, V.1970, Caballero Ruano, A., 2ª Prom. (TFC 4308); Las Carboneras, Anaga, 20.V.79, del Arco Aguilar, M., (TFC Duplic. 8626); Las Mercedes, VI.79, Viera, A., G. Rivero, (TFC 20573); Ctra. Forestal del Roque de los Pasos. Anaga, 2.IX.79, Hernández Padrón, C., P.L. Pérez de Paz, (TFC Duplic. 21662); Madre del Agua. Agua García. 14.VIII.1979, Wildpret de la Torre, W., A. García Gallo, L. Negrín (TFC 10393); Madre del Agua, Agua García, 27.VI.1980, Wildpret de la Torre, W., A. García Gallo, (TFC 10394); Bco. de Badajoz, 4.V.84, Charpin, A., O. Rodríguez Delgado, (TFC 27766); Masca, 3.V.84, Lecuona, R, (TFC 13216 P-TFC 377); Anaga, Bco. de Chinobre, 800 m.s.n.m., 25.V.84, Pérez de Paz, P.L., C. Hernández Padrón, (TFC 13251, P-TFC 412); Laurisilva del Monte de Las Mercedes, 15.II.92, Pérez Hernández, D., Alumnos de 5° curso., P.L. Pérez de Paz, (TFC 35120); Junto Charcos de Erjos, Los Silos, 23.VII.94, Lucía Sauquillo, V., C. Acosta, (TFC 38195); La Laguna, Monte de Anaga, 920 m.s.n.m., 18.IX,96, Acebes, J.R., (TFC 40893, Duplic. 40304); Bco. Ijuana, Ladera N, 650 m, 11.VI.96, Racca, R., (TFC 39012); Pedro Álvarez, (Test Fitoquímico), 2.VI.99, Sánchez Mateo, C., (TFC 42213); Icod Alto, El Reventón, (El Amparo), (M. Test), 8.V.99, Sánchez Mateo, C.C., (TFC 41953); Bco. de Tahodio, 17.IV.99, Becher, S., T. Himsted, (TFC 42256); La Esperanza, C/El Sol n°39, 7.V.2002, La Serna Ramos, I., J. Afonso Vargas, (TFC 43601). Inventario de la Flora y Vegetación del Parque Nacional del Teide (2001–2003) V.E. Martín Osorio, W. Wildpret de la Torre & B. Hernández Bolaños: Guajara, Nacimiento del Barranco del Río (42C 23 a), 09/06/2003, F. Mesa (TFC 44221).

Como *Hypericum inodorum*: 26.VI.69, Bramwell (TFC 247); Arafo, Altos de Hoya Fría, 27.VI.71, Barquín Díez, E., (TFC 21767); Bco. de Badajoz, Güímar, 31.V.73, La Serna Ramos, I., (TFC 35767); El Tablado, 5.VII.73. Beltrán Tejera, E., (TFC 27183); Bco. del Cuervo (Trevejos), 25.VII. 73, La Serna Ramos, I., (TFC Duplicatas 27184); Anaga, Pico del Inglés, 16.IV.79, Fernández Casas, J., (TFC 20731); Las Carboneras, 8.VIII.79, Bañares, A., (TFC 21131); Bco. Seco, Adeje, 14.IV.83, Pérez de Paz, P.L., L. Negrín, (TFC Duplicc. 29387); Bco. de los Riachuelos, 1.VI.83, Socorro, J.S., (TFC 18464); Icod, Palopique, 500 m., 29.VI.84, Ardévol González, F., (TFC 19174); Las Mercedes, El Moquinal, 2.VI.85, Pérez de Paz, P.L., (TFC 24484 Duplicatas); Las Cañadas, Topo de la Grieta, 20.VI.85, Pérez de Paz, P.L., I. La Serna Ramos, IMM, (TFC 24479 Duplicatas); Pico de las Flores. Cumbre central, 20.VI.85, Pérez de Paz, P.L. et al. (TFC 24487 Duplicatas); Pico del Inglés. Las Mercedes, 12.VII.85, IMM, (TFC 24485 Duplicatas); Icod, Los Márquez, 8.VII.1985, Ardévol González, F., (TFC 37990); Icod, 19.XI.86, Ardévol González, F., (Ruiblás. TFC 38317); Bco. de Araba, 550 m.s.n.m., 30.X.88, MMG., (TFC 29764); Los Realejos, Galería "Toste" 470 m.s.n.m. Pared izqda. entrada, 8.VI.90, Hernández, C.D., (TFC 33306), incomplet Herbar. material citado en la tesina de C.D. Hernández; Pista Casa Forestal Los Realejos, Altos de la Orotava, 26.VI.90, Pérez de Paz, P.L., D. Pérez, (TFC Duplic. 32595); Entrada galería "Florida baja" Los Realejos, 19.V.91, Hernández, C.D., (TFC 33420).

Uses, Benefits and/or Medicinal Properties

> "*Cicatricial, antihysterical, sedative, hypocholesterolemic, pectoral. Useful part: whole plant, flowers, fruits. Instructions for use: juice, infusion, poultice.*" (Translated from: Pérez de Paz and Hernández Padrón 1999)

3.5.59 Juniperus cedrus *Webb & Berthel. ssp.* cedrus

Phyt. Canar., 3(2/3), 277 (1844)

Endemism, Certain native
Family: *Cupressaceae*
Common name: Cedro, Cedro canario o Sabina (ES); Canary Island Juniper (EN)
Biotype: tree (phanerophyte—mesophanerophyte)
Flowering: May–July (flower ♂: February–March)
Fruiting: July–September
Pollination: wind pollination (anemogamy)
Dispersal: self and animal dispersal (barochory, endozoochory)
Chromosome No: 2n = 22

Description

Tree that can reach 5 m in height; arched, hanging branches, dark flexible stems, the youngest 3-gonous, glaucous (Fig. 3.183).

Leaves distributed in threes, 7–8 mm long, 0.5 mm wide, stacked or subspaced, glaucous or pale green, ascending or spreading, convex carinate below, canaliculate above, prominent midvein two crowded glaucous lines, sessile, subdilated base, obtuse or acute apex, sharp.

Fig. 3.183 (**a**) Risco Verde, May 2004. (Photo credit: Beatriz Hernández) (**b**) Alto de La Fortaleza. (Photo credit: Victoria Eugenia Martín Osorio)

Male axillary catkin ovate, obtuse, scales distributed in threes whorled, rhomboid-rounded yellow, upper stamens purple, at the base 2 or 3 fovent. Female catkin 3-ovulate, elongated ovules, naked horny apex.

Orbicular galls, the youngest bluish, oval apex exerted tricorne, the mature orbicular baccata orange, monospermous. Seed subtriangular or compressed-orbicular, acute margin often covered by a deciduous membrane, testa bony, tegmen very thin, cylindrical, surrounded by a thick perisperm. Embryo in elongated perispermic axis, spatulate, thick plane of the cotyledon elongated-ovate, radicle cylindrical, ending in a suspended funiculus. Reproductive biology: Dioecy and, exceptionally, sometimes Monoecy. The galls mature the second year during the months of March to June. For their germination to occur, they must pass through the digestive tract of birds, especially corvids.

According to Rumeu et al. (2015): The study of the seed dispersal system revealed that the winter visitor ***Turdus torquatus*** is the most effective dispersing agent, and that the endemic lizards ***Gallotia galloti*** play an important quantitative role, especially in the warmer months. The extinction of the raven Corvus corax in Teide National Park implies a strong dependence of ***J. cedrus*** on the visit of the wintering population of ***T. torquatus*** for medium- or long-distance dispersal of its seeds, which denotes a certain fragility in its dispersal system. The molecular approach carried out in the Riscos de la Fortaleza population indicated that the gene flow of ***J. cedrus*** does not have a panmictic dynamic, but that there is a spatial genetic structure and the population is subdivided into 13 groups or neighborhoods of related individuals. The inbreeding rate of the population is significant and, furthermore, within the detected neighbourhoods, inbreeding has a significant and negative influence on the fertility of the mother plants.

Several cedar specimens from the Teide National Park have been dated, resulting in the so-called “Patriarch” of Montaña Rajada being dated to 1050 (±30 years BP [Before Present]), which is 1120 ± 30 years old. Another specimen of Canary cedar (***Juniperus cedrus***) located in the Teide National Park is also considered to be the oldest tree in the European Union. Thanks to the radiocarbon technique, a dating of 1481 years of age has been obtained. The 1481-years-old juniper as well as “The Patriarch” are both females and they are still producing viable fruits, suggesting that there are no age limitations to fruit production (Sangüensa-Barreda et al. 2022).

Distribution

Endemic species of the islands of La Palma, La Gomera, Tenerife and Gran Canaria (Figs. 3.184 and 3.185). On the island of Tenerife it is distributed from 900 m above sea level to the edge of the Pared de Las Cañadas, at about 2675 m above sea level (Montaña Guajara).

In the Teide National Park, it develops along the rocky areas of the entire escarpment and slopes of Las Cañadas (from Montaña del Cerrillar to Cañada del Cedro), including the Roques de García. Outside of La Pared, it also colonizes the escarpments and slopes of La Fortaleza and, in a more dispersed form, the later lava fillings of the eastern interior of Las Cañadas (area of Montaña Mostaza, Minas de San

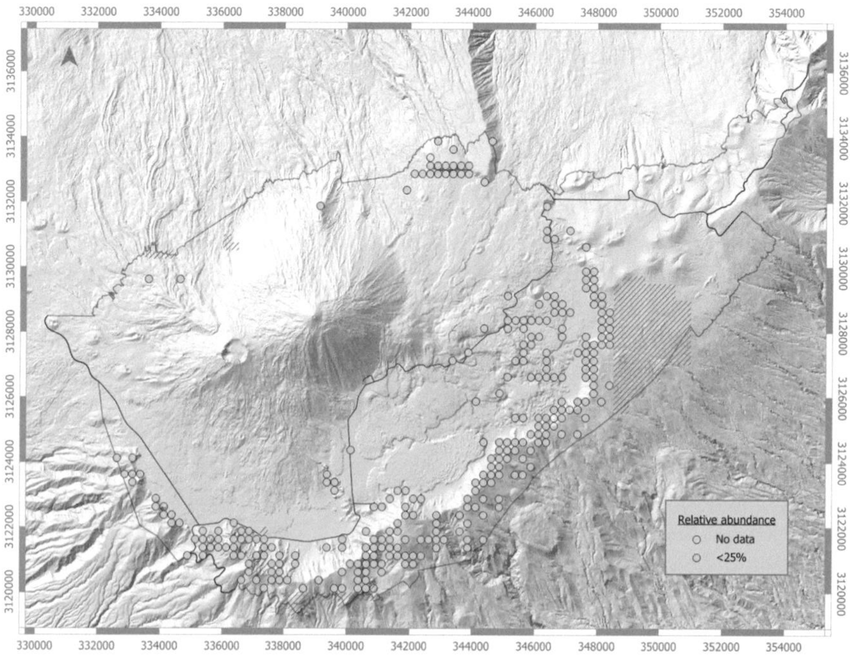

Fig. 3.184 *Juniperus cedrus* subsp. *cedrus* distribution, N° Subgrids (250 × 250 m): **288**/3741

Fig. 3.185 Endemism of the central and occidental islands

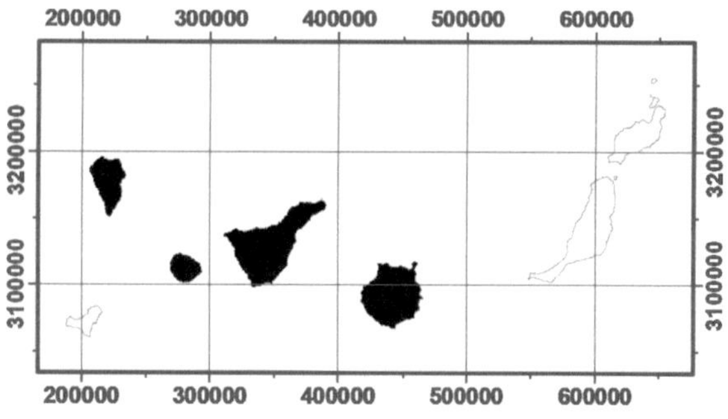

José, Valle de Las Piedras Arrancadas and Malpaís de Las Pilas). Very occasionally, individuals were also found in the lavas of Cerro Las Chajoras (NW-Pico Viejo Volcano) and Echisere (N-Teide Slopes). Also, in the Wolfredo Wildpret Botanical Garden of the Teide National Park.

Habitat and Phytosociology

Cedral of the high Tenerife summit and La Palma, where it appears as a companion in communities of broom and pine forests. On the islands of Gran Canaria and La Gomera, it forms part of communities and complexes of rock vegetation.

It is present in communities of the ***Chamaecytiso-Pinetea canariensis*** class, ***Cisto-Pinion canariensis*** alliances and ***Spartocytision supranubii***.

It defines the ***Juniperion cedri*** alliance, which brings together the formations of Canary cedars installed on escarpments, cliffs and rocks of the wall or precaldera building of Las Cañadas del Teide in Tenerife and the escarpments of the Caldera de Taburiente in La Palma.

It is a characteristic species of ***Adenocarpo viscosi-Juniperetum cedri***, the head of the Tenerife edapho-oxerophilous series of Mediterranean lithosolic pluviseasonal oceanic Supra-Mediterranean dry-subhumid euoceanic ***Juniperus cedrus*** and ***Adenocarpus viscosus*** with ***Echium wildpretii*** and ***Pterocephalus lasiospermus***. Its current distribution is associated with some of the oldest volcanic materials in the territory. It is a cedar forest that in the past may have been more widely distributed and that currently represents the potential vegetation of the high Canary summit (Martín Osorio et al. 2007, 2015)

On the island of La Palma, the association of which it is characteristic takes the name of ***Adenocarpo spartioidis-Juniperetum cedri***, a humid Supra-Mediterranean edaphoxerophilous series, also developed on rocky escarpments and cliffs of materials prior to the gravitational landslides of the high summit.

Conservation Status

According to the data collected during the sampling, in 94% (271) of the subgrids in which this subspecies was found, no damage of any kind was observed. Only in 3.5% (10), damage was detected (browsing, high mortality without regeneration, etc.), but this did not affect more than 25% of the specimens. Damage was found on more than 25% of the specimens in only 1% (3) of the subgrids where this taxon appeared. In the rest (4), this variable was not specified at the time of sampling.

This species is relatively well preserved on all the islands, except in Gran Canaria, where its populations are very fragmented and with very few individuals. On the rest of the islands, the populations tend to consist of adult specimens (often centuries old and with abundant fruiting), and few juveniles. In the case of the Teide National Park, the populations show symptoms of predation by rabbits and mouflons, although there are also other threat factors such as adverse weather events, landslides, fires, absence of birds that can promote the dispersal and germination of seeds, and absence of space-time microenvironments for germination (Martín Osorio et al. 2010).

> "*The rarity of this species has often been attributed to the belief that its 'fruits' (arcestidae) must be eaten by birds (crows, rooks, etc.) so that the seeds they contain germinate. This is only partly true, since the acidic pH of the stomach of these birds dilutes the resins and favors germination, a circumstance that can also be achieved by appropriate treatments.*" (Translated from: Pérez de Paz and Hernández Padrón 1999)

> "*It is cited as 'in danger of extinction' in the National Catalogue of Threatened Species, Law 4/1989, of March 27 (B.O.E. No. 74, March 28, 1989) on the Conservation of Natural Spaces and of Wild Flora and Fauna.*" (Translated from: García Gallo et al. 2003)

Threat Category

Anexo II: Protegidas (Gobierno de Canarias 1991)
LR ca (Bañares et al. 1998)
E (Walter and Gillett 1998)
VU D1 (VVAA 2000)
VU (Bañares et al. 2004)
NT (Bañares et al. 2008)
Vulnerable (Ministerio de Medio Ambiente, Medio Rural y Marino 2011)

Germplasm Banks

Jardín Botánico Canario Viera y Clavijo.

Exsiccata (Herbarium Specimens)

Tenerife: 18.07.1969, Bramwell 1797 (TFC 259); Jardín de la Universidad de La Laguna, 3.11.1969, E. Beltrán Tejera (TFC 35427); Ibid., 21.11.1969, ejusd. (TFC 35419); Las Cañadas, 28.07.1971, A. Acuña & F. Estévez (TFC 21753); El Sombrero, Boca Tauce, Cañada de Diego Hernández, 28.07.1971, ejusd. (TFC 21943, 21944 + Dupl.); Mña. Guajara, 15.11.1972, P.L. Pérez de Paz & J.R. Acebes (TFC 23299); Cmo. Las Peras-La Laguna, 15.03.1973, I. La Serna (TFC 23610); Mña. La Fortaleza-Las Cañadas, 19.05.1973, W. Wildpret, AAG, E. Beltrán Tejera, P.L. Pérez de Paz, LG, CGR, H. Metlesics & E. Barquín Díez (TFC 16927); 18.08.1974, W. Wildpret, BM & J.R. Acebes (TFC 4530); Bco. de Badajoz, 17.04.1975, A.D. Benítez (TFC 42878); Mña. La Fortaleza-Las Cañadas, 19.05.1979, W. Wildpret, AA, E. Barquín, P.L. Pérez de Paz, L. Gallo, CG, H. Metlesics & E. Beltrán (TFC 16927—Parasitada); Cmo. Las Peras-La Laguna, 3.03.1980, C. León Arencibia & I. La Serna (TFC 15982 + Dupl.); Campus Universitario, 3.03.1980, C. León Arencibia (TFC 23246 + Dupl.); Ibid., 3.03.1980, C. León Arencibia & I. La Serna (TFC 30096); La Fortaleza, 13.06.1980, E. Beltrán Tejera (TFC 20907 + Dupl.); Ctra. Icod-La Guancha, 6.07.1983, F. Ardévol González (TFC 38399); Hoya Nadía, 350 m.s.n.m., 7.07.1983, ejusd. (TFC 19011); Cañada de Diego Hernández, 12.05.1984, A. Charpin & W. Wildpret (TFC 20065); Bco. del Río-Cmo. de la Hidro, 850 m.s.n.m., 14.03.1989, MS & MM (TFC 29874); Las Cañadas, 1992, J.R. Acebes, P.L. Pérez de Paz & C. León Arencibia (TFC 36412 + Dupl.); Las Mercedes, 15.02.1992, D. PH, Alumnos 5 °C & P.L. Pérez de Paz (TFC 35123); Sabinar de Afur, Marzo 1994, O. Rodríguez Delgado & E. Beltrán Tejera (TFC 36429); Ucanca-Las Cañadas, 13.07.1994, P.L. Pérez de Paz & DPH (TFC 36784 + Dupl.—Test Fitoquímico). Inventario de la Flora y Vegetación del Parque Nacional del Teide (2001–2003) V.E. Martín Osorio, W. Wildpret de la Torre & B. Hernández Bolaños: Pared de Las Cañadas—Risco Verde, 29/08/2001, W. Wildpret de la Torre (TFC 44889 + Dupl.); Bares de El Portillo en el Parque Nacional del Teide, 13/11/2003, V.E. Martín Osorio, W. Wildpret de la Torre & B. Hernández Bolaños (TFC 44933 + Dupl.); Pie de Risco Verde, Siete Cañadas,

12/05/2004, B. Hernández Bolaños, V.E. Martín Osorio & W. Wildpret de la Torre (TFC 45431).

La Palma: Morro Negro-Caldera, 25.08.1979, W. Wildpret, P.L. Pérez de Paz, M. del Arco Aguilar & I. La Serna (TFC 16869 + Dupl.); Ibid., 9.08.1985, P.L. Pérez de Paz & C. Hernández Padrón (TFC 24619 + Dupl.); Los Andenes-Garafía, 16.06.1995, P.L. Pérez de Paz (TFC 37956); Morro Negro, 2100 m.s.n.m., 16.06.1995, ejusd. (TFC 37954); Ibid., 2000 m.s.n.m., 16.06.1996, ejusd. (TFC 37955).

La Gomera: Bajo Montaña del Cedro, 2100 m.s.n.m., 3.05.1990, M. del Arco Aguilar & M. Salas Pascual (TFC 38396); Tajaque, 1000 m.s.n.m., 17.06.1995, P.L. Pérez de Paz & A. Fernández (TFC 37947—Anál. Fitoquímico).

Uses, Exploitation and/or Medicinal Properties

In the past, its wood was highly valued for furniture and cabinet making. This caused that, in the middle of the twentieth century, its populations were reduced only to those areas that were very difficult to access for its extraction.

> "*Antiseptic, parasiticide, anti-inflammatory: cure internal blows, vulnerary; crafts; cabinet making, forestry interest: wood, firewood; fodder. Useful part: wood and acerboides. Instructions for use: cedar oil, liniment, ointment, salve, 'cedar wine'; browsing, 'dairy' fodder.*" (Translated from: Pérez de Paz and Hernández Padrón 1999)

3.5.60 Lamarckia aurea *(L.) Moench*

Meth., 201 (1794)

Achyrodes aureum (L.) Kuntze *in* Revis. Gen. Pl. 2: 758 (1891)
Chrysurus aureus (L.) Besser *in* Cat. Jard. Bot. Krzemieniec: 30 (1810)
Chrysurus cynosuroides Pers. *in* Syn. Pl. 1: 80 (1805), nom. superfl.
Cynosurus aureus L. *in* Sp. Pl.: 73 (1753)

Not endemic, Probable native
Family: *Poaceae*
Common name: Cepillito dorado (ES); Golden dog's tail, goldentop grass (EN)
Biotype: therophyte—biennial
Flowering: February–June
Fruiting: February–June
Pollination: wind pollination (anemogamy)
Dispersal: wind or animal dispersal (anemochory—zoochory)
Chromosome N°: 2n = 14

Description

Annual plant with stems up to 50 cm, striate, glabrous.

Leaves with ligule of 4–12 mm, oblong-obtuse; blade up to 5–9 × 1.1 cm, caniculate, smooth or with slightly antrorse-scabrid margin. Panicle 2 cm oblongoid, greenish at first then golden-yellow and at the end of maturity slightly violet (Fig. 3.186).

Spikelets with villous pedicel, the sterile ones 5–8 mm, linear, with lanceolate glume, uninervated, more or less antrorse-scabrid, with trinervated lemmas, denticulate at the apex, more or less antrorse-scabrid; the fertile ones with glumes of 3–4.5 mm, similar to the sterile ones; fertile flower lemma 2.4–3 mm, lanceolate, antrorse scabrid, with bidentate membranous apex and subterminal awn 5–9 mm; fertile flower palea 2.4–3 mm, lanceolate, with antrorse-scabrid keels.

Distribution

Species distributed throughout the Mediterranean, Macaronesia (Madeira and the Canary Islands), Africa and Central and Western Asia. Introduced to various parts of the globe. Present on all the islands of the Canary Archipelago (Fig. 3.187).

In the Teide National Park (Fig. 3.188) it was found in Montaña Echisere (N-Teide), lava fills in Cañada de La Grieta, Cañada Blanca, near the Parador Nacional, Llanos de la Escoronada (S-Pico Viejo) and southern flanks of Montaña Guajara; between 1700 and 2200 m a.s.l.

Fig. 3.186 Detail of *Lamarckia aurea*, January 2015. (Photo credit: Victoria Eugenia Martín Osorio)

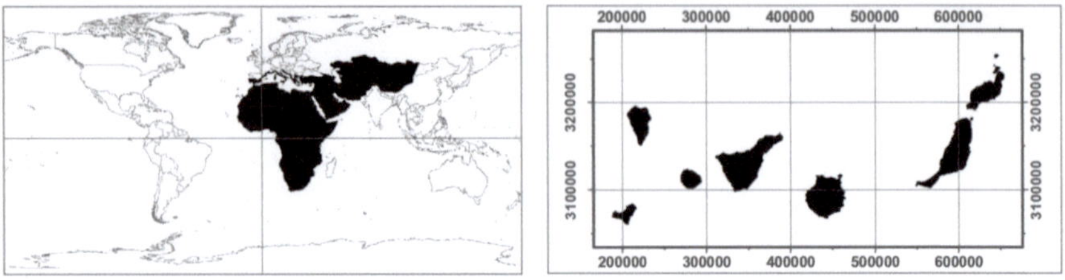

Fig. 3.187 Present on all the Canary Islands

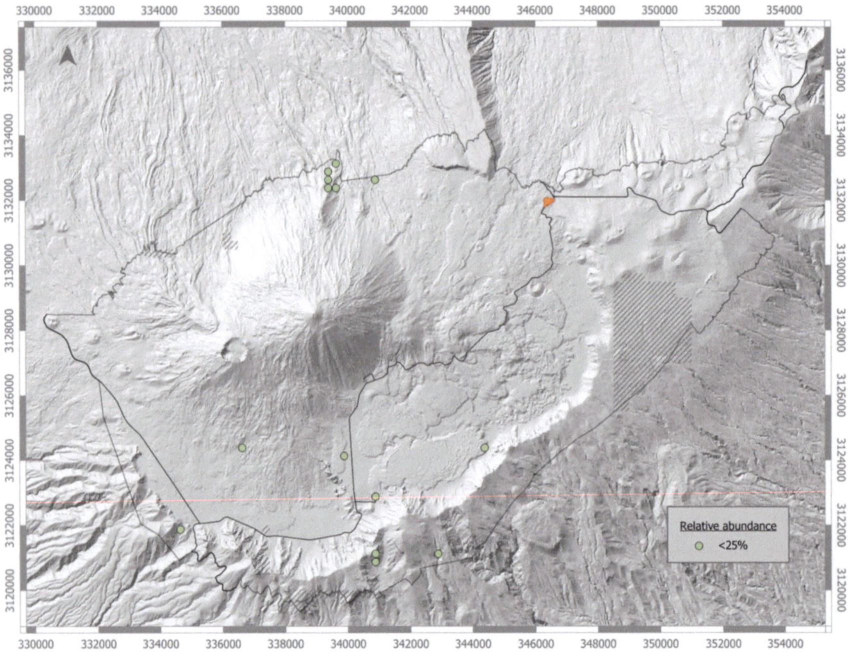

Fig. 3.188 *Lamarckia aurea* distribution, N° Subgrids (250 × 250 m): **14**/3740

Habitat and Phytosociology

It usually colonizes grasslands and ruderal-nitrophilous environments, quite common in abandoned farms, roadsides, roofs and uncultivated land. Frequent in the Infra and Mesomediterranean bioclimatic zones.

It occurs in communities of the order ***Sisymbrietalia officinalis***, moderately nitrophilous ruderal and roadside vegetation, with vernal or summer phenology, widely distributed throughout the Mediterranean region and other Holarctic territories with a Mediterranean bioclimate.

Conservation Status

In 57% (8) of the subgrids in which this species was found, no damage of any kind was observed. In the remaining 43% (6), damage was found on more than 25% of the specimens, and these were completely dry, according to the notes of the field team.

Exsiccata (Herbarium Specimens)

Tenerife: Campus universitario, 2.3.1970, E. Beltrán Tejera (TFC 35510); Puerto de la Cruz, 25.3.1973, I. La Serna (TFC 35572); La Laguna, 3.1975, CGF (TFC 33433); Santa María del Mar, 3.1979, W. Wildpret & M. del Arco Aguilar (TFC 8665); Santa María del Mar, ?, W. Wildpret & M. del Arco Aguilar (TFC 24941); Santa Cruz, Presa de Los Campitos, 24.3.1980, W. Wildpret, M. del Arco Aguilar, J.R. Acebes & M.C. León Arencibia (TFC 8856); El Escobonal, Bco. de Erques, 6.3.1981, O. Rodríguez Delgado & PGC (TFC duplic. 12624); El Escobonal, Bco. de Laja de Erques, 23.3.1981, O. Rodríguez Delgado (TFC duplic. 12625); La Laguna, 17.3.1982, W. Wildpret & A. García Gallo (TFC duplic 22530); Icod, Hoya Blanca, 425 m.s.n.m., 10.6.1983, F. Ardévol González (TFC 19256); Santa Cruz, Bco. Tahodio, 21.2.1985, P.L. Pérez de Paz (TFC 22999 duplic.); La Laguna, 6.3.1985, W. Wildpret & A. García Gallo (TFC 22531); La Laguna, Plaza del Cristo, 26.3.1985, A. García Gallo (TFC 22532; 22533); Güímar, Arrastradero de Arriba, Lomo de Mena, 21.2.1986, O. Rodríguez Delgado (TFC 28099); Los Silos, sobre Las Canteras, 23.6.1989, V. Lucía Sauquillo & C. Acosta (TFC 38272); El Socorro, La Padilla, 30.1.1993, W. Wildpret & T. Jiménez Felipe (TFC duplic. 36322); Bco. de Ijuana, Las Casillas, ladera N, 1.3.1996, R. Racca (TFC 39022); Sitio de Interés Científico del Acantilado de la Hondura, SICAH, Fasnia, 30.3.1996, GMCT (TFC 39505); Bco. de Tahodio, Barrio de la Alegría, 22.2.1999, S. Becker & T. Himstedt (TFC 42112); Ctra. General de El Sauzal, TF-1224, 250 m.s.n.m. UTM 28RCS 933504, 30.3.2002, V. Lucía Saquillo & M.L. Rodríguez Navarro (TFC 44666).

La Gomera: Parque Nacional Garajonay, 8.5.1980, A. Charpin & P. Ranero (TFC 20008).

Gran Canaria: Moya, Bco. Oscuro, 5.1980, P.L. Pérez de Paz & C. Suárez (TFC 11133, duplic. In Juc.); Bco. del Cernícalo, primavera 1989, S. Ortega González & E. Fraile Bonafonte (TFC 30442); Parque Natural Bco. de Guayadeque, Caldera de los Marteles, 21.2.1990, F. Amor & P.L. Pérez de Paz (TFC 30606); Parque Natural Bco. de Guayadeque, cauce del barranco a la altura del cruce de la ctra., 8.3.1990, F. Amor & P.L. Pérez de Paz (TFC 30797); Parque Natural Bco. de Guayadeque, Bco. de Guayadeque, 31.3.1990, F. Amor & P.L. Pérez de Paz (TFC 30650).

Lanzarote: Tahiche, 17.2.1978, E. Beltrán, J.R. Acebes, M.C. León Arencibia & M. del Arco Aguilar (TFC 28962); Teguise, Nazaret, 17.2.1978, ejusd. (TFC 29078); Tabaibal cerca de los Jameos, Punta de Mujeres, 17.2.1978, ejusd. (TFC 34859); Arrecife, llanos pedregosos por detrás del polideportivo de Arrecife, 4.4.1993, J.A. Reyes-Betancort (TFC 37179); Arrecife, 4.4.1993, ejusd. (TFC 37481); Arrecife, 4.4.1993, ejusd. (TFC 37492); Arrecife, Playa del Reducto, 6.4.1993,

J.A. Reyes-Betancort & M.C. León Arencibia (TFC duplic. 37488); Teguise, Vega de Tahiche, 12.3.1996, J.A. Reyes-Betancort, A. Losada Lima & M.C. León Arencibia (TFC 39687).

3.5.61 Laphangium teydeum *Wildpret & Greuter*

In Greuter, W., *Willd.*, 33, 239–244 (2003)

Gnaphalium teydeum Knapp. *Oberhessische Naturwissenschaftliche Zeitschrift*, 42: 77–95 (1976).
Helichrysum teydeum (Wildpret & Greuter) Raus Willdenowia 45: 452 (2015)

Endemism, Certain native
Family: *Asteraceae*
Nombre vernáculo: Borriza de cumbre o del Teide (ES); Teide cudweed (EN)
Biotype: therophyte
Flowering: May–June
Fruiting: June–July
Pollination: –
Dispersal: wind dispersal (anemochorous)

Description

Perennial plant. Flowering stems 3–10 cm tall, tomentose (Fig. 3.189).

Leaves simple, undivided, soft, surface tomentose on both sides, ingrown, protected by white hairs. The leaves at the base are spatulate, often acute, 7–12 mm long, 3–5 mm wide. The stem leaves are alternate, sublinear, frequently with the margin revolute, retrorse and recurved, narrower than the basal ones (7–13 mm long, 1–2 mm wide).

Fig. 3.189 (**a**) Teide National Park, June 2004. (Photo credit: Beatriz Hernández) (**b**) May 2004. (Photo credit: Victoria Eugenia Martín Osorio)

Several **flower** heads per peduncle or flowering stem; synflorescences are spike-shaped, globose or subglobose, ca. 8–20 mm wide. Involucre with a cottony base, scales (bracts) 3–4 mm long, clear, slaty, the outer ones oval, the inner ones oblong-lanceolate, scarious and transparent at the apex.

Remarks:

Gnaphalium teydeum Knapp (*Laphangium teydeum*) is a taxon close to ***G. supinum*** L., from which it differs fundamentally in the morphology of the basal leaves and in the number of capitula that make up the synflorescence. Little is known about the different aspects of its biology. It occurs in perennial forms and apparently does not show special problems of fertilization and germination.

Distribution

It exists only on the island of Tenerife: in the Teide National Park, specifically, between 3375 and 3575 m a.s.l. During the sampling, apart from locating it planted in the Botanical Garden, the population was only found naturally located in areas of Pico del Teide, however, Marrero Gómez et al. (2004) also mention another population in Montaña Guajara (Figs. 3.190 and 3.191).

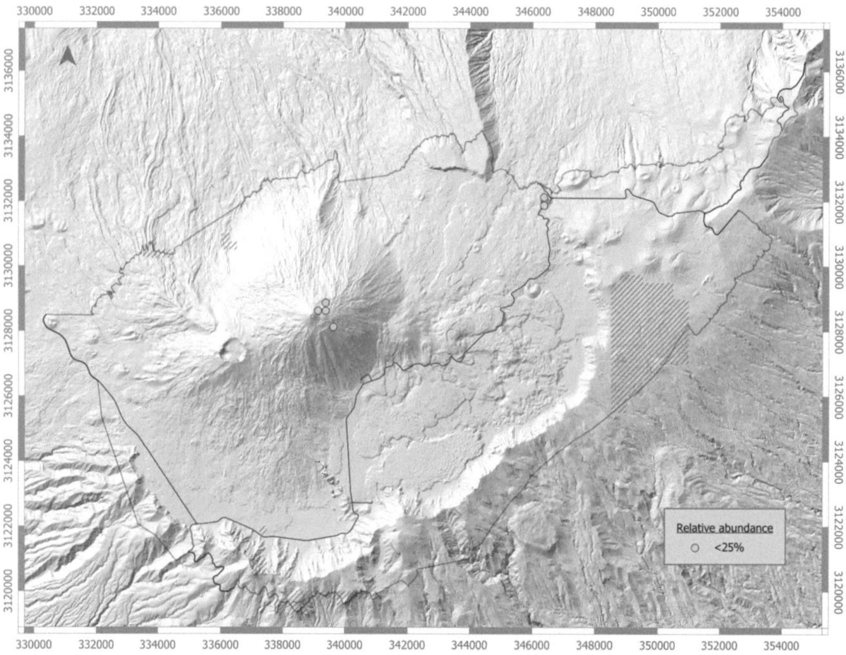

Fig. 3.190 *Laphangium teydeum* distribution, N° Subgrids (250 × 250 m): **6**/3741

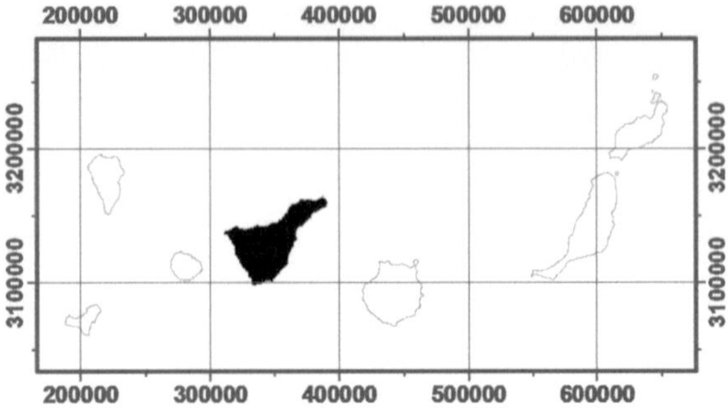

Fig. 3.191 Endemism of the island of Tenerife

Habitat and Phytosociology

It develops above 2000 m a.s.l., in humid places (seepage or springs), which include the fumaroles at the top of Pico del Teide.

Characteristic species of the ***Vulpio myuri-Laphangietum teydei*** microassociation, a terophytic micrograssland that develops in the warm soils near the Rambleta fumaroles, at the base of the cone at 3550 m a.s.l.

It is also found in ***Violetum cheiranthifoliae***, a geopermaseries of Tenerife Mediterranean pluviseasonal oceanic Oromediterranean dry euoceanic of the open glareose dwarf shrubs of ***Viola cheiranthifolia*** with ***Argyranthemum tenerifae*** and ***Silene nocteolens*** with ***Laphangium teydeum***, these two associations represent the complex of Canarian Oromediterranean vegetation of the highest zone of the Teide stratovolcano.

Conservation Status

As threat factors we can highlight: the scarcity of its population, its low ecological plasticity (humid enclaves) and the eminent tourist pressure of the path that is close to its distribution area. Even so, it seems that the populations present a certain stability. Conservation measures include: accessions to germplasm banks, the translocation of specimens to less visited wet areas, undertaking studies on their population dynamics, continuing with genetic studies, proposing their inclusion in the Habitats Directive and addressing the reinforcement of existing natural populations (Marrero Gómez et al. 2004).

According to the data collected during sampling, in the natural population located on Pico del Teide, in 50% of the sub-grids in which this species was found, no damage of any kind was observed. In the other 50%, damage was detected (browsing, high mortality without regeneration, etc.), but this did not affect more than 25% of the specimens.

Threat Category

CR (U.I.C.N., 1994)
CR D (Bañares et al. 1998)
CR D (VVAA 2000)

Sensible a la alteración de su hábitat (Gobierno de Canarias 2001)
EN (Bañares et al. 2004)
EN D (Moreno 2008)

Germplasm Banks

E.T.S. Ingenieros Agrónomos de Madrid.
Jardín Botánico Canario Viera y Clavijo.
Real Jardín Botánico de Madrid.

Exsiccata (Herbarium Specimens)

Tenerife: Fumarolas de la Rambleta-Las Cañadas del Teide, 3500 m.s.n.m., 22.06.1989, W. Wildpret & M. León Arencibia (TFC 31002). Inventario de la Flora y Vegetación del Parque Nacional del Teide (2001–2003) V.E. Martín Osorio, W. Wildpret de la Torre & B. Hernández Bolaños: Fumarola de la Rambleta del Cono del Teide, 27/11/2003, V.E. Martín Osorio & W. Wildpret de la Torre (TFC 44942); Jardín Botánico (P.N. Teide) (P23) (30D 11 b), 17/06/2004, B. Hernández Bolaños (TFC 45399).

3.5.62 Lotus campylocladus *Webb & Berthel. subsp.* campylocladus

Phyt. Canar., 3 (II/A): 83 (1842)

Endemism, Certain native
Family: *Fabaceae*
Common name: Corazoncillo de Tenerife (ES), Tenerife bird-foot trefoil (EN)
Biotype: subshrub (chamaephyte)
Flowering: March–June
Fruiting: April–August
Pollination: insect pollination (entomogamy)
Dispersal: –
Chromosome N°: 2n = 1; n = 7

Description

Chamaephyte with a procumbent, hairy, woody base. Branched stems provided with leaves with five leaflets and ovate, lanceolate or oblanceolate, obtuse, 4–10 mm long, with dense, tightly packed hairs, short peduncles, 1.5 cm or less.

Yellow flowers (Fig. 3.192) arranged in a raceme of three to six, approximately 1–1.5 cm long. Calyx hirsute, bilabiate, 5-fissured, two of the upper parts elongated, lower part turbinate. Corolla long; ovate, carinate vexil (imbricated pre-flowering); the apex of the wing free, upper margin connivent, in front of the nail

Fig. 3.192 (**a**) Cañada del Cedro, January 2004 (**b**, **c**) La Fortaleza, June 2004. (Photo credit: Victoria Eugenia Martín Osorio)

saccato-swollen, often also coherent; the carinae of the wings longitudinal, ending in a narrow, curved apex. Stamens diadelphous. Longitudinal filament that goes and comes after the anthesis, elongated apex. Anthers equal. Style attenuated at the base provided with capitate stigma.

Legume subcylindrical, subarched, adapted, often torulose (with regular constrictions), acute, 1-loculate, or septal between the divisions of the seeds, polyspermous, 2-valve.

Distribution

Endemism of the island of Tenerife and La Palma; doubtful on the island of El Hierro. In Tenerife we can find it from 600 to 2300 m a.s.l.

The two most important localities within the Teide National Park are in La Fortaleza and Los Roques de García. Occasionally, it is also found in certain areas of the Wall: Boca Tauce, S-Roque del Almendro, Lomo Corto in the Cañada del Corral towards Llano Ucanca, El Riachuelo, Colorado de Guajara (W-Barranco del Río), Roque del Pino (Cañada de La Mareta), Cañada de La Grieta, Risco El Colmenar (southern slopes of Topo de La Grieta), Risco Verde and areas close to Portillo Bajo. Also, in the Wolfredo Wildpret Botanical Garden of the Teide National Park (Figs. 3.193 and 3.194).

Habitat and Phytosociology

It is usually found preferably in nitrophilous environments on the edge of the road. Characteristic and bioindicator species of ***Cisto symphytifolii-Pinion canariensis***, which groups the closed climatophilous and edaphoxerophilous meso-macroforests, well characterized by the Canary Island Pine, dry-humid

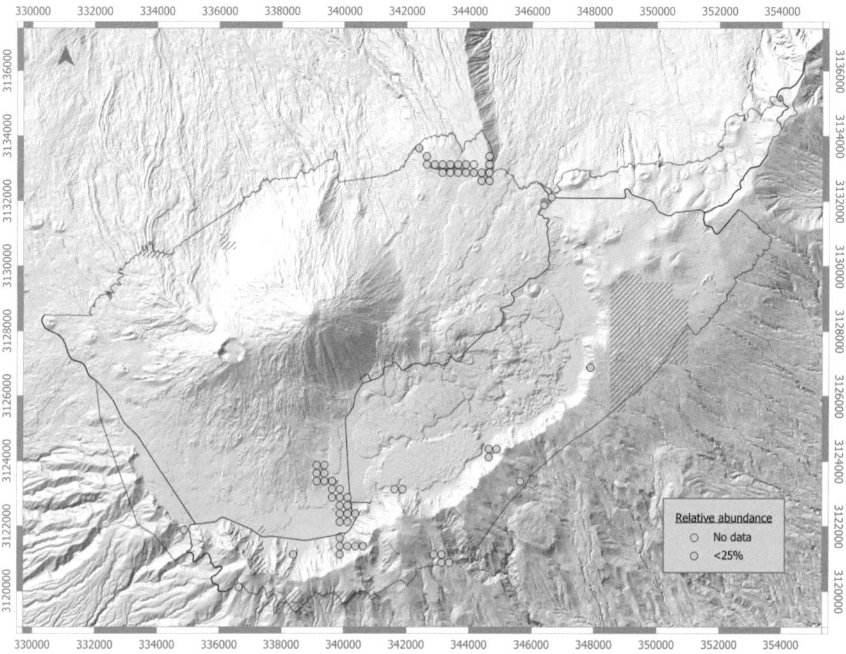

Fig. 3.193 *Lotus campylocladus* distribution, N° Subgrids (250 × 250 m): **61**/3741

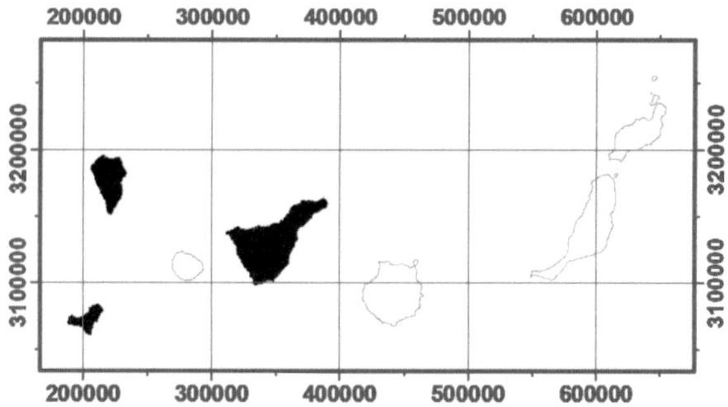

Fig. 3.194 Endemism of the island of Tenerife, la Palma and El Hierro

Meso-Supra-Mediterranean, rarely sub-cloudy, of western Canary Island distribution. It also appears within ***Spartocytision supranubii***, an alliance that groups the high-growing high summit broom and dwarf forests, climatophilous or edaphoxerophilous, sometimes catenal or serial of the Canary Island pine forests; of a dry-humid Meso-Oro-Mediterranean distribution in La Palma and Tenerife.

Conservation Status

In practically most of the subgrids in which this species was found, no damage of any kind was observed.

Threat Category

R (Barreno et al. 1984)
LR pm (Bañares et al. 1998)

Germplasm Banks

E.T.S. Ingenieros Agrónomos de Madrid.
Jardín Botánico Canario Viera y Clavijo.

Exsiccata (Herbarium Specimens)

Tenerife: mayo 1969, Bramwell 1408 (TFC 297); 4.12.1971, E. Beltrán Tejera (TFC 4461); 19.05.1973, W. Wildpret, E. Barquín, AA, E. Beltrán, P.L. Pérez de Paz, LG, MCG & H. Metlesics (TFC 4514); Mña. La Fortaleza, 19.05.1973, W. Wildpret, AA, E. Barquín, H. Metlesics, L. Gallo, P.L. Pérez de Paz, C. Gil & E. Beltrán (TFC 16024); Ibid., 19.05.1973, P.L. Pérez de Paz (TFC 23296 + Dupl.); 24.05.1975, E. Rodríguez Henríquez (TFC 4412); Las Cañadas, 29.05.1975, A. Delgado Benítez (TFC 42903); Altos de Igueste de Candelaria, 26.03.1976, M. del Arco Aguilar & E. Barquín Díez (TFC 25880); Bco. de Erques, 3.04.1976, J.R. Acebes, P.L. Pérez de Paz, C. León Arencibia & M. del Arco Aguilar (TFC 32682); Monte del Escobonal, 26.12.1976, O. Rodríguez Delgado (TFC 27822); Las Arenas-San José de los Llanos-Erjos, 29.08.1979, W. Wildpret, I. La Serna, L. Negrín, A. Losada Lima & A. García Gallo (TFC 24706 + Dupl.); Bco. Albarderos-Arico, 1300 m.s.n.m., 1.04.1980, W. Wildpret, J.R. Acebes, M. del Arco Aguilar, C. León Arencibia, L. Negrín & AVR (TFC 25142); Valle del Volcán de las Arenas-Cumbres de Arafo, 9.05.1981, W. Wildpret, M. del Arco Aguilar & P.L. Pérez de Paz (TFC 23709, 23736 + Dupl.); Las Cañadas, 2.05.1982, R. Lecuona (TFC 13114 + Dupl.); 12.05.1984, A. Charpin & W. Wildpret (TFC 20063); Pista entre Mña. Colorada y Bco. de Guaco, 17.10.1984, O. Rodríguez Delgado (TFC 27831 + Dupl.); La Laguna, 19.02.1988, A. García Gallo & J.M. González (TFC 26087, 26088, 26089); Lomo de la Florida-Arico, 800 m.s.n.m., 9.03.1990, P.L. Pérez de Paz, M. del Arco Aguilar & M. Salas Pascual (TFC 38366 + Dupl.); Altos de la Mesa Araya-Candelaria, UTM 3613141, 1993, W. Wildpret, MMG, ECA & O. Rodríguez Delgado (TFC 36126 + Dupl.). Inventario de la Flora y Vegetación del Parque Nacional del Teide (2001–2003) V.E. Martín Osorio, W. Wildpret de la Torre & B. Hernández Bolaños: Al NW de la Cañada de Los Guancheros—La Fortaleza (30A 24 a), 09.07.2001, Y. de Rioja (TFC 44921); La Fortaleza (30A 14), 06/06/2004, V.E. Martín Osorio & W. Wildpret (TFC 45428).

3.5.63 Melica canariensis *W. Hempel*

First published in Feddes Repert. Spec. Nov. Regni Veg. 75: 109 (1967)

Melica ciliata L. ssp. *magnolii* (Green & Godr.) K. Richt *Pl. Eur.*, 1, 78 (1890)

Macaronesian endemism, Certain native
Family: *Poaceae*
Common name: Triguerilla de Canarias (ES); Canary Melic Grass (EN)
Biotype: hemicryptophyte
Flowering: May–June
Fruiting: May–June
Pollination: wind pollination (anemogamy)
Dispersion: wind dispersal (anemochory)
Chromosome N°: –

Description

Perennial plant (Fig. 3.195). **Stem** 50–120 cm tall, the lower part almost woody, smooth glabrous with nodular branches truncated by the leaf sheaths. With numerous panicles. Leaf sheaths somewhat rough, glabrous and fused on the lower part. Edge with a large scaly extension of the ligule.

Leaves flat or slightly rolled (6–) 8–13 (17) cm. long and 1.5–3 mm wide, the upper part glabrous, rough or more or less hairy. 15–17 (21) veined, the veins protrude somewhat on the lower part. Ligule elongated blunt, bifid laciniate, somewhat scabrid 2–3 mm long. Branched panicle 9–21 cm long, often interrupted basally. Primary branchlets of the panicle provided with 6 (9–15)–20 primary spikelets, 4–8 secondary spikelets and 2–5 tertiary spikelets (Fig. 3.196).

Pale uniflorous spikelets 4.5–6 mm long. Spikelet peduncles hairy at the top. **Glumes** unequal, the lower ones 5 mm long, short, lateral veins short (1/2), median veins 5-veined to the apex, upper glume 5.5–6 mm long, lanceolate, acute. Median vein to the apex and lateral vein up to 2/3 of the length. Glumelle acute 4.5–5.5 mm long, lanceolate, glabrous dorsum 7-veined. Median veins reach the apex, lateral veins reach up to 2/3 of the length of the glumelle. Palea carinate, apex ciliolate and bifid, 3–4 mm long. Lower apicular glume 2.0–2.5 mm long, cylindrical, glabrous to more or less scabrid.

Fig. 3.195 Cañada del Capricho, July 2003. (Photo credit: Victoria Eugenia Martín Osorio)

Fig. 3.196 Spikelet of *Melica canariensis*. (Photo credit: Victoria Eugenia Martín Osorio)

Remarks:

The type sheet was taken in Bajamar on 13/5/1901 by Bornmüller, Plantae exsic. Canarienses no. 2950 JE.

Distribution

Macaronesian endemism. It appears throughout the Canary Islands, except in the westernmost islands: El Hierro and La Palma (Fig. 3.197).

In the Teide National Park (Fig. 3.198) it was found in Cañada de Diego Hernández (very sporadically), near the base of the Cable Car, Montaña Majúa and Corredor de La Corbata, Cañada del Capricho (where it has been found in abundance), between the Parador Nacional and Roques de García and, more dispersed in Llano Ucanca and Pared, between 2025 and 2400 m a.s.l.

Habitat and Phytosociology

It is a characteristic species of ***Brachypodietalia phoenicoidis***, of the ***Festuco-Brometea*** class, of anthropogenic, slightly xerophilous grasslands of perennial herbs, sub-Meso-Supratempered and Meso-Supra-Mediterranean, dry-humid. In the Teide National Park, it also appears as a characteristic element of the

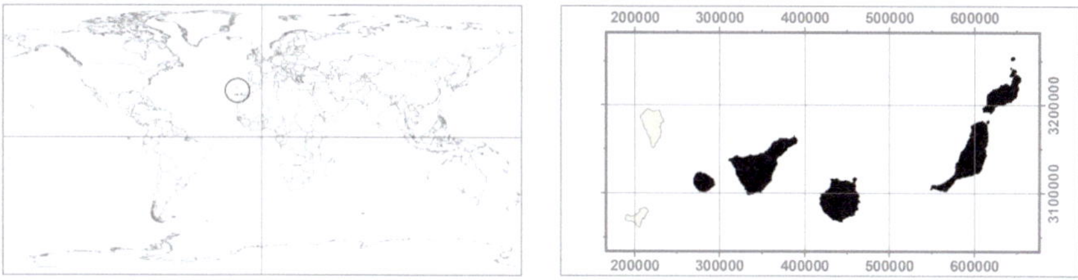

Fig. 3.197 Present on all the Canary Islands except in the westernmost islands: El Hierro and La Palma

Fig. 3.198 *Melica canariensis* distribution, N° Subgrids (250 × 250 m): **23**/3740

Cerrillar—Crespar, which colonizes the slopes of gelifract debris inside the Pared de Las Cañadas: ***Arrhenathero calderae-Plantaginetum webbii***.

Conservation Status

In most of the subgrids in which this species was found (22), no damage of any kind was observed. Only in one (1) of them, damage was detected (browsing, high mortality without regeneration, etc.), but this did not affect more than 25% of the specimens.

Threat Category

LR pm (Bañares et al. 1998)
VU (Bañares et al. 2004)
VU D2 (Moreno 2008)

Germplasm Banks

E.T.S. Ingenieros Agrónomos de Madrid.
Jardín Botánico Canario Viera y Clavijo.

Exsiccata (Herbarium Specimens)

Tenerife: Inventario de la Flora y Vegetación del Parque Nacional del Teide (2001–2003) V.E. Martín Osorio, W. Wildpret de la Torre & B. Hernández Bolaños: Cañada del Capricho (42A 32 c), 02.07.2003, F. Mesa, A. Montañés & Y. de Rioja (TFC 44863).

3.5.64 **Mentha longifolia** *(L.) L. var.* **teydea** *Bolle*

Bonplandia, 8, 286 (1860)

Mentha sylvestris var. *teydae* Bolle *in* Bonplandia (Hannover) 8: 286 (1860)

Endemism, Certain native
Family: *Lamiaceae*
Common name: Menta o Mastranto (ES) Teide Horse Mint (EN)
Growth form: subshrub/woody herb (chamaephyte)
Flowering: May–July
Pollination: insect pollination (entomogamy)
Dispersal: Shortrange self-dispersal (alatelochory)
Fruiting: July–August.
Chromosome N°: –

Description

Perennial plant of about 20–30 cm, pubescent quadrangular stem, densely tomentose and short.

Leaves decussate, greyish-green, lanceolate to ovate-lanceolate, sessile, with serrated to serrulate edges, densely tomentose both on the upper and lower surface; tomentum denser and longer on the lower surface; bracts ovate-lanceolate to linear, tomentose on the upper surface and on the margins, hirsute on the lower surface.

Inflorescences in terminal spikelets (Fig. 3.199), solitary or branched, terminal and from basal branches or solitary or in number of three, generally, in the latter case, the central one being longer, rarely shorter. Only occasionally, the number of spikelets is greater than three and they can be: long (up to 30 cm) with very

Fig. 3.199 Lagunazo Siete Cañadas, (**a**) May 2004. (Photo credit: Beatriz Hernández) (**b**) June 2005. (Photo credit: Victoria Eugenia Martín Osorio)

separated verticilasters, longibracteate; congested (0.5–6 cm); and mixed, with dense whorls at the base and rosary-like towards the apex.

Pedunculated flowers, peduncles up to 3 mm, tomentose; calyx very variable depending on whether it belongs to one or another of the previously mentioned spicastres, with calyxes of 1.5–3 mm, externally hairy and glandular, with large yellow glands and internally glabrous, membranous green to greyish-purple, narrowly bell-shaped, subbilabiate to bilabiate, actinomorphic to subactinomorphic, subulate teeth subequal to differentiated into two lips, the upper one tridentate, deltoid, and the lower bidentate, long subulate. Tetralobed corolla, upper lobe wider and notched, exclusive or included depending on the calyx, externally hairy, internally glabrous. Stamens subequal, armies. Style more or less long with bilaciniate stigma of subequal curved laciniae. Nutlets 0.6–0.7 mm.

Remarks:

> Socorro (1985) *describes an ecoform that grows in the "Cañadas" biotope, which he calls "longespicata"; the specimens have a dense, congested picastre that may correspond to a first flowering in spring-summer, followed by a late flowering (summer-autumn) that produces abnormally long picastres, together with other dense ones. He also points out that it seems important to follow the phenological cycle more closely in its natural habitat, as well as to try to germinate the plants in "standard" conditions to observe the constancy of these characteristics.*

Bolle *(*1860*) in Bonplandia (Hannover) 8: 286 cita: var. teydea, C. Bolle: statura summopere humili, foliis angustis valde acutis. Hab. haec insignis varietas, eui ulterios studeatur, ut melius innotescat, in regione excelsa summi montis Pico de Teyde ad scaturigines paucas parce inter pumices fluentes. Col de la Montaña de Trigo (Borgeau!).*

(=Extremely) short in stature, with narrow and very sharp leaves. This remarkable variety, which I studied more thoroughly, to get to know it better, is found in the high region of the highest Pico de Teyde, in some springs that spring up sparsely between the pumice stone. Col de la Montaña de Trigo (¡Borgeau!). (Translation:V Martín Osorio)

Distribution

Taxon exclusive to Las Cañadas del Teide, whose distribution area extends from 2025 m a.s.l. in the Llano de Ucanca, where it appears very occasionally, to 2200 m a.s.l. in the Cañada de La Mareta and Cañada del Montón de Trigo (Figs. 3.200 and 3.201).

Habitat and Phytosociology

It normally occurs in endorheic environments, such as those of the Ucanca Valley and Siete Cañadas, established in the cracks formed by the drying of the clays at the bottom of the ephemeral lagoons formed by the thawing of ice and rain.

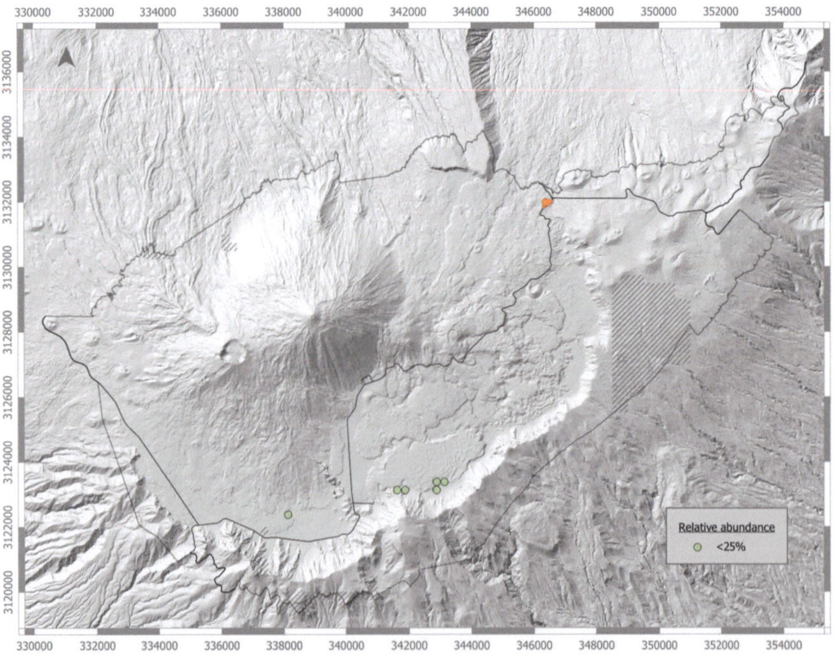

Fig. 3.200 *Mentha longifolia* var. *teydea* distribution, N° Subgrids (250 × 250 m): **6**/3741

Fig. 3.201 Endemism of Teide National Park, Tenerife

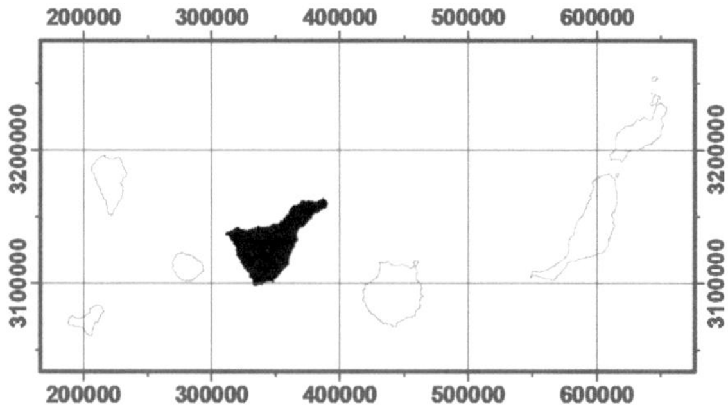

It is found within the ***Magnocaricion elatae*** alliance. Within this, it is a characteristic species of the ***Mentho longifoliae-Caricetum calderae*** association, typical of environments subject to temporary flooding, springs or stream edges in the Supra-Mediterranean region.

Conservation Status

In all the subgrids in which this variety was found, the conservation status was high, so no damage of any kind was observed.

Threat Status

Not Available.

Germplasm Banks

Jardín Botánico Canario Viera y Clavijo.

Exsiccata (Herbarium Specimens)

Tenerife: Fte. Cañada de la Grieta, 2175 m.s.n.m., 2.10.1982, S. Socorro (TFC 19975); Bco. del Riachuelo, 2100 m.s.n.m., 3.10.1982, S. Socorro (TFC 19974); C. del Mt. de Trigo, 2160 m.s.n.m., 12.10.1982, S. Socorro, W. Wildpret & E. Beltrán (TFC 18498, 18499, 19976); Ibid., 12.10.1982, S. Socorro (TFC 19973); Ibid., 10.12.1982, S. Socorro, W. Wildpret & E. Beltrán (TFC 19977); Ibid., 12.10.1982, S. Socorro, W. Wildpret & E. Beltrán (TFC 19978); Ibid., 2.10.1982, S. Socorro (TFC 19979); Fte. C. de la Grieta, 2160 m.s.n.m., 2.10.1982, S. Socorro (TFC 19980); Ibid., 2.10.1982, S. Socorro (TFC 19981); Ibid., 2.10.1982, S. Socorro (TFC 19982); Bco. del Río, 2100 m.s.n.m., 15.05.1983, S. Socorro (TFC 19983). Inventario de la Flora y Vegetación del Parque Nacional de Teide (2001–2003) V.E. Martín Osorio, W. Wildpret & B. Hernández Bolaños: Llano de Ucanca, en la parte superior, al pie de la base del Teide, 04/06/2003, A. Montañés Rodríguez (TFC 44252).

3.5.65 Micromeria lachnophylla *Webb & Berthel.*

Phyt. Canar., 3, 73 t. 156 A (1844)

***Micros** (griego): pequeño -**meros** (griego): parte*
Micromeria julianoides Webb & Berth.
Ibíd., 3:78. t. 160 B. (1844).
Micromeria teydensis Bolle.
Bonplandia, 8: 282 (1860).
Micromeria varia f. lachnophylla Christ.
Bot. Jahrb., 9: 133 (1888).
Satureja lachnophylla (Webb & Berth.) Briq., Engler et Prantl. *Natürl. Pflanzenfam.*, 4(3a): 299 (1896).
Satureja julianoides (Webb) Briq.
Ibíd. (1896).
Satureja teydensis (Bolle) Briq.
Ibíd. (1896).
Clinopodium julianoides (Webb & Berth.) Kuntze, O. Rev. Gen., 3: Lab. 511–531 (1898).
Micromeria varia v. lachnophylla (Webb et Berth.) Bornm. *Feddes Repert.*, 19: 198 (1924); Ceballos et Ortuño, *Veg. Fl. Forest. Canar. Occ.*: 421 (1951).
Micromeria varia v. *julianoides* (Webb et Berth.) Bornm. l.c. (1924); Burchard, *Biblioth. Bot.* (Stutgartt), 98: 182 (1929); Ceballos et Ortuño, l.c. (1951).
Thymus polimorpha v. *pubescens* Webb & Berth. MSS., nom. *in* schaed. (Fl!)

Endemism, Certain native
Family: *Lamiaceae*
Vernacular name: Tomillo de Las Cañadas o del Teide, Tomillo burriquero (ES); Teide whitweed (EN)
Biotype: chamaephyte
Flowering: May–September
Fruiting: May–September
Pollination: insect pollination (entomogamy)
Dispersal: animal dispersal (zoochory)
Chromosome N°: –

Description

Chamaephyte 10–30 (40) cm tall, usually erect; woody stem, branched from the base; erect or ascending branches, graceful; flexible; the older ones subcylindrical, with peeling bark, glabrescent and brown; the younger ones subquadrangular, tomentose; very small hairs, often deflexed; greenish or tinged with purple; internodes generally 0.1–2.0 cm long.

Leaves revolute, sessile, the larger ones sublanceolate, attenuated towards the apex, the smaller ones arranged in fascicles in the axil of the previous ones, ericoid or linear; greenish-grey or yellowish; finely tomentose, varying greatly in the density and size of the tomentum; in the young shoots and lower branches imbricated, closely decussated, giving them a characteristic spiciform appearance.

Persistent cymes; 3–10 (20) flowers, subsessile or pedunculate; peduncles between 0.5 and 10 mm long, tomentose; bracts linear-lanceolate, subulate, greenish, often tinged with purple at the apex, hairy 2–4 mm long; flowers with confluent pedicels or arranged in barely birambed cymes. Calyx subcylindrical, attenuate at the base (3–4 mm long), purple colored in its upper half, covered with a very fine and small tomentum; with (13–14) 15 ribs; teeth subequal, straight, subulate, hairy on its inner surface; calyx tube 2/3–3/4 of its length. Corolla exserted (Fig. 3.202), white or lightly tinged with lilac; limb 2.5–3.0 (3.5) mm in diameter; bilabiate; Upper lip bifid, lower lip trifid, lateral lobes rounded, the central one larger and subflaveliform; hairy on the outside. Stamens of the upper lip inserted, those of the lower lip slightly exserted. Anthers small; purple; frequently sterile, subparallel. Stigma bifid, branches purple coloured. Nuts oblong, chestnut coloured.

Remarks:

When flowering occurs outside of spring or even within it, specimens are frequently found with totally or partially androsterile flowers, while the size of the corolla decreases noticeably and frequently the whole plant shrinks, allowing these plants to be distinguished from fertile ones at a glance. Fruiting is abundant and seems to be the only way to ensure the progress of the plant, which reproduces easily by seeds. As with other species of the genus, its transplant is doomed to failure.

Occasionally, in various localities of debris at the foot of the cliffs of La Fortaleza, Cañada de Diego Hernández, between Topo de la Grieta and Guajara, etc., other clearly hairier specimens have been observed coexisting with these specimens

Fig. 3.202 Teide National Park, June 2004. (Photo credit: Beatriz Hernández)

considered to be typical, without there apparently being an ecological cause that motivates it, and which are not considered appropriate to give a different taxonomic treatment since they fit, due to the rest of their characteristics, with those described for the other specimens (Pérez de Paz 1978).

Distribution

Endemism of the island of Tenerife. Its optimal distribution area is between 2000 and 2400 m a.s.l., although it occasionally reaches 3000 m a.s.l. on the southern slopes of Teide (Figs. 3.203 and 3.204).

In Pérez de Paz (1978), the following was noted: "it is quite abundant in the clearings of broom and flatpod formations throughout the Circo de Las Cañadas, a place considered to be the area where the genuine form of the species grows. It has been frequently observed in the vicinity of the Fortaleza, Montaña de los Tomillos, Arenas Negras, Cañada de Diego Hernández, etc. It also reaches beyond the summits of the Pared, descending along the outer slopes until it comes into contact with the pine forests, where it mixes with ***M. hyssopifolia*** Webb et Berthel, from which it is sometimes difficult to separate. Gradation between the two species has been observed in Altos del Pinar de Los Realejos, the upper part of the Ladera de Güímar, Altos de la Sabinita (Arico), Ladera de Tamaimo, Santiago del Teide, and in general in all the places where the transition from the middle and montane floor to the Filo de las Cañadas occurs in a smooth manner."

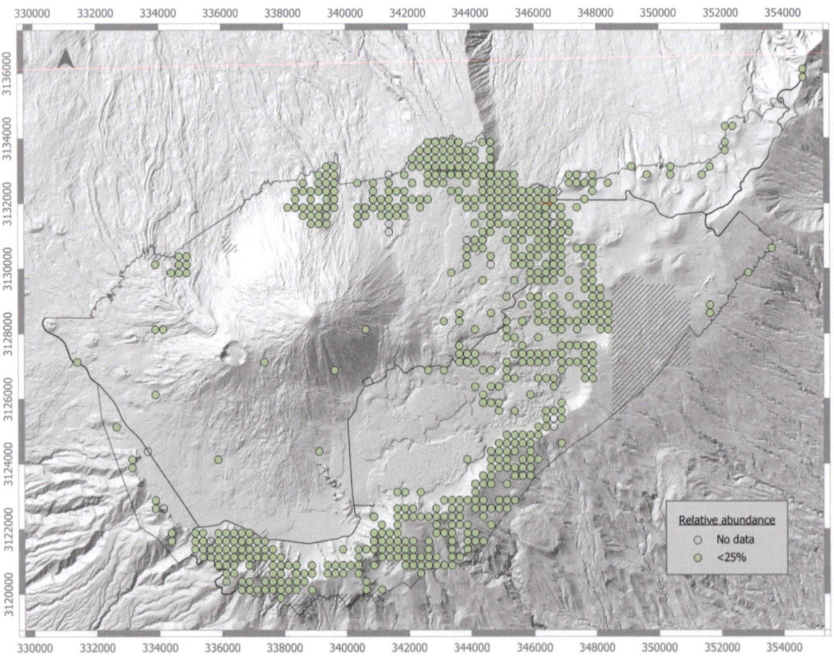

Fig. 3.203 *Micromeria lachnophylla* distribution, N° Subgrids (250 × 250 m): **642**/3741

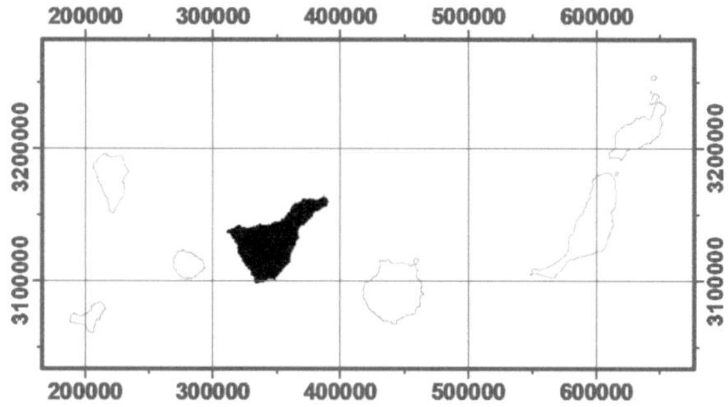

Fig. 3.204 Endemism of the Teide National Park on the island of Tenerife

During the sampling carried out, it was located in a significant way from Montaña Echisere, towards La Fortaleza and up to the Portillo area, to continue extending continuously along the entire edge of the Circo de Las Cañadas up to Boca Tauce. From this point and towards Cañada del Cedro, it is much less abundant.

It colonizes both the interior and exterior slopes of the Pared, except for the areas towards the Dorsal, where it disappears more or less categorically. In the latter, in fact, it is distributed in a very punctual and dispersed way only in the heights of the Valle de La Orotava and Arico and Fasnia on the other slope. In the interior areas of Las Cañadas, its distribution is much more dispersed (flow-fills in the eastern half), appearing only very occasionally towards the central and western areas.

Habitat and Phytosociology

It is a xerophilous and heliophilous plant, although it can be found in partially shady situations: cracks in lava flows, cliffs of Las Cañadas and the head of ravines that originate on its outer edge. Occasionally, it can intervene in the undergrowth of the last manifestations of the pine forest (ecotone zones between this and the formations of broom). It can also sporadically ascend the slopes of Pico del Teide, where some small specimens have been observed, growing among pebbles and lava cliffs above 2700 m a.s.l. (Pérez de Paz 1978).

Puppo (2015) points out that both ***M. lachnophylla*** and ***M. lasiophylla*** are very restricted to the central area of the island of Tenerife, geologically younger than the paleoislands of Anaga, Teno and Adeje, where other species and/or subspecies of this same genus can be found.

It is a characteristic species and bioindicator of ***Pterocephalenion lasiospermi***, a suballiance that groups permanent and serial dwarf shrub associations, which colonize volcanic deposits and denuded soils of pine forests and broom shrublands, dry-subhumid Meso-Supra-Mediterranean, endemic to the island of Tenerife.

It also defines or is present in the following communities:

Community described as Cerrillar-Crespar, where there is a phytosociological association of the National Park that preferably colonizes the slopes of gelifract debris of the Pared de las Cañadas:

Arrhenathero calderae-Plantaginetum webbii
Cheirolopho teydis-Carlinetum xeranthemoides
Tolpido webbii-Micromerietum lachnophyllae
Sileno berthelotianae-Cerastietum sventenii

Conservation Status

In 96% (614) of the subgrids in which this species was found, no damage of any kind was observed. Only in 3% (21), damage was detected (browsing, high mortality without regeneration, etc.), but this did not affect more than 25% of the specimens. In the rest (7), this variable was not specified at the time of sampling.

Threat Category

nt (Barreno et al. 1984)
Anexo II: Protegidas (Gobierno de Canarias 1991)
LR pm (Bañares et al. 1998)

Exsiccata (Herbarium Specimens)

Tenerife: Filo de Las Cañadas, 1845, E. Bourgeau 255 (FI!; Z); Ibid., s.d., P. Barker-Webb (FI!); Cañadas del Teide, s.l., 20.05.1846, E. Bourgeau, 915 (FI!); Ibid., Julio 1855, ejusd. (Z!); Ibid., 7.07.1855, ejusd. 1464, 1465 (FI!); Ibid., 25.08.1880, Hillebrand (Z!); La Fortaleza, Abril 1884, Christ (Z); Cañadas, s.l., 28.06.1900, Bornmüller, 1131 (Z!); La Fortaleza, 18.07.1969, D. Bramwell, 2100 (TFC 342); El Cabezón-Las Cañadas, 1.08.1969, ejusd. 2130 (TFC 336); El Portillo-Las Cañadas, 1900 m.s.n.m., 28.07.1971, FECh & AAG (TFC 15991); Ca. de la Mña. de los Tomillos-El Portillo-Las Cañadas, 15.11.1972, P. Pérez (TFC 5696, Duplic. MA); El Portillo-Las Cañadas, P.L. Pérez de Paz & J.R. Acebes (TFC 5697); Arenas Negras-Las Cañadas, Mayo 1973, P.L. Pérez de Paz (TFC 5695, Dupl. in BM, FI); La Fortaleza, Mayo 1973, ejusd. (TFC 5700, Dupl. in P, B); Ibid., Mayo 1973, ejusd. (TFC 5698, Dupl. K); La Fortaleza, 4.06.1973, H. Metlesics (TFC 4392; 4383); La Fortaleza, 19.05.1973, P. Pérez (TFC 5698, Duplic. K; TFC 5700, Duplic. B, P.); Arenas Negras-Las Cañadas, ejusd. (TFC 5695, Duplic. BM, FI); Alrededores del Topo de la Grieta-Las Cañadas, 2.06.1973, ejusd. (TFC 5699, Duplic. MAC); La Fortaleza-Las Cañadas, Julio 1973, H. Metlesics (TFC 4393); Las Cañadas, Julio 1973, P.L. Pérez de Paz (TFC 5699, Dupl. in MAC); Cañada de Diego Hernández-Las Cañadas, 3.08.1974, W. Wildpret & J.R. Acebes (TFC 5701); El Portillo-Las Cañadas, 13.97.1994, P.L. Pérez de Paz & D.P. Hernández (TFC 36785 + Dupl.). Inventario de la Flora y Vegetación del Parque Nacional de Teide (2001–2003) V.E. Martín Osorio, W. Wildpret & B. Hernández Bolaños: Al W de la Montaña del Arroyo o de la Zarza (31A 19), 23.04.2001, L. Afonso, A. Hernández, Y. de Rioja & F. Tapia (TFC 44879); Roques de Chavao (41A 34 c), 07/05/2001, L. Afonso, A. Hernández, Y. de Rioja & F. Tapia (TFC 44885); Roques de Chavao (41A 34 c), 07.05.2001, L. Afonso, A. Hernández, Y. de Rioja & F. Tapia (TFC 44888); Pared de Las Cañadas—Risco Verde, 29.08.2001, W. Wildpret de la Torre (TFC 44892);

Al N de la Cañada de Pedro Méndez (41A 28), 28.04.2003, A. Montañés Rodríguez (TFC 44849); Jardín Botánico (P.N. Teide) (E26) (30D 2 c), 22/06/2004, B. Hernández Bolaños (TFC 45393).

Uses and/or Medicinal Properties

> "*Antiseptic, preservative, aromatic, pectoral; bad forage. Useful part: whole plant, upper shoots. Use: teas, bouquet, browsing.*" (Translated from: Pérez de Paz and Hernández Padrón 1999)

3.5.66 Micromeria lasiophylla *Webb & Berthel.* *ssp.* lasiophylla

Phyt. Canar., 3, 74. t. 156 B. (1844)

Micromeria oreothymbra Webb. *MSS. Herbarium Webbianum*, nom. *in* schaed. (s.d.).
Micromeria varia Bentham. DC., *Prodr.*, 12: 216 (1848); pro parte.
Satureja lasiophylla (Webb et Berth.) Willemse. Hansen, A. & P. Sunding (1993). Flora of Macaronesia. Checklist of vascular plants, 4. rev. ed., *Sommerfeltia*, 17: 1–295. Oslo
***Micros** (griego): pequeño -**meros** (griego): parte*

Endemism, Certain native
Family: *Lamiaceae*
Common name: Tomillo de cumbre tinerfeño (ES); Tenerife summit whitweed (EN)
Biotype: camophyte
Flowering: May–September
Fruiting: May–September
Pollination: insect pollination (entomogamy)
Dispersal: animal dispersal (zoochory)
Chromosome No: –

Description

Chamaephyte. Chasmophyte 10–25 cm tall, with a short, abundantly branched stem; ascending or procumbent, tangled branches; the older ones have broken bark; the younger ones subquadrangular, tinged with purple, hairy; internodes generally 0.1–1.0 cm long.

Leaves sessile, sometimes shortly petiolate, the largest up to 12 × 3 mm, the smallest arranged in fascicles in the axil of the largest, dark green, densely hairy on both sides, with long and short hairs interspersed.

Cymes usually of 5–12 **flowers** (Fig. 3.205), subsessile or shortly pedunculated, forming dense and well-defined verticillasters; peduncles usually no more than 0.5 cm, filiform, hairy like the leaves. Bracts linear-lanceolate, subulate, hairy

Fig. 3.205 Montaña Guajara, June 2003. (Photo credit: Victoria Eugenia Martín Osorio)

(1–2 mm long). Calyx tubular-cylindrical, sometimes sub-bell-shaped when mature (2.5–3.5 mm long), intensely purple in the upper 2/3 of its length; hairy; with (13–14) 15 ribs, with subequal, straight, lanceolate teeth, hairy on its inner surface; tube 1/2–2/3 of its length. Corolla exerted, 1–2 mm larger than the calyx; limb with revolute lobes, (3) 3.5–4.0 mm in diameter, white, sometimes with purple streaks on the throat; upper lip bifid; lower lip trilobed, lateral lobes subquadrangular, the central one emarginate. Stamens of the lower lip exerted, those of the upper included. Anthers deep purple, large oval, subdivaricate. Style hyaline, ending in a stigma with subequal branches, coloured like the purple anthers. Nutlets golden-brown, subtrigone, obtuse (0.7–0.8 to 0.3–0.5 mm).

Distribution

Endemism of the high summit of the island of Tenerife. Puppo (2015) points out that both ***M. lachnophylla*** and ***M. lasiophylla*** are very restricted to the central area of the island of Tenerife, geologically younger than the paleoislands of Anaga, Teno and Adeje, where other species and/or subspecies of this same genus can be found (Figs. 3.206 and 3.207).

In the Teide National Park during sampling, it has been found very punctually, between 2150 and 2450 m a.s.l.: on the Ladera La Mesa (N-Izaña), Alto de la Pared in the Cañada de Las Pilas, Topo de la Grieta, western sector of Montaña Guajara and Morro and Risco del Tiro del Guanche, near Boca Tauce.

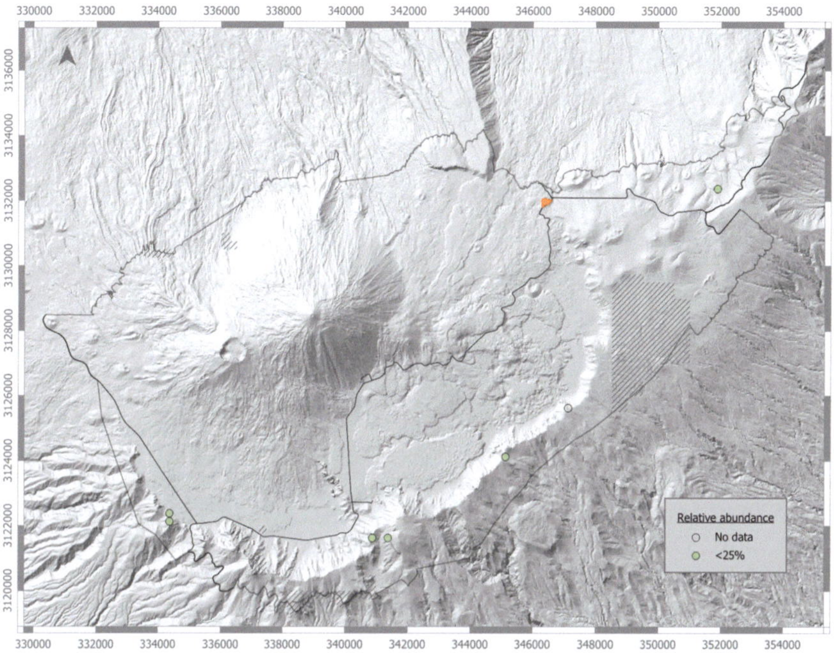

Fig. 3.206 *Micromeria lasiophylla* distribution, N° Subgrids (250 × 250 m): **7**/3741

Fig. 3.207 Endemism of the island of Tenerife

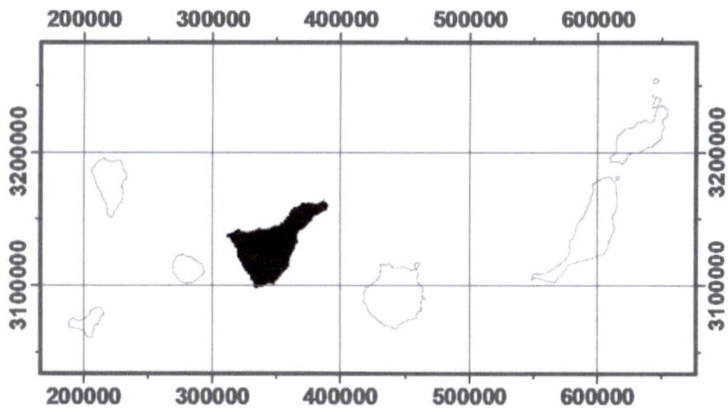

Habitat and Phytosociology

It usually colonizes cracks and fissures in cliffs or crags. Species found within the domain of the ***Chamaecytiso-Pinetea canariensis*** class, which defines the Canarian coniferous forests (***Chamaecytiso-Pinetalia canariensis***), mountain retamoid shrublands and Oromediterranean dwarf shrublands of the Teide volcanic cone (***Spartocytisetalia supranubii***), climatophilous, edaphoxerophilous, pyrophilous or serial, endemic to the midlands and summits of the Central-Western Canary Islands, oceanic pluviseasonal Mediterranean, dry-humid Thermo-Oromediterranean, eu-semi-hyperoceanic and euceanic of the Western Canary biogeographic subprovince.

Conservation Status

In practically all the subgrids in which this species was found, no damage of any kind was observed, so its specimens were well preserved.

Threat Category

R (Barreno et al. 1984)
VU D2 (Bañares et al. 1998)
VU (Bañares et al. 2004)
VU D2 (Moreno 2008)

Exsiccata (Herbarium Specimens)

Tenerife: Las Cañadas, s.l., Marzo 1844, H. Christ (Z!); Acantilados de la Mña. de Guajara, 15.11.1972, P. Pérez & J.R. Acebes (TFC 5706, Duplic. B, FI, LPA; TFC 5707; 5708); Topo de la Grieta, 2.06.1973, P. Pérez (TFC 5709; 5710, Duplic. BM, K, MA, MAF, O; TFC 5711, Duplic. FI, MAC; TFC 5712; 5713); El Sombrero, 2.07.1977, H. Metlesics (TFC 6652); Boca de Tauce, 2100 m.s.n.m., W. Wildpret & S. Socorro (TFC 34261 + Dupl.). Inventario de la Flora y Vegetación del Parque Nacional de Teide (2001–2003) V.E. Martín Osorio, W. Wildpret & B. Hernández Bolaños: Al W de Izaña—Montaña de El Cabezón (31C 2 b), 24/04/2001, L. Afonso, A. Hernández, Y. de Rioja & F. Tapia (TFC 44917); Siete Cañadas (36D 30 c), 22/06/2001, L. Afonso, A. Hernández, Y. de Rioja & F. Tapia (TFC 45406); Cerca del camino de El Capricho, hacia el Oeste, en lo alto (41D 20 d), 23/06/2003, A. Montañés Rodríguez (TFC 44924).

3.5.67 Monanthes brachycaulos *(Webb* in *Webb & Berthel.) Lowe var.* nivata *Svent.*

Addit. Fl. Canaries, I, 17 (1960)

Endemism, Certain native
Family: *Crassulaceae*
Common name: Pelotilla isleña (ES)
Biotype: hemicryptophyte
Flowering: May–June
Fruiting: June–July
Pollination: –
Dispersal: wind dispersal (anemochory)

Description

It differs from the typical species by presenting an elongated, hardened trunk, covered with old leaves. Dense **rosette** (Fig. 3.208) with subteretiform-spatulate (partially cylindrical) leaves, rigid, with angular apex; yolk globular-stained.

Calyx divided, lanceolate, blunt; petal glabrous, linear, spiked; scale with subcuneiform unguiculus, 2/3 smaller than the subflabellate limb, emarginate limb, approximately 2 mm long, pointed, rubristriate and crenulate margin. Oblong-elliptic seed. Sventenius (1960): It differs from the type with a long trunk, hard, covered with thick leaves; dense rosette with subterreform-spatulate leaves, rigid, angled at the apex; globular-depressed gem; calyx with lanceolate divisions, obtuse; petals glabrous, linear, apiculate: scale with a subcuneiform claw, 2/3 shorter than the subflabelliform limbus, slightly emarginate, about 2 mm. lat., puncticulata, red-striated and with a crenulated margin; oblong-elliptic seed. It lives in the sandy crevices of rocky cliffs.

The original place of var. *nivata*; Las Cañadas, on the promontory called "La Fortaleza" at 2200 m above sea level. More or less abundant.

Remarks:

According to Praeger (1967) this plant is similar to *M. pallens*, having, like it, a singular rosette (usually), and a very thick, bulbous or cylindrical base of the trunk. It can be distinguished during growth periods by its longer and much wider leaves

Fig. 3.208 Teide National Park. (Photo credit: Victoria Eugenia Martín Osorio)

(4–5 mm wide instead of 2 mm), which are much fewer in number (about 20 to the rosette instead of two or four times as much).

This taxon is very variable depending on the habitat, which affects its appearance, size of the leaves and flowering stem, depending on the ecological conditions of the different localities.

Nyffeler (1992) synonymizes *Monanthes niphophila* as *M. brachycaulos*; but there are still some doubts about whether *M. niphophila* would definitely be a diminutive form of *M. brachycaulos* or, on the contrary, its status should be conserved as an independent species.

Distribution

Endemism exclusive to the island of Tenerife (Figs. 3.209 and 3.210). In the Teide National Park, it was found occasionally in the surroundings of La Crucita (Morra La Casilla) and more importantly, on the western flank of La Fortaleza, between 1950 and 2125 m above sea level.

Habitat and Phytosociology

It usually colonizes humid and sunny rock fissures. ***M. brachycaulos*** is found within ***Greenovio aureae-Aeonietea***, a class of perennial vegetation, which groups comophytic and chasmo-comophytic permanent perennial communities, endemic to the Canary-Madeirense biogeographic subregion.

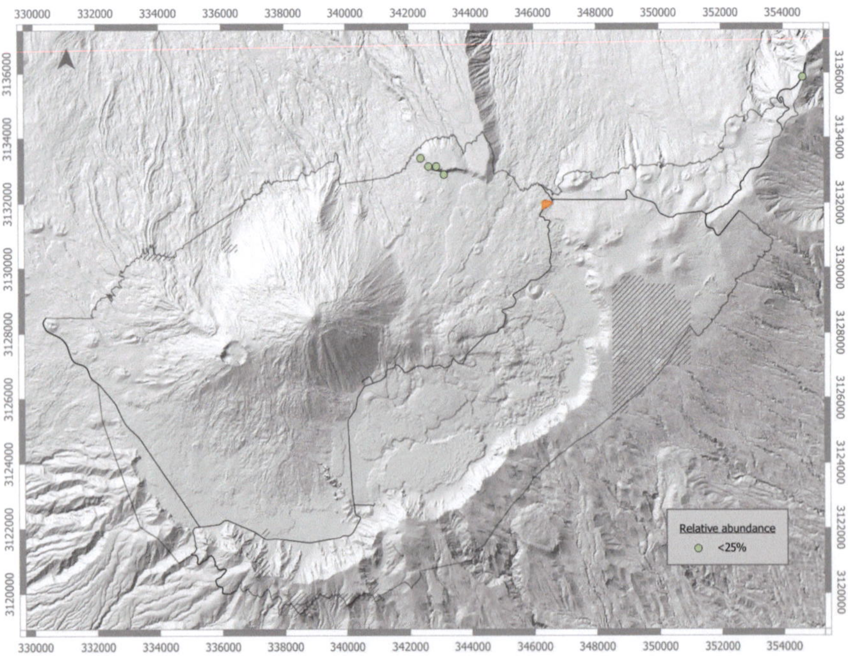

Fig. 3.209 *Monanthes brachycaulos* var. *nivata* distribution. N° Subgrids (250 × 250 m): **5**/3741

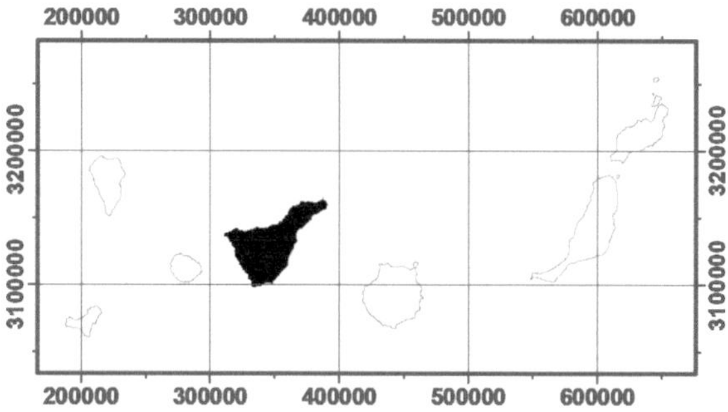

Fig. 3.210 Endemism of the island of Tenerife

Conservation Status

In all the subgrids in which this variety was found, no damage of any kind was observed, so its conservation status was high.

Category of Amenaza

nt (*M. brachycaulos*—Barreno et al. 1984)
Anexo II: Protegidas (Gobierno de Canarias 1991)
LR pm (Bañares et al. 1998)

Exsiccata (Herbarium Specimens)

Tenerife: Bco. del Agua—Güímar, 4.01.1989, A. Bañares & R. Mesa (TFC 29992). Inventario de la Flora y Vegetación del Parque Nacional del Teide (2001–2003) V.E. Martín Osorio, W. Wildpret & B. Hernández Bolaños: La Fortaleza (30A 14), 06/06/2004, V.E. Martín Osorio & W. Wildpret de la Torre (TFC 45429)—como *M. brachycaulos.*

3.5.68 Nepeta teydea *Webb & Berthel. var.* albiflora *Svent.*

Add. Fl. Canar.: 56 (1960)

Endemism, Certain native
Family: *Lamiaceae*
Common name: Hierba gatera, Nébeda, Tonática, Salvia de cumbre, Yerba del Teide (ES); Teide catnip (EN)
Biotype: hemicryptophyte
Flowering: April–June
Fruiting: May–July
Pollination: insect pollination (entomogamy)

Description

It differs from the type species in the shape of the main bracts, which are mutilated: short calyx, longer upper part of the tube and less ciliated divisions; white, immaculate corolla (Fig. 3.211).

Sventenius (1960):

Differt a typo bracteis principalibus muticis; calyce breviori, tubo parte superiori plus prolato et a divisionibus minus ciliatis; corolla alba, immaculata. Floret et fructificat ut in typo. Habitat inter scorias vulcanicas et in clivis petrosis.

Locus originis: Nivaria (Tenerife); in magno circo montano "Las Canadas" dicto, in loco nominato "Arenas negras" ad 2100 m. supra mare, ubi invenitur sat abundanter.

Lecta die 12 Aprilis 1944. Etiam visa est in loco dicto "El Portillo" atque in circunstantibus locis; parum frequens.

Distribution

Endemism of the island of Tenerife exclusive to Las Cañadas. We can find it in areas of El Portillo, Montaña El Cerrillar, Llano de Maja and Montaña de Las Piedritas or del Pedregal and southeastern slopes of Morro El Río (ridge of Barranco del Río); from 2000 to 2350 m a.s.l. (Figs. 3.212 and 3.213).

Fig. 3.211 *Nepeta teydea* var. *albiflora*. (Photo credit: Victoria Eugenia Martín Osorio)

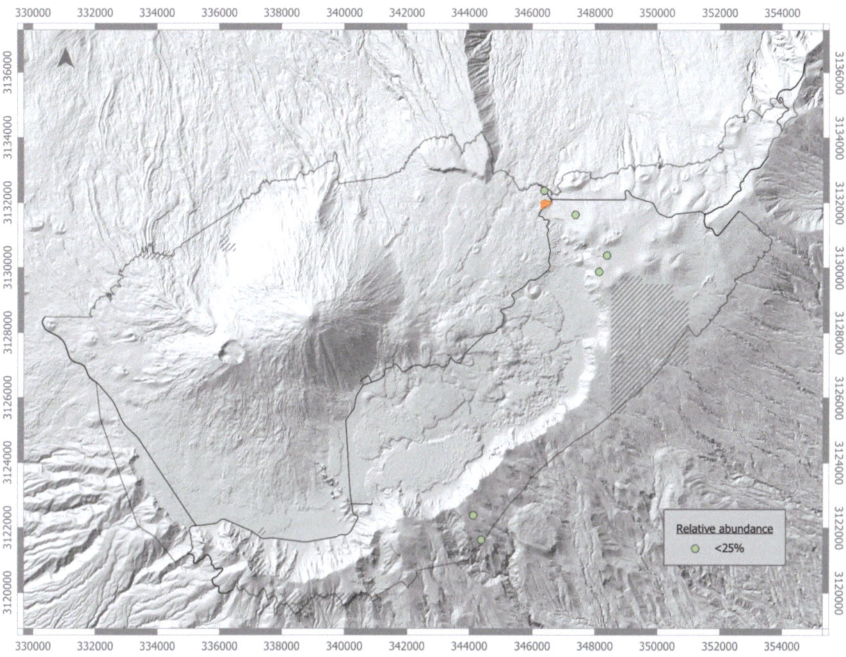

Fig. 3.212 *Nepeta teydea* var. *albiflora* distribution, N° Subgrids (250 × 250 m): **6**/3741

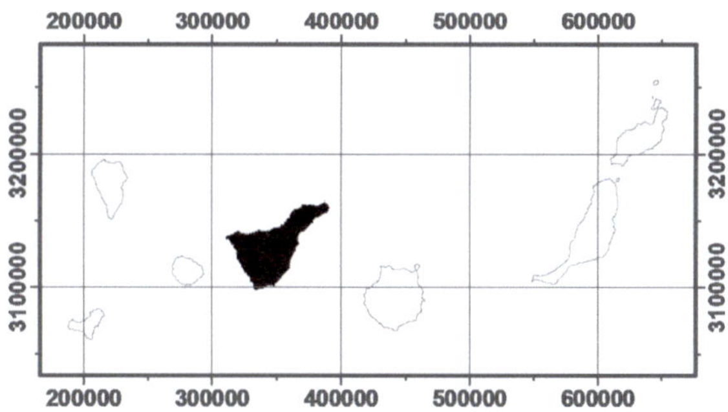

Fig. 3.213 Endemism of the Teide National Park on the island of Tenerife

Habitat and Phytosociology

Very rare variety in environments like the type species (rocky and stony). It is found within the ***Spartocytision supranubii*** alliance, which groups the summit shrublands of brooms, flatpods and summit sweet broom, climatophilous or edaphoxerophilous, sometimes catenal or serial of the Canary pine forests, with a distribution in La Palma and Tenerife, dry-humid Meso-Oromediterranean.

Conservation Status

In all the subgrids where this variety was found, its conservation status was high, so no damage of any kind was observed.

Exsiccata (Herbarium Specimens)

Tenerife: 28.05.1970, T. Bravo (TFC 350); Cañada de Diego Hernández—Las Cañadas, 28.05.1970, T. Bravo (TFC 10030); Arenas Negras, 2.06.1973, P.L. Pérez de Paz (TFC 10029, 10032); Arenas Negras—Las Cañadas, 1.04.1974, J.R. Acebes (TFC 6714); El Portillo, 18.05.1980, L. Negrín (TFC 10014); Ibid., 13.06.1980, E. Beltrán Tejera & L. Negrín (TFC 10027); El Portillo, 20.06.1982, L. Negrín (TFC 17222 + Dupl., 17223 + Dupl., 17224 + Dupl.); Ibid., 12.06.1983, ejusd. (TFC 17385). Inventario de la Flora y Vegetación del Parque Nacional del Teide (2001–2003) V.E. Martín Osorio, W. Wildpret & B. Hernández Bolaños: Guajara, Nacimiento del Barranco del Río (42C 14 b), 22/05/2003, A. Montañés Rodríguez (TFC 44274).

Uses, Exploitation and/or Medicinal Properties

> "*Anticatarrhal, diuretic, pectoral, hypoglycemic, aphrodisiac; poor quality fodder. Useful parts: leaves, tops. Instructions for use: decoction; browsing.*" (Translated from: Pérez de Paz and Hernández Padrón 1999)

3.5.69 Nepeta teydea *Webb & Berthel. var.* teydea

Hist. Nat. Iles Canaries Phyt. Canar., 3(2/3), 85 (1844–1850)

Glechoma teydea (Webb & Berthel.) Kuntze *in* Revis. Gen. Pl. 2: 519 (1891)

Endemism, Certain native
Family: *Lamiaceae*
Common name: Hierba gatera, Nébeda, Tonática, Salvia de cumbre, Yerba del Teide (ES); Teide catnip (EN)
Biotype: hemicryptophyte
Flowering: April–June
Fruiting: May–July
Pollination: insect pollination (entomogamy)
Dispersal: –
Chromosome N°: 2n = 16; n = 8

Description

Branches erect, tetragonous, hollow, hairy, very soft white hairs, articulated. **Leaves** petiolate, the upper ones subsessile, lanceolate or ovate-lanceolate, acute, cordate base, finely crenate-toothed, hairy on both sides, pale.

Raceme simple or with a branchy base, elongated, leaves of the flowers lanceolate entire, bracts linear-lanceolate, veined, mottled; calyx elongated or short; cyme pedunculated below, subsessile in aggregate spikes above; calyx short,

Fig. 3.214 Siete Cañadas, May 2004. (Photo credit: Beatriz Hernández)

campanulate-tubulous, 15-striated, teeth subequal or extremely subelongate, lanceolate base, subulate apex, smooth jaws, fruit little dilated; **Corolla** externally hirtulate, interiorly smooth, calyx protruding, one and a half times; tuberculate-scabrous semimericarp, blackish (Fig. 3.214).

Distribution

Endemism of the islands of La Palma and Tenerife. Within the Teide National Park it is found between 1900 and 3400 m above sea level; widely distributed throughout the NE-SW half of the park itself: from La Crucita, areas of La Dorsal, Altos de La Orotava and Izaña, extending more or less continuously to Boca Tauce, Cañada del Cedro and Narices del Teide. It also appears in La Fortaleza (Figs. 3.215 and 3.216).

Habitat and Phytosociology

It is usually found preferably in rocky and stony environments. Much less abundant in sandy or silty substrates of endorheic plains.

It is a characteristic species and bioindicator of ***Chamaecytiso-Pinetea canariensis***, a type of vegetation that includes coniferous forests, rhetamoid shrublands and colonizing dwarf mountain shrublands serial of Tenerife and La Palma.

It participates in the following associations:

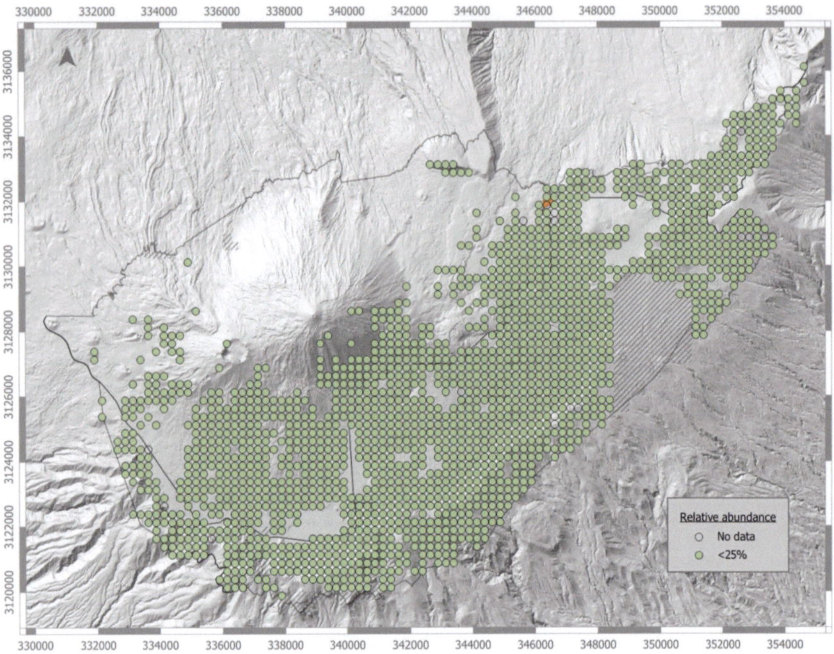

Fig. 3.215 *Nepeta teydea* var. *teydea* distribution, N° Subgrids (250 × 250 m): **1846**/3741

Fig. 3.216 Endemism of the island of Tenerife and La Palma

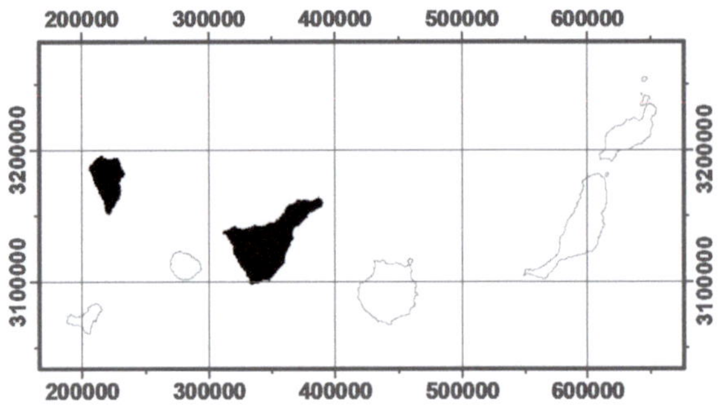

Spartocytisetum supranubii
Spartocytisetum supranubii-descurainietosum lemsii
Descurainio bourgeauanae-Spartocytisetum supranubii erigerontetosum calderae
Arrhenathero calderae-Plantaginetum webbii
Cheirolopho teydis-Carlinetum xeranthemoides
Tolpido webbii-Micromerietum lachnophyllae
Arrhenathero calderae-Adenocarpetum viscosi

Conservation Status

In 92% (1692) of the subgrids in which this species was found, no damage of any kind was observed. Only in 7% (131), damage was detected (browsing, high mortality without regeneration, etc.), but this did not affect more than 25% of the specimens. Damage was only found on more than 25% of the specimens in less than 1% (8) of the subgrids where this taxon occurred. In the rest (15), this variable was not specified at the time of sampling.

Threat Category

R (Barreno et al. 1984—*N. teydea*)
Anexo II: Protegidas (Gobierno de Canarias 1991)

Germplasm Banks

Jardín Botánico Canario Viera y Clavijo.

Exsiccata (Herbarium Specimens)

Tenerife: La Fortaleza, 19.05.1973, P.L. Pérez de Paz (TFC 10028); Bco. del Río, 16.05.1978, E. Beltrán Tejera (TFC 10031); El Portillo, 18.05.1980, L. Negrín (TFC 10013); Cercanías del Parador de Las Cañadas, 18.05.1980, L. Negrín (TFC 10015); El Portillo, 13.06.1980, E. Beltrán Tejera & L. Negrín (TFC 10016); Ibid.-Borde del sendero turístico, 13.06.1980, ejusd. (TFC 10017); La Fortaleza, 13.06.1980, ejusd. (TFC 10018, 10019 + Dupl., 10020, 10021, 10022, 10023, 10024, 10025, 10026); Cercanías del Parador de Turismo de Las Cañadas, 20.06.1982, L. Negrín (TFC 17236); El Portillo, 20.06.1982, L. Negrín (TFC 17234 + Dupl., 17235, 17236, 17237 + Dupl.); Izaña, 20.06.1982, L. Negrín (TFC 17225 + Dupl., 17226 + Dupl., 17227, 17228, 17229 + Dupl., 17230+ Dupl., 17232 + Dupl.,17233 + Dupl.); Boca de Tauce, 12.06.1983, L. Negrín (TFC 17384 + Dupl.); Las Benjaminas-Icod, 8.06.1989, F. Ardévol González (TFC 19146); Bco. de Erques-Adeje, 1800 m.s.n.m., J.M. Arteaga & P.L. Pérez de Paz (TFC 40657—Test Fitoquímico). Inventario de la Flora y Vegetación del Parque Nacional del Teide (2001–2003) V.E. Martín Osorio, W. Wildpret & B. Hernández Bolaños: Roques al Sur del Llano Ucanca (41D 35 d), 14/05/2003, A. Montañés Rodríguez (TFC 44297); Ladera Sur del Teide, hacia Pico Viejo (35D 31 c), 16/06/2003, F. Mesa (TFC 44205).

Uses, Exploitation and/or Medicinal Properties

> "*Anticatarrhal, diuretic, pectoral, hypoglycemic, aphrodisiac; poor quality fodder. Useful parts: leaves, tops. Instructions for use: decoction; browsing.*" (Translated from: Pérez de Paz and Hernández Padrón 1999)

3.5.70 Pimpinella cumbrae *Link*

in Buch., *Phys. Beschr. Can. Ins.*, 152 (1825)

Pimpinella buchii Webb. *Phyt. Canar.*, 3(2/2): 151–152.
Tragium buchii Webb. icon. 73.
Pimpinella cumbrae Buch. ubers. can., pág. 29 et *Phys. Beschr. Can. Ins.*, pág. 152; DC., *Prodr.*, vol. 4, pág. 120.
Tragium incanum Choisy. *in* DC., *Prodr.*, vol. 4, pág. 120.

Endemism, Certain native
Family: *Apiaceae*
Common name: Perejil de cumbre (EN); Summit parsley (EN)
Biotype: hemicryptophyte
Flowering: May–July
Fruiting: June–August
Pollination: insect pollination (entomogamy)
Dispersal: short range self-dispersal (atelochory)

Description

Hemicryptophyte with erect stems, 60–90 cm tall, rounded, pubescent, striated, subaphyllous with the upper part branchy. Pubescent branches, very furrowed; the lower ones solitary and distant, the upper ones very trichotomous; apex in terminal umbel.

Radical **leaves** numerous, very petiolate, with subangular petioles, short and dilated on the lower part, at the base of the stems unipinnate, scarcely vaginalized, with a thickly ciliated vaginal margin. Pinnatisect leaves (Fig. 3.217), the surrounding ones oblong, with seven fleshy segments, ovate to rotund, coarsely toothed; the lateral ones obovate-cuneate, incised-serrated, white-pubescent, with a trilobed terminal segment with incised-serrated cuneate lobes. Cauline leaves diverse. The leaves of the lower stems are sometimes like the radicals; those of the upper stems are very short, branched, sparsely petiolated, with vaginiform petioles, ciliate margins, pinnatisect, segments narrowly linear, acute, incised, but with entire margins; the upper leaves are simple and linear.

Involucre and involucellus absent. Umbels with seven rays, sometimes 6–8, rounded, subequal. Umbelule with 9–14 rays, longitudinally unequal, 2–4 mm long. Flowers few. Corolla whitish, petals ovate, concave, hairy back, canaliculate and emarginate at the top, narrow apex, slightly laciniate; lacinia inflexa. Stamens five alternate, whitish, long filaments, incurved during the summer period. Anthers bilobed, glabrous, yellowish, turgid. Two reflexed, yellowish styles with capitate apex. Pulvinular stylopodium yellowish. Fruits subglobular, hairy-velvety, the younger ones tomentose. Seeds convex and flat.

Fig. 3.217 (**a**) Teide National Park, Aug. 2008 (**b**) Route to Vilaflor, June 2006. (Photo credit: Beatriz Hernández)

Distribution

Endemism of the islands of La Palma and Tenerife. It extends along the high summit of Tenerife, between 1900 and 2700 m a.s.l., especially related to the steep areas of the inner slope of the Pared de Las Cañadas, Barranco del Río and Roques de García. It also appears very occasionally in areas of La Dorsal or Altos de La Orotava. Also, in the Teide Botanical Garden (Figs. 3.218 and 3.219).

Habitat and Phytosociology

It preferably colonizes steep areas and rocky-stony slopes. It is a characteristic species and bioindicator of ***Spartocytisetalia supranubii***.

It also characterizes or intervenes in the following associations:

Adenocarpo viscosi-Juniperetum cedri
Arrhenathero calderae-Plantaginetum webbii
Cheirolopho teydis-Carlinetum xeranthemoides
Sileno berthelotianae-Cerastietum sventenii

Conservation Status

In 95% (280) of the subgrids in which this species was found, no damage of any kind was observed. Only in 4% (12), damage was detected (browsing, high mortality without regeneration, etc.), but this did not affect more than 25% of the specimens. In the rest (4), this variable was not specified at the time of sampling.

Threat Category

nt (Barreno et al. 1984)
VU-D2 (Bañares et al. 1998)
VU (Bañares et al. 2004)
NT (Moreno 2008)

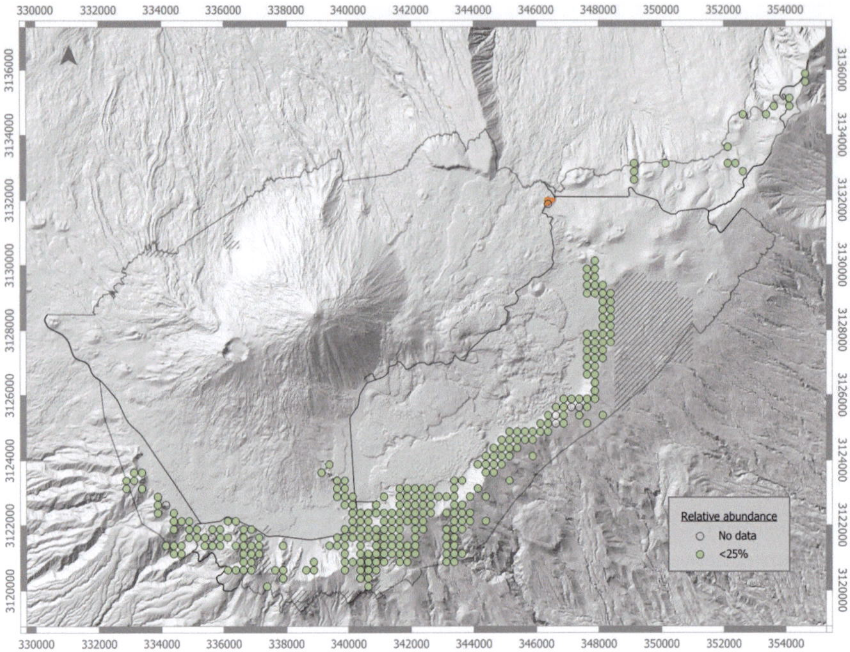

Fig. 3.218 *Pimpinella cumbrae* distribution, N° Subgrids (250 × 250 m): **296**/3741

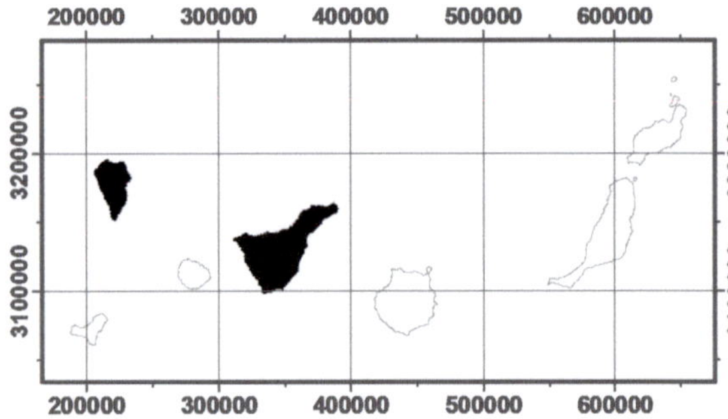

Fig. 3.219 Endemism of the island of Tenerife and La Palma

Germplasm Banks

E.T.S. Ingenieros Agrónomos de Madrid.
Jardín Botánico Canario Viera y Clavijo.

Exsiccata (Herbarium Specimens)

Tenerife: 29.06.1969, Bramwell 2126 (TFC 387); Las Cañadas, 28.07.1971, F. Estévez & A. Acuña (TFC 15993 + Dupl.); Cañada de Diego Hernández-Topo de la Grieta, 2.06.1973, P.L. Pérez de Paz (TFC 23333); Cañada de Diego Hernández, 2.06.1973, M.C. Gil (TFC 24089 + Dupl.); Las Cañadas, 31.06.1973, W. Wildpret,

PP, MG, H. Metlesics, CG & E. Beltrán Tejera (TFC 3886); 3.07.1974, J.R. Acebes (TFC 4537); Las Cañadas, 2.07.1977, W. Wildpret, M. del Arco Aguilar, C. León Arencibia & I. La Serna (TFC 25094); Las Cañadas, 7.07.1977, SMG & LD (TFC 23228 + Dupl.); Las Cañadas, 13.06.1981, CRP (TFC 24496); Cañada de Diego Hernández, 20.06.1985, P.L. Pérez de Paz, IMM & I. La Serna (TFC 24757 + Dupl.); Llano de Ucanca-Risco del Cedro, 2100 m.s.n.m., 2.08.1998, P.L. Pérez de Paz, C. Hernández Padrón & DPH (TFC 41623 + Dupl.); Degollada de Guajara-Las Cañadas, 2300 m.s.n.m., 29.08.2000, P.L. Pérez de Paz & DPH (TFC 42811 + Dupl.—Test. Fitoquímico—A. Velasco, Madrid). Inventario de la Flora y Vegetación del Parque Nacional del Teide (2001–2003) V.E. Martín Osorio, W. Wildpret & B. Hernández Bolaños*:* Guajara (42C), 17/06/2003, Y. de Rioja Santos (TFC 44257); Guajara, cerca de El Capricho (42C 10 a), 17/06/2003, A. Montañés Rodríguez (TFC 44270).

3.5.71 Pinus canariensis *C. Sm.* ex *DC.* in *Buch*

C.L.von Buch, Phys. Beschr. Canar. Ins.: 159 (1825)

Endemism, Certain native
Family: *Pinaceae*
Common name: Pino, Pino canario, Pino tea, Pino padre (ES); Canary Pine Tree (EN)
Biotype: tree (megaphanerophyte)
Flowering: March–April
Fruiting: July–September
Pollination: wind pollination (anemogamy)
Dispersal: wind—self dispersal (anemochory—barochory)
Chromosome No: 2n = 24; n = 12

Description

Tree up to 60 m tall and 2.5 m in diameter (Fig. 3.220), although it normally does not exceed 40 m. Straight and cylindrical trunk. Bark is almost smooth during the first years of growth, then thickens rapidly, cracks and takes on a reddish brown colour. In older trees the bark forms smooth plates taking on an ashy colour. Pyramidal-conical crown in young individuals, which in adults becomes rounded. The presence of adventitious shoots on the trunk, with dense, bluish primary leaves, is one of the characteristics of the Canary Island pine.

The **leaves** are needle-like, always grouped in groups of three, quite long, light green, very thin and flexible, 20–30 cm long by 1 mm thick, wrapped at their base by a membranous sheath 10–15 mm long. The leaves of the first year are short and very glaucous; Normal ones are pointed at the end and finely serrated at the margins. Thick, non-resinous buds.

Fig. 3.220 La Crucita, August 2008. (Photo credit: Beatriz Hernández). Male cone, August 2015. (Photo credit: Victoria Eugenia Martín Osorio)

Reproductive structures of male flowers ovate—oblong, with the appearance of catkins, made up of numerous stamens. They appear at the end of the branches of the year before flowering, grouped in strobili 5–10 cm long, greenish-yellow in colour. Solitary **cones**, rarely whorled in groups of two or three, oblong—fusiform, reddish-brown and shiny, 10–18 cm long and about 5 cm thick, pedunculated. Rhomboid shields wider than high, very lignified and tight. Obovoid pine nuts 1 cm long and 6 mm thick, with a hard shell, blackish on one side and mottled on the other.

Remarks:

> *It flowers and matures alternately every 2 years. "Natural regeneration from undisturbed seed in good soil is abundant and easy. Immediately after germination, the main root develops rapidly; the young stem, before branching, is whitish and covered with very glaucous and finely serrated primordial leaves. Growth, slow during the stage following germination, gradually increases until the plant is in full possession of the soil, becoming quite rapid from the fifth year onwards; it is not unusual to see 10 to 15 year old specimens that lengthen their vine by 1 to 1.50 m during the vegetative period; this speed slows somewhat in the following years, when, on the other hand, growth in thickness is more active. From 30 years of age (15–20 m), growth decreases, although it continues, increasingly slower, until 70 or 75 years of age (30–40 m), when it becomes null or almost null (…), then beginning a period of precarious life, which can last even centuries without the tree entering into clear decrepitude."* (Translated from: Ceballos and Ortuño 1976)

Distribution

Endemism of the central-western Canary Islands (Fig. 3.221). In Tenerife, it can occasionally be found from 100 m above sea level, reaching altitudes of up to 2675 m above sea level (near Las Narices del Teide), although its optimal altitude range extends from 700 to 1500 m above sea level.

It forms important natural forests in Tenerife, La Palma, El Hierro and Gran Canaria, and also exists as an isolated native species in La Gomera. This species has

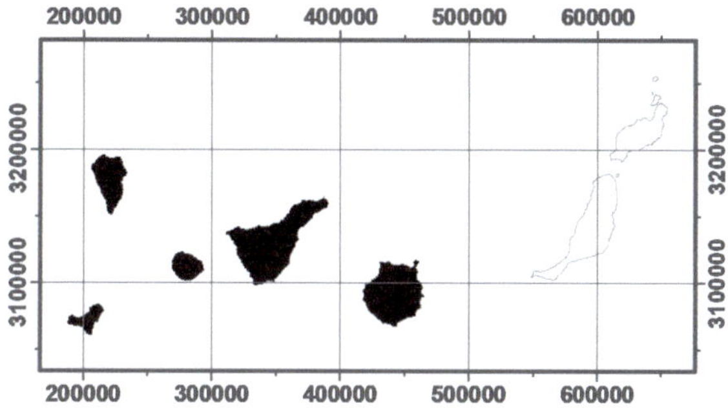

Fig. 3.221 Endemism of the central and occidental islands

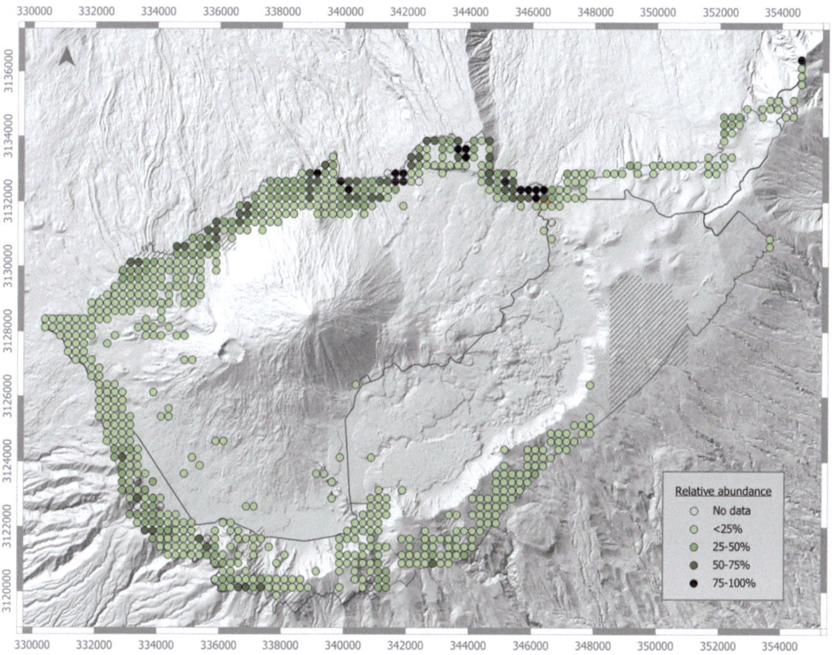

Fig. 3.222 *Pinus canariensis* distribution, N° Subgrids (250 × 250 m): **723**/3741

been the main element of several important plantation and repopulation projects, generally in all the central and western islands.

In the Teide National Park (Fig. 3.222), it can be found more or less continuously along the entire boundary of the National Park itself (Altos de La Orotava, La Fortaleza, northern slopes of the Teide-Pico Viejo edifice, Montaña Samara, Cañada del Cedro and Boca Tauce). Precisely, from the Llano de Ucanca cirque its formation becomes much more discontinuous in interior areas of the Pared de Las Cañadas (except for the vicinity of Riachuelo and Cañada del Capricho).

Practically, from La Cañada de Las Pilas and up to El Portillo and higher areas of La Dorsal, the upper limit of the Pinar is outside the area of influence of this protected area.

Near the National Park the ***Pinus canariensis*** pine forest was seriously affected in 2023 by a fire that spread over more than 15,000 ha, being the most serious on the island in recent decades.

Habitat and Phytosociology

It commonly colonizes rocky and stony volcanic substrates. It is a characteristic species and bioindicator of the ***Chamaecytiso-Pinetea canariensis*** class, which includes coniferous forests, rhetamoid shrublands and colonizing mountain dwarf shrublands serial of Tenerife and La Palma.

It appears as a protagonist or also intervening in the following units:

Sideritido solutae-Pinetum canariensis
Myrico fayae-Pineto canariensis, beyond the limits of the National Park
Adenocarpo viscosi-Chamaecytisetum angustifolii
Tolpido webbii-Micromerietum lachnophyllae
Arrhenathero calderae-Plantaginetum webbii

Conservation Status

In 89% (642) of the subgrids in which this species was found, no damage of any kind was observed. Only in almost 6% (40), damage was detected (browsing, high mortality without regeneration, etc.), but this did not affect more than 25% of the specimens. Damage was only found on more than 25% of the specimens in 4% (32) of the sub-grids where this taxon occurred. In the rest (9), this variable was not specified at the time of sampling.

Threat Category

nt (Barreno et al. 1984)
Anexo III (Gobierno de Canarias 1991)
LR pm (Bañares et al. 1998)

Germplasm Banks

E.T.S. Ingenieros Agrónomos de Madrid.
Jardín Botánico Canario Viera y Clavijo.

Exsiccata (Herbarium Specimens)

Tenerife: Pinar de La Esperanza, A. Santos (TFC 22910); San Miguel de Abona, 17.12.1969, E. Beltrán Tejera (TFC 35418); Cmo. Las Peras-La Laguna, 15.03.1973, I. La Serna (TFC 35697); La Laguna, 20.03.1975, B. Castro (TFC 33490); La

Laguna, 31.03.1985, P.L. Pérez de Paz (TFC 456, 13291 + Dupl.); Pinar de Arico-Contador y cumbre, 9.03.1990, M. del Arco Aguilar, P.L. Pérez de Paz & M. Salas Pascual (TFC 38365 + Dupl.); Bco. Cuevas Negras-Los Silos, 9.03.1997, V. Lucía Sauquillo & C. Acosta (TFC 40061 + Dupl.). Inventario de la Flora y Vegetación del Parque Nacional del Teide (2001–2003) V.E. Martín Osorio, W. Wildpret & B. Hernández Bolaños: Casa al lado del Bar de El Portillo en el Parque Nacional del Teide (30D 32 a), 13/11/2003, V.E. Martín Osorio, W. Wildpret de la Torre & B. Hernández Bolaños (TFC 44936 + Dupl.)

Gran Canaria: Risco del Hurón-Bco. Oscuro-Valleseco, 3.05.1980, P.L. Pérez de Paz & C. Suárez (TFC 11087).

Uses, Exploitation and/or Medicinal Properties

> "*Canarian pine wood is of good quality among those of the genus (…). This wood, very resinous, acquires great density and resistance when it is tanned, becoming practically incorruptible, due to the impregnation of resin which, at the same time as enhancing its beautiful colour, makes it translucent in the thickness of common boards; for all this reason the teak of P. canariensis is highly appreciated and sought after today, which is almost considered in the category of precious wood (…). We have frequently seen this wood used in skirting boards, balustrades, lattices and even furniture for the interior decoration of buildings and stately mansions in the Canary Islands, apart from its main application since ancient times as construction wood.*" (Ceballos and Ortuño 1976)

> "*Analgesic, anti-catarrhal, hair tonic, antiscorbutic, odontalgic, antirheumatic, antiasthmatic, anti-inflammatory; Crafts: cabinetmaking, sculpture; forestry interest: wood, firewood, construction, pine needles or pine base; ornamental. Useful parts: buds and shoots; pine cones, seeds; resin; tar or black pitch; wood: tea; whole plant. Instructions for use: decoction (external use); infusion; food; torches: light; construction: beams, floors, roofs, balconies; crafts: boxes, vessels, spoons, tools, walking sticks; manure.*" (Translated from: Pérez de Paz and Hernández Padrón 1999)

3.5.72 Pinus radiata *D. Don*

Trans. Linn. Soc. London, 17, 442 (1836)

Pinus insignis var. *radiata* (D.Don) Lemmon *in* Bienn. Rep. Calif. State Board Forest. 2: 76, 114 (1888), nom. illeg.

Not endemic, Certain introduced
Family: *Pinaceae*
Common name: Pino de Monterrey, Pino americano o Pino insigne (ES) Monterey Pine (EN)
Biotype: tree (phanerophyte)
Flowering: March–April
Fruiting: the cones open in the second autumn
Pollination: wind pollination (anemogamy)

Dispersal: wind and self-dispersal (anemochory—barochory)
Chromosome N°: 2n = 24; n = 12

Description

Tree up to 30 (40) m (Fig. 3.223). Erect trunk, with thick, reddish-brown rhytidome. Horizontal or erect-patent branches, ascending; reddish-brown twigs.

Leaves 7–15 × 0.12–0.17 cm, in fascicles of 3 (29) flexible, bright green. Strobili 7–14 × 5–8 cm, ovoid, subsessile, in pairs or whorls of 3–5, strongly asymmetrical, with very prominent apophyses of the external scales. The cones can remain closed for several years without the germinative capacity of the pine nut being altered. Seeds of 5–8 mm.

Distribution

Originally from the coasts of southern California, species introduced in northern Spain, northern and central Portugal, and the Canary Islands. Present on all the islands of the Canary Archipelago, except Lanzarote and Fuerteventura (Fig. 3.224).

In the Teide National Park (Fig. 3.225), it was found occasionally in Montaña de Igueque (Güímar highlands), in Hoya el Peñón (N-Izaña), west of La Fortaleza and near the Roques de García. In greater numbers, in Portillo Bajo and Portillo Alto; between 1900 and 2200 m a.s.l.

Habitat and Phytosociology

Generally, it has been cultivated mainly in non-calcareous soils and areas with a temperate and humid winter climate, between 0 and 800 m a.s.l. It was planted in the 1950s on the islands of Tenerife, La Gomera, El Hierro, La Palma, and Gran

Fig. 3.223 El Portillo Alto, Teide National Park, June 2005. (Photo credit: Victoria Eugenia Martín Osorio)

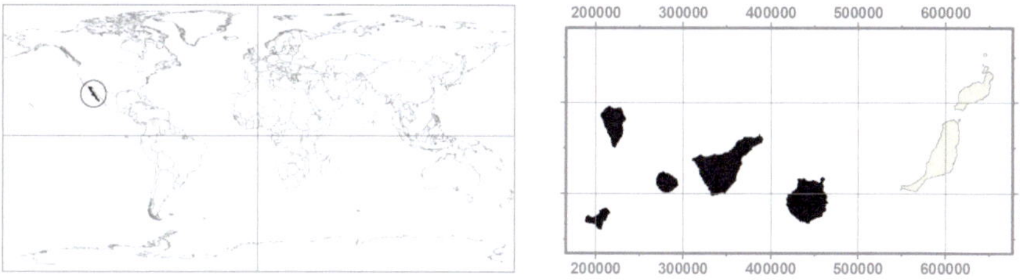

Fig. 3.224 Present on all the Canary Islands except Lanzarote and Fuerteventura

Fig. 3.225 *Pinus radiata* distribution, N° Subgrids (250 × 250 m): **14**/3740

Canaria by the Institute for the Conservation of Nature (ICONA) to repopulate the areas between 1000 and 1500 m above sea level, preferably facing north.

Conservation Status

In most of the subgrids where this species was found (13), no damage of any kind was observed. Only in one (1), damage was detected (browsing, high mortality without regeneration, etc.), but this did not affect more than 25% of the specimens.

Exsiccata (Herbarium Specimens)

Tenerife: La Laguna, Mesa Mota, 12.3.1988, M. Salas Pascual (TFC duplic. 38364); La Laguna, Campus Central, 3.1994, M. Sansón & M.C. León Arencibia & E. Beltrán Tejera (TFC duplic. 36431); Los Silos, La Juncia, 1.2.1997, V. Lucía Sauquillo (TFC 40024 + duplic.).

Uses, Exploitation and/or Medicinal Properties

> "*Pectoral, balsamic, antirheumatic; wood: construction, carpentry, cabinetmaking; forestry interest. Useful part: buds, resin. Instructions for use: exceptional: inhalation, infusion.*" (Translated from: Pérez de Paz and Hernández Padrón 1999)

3.5.73 Piptatherum coerulescens *(Desf.) P. Beauv.*

Ess. Agrost., 18, 173 (1812)

Agrostis coerulescens (Desf.) Poir. *in* J.B.A.M.de Lamarck, Encycl., Suppl. 1: 258 (1810)
Milium coerulescens Desf. *in* Fl. Atlant. 1: 66 (1798)
Oryzopsis coerulescens (Desf.) Hack. *in* Denkschr. Kaiserl. Akad. Wiss., Wien. Math.-Naturwiss. Kl. 50: 75 (1885)
Stipa coerulescens (Desf.) Raspail *in* Ann. Sci. Nat. (Paris) 5: 449 (1825)
Urachne coerulescens (Desf.) Trin. *in* Fund. Agrost.: 110 (1820)

Not endemic, Probable native
Family: *Poaceae*
Common name: Cerrillón granudo (ES); Ricegrass (EN)
Biotype: hemicryptophyte
Flowering: April–June
Fruiting: April–June
Pollination: wind-pollination (anemogamy)
Dispersal: wind dispersal (anemochory)

Description

Perennial cespitose plant (Fig. 3.226). **Stems** up to 80 cm, striated, glabrous. **Leaves** with glabrous sheath; ligule 2.5–9 mm, acute, slightly pubescent; blade up to 20 (−40) × 0.05–0.3 cm, flat or convolute, glabrous, with markedly striated and scabrid upper surface with scabrid margin.

Panicle 8–15 cm, lax, with smooth or very slightly scabrid axis; smooth or slightly scabrid branches, flexuous; lower node with 1–2 branches.

Spikelets 5.8–8 mm, usually violet or purple. Glumes longer than the flower, with 3–5 nerves, ovate-lanceolate, acuminate; the lower one 5.8–8 mm; the upper

Fig. 3.226 Wet walls in Teide National Park, June 2006. (Photo credit: Victoria Eugenia Martín Osorio)

one 5.5–7.7 mm. Lemma 3.2–4 mm, generally covered by glumes, glabrous, hardened and blackish when mature, with terminal awn 2.3–5 mm. Palea as long as the lemma. Anthers ca. 3 mm.

Distribution

Distributed throughout the Mediterranean region and southern Portugal, N Africa, Central and SW Asia and, in Macaronesia, only in the Canary Islands. Present on the islands of La Gomera, Tenerife, Gran Canaria and Lanzarote (Figs. 3.227 and 3.228).

Piptatherum coerulescens is considered introduced according to Plants of the World Online (https://powo.science.kew.org/taxon/urn:lsid:ipni.org:names:416076-1) and in Martín Osorio et al. (2006). According to the sampling carried out, it is widely spread and was located in: La Tarta, near the Fasnia volcanoes, areas of El Capitán (highs of the Orotava Valley), La Fortaleza and along practically the entire wall of Las Cañadas, including the Roques de García; between 1900 and 2600 m a.s.l.

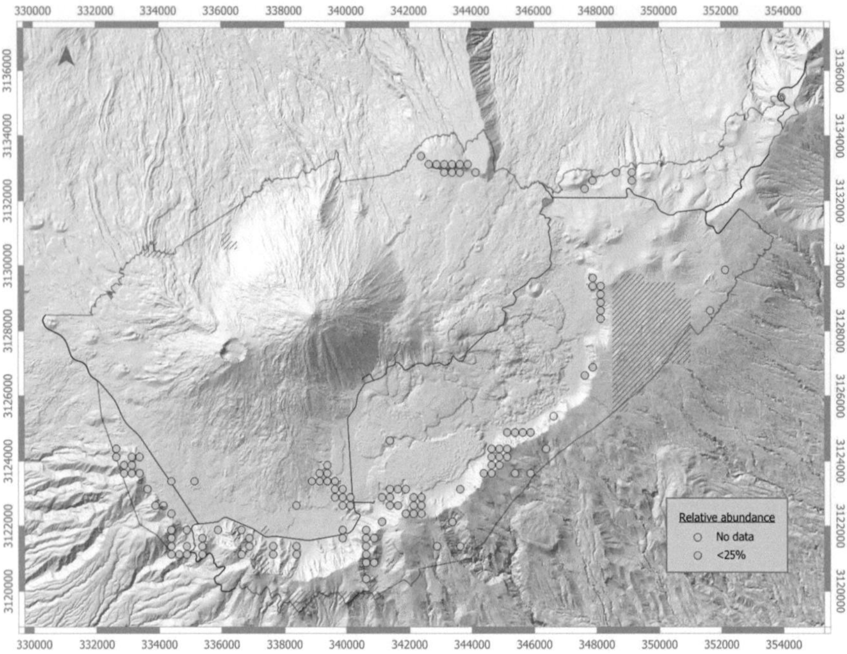

Fig. 3.227 *Piptatherum coerulescens* distribution, N° Subgrids (250 × 250 m): **126**/3741

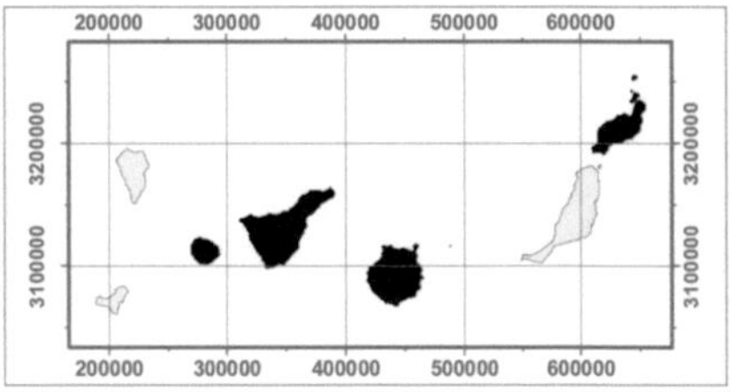

Fig. 3.228 Present on the islands of La Gomera, Tenerife, Gran Canaria and Lanzarote

Habitat and Phytosociology

It lives in rocky areas and grasslands on stony soils. Widely distributed from the infra to the Supra-Mediterranean floor.

It is a characteristic species and bioindicator of the ***Lavaterion maritimae*** alliance, which groups the western and North African Mediterranean rupicolous chasmophytic and chasmocomophytic associations, nitrophilous, subnitrophilous and ornithocoprophilous littoral, with analogous interior thermophilic stations, typical of rocky stations visited by herbivores and by nesting birds; semi-arid-dry Thermo-Meso-Mediterranean. It is also typical of communities of the ***Lygeo sparti-Stipetea tenacissimae*** class, vegetation that develops on soils with a loamy or clayey texture, generally rich in bases, constituting relatively evolved serial stages and favoured by fire and extensive grazing practices, whose optimum corresponds to warm and dry climates.

Conservation Status

In practically all the subgrids in which this species was found (120), no damage of any kind was observed. Only in four (4) of them, damage was detected (browsing, high mortality without regeneration, etc.), but this did not affect more than 25% of the specimens. In the rest (2), this variable was not specified at the time of sampling.

Remarks:

Martín Osorio et al. (2006), taking as criteria the number of specimens, the surface area occupied and the colonization power of the species, determined as well-established transforming invaders within the Teide National Park, the taxa ***Avena barbata***, ***Avena sterilis***, ***Bromus tectorum***, ***Dittrichia viscosa***, ***Piptatherum coerulescens*** and ***Festuca myuros***.

Germplasm Banks

E.T.S. Agricultural Engineers of Madrid.
Royal Botanical Garden of Madrid.

Exsiccata (Herbarium Specimens)

Tenerife: Arico, La Sabinita, 27.1.1973, P.L. Pérez de Paz (TFC 32967); Masca, 22.7.1975–1.1994, E. Beltrán Tejera, M. del Arco Aguilar, M.C. León Arencibia, I. La Serna & JBD (TFC duplic. 36422); Arico, Bco. del Río, 1976, E. Beltrán Tejera, M. del Arco Aguilar & M.C. León Arencibia (TFC 36421); Masca, aprox. 800 m.s.n.m., 15.4.1976, E. Beltrán Tejera, I. La Serna, M. del Arco Aguilar, H. Metlesics et al. (TFC 36423); Bco. Ruiz, altos, 160 m.s.n.m., 21.1.1977, E. Beltrán Tejera, M. del Arco Aguilar, M.C. León Arencibia et al. (TFC duplic. 34564); Arona, Ifonche, 24.3.1979, W. Wildpret, E. Beltrán Tejera, J.R. Acebes, M. del Arco Aguilar & M.C. León Arencibia (TFC duplic. 35266); Las Cañadas, La Fortaleza, 13.6.1980–12.1992, E. Beltrán Tejera & M.C. León Arencibia (TFC duplic. 35257); El Escobonal, Bco. de Erques, 30.10.1980, E. Beltrán Tejera & O. Rodríguez Delgado (TFC duplic. 12639); Arico, Bco. del Río, 1.4.1980, W. Wildpret, J.R. Acebes, M. del Arco Aguilar, M.C. León Arencibia, L. Negrín & AVR (TFC 25132 + 2 duplic.); Punta del Hidalgo, 10.5.1980, A. Hardisson (TFC 23037); Agache, Bco. del Espigón, 2.4.1981, O. Rodríguez Delgado (TFC 12640); Masca, 11.6.1983, P.L. Pérez de Paz (TFC duplic. 33779); Bco. el Riachuelo, 12.5.1984, A. Charpin & W. Wildpret (TFC 20046; 20082); Adeje, Bco. del Infierno, 400 m.s.n.m., W. Wildpret, M. del Arco Aguilar et al. (TFC 9544); Izaña, 25.5.1987, W. Wildpret & O. Rodríguez Delgado (TFC 29207); Agache, Güímar, La Ladera, 6.3.1988, W. Wildpret & O. Rodríguez Delgado (TFC duplic. 28109); La Laguna, Valle Colín, 28.4.1988, W. Wildpret, A. García Gallo & ECA (TFC 26816; 26817; 26818; 26819 + duplic. 26820); Mesa del Sabinal, 480 m.s.n.m., 18.6.1996, R. Racca (TFC 39151); Ladera de Güímar, 9.3.1990, P.L. Pérez de Paz, M. del Arco Aguilar & M. Salas Pascual (TFC 38360); Bco. de Tahodio, cerca de Jardina, 8.5.1999, S. Becker, O. Rodríguez Delgado & T. Himstedt (TFC 42146). *Inventario* *de la Flora y Vegetación del Parque Nacional del Teide (2001–2003) V.E. Martín Osorio, W. Wildpret & B. Hernández Bolaños*: Zapatito de la Reina (41D 25 b), 28/05/2002, L. Afonso & A. Hernández (TFC 45424); Cañada del Capricho, 02/07/2003, F. Mesa, A. Montañés & Y. de Rioja (TFC 44864); La Fortaleza, Parque Nacional del Teide, 06/06/2004, V.E. Martín Osorio & W. Wildpret de la Torre (TFC 45441 + Dupl.); Al sur de la Mña. Angostura (42B 5 b), 06.06.2003, F. Mesa (TFC 44216); La Fortaleza, 05.06.2004, V.E. Martín Osorio & W. Wildpret (TFC 44960); Guajara, Nacimiento del Barranco del Río (42C 23 a), 09.06.2003, F. Mesa (TFC 44220).

Gran Canaria: Subida a Fataga, 15.4.1976, E. Beltrán Tejera, J.R. Acebes, M. del Arco Aguilar & B. Méndez Pérez (TFC 25863 + duplic.); San Bartolomé, Risco Blanco, 15.4.1976, ejusd. (TFC 25865); Guía, Bco. La Colmenilla, 18.5.1986, C. Suárez (TFC 31674; 31675; 31676); Arucas, Lomo Riquianez, 22.5.1986, C. Suárez (TFC 31702); Valsequillo, Bco. del Agua, 18.6.1986, C. Suárez (TFC 31715); Parque Natural de Guayadeque, cauce del Bco. de Guayadeque, 23.6.1989, P.L. Pérez de Paz & F. Amor (TFC 30532); Ibid, 8.3.1990, ejusd. (TFC 30793); Tejeda, Pajonales, Presa de las Niñas, 10.4.1992, M. Salas Pascual (TFC 37082); San Bartolomé, Camino del Pinar, 900 m.s.n.m., 18.4.1992, M. Salas Pascual (TFC 37081); Pinar de Inague, 1.10.2003, S. Rivas-Martínez, M. del Arco Aguilar, J.R. Acebes & M.L. López (TFC 44413); Pinar de Inague, 1.10.2003, S. Rivas-Martínez (TFC 44414; 44415).

Lanzarote: Malpaís de la Corona, 3.1.1994, J.A. Reyes-Betancort (TFC 38550).

3.5.74 Piptatherum miliaceum *(L.) Coss.*

Not. Pl. Crit., 129 (1851)

Oloptum miliaceum (L.) Röser & Hamasha First published *in* Pl. Syst. Evol. 298: 365 (2012)
Achnatherum miliaceum (L.) P.Beauv. *in* Ess. Agrostogr.: 19 (1812)
Agrostis miliacea L. *in* Sp. Pl.: 61 (1753)
Milium multiflorum Cav. *in* Descr. Pl.: 36 (1801), nom. superfl.
Nassella multiflora Druce *in* List Brit. Pl., ed. 2: 125 (1928), nom. superfl.
Oryzopsis miliacea (L.) Asch. & Schweinf. *in* Mém. Inst. Égypt. 2: 169 (1887)
Oryzopsis multiflora Druce *in* Rep. Bot. Exch. Club Soc. Brit. Isles 3: 180 (1912 publ. 1913), nom. superfl.
Piptatherum frutescens Schult. *in* Mant. 2: 183 (1824), nom. superfl.
Piptatherum miliaceum (L.) Coss. *in* Notes Pl. Crit.: 129 (1812)
Stipa miliacea (L.) Hoover *in* Leafl. W. Bot. 10: 340 (1966)

Not endemic to the Canary Islands, Probably native
Family: *Poaceae*
Common name: Cerrillón o Cerrillón fino (ES); Narrow ricegrass (EN)
Biotype: hemicryptophyte

Flowering: April–November
Fruiting: April -Novembe
Pollination: wind pollination (anemogamy)
Dispersion: wind—animal dispersal (anemochory—zoochory)

Description

Perennial cespitose plant with stems up to 100 cm (Fig. 3.229), erect or ascending, striate, glabrous.

Leaves with glabrous sheath; ligule 1–3 mm, pubescent. Blade up to 50 × 0.2–1.2 cm, flat or convolute when dry, with markedly striate and scabrid upper surface and smooth underside.

Panicle 20–50 cm, pyramidal, with pendulous upper part, very lax; axis generally smooth; branches whorled, erect-patent or patent, flexuous, capillary, scabrid; lower node with 4–10 (−17) branches, with all spikelets fertile.

Spikelets 2.5–3.6 mm, frequently violet. Glumes longer than the flower, lanceolate or ovate-lanceolate, three-veined, acuminate, glabrous; the lower one 2.5–3.6 mm; the upper one 2.3–3.4 mm. Lemma 1.7–2.5 mm, three-veined, obtuse, glabrous, with a terminal awn of 2.5–5 m. Palea as long as the lemma. Anthers 1–1.5 mm.

Fig. 3.229 Detail of *Piptatherum miliaceum*. (Photo credit: GG Casanova)

Distribution

It extends throughout the Mediterranean region, Africa, and Central and SW Asia. Very common throughout the Iberian Peninsula and in the Azores, Madeira and the Canary Islands. Present on all the islands of the Canary Archipelago (Fig. 3.230).

In the Teide National Park (Fig. 3.231) it has been found very punctually in La Crucita (2000 m a.s.l.), La Fortaleza and the southeastern slopes of Roque El Almendro; more continuously, from the Barranco del Río, Guajara Mountain, Ucanca Tableland and the flank of La Pared towards Llano de Ucanca, reaching 2600 m above sea level.

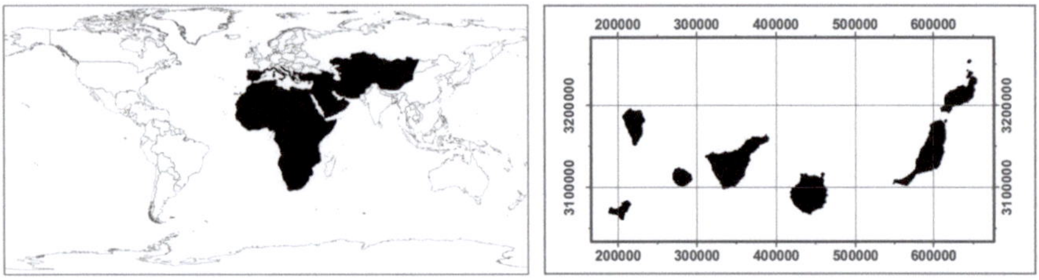

Fig. 3.230 Present on all Canary Islands

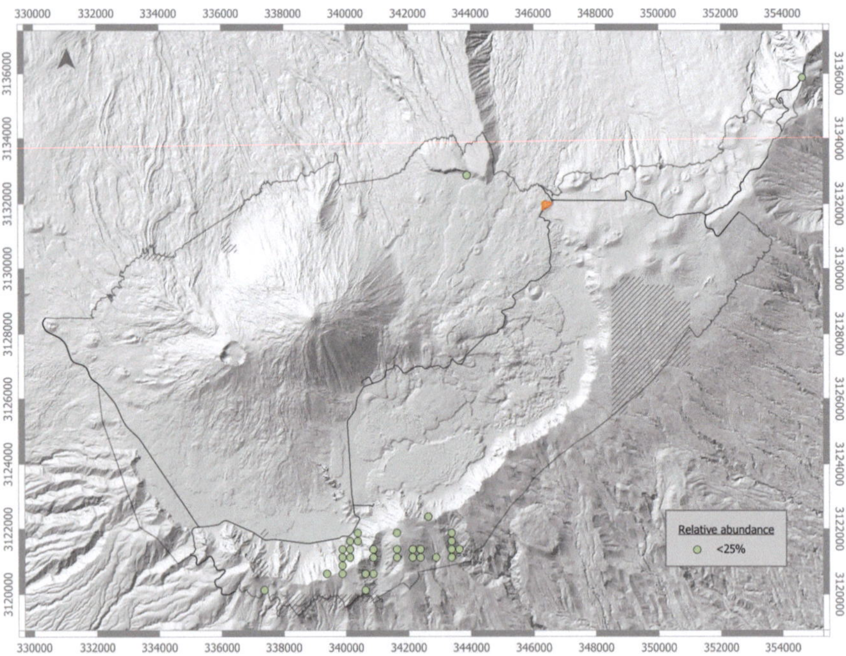

Fig. 3.231 *Piptatherum miliaceum* distribution, N° Subgrids (250 × 250 m): **32**/3741

Habitat and Phytosociology

It usually colonizes areas of grasslands and herbaceous lands, roadside slopes and disturbed and uncultivated lands. It is a characteristic species and bioindicator of ***Bromo madritensis-Piptatherion miliacei***, an alliance that groups the associations of perennial grasses-herbs, which develop on deep soils in ruderal-nitrophilous environments and abandoned crops; of circum-Mediterranean distribution with Cantabrian-Atlantic and Canary-Madeiran infra-Meso-Mediterranean semi-arid-subhumid disjunctions.

Conservation Status

In most of the subgrids in which this species was found (28), no damage of any kind was observed. Only in the remaining four (4), damage was detected (browsing, high mortality without regeneration, etc.), but this did not affect more than 25% of the specimens.

Germplasm Banks

E.T.S. Ingenieros Agrónomos de Madrid.
Jardín Botánico Canario Viera y Clavijo.

Exsiccata (Herbarium Specimens)

Tenerife: Cuchillos del Saltadero, sobre San Andrés, 12.3.1973, W. Wildpret, E. Beltrán Tejera, AS, P.L. Pérez de Paz & R. Afonso (TFC duplic. 4065); La Orotava, A1, Cuartel de Bomberos, 260 m.s.n.m., 10.11.1977, E. Beltrán Tejera & SMG (TFC 34563); La Laguna, La Manzanilla, 4.5.1981, M. Arco Aguilar & J.R. Acebes (TFC 24955 + 5 duplic.); La Laguna, 21.1.1982, A. García Gallo (TFC 22559; 22560; duplic. 22561); Icod, Los Castañeros, 550 m.s.n.m., 21.7.1984, F. Ardévol González (TFC 19264); Masca, 27.5.1990, C.D. Hernández (TFC 33428); Santiago del Teide, entrada a galería de Tamaimo, 660 m.s.n.m., 27.5.1990, ejusd. (TFC 33327); Tamaimo, entrada al túnel de Tamaimo, 9.6.1990, C.D. Hernández (TFC 33411); Los Silos, Pina, 27.8.1991, V. Lucía Sauquillo (TFC 38141); Tegueste, pueblo, 17.10.1992, T. Jiménez Felipe (TFC 36324); Güímar, Bco. de Badajoz, 31.5.1993–1.1994, I. La Serna & M.C. León Arencibia (TFC 33338); La Orotava, Punta del Fraile, 15.3.1994, F. Ramallo (TFC duplic. 37540); Los Silos, La Costa, 15.6.1995, V. Lucía Sauquillo & C. Acosta (TFC 39950 + duplic.); Inventario de la Flora y Vegetación del Parque Nacional del Teide (2001–2003) V.E. Martín Osorio, W. Wildpret de la Torre & B. Hernández Bolaños: Al W de Izaña—Montaña de El Cabezón (31C 2), 24.04.2001, L. Afonso, A. Hernández, Y. de Rioja & F. Tapia (TFC 44914); Guajara, Nacimiento del Barranco del Río (42C 14 b), 22.05.2003, A. Montañés Rodríguez (TFC 44279); La Fortaleza (30A 25 b), 03.04.2001, L. Afonso, A. Hernández, Y. de Rioja & F. Tapia (TFC 44963).

Gran Canaria: San Nicolás de Torrentino, Las Tabladas, 14.11.1975, J.R. Acebes (TFC 25859); Agaete, Bco. de la Palma, 3.6.1979, J.R. Acebes (TFC duplic. 32498).

Lanzarote: Tinajo, 24.4.1996, J.A. Reyes-Betancort & C. García (TFC duplic. 40185); Valle de Malpaso, 2.7.1994, J.A. Reyes-Betancort (TFC 38510 + duplic.).

?, 2.11.1974, E. Beltrán Tejera, M. Arco Aguilar, B. Méndez Pérez & J.R. Acebes (TFC 4431).

3.5.75 Plantago webbii *Barnéoud*

Monogr. Pl., 50 (1845); Dcne., *in* D.C., *Prodr.*, XII, 733

Psyllium webbii (Barnéoud) Soják *in* Čas. Nár. Mus., Odd. Přír. 140: 131 (1972)
Plantago seetzenii Decne. *in* A.P.de Candolle, Prodr. 13(1): 733 (1852)

Endemism, Certain native
Family: *Plantaginaceae*
Common name: Pinillo blanco o Crespa (ES); Teide plantain (EN)
Biotype: shrub (chamaephyte)
Flowering: May–June
Fruiting: June–July
Pollination: wind pollination (anemogamy)
Dispersal: animal dispersal (zoochory)
Chromosome N°: 2n = 12

Description

Chamaephyte with branches separated by 1–1.5 cm (Fig. 3.232). **Leaves** slightly thick, linear and curved, obtuse at the apex, dilated base with short sheath 1.5–3 mm long and 1 mm wide. Erect spike peduncle from 1 cm long to 5–6 cm, whitish and hairy. Erect spike cylindrical with whorls of few flowers, tomentose. Acute bracts with a membranous-dilated base, 4–5 mm long, bristly superficially with greater abundance on the keel. **Corolla** tube glabrous and wrinkled with acute lanceolate lobes. Ovoid capsule provided with two shiny black seeds.

Distribution

It is found in the high peaks of Tenerife, La Palma and Gran Canaria (Fig. 3.233), reaching altitudes of 2300 m. Within the Teide National Park (Fig. 3.234), it is found between 2000 and 2400 m above sea level, mainly on the interior slopes of the Las Cañadas Wall: Cañada de Diego Hernández, Risco de Las Pilas, Morro La Angostura, Cañada del Montón de Trigo, Cañada del Capricho, Roques de García, and Cañada del Corral and Cañada del Pinito in the Ucanca Valley. It also appears in La Fortaleza, very punctually and dispersed in interior areas of the Circo de Las Cañadas and in the Teide Botanical Garden.

Fig. 3.232 Cañada de Diego Hernández, June 2006. (Photo credit: Victoria Eugenia Martín Osorio)

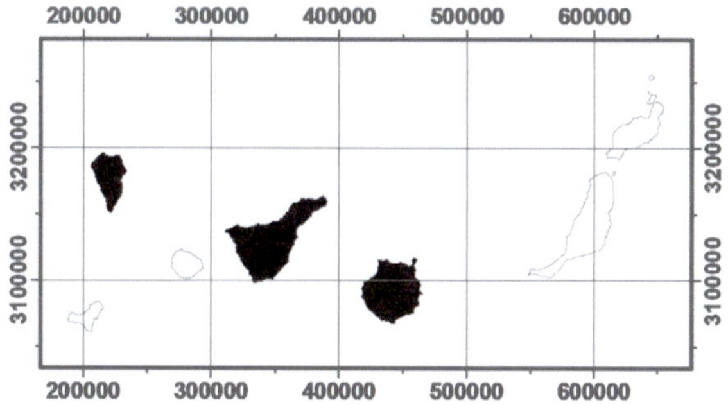

Fig. 3.233 Endemism the summits of Tenerife, La Palma & Gran Canaria

Habitat and Phytosociology

Rock and pebble environments or gelifracts. It is a characteristic species of the communities integrated in the phytosociological alliance ***Plantaginion webbii***, which groups the teide plantain (*Plantago webbii*) mountain dwarfs shrublands, prime colonisers of lava flows, volcanic deposits and periglacial rubble and gelifracts; as well as serials on eroded soils of Canary Island pine forests, cedar forests or Teide broom shrublands; with a distribution in La Palma-Tenerife, dry-humid Meso-Supra-Mediterranean.

In addition, it is also characteristic or intervenes in the following associations:

Arrhenathero calderae-Plantaginetum webbii community described as Cerrillar-Crespar, a phytosociological association of the National Park that preferably colonizes the slopes of gelifract rubble of the Pared de las Cañadas:

Cheirolopho teydis-Carlinetum xeranthemoides

Tolpido webbii-Micromerietum lachnophyllae

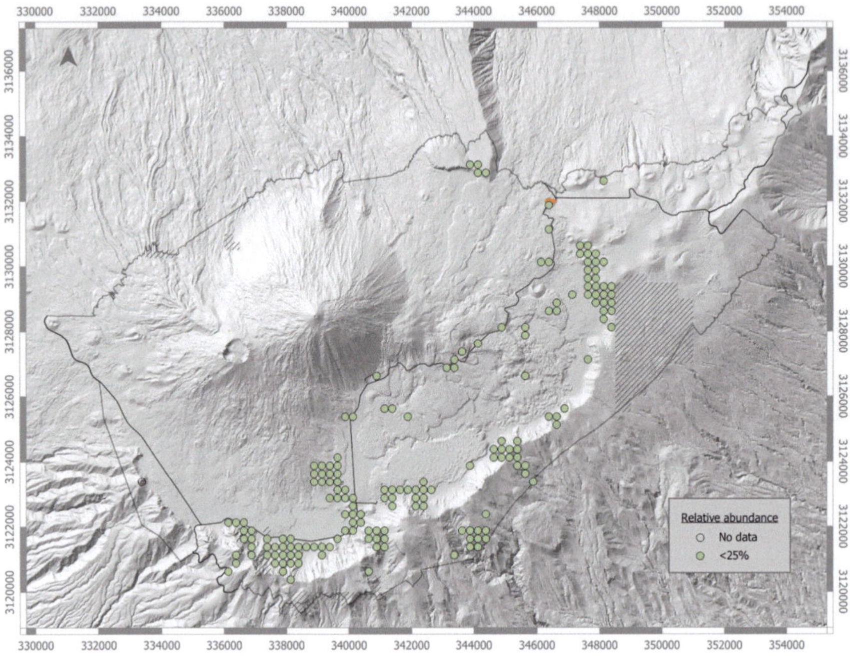

Fig. 3.234 *Plantago webbii* distribution, N° Subgrids (250 × 250 m): **175**/3741

Conservation Status

In almost all the subgrids in which this species was found, no damage of any kind was observed. Only in 1% (2), damage was detected (browsing, high mortality without regeneration, etc.), but this did not affect more than 25% of the specimens.

Threat Category

nt (Barreno et al. 1984)
LR pm (Bañares et al. 1998)

Germplasm Banks

E.T.S. Ingenieros Agrónomos de Madrid.
Jardín Botánico Canario Viera y Clavijo.

Exsiccata (Herbarium Specimens)

Tenerife: Roque junto al Parador de Las Cañadas, 10.05.1969, A. Santos (TFC 1508, 1550, 1574); Cañada de Diego Hernández, Junio 1969, Bramwell (TFC 1701); La Fortaleza-Las Cañadas, 18.07.1969, Bramwell 1794 (TFC 1794); Bco. del Infierno, Marzo 1971, A. Santos Guerra (TFC 1507); Roques de García-Las Cañadas, 1.07.1971, AC & L. Gallo (TFC 1502 + Dupl.); Mña. Guajara, 15.11.1972,

J.R. Acebes & P.L. Pérez de Paz (TFC 23341); Base Mña. Guajara, 15.11.1972, J.R. Acebes (TFC 6779); Las Cañadas del Teide, 2.05.1975, ERH (TFC 33511); Las Cañadas, 2.07.1977, W. Wildpret, M. del Arco Aguilar, C. León Arencibia & I. La Serna (TFC 25080); Las Cañadas, 7.07.1977, SMG & LD (TFC 23231); La Fortaleza, 13.06.1980, E. Beltrán Tejera (TFC 20928 + Dupl.); Cañada de la Grieta-Las Cañadas, Junio 1981, CRP (TFC 24490); Topo de la Grieta, 20.06.1985, P.L. Pérez de Paz, IMM & I. La Serna (TFC 24748 + Dupl.). Inventario de la Flora y Vegetación del Parque Nacional del Teide (2001–2003) V.E. Martín Osorio, W. Wildpret & B. Hernández Bolaños: Zapatito de la Reina (41D 25 b), 14/04/2003, V.E. Martín Osorio & W. Wildpret de la Torre (TFC 44877 + 2 Dupl.); Roques al Sur del Llano Ucanca (41D 35 d), 14/05/2003, A. Montañés Rodríguez (TFC 44298); Guajara, por Montaña Colorada (42C 16 a), 25/06/2003, A. Montañés Rodríguez (TFC 44284); Cañada del Capricho, 02/07/2003, V.E. Martín Osorio & W. Wildpret de la Torre (TFC 44201); Cañada de Diego Hernández (36B 30 a), 12/05/2004, B. Hernández Bolaños, V.E. Martín Osorio & W. Wildpret de la Torre (TFC 45432)

La Palma: Nambroque, A. Santos (TFC 1548 + Dupl.); Roque de Los Muchachos, 8.07.1973, P.L. Pérez de Paz (TFC 16013); Laderas de Birigoyo-Sobre lapillis rojos, 1425 m.s.n.m., 29.07.1977, M. del Arco Aguilar, I. La Serna & C. León Arencibia (TFC 32907).

Gran Canaria: Degollada de Cruz Grande, 28.03.1997, A.D. Hernández (TFC 40763).

3.5.76 Poa infirma *Kunth*

First published *in* F.W.H.von Humboldt, A.J.A.Bonpland & C.S.Kunth, Nov. Gen. Sp. 1: 158 (1816)

Eragrostis infirma (Kunth) Steud. *in* Nomencl. Bot., ed. 2, 1: 568 (1840)
Megastachya infirma (Kunth) Roem. & Schult. *in* Syst. Veg., ed. 15bis. 2: 585 (1817)
Ochlopoa infirma (Kunth) H.Scholz *in* Ber. Inst. Landschafts Pflanzenökol. Univ. Hohenheim Beih. 16: 59 (2003)

Not endemic, Probable native
Family: *Poaceae*
New record for Teide National Park
Common name: Poa frágil (ES); early meadow-grass, weak bluegrass (EN)
Biotype: therophyte
Flowering: March–June
Fruiting: March–June
Pollination: wind pollination (anemogamy)
Dispersal: wind dispersal (anemochory)
Chromosome N°: 2n = 14

Description

Annual plant (Fig. 3.235) with stems up to 25 cm, erect, striated, glabrous.

Leaves with 0.5–2 (−2.5) mm truncated ligule; blade up to 6 (−10) cm long and 1–2.3 (−3.5) mm wide, flat, with striated upper surface, glabrous or with scabrid margin.

Panicle 1–4 cm, ovate or lanceolate in outline, lax, with 1–3 branches per node, more or less erect-patent at anthesis.

Spikelets 2–4 mm, ovate-lanceolate, with (1–) 2–4 flowers. Unequal glumes with scarious margin and smooth or slightly scabrid veins; the lower one 1–1.5 (−2) mm, ovate, uninervated; the upper one (1.2–) 1.5–2.2 (−2.3) mm elliptic three-veined, lemma 1.8–2.7 mm, oblong, with 5 generally hairy nerves and with scarious margin. Palea somewhat shorter than the lemma. Anthers 0.2–0.4 (−0.5) mm.

Distribution

It extends throughout Europe, Irano-Turanian region, Mediterranean region, Madeira and the Canary Islands. Present in all the Canary Islands, except in El Hierro (Fig. 3.236). In the Teide National Park (Fig. 3.237) it was found punctually in the area around Pico del Teide, between 3425 and 3600 m a.s.l. Possible anthropic influence on the distribution of this species in the National Park.

Fig. 3.235 *Poa infirma*, 30 May 2007 Teide cone at 3600 m above sea level. (Photo credit: Victoria Eugenia Martín Osorio)

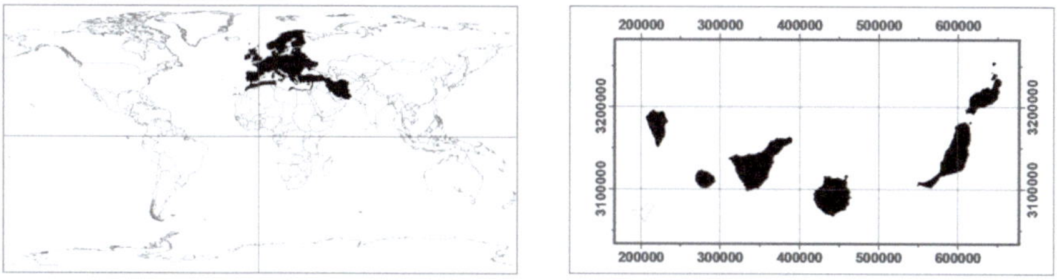

Fig. 3.236 Present on all the Canary Islands except El Hierro

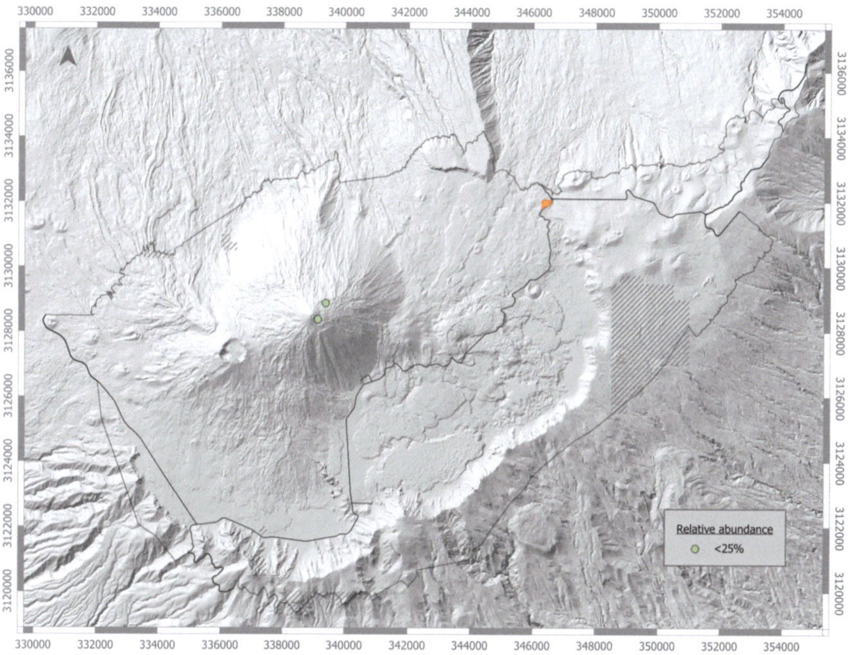

Fig. 3.237 *Poa infirma* distribution, N° Subgrids (250 × 250 m): **2**/3740

Habitat and Phytosociology

In general, it is usually found in meadows, rocky areas and ruderalized places.

It is a characteristic species of the ***Polygono-Poetea annuae*** class communities, of nitrophilous, road-dwelling and first-colonizing vegetation, which thrives on soils compacted by trampling both on paths and roads and in urban areas.

It is also a characteristic species of the association given in the Teide National Park, ***Poo infirmae-Saginetum stoloniferi***, an oligospecific, thermos edaphohygrophilous and somewhat sciophilous community that occupies only a few square decimetres of surface, in a microclimatic environment of permanently humid and warm lands (35 °C), from the top of the peak to La Rambleta del Teide (3650–3700 m a.s.l.); it generally appears in trampled places on the edge of the Teide crater.

Conservation Status

In the two (2) only subgrids in which this species was found, no damage of any kind was observed.

Exsiccata (Herbarium Specimens)

Tenerife: Santa Cruz, Plaza de España, 5.5.1985, P.L. Pérez de Paz (TFC 13327 + duplic.); La Orotava, Punta del Fraile, 15.3.1994, F. Ramallo (TFC 37539); Los Silos, Erjos, V. Lucía Sauquillo & J.A. Reyes-Betancort (TFC 40787); Bco. de Tahodio, cerca de Jardina, 6.5.1999, S. Becker, T. Himstedt & O. Rodríguez Delgado (TFC 42149). Inventario de la Flora y Vegetación del Parque Nacional del Teide (2001–2003) V.E. Martín Osorio, W. Wildpret & B. Hernández Bolaños: Cono del Teide (35B 37 b), 23/04/2003, A. Montañés Rodríguez (TFC 44845).

Lanzarote: Teguise, UTM 28RTF 402152, 20.3.1997, J.A. Reyes-Betancort (TFC 40479).

3.5.77 Polycarpaea tenuis *Webb* ex *Christ*

In Christ, Bot. Jahrb. Syst., 103 (1887)

Polycarpon tenue (Webb ex Christ) Kuntze *in* Revis. Gen. Pl. 1: 51 (1891)
Polycarpaea aristata Webb. l. cit. nec Chr. Smith.

Endemism, Certain native
Family: *Caryophyllaceae*
Common name: Pataconejo de cumbre (ES)
Biotype: hemicryptophyte
Flowering: June
Fruiting: July–August
Pollination: self-pollination (autogamy)
Dispersal: wind dispersal (anemochory)

Description

Perennial **herb**. Dense, very branched, woody at the base. **Leaves** in whorls, linear or subspatulate, silvery-white, pubescent, awned. Axillary and terminal cymes, dense. **Stipules** fleshy (Fig. 3.238). **Flowers** 5-partite. Petals (staminodes) present. Style trifid at the tip. Capsule with 3 valves.

Fig. 3.238 Wolfredo Wildpret Botanical Garden of Teide National Park, June 2004. (Photo credit: Beatriz Hernández)

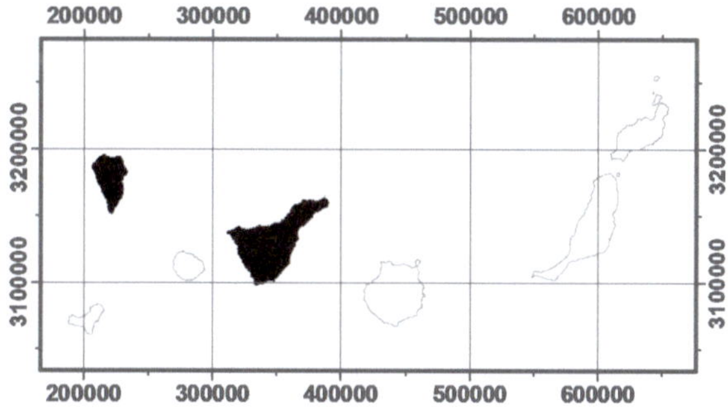

Fig. 3.239 Endemism of the island of Tenerife and La Palma

Distribution

Endemism of the islands of La Palma and Tenerife (Fig. 3.239). In the Teide National Park (Fig. 3.240), it is widely distributed throughout the interior of Las Cañadas, except for the Teide—Pico Viejo complex; in a more dispersed and punctual way, along the Wall, even reaching beyond it towards the southern slope. Altitudinally, it is found between 1650 and 2700 m a.s.l.

Habitat and Phytosociology

It usually appears in rocky or recent lava soils. It is a characteristic species and bio-indicator of ***Plantaginion webbii***, an alliance that groups the white "pinillo" (***Plantago webbii***) oyster mushrooms, first colonizers of lava flows, lapillis and other volcanic deposits. We can also find it in the following associations:

Arrhenathero calderae-Plantaginetum webbii,
Tolpido webbii-Micromerietum lachnophyllae

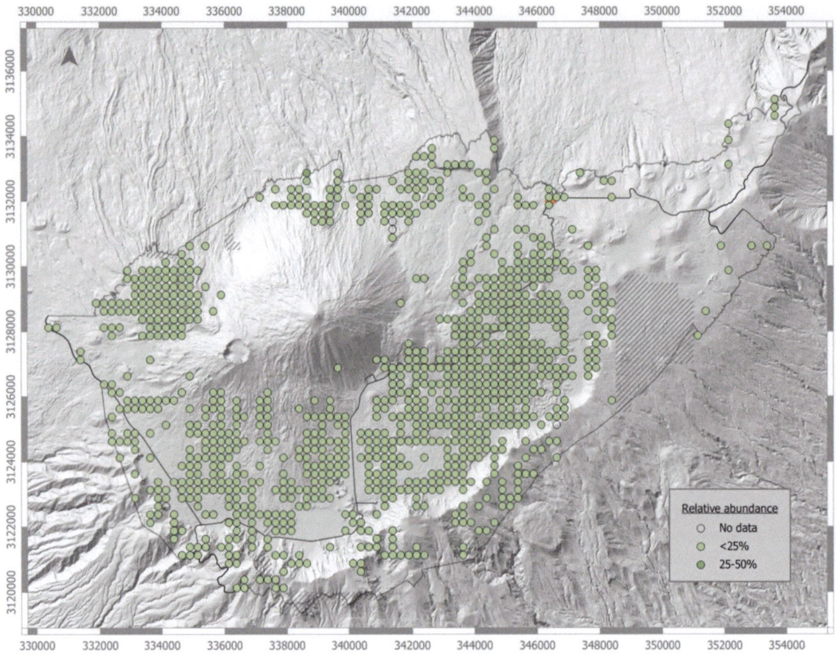

Fig. 3.240 *Polycarpaea tenuis* distribution, N° Subgrids (250 × 250 m): **1037**/3741

Conservation Status

In almost 99% (1023) of the subgrids in which this species was found, no damage of any kind was observed. Only in 1% (11), damage was detected (browsing, high mortality without regeneration, etc.), but this did not affect more than 25% of the specimens. Damage was only found on more than 25% of the specimens in 2 of the sub-grids where this taxon appeared. In the rest (1), this variable was not specified at the time of sampling.

"*Not very abundant in Las Cañadas, unique locality.*" (Sventenius 1946)

Threat Category

R (Barreno et al. 1984)
LR pm (Bañares et al. 1998)

Germplasm Banks

E.T.S. Ingenieros Agrónomos de Madrid.

Exsiccata (Herbarium Specimens)

Tenerife: Las Cañadas, 28.7.1971, F. Estévez & A. Acuña (TFC 8407); Mña. de Guajara, 11.11.1972, P.L. Pérez de Paz & J.R. Acebes (P-TFC: 264) (TFC 13147); La Fortaleza, 19.5.1973, H. Metlesics (TFC 4388); Bco. de la Orchilla, 1.6.1973, E. Beltrán Tejera (TFC 3829); Cañada de Diego Hernández, Topo de la Grieta, 2.6.1973, P.L. Pérez de Paz (P-TFC:262) (TFC 13145); La Fortaleza, 19.5.1975, W. Wildpret, P.L. Pérez de Paz, A. Acuña, E. Beltrán Tejera, C. Gil, E. Barquín Díez, H. Metlesics (TFC 3887); Topo de la Grieta, 31.5.1975, H. Metlesics (TFC 3899; 3900); Anocheza, Ladera de Güímar, 28.1.1976, M.C. León Arencibia, M. Arco Aguilar, E. Beltrán Tejera & J.R Acebes (TFC 4910); Altos de El Escobonal, 20.12.1976, O. Rodríguez Delgado (TFC 27628); La Fortaleza, 13.6.1980, E. Beltrán Tejera (TFC duplic. 8426); Caldera de Pedro Gil, Volcán de las Arenas, 16.5.1981, P.L. Pérez de Paz, M. Arco Aguilar & W. Wildpret (P-TFC 128) (TFC 13046); Los Órganos de Aguamansa, 23.5.1981, M. Arco Aguilar & W. Wildpret (P-TFC 131) (TFC 13049 + 3 duplic.); Las Cañadas, 30.5.1981, W. Wildpret, E. Beltrán Tejera, A. Losada Lima & A. García Gallo (P-TFC 133) (TFC duplic. 13051); Arenas Negras, 20.6.1982, W. Wildpret, E. Beltrán Tejera, A. Losada Lima & A. García Gallo (TFC 7886); sur de Guajara, 2300 m.s.n.m., 28.5.1983, W. Wildpret & J. S. Socorro (TFC 34250); Las Cañadas, ctra. de Guajara, 25.4.1987, P.L. Pérez de Paz, W. Wildpret & Biondi (TFC 24081); Icod, Los Castillos, 1600 m.s.n.m., 15.6.1988, F. Ardévol González (TFC 19323); La Ladera, 5.8.1988, O. Rodríguez Delgado (TFC 29192); Las Cañadas, 1992, J.R. Acebes, P.L. Pérez de Paz, M.C. León Arencibia et al. (TFC duplic. 36415). Inventario de la Flora y Vegetación del Parque Nacional del Teide (2001–2003) V.E. Martín Osorio, W. Wildpret & B. Hernández Bolaños: Al SW de la Tarta (31A 4), 5/04/2001, F. Tapia & Y. de Rioja Santos (TFC 44904); Pared de Las Cañadas—Risco Verde, 29/08/2001, W. Wildpret de la Torre (TFC 44890); Por debajo de El Roque Los Almendros, hacia el SW, al Oeste de Los Retamares (41D 43 b), 28/04/2003, A. Montañés Rodríguez (TFC 44850); Entrada al Jardín Botánico (30D 11 b) (T19), 17/06/2004, B. Hernández Bolaños (TFC 45403); Jardín Botánico del Parque Nacional del Teide,13/11/2003, V.E. Martín Osorio, W. Wildpret de la Torre & B. Hernández Bolaños (TFC 44931).

La Palma: Los Sauces, 11.8.1986, P.L. Pérez de Paz & J. Leal (TFC 24188 + duplic.).

Gran Canaria: Bajo Montaña del Cedro, 2100 m.s.n.m., 3.5.1990, M. del Arco Aguilar & M. Salas Pascual (TFC 38401); ?, 18.7.1969, 1795, Bramwell (TFC 400).

3.5.78 Pterocephalus lasiospermus *Link* ex *Buch*

C.L.von Buch, Phys. Beschr. Can. Insl.: 150 (1825)

Pterocephalus dumetorum ß *pusilla* Court. Dips., pág. 32.
Scabiosa lasiosperma (Link) D.Dietr. *in* Syn. Plant. 1: 448 (1839)
Scabiosa fruticulosa C.Sm. *ex* DC. *in* Prodr. 4: 654 (1830)

Endemism, Certain native
Family: *Dipsacaceae*
Common name: Hierba conejera, Rosalito o Rosalillo de cumbre (ES); Summit rosebush (EN)
Biotype: shrub (chamaephyte)
Flowering: May–August
Fruiting: July–September
Pollination: insect pollination (entomogamy)
Dispersal: wind or animal dispersal (anemochory—zoochory)
Chromosome N°: 2n = 18

Description

Chamaephyte up to 1 m high, multiple, greenish-greyish fruit-bearing stems. **Leaves** greenish-greyish, somewhat pale, hirsute-silver, oblong-lanceolate, acute, sub-obtuse, lower part mononerved, reticulate-veined, very hairy and densely tomentose; whitish tomentum; short petiole. Great abundance of pale pinkish **flowers** arranged in solitary, long-pedunculated capitula (Fig. 3.241).

Distribution

Endemism of the island of Tenerife. In the Teide National Park (Fig. 3.242) it is found between 1650 and 3400 m of altitude, and can exceptionally descend, outside the limits of the National Park, to much lower altitudes. Widely distributed throughout the territory of Las Cañadas, except on slopes and lava flows of the Pico Viejo—Teide—Montaña Blanca complex, and higher areas of La Dorsal (Llano de Maja and surrounding areas) (Fig. 3.243).

Fig. 3.241 National Park towards Vilaflor, June 2006. (Photo credit: Beatriz Hernández)

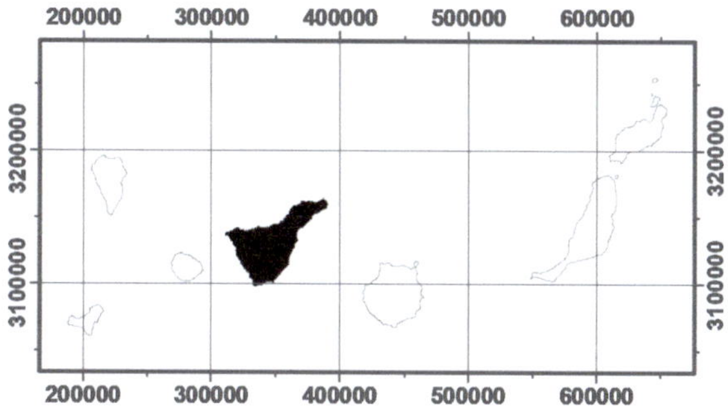

Fig. 3.242 Endemism of the Teide National Park on the island of Tenerife

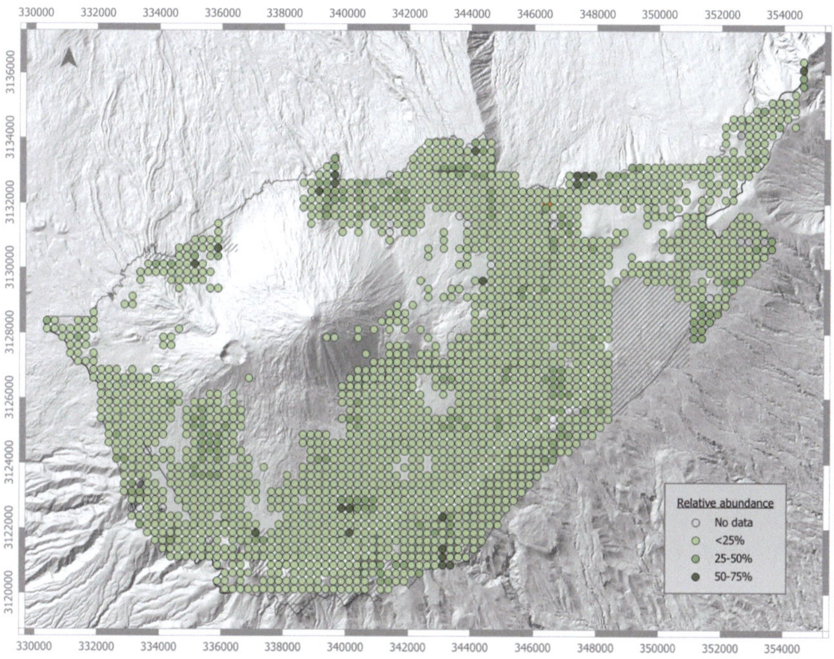

Fig. 3.243 *Pterocephalus lasiospermus* distribution, N° Subgrids (250 × 250 m): **2184**/3741

Habitat and Phytosociology

It grows on rocky slopes and on volcanic substrates, preferably of a lava nature.

It is a characteristic species and bioindicator of ***Pterocephalenion lasiospermi***, forming part of the community of replacement of the summit brooms. ***Erysimo scoparii-Pterocephaletum lasiospermi***.

It is also found in the following communities:

Adenocarpo viscosi-Chamaecytisetum angustifolii
***Adenocarpo viscosi-Juniperetum cedri*,**
Arrhenathero calderae-Plantaginetum webbii, that colonizes the slopes of gelifractic debris of the Las Cañadas Wall.

Cheirolopho teydis-Carlinetum xeranthemoides
Tolpido webbii-Micromerietum lachnophyllae
Arrhenathero calderae-Adenocarpetum viscosi

Conservation Status

During the sampling carried out, in 63% (1365) of the subgrids in which this species was found, no damage of any kind was observed. In 35% (773), damage was detected (browsing, high mortality without regeneration, etc.), but this did not affect more than 25% of the specimens. Damage was only found on more than 25% of the specimens in just over 1% (30) of the sub-grids where this taxon appeared. In the rest (16), this variable was not specified at the time of sampling.

> "*Beautiful shrub, with shaggy-silver foliage and great abundance of pinkish-purple flowers, from the slopes and rocky plains in the eastern and western part of the Cañadas. Rare species and with a tendency to be very scarce. Unique locality.*" (Sventenius 1946)

> "*It was considered rare in many studies at the beginning of the century but now, at present, it is a very abundant species, especially on the southern slope of the Teide National Park and does not seem to be threatened.*" (García Gallo et al. 1993)

Threat Category

nt (Barreno et al. 1984)
LR pm (Bañares et al. 1998)

Germplasm Banks

E.T.S. Ingenieros Agrónomos de Madrid.
Jardín Botánico Canario Viera y Clavijo.

Exsiccata (Herbarium Specimens)

Tenerife: 11.06.1969, Bramwell 1696 (TFC 416); 18.07.1969, Bramwell 1790 (TFC 415); 18.01.1971, P.L. Pérez de Paz (TFC 680); Altos de Vilaflor, 12.06.1971, E. Beltrán Tejera (TFC 21630); Corral del Niño, El Portillo, Cañada Blanca, Roques de García, 28.07.1971, F. Estévez & A. Acuña (TFC 15992); 4.12.1972, E. Beltrán Tejera (TFC 4460); Las Cañadas, 19.05.1973, E. Beltrán Tejera & LG (TFC 3881, 3882, 3883); Cañada de Diego Hernández, 2.06.1973, M.C. Gil Rodríguez (TFC 24095); Las Cañadas, 31.06.1973, W. Wildpret, PP, M. Galván, CG, H. Metlesics & E. Beltrán Tejera (TFC 3880); Ibid., 2.07.1977, W. Wildpret, M. del Arco Aguilar, C. León Arencibia & I. La Serna (TFC 25079); Ibid., 28.05.1979, E. Beltrán Tejera (TFC 7944, 7945, 7946 + Dupl.); Cañada de la Grieta, Junio 1981, CRP (TFC 24492); Las Cañadas, 27.06.1982, RLN (TFC 25451); Topo de la Grieta, 20.06.1985, P.L. Pérez de Paz, I. La Serna & IMM (TFC 24743 + Dupl.); E. Bibanago-Bco. de Castro-Icod, 500 m.s.n.m., 15.07.1985, F. Ardévol González (TFC 19184); Boca

Tauce-Las Cañadas, 30.07.1988, P.L. Pérez de Paz (TFC 30047); Chimoche-La Orotava, 1440 m.s.n.m., 28.07.1997, A. Santos (TFC 41565—AHIM. NÚM.: 172. Centuria II-1997); Inmediaciones de Boca Tauce, 2200 m.s.n.m., 2.08.1998, P.L. Pérez de Paz, C. Hernández Padrón & D.P. Hernández (TFC 41621 + Dupl.). Inventario de la Flora y Vegetación del Parque Nacional del Teide (2001–2003) V.E. Martín Osorio, W. Wildpret & B. Hernández Bolaños: Roques al Sur del Llano Ucanca (41D 35 d), 14/05/2003, A. Montañés Rodríguez (TFC 44840); Guajara, cerca de El Capricho (42C 1 a), 15/05/2003, A. Montañés Rodríguez (TFC 44854); Cañada Blanca, Cueva de Diego Hernández, 25/06/2003, F. Mesa, A. Montañéz & Y. de Rioja (TFC 44865).

3.5.79 Ranunculus cortusifolius *Willd.*

Enum. Pl. Hort. Berol., 588 (1809)

Ranunculus spicatus subsp. *cortusifolius* (Willd.) Maire *in* É.Jahandiez & al., Cat. Pl. Maroc 2: 250 (1932)
Ranunculus megaphyllus Steud. *in* Nomencl. Bot., ed. 2, 2: 434 (1841)
Ranunculus teneriffa Pers. *in* Syn. Pl. 2: 103 (1806)

Macaronesian endemism, Certain native
Family: *Ranunculaceae*
New record for Teide National Park
Common name: Morgallana, Morgallón, Murgallana, Botón de oro o Burguñuela (ES); Canary buttercup (EN)
Biotype: hemicryptophyte
Flowering: February–June
Fruiting: February–June
Pollination: insect pollination (entomogamy)
Dispersal: wind dispersal (anemochory)
Chromosome N°: n = 8; 2n = 16

Description

Herbaceous, perennial plant. Robust, hairy, with tubers. **Alternate leaves**, simple or compound, the basal ones up to 30 cm in diameter, orbicular-cordate, superficially lobed (Fig. 3.244).

Inflorescence subcorymbose; regular flowers, hermaphrodite, in cymose panicles, flowers up to 5 cm in diameter; bright yellow petals, 5. Many stamens. Numerous achenes, smooth, black when mature.

Fig. 3.244 At the foot of the columnar disjunction wall to the N of Izaña, May 2005. (Photo credit: Beatriz Hernández)

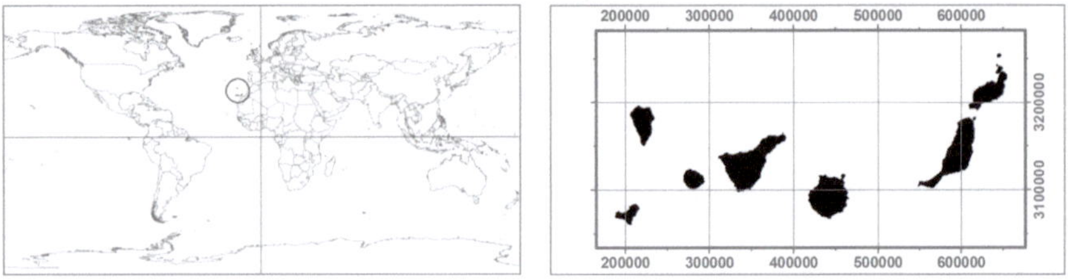

Fig. 3.245 Present on all the Canary Islands

Distribution

It is found in the Macaronesian region, except Cape Verde. Abundant on the island of Tenerife (Teno, Masca, Anaga, Adeje, Icod, etc.) and also present in the rest of the Canary Islands (Fig. 3.245).

In the Teide National Park (Fig. 3.246) it has been found very punctually only in areas of La Dorsal: at the head of Barranco La Vera (Montaña Los Arrastraderos), areas of Salto Frío (N-Izaña) and Colorado Chico (between Montaña Los Atajos and Montaña de Caramujo); between 2025 and 2100 m a.s.l.

Habitat and Phytosociology

It normally occurs between 200 and 1500 m a.s.l., being abundant in green and fayal-brezal scrubland, and in humid rocks in the low areas; Thermo and Mesomediterranean bioclimatic zones. It is a characteristic species of communities of the alliance ***Ranunculo cortusifolii-Geranion canariensis***, which groups the megaphorbic communities of the mesophytic edges of the laurel forests and the humid, Thermo-Meso-Mediterranean subhumid-humid Canary Island pine forests of the Canary Islands.

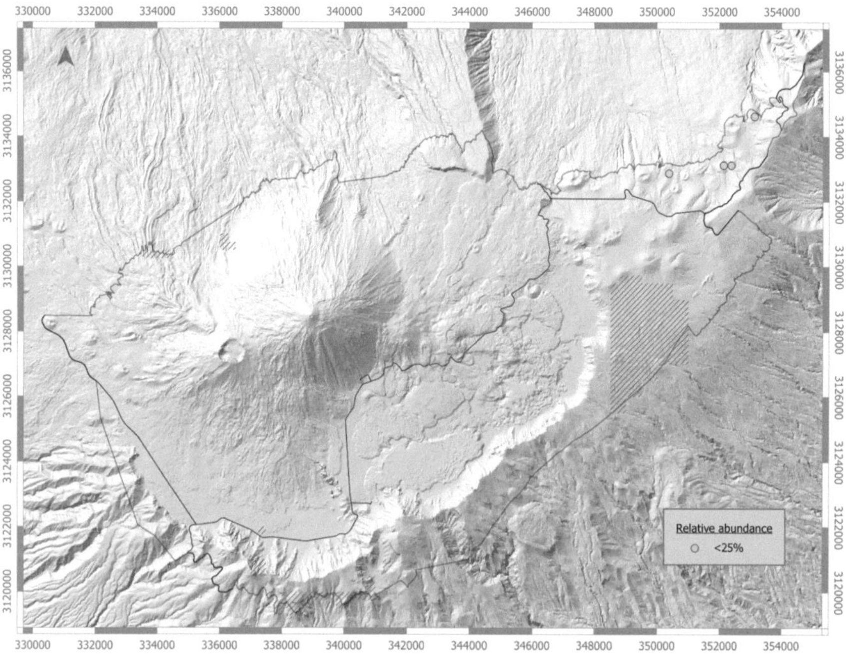

Fig. 3.246 *Ranunculus cortusifolius* distribution, N° Subgrids (250 × 250 m): **4**/3740

Conservation Status

In all the subgrids in which this species was found (4), no damage of any kind was observed.

Threat Category

LR pm (Bañares et al. 1998)

Germplasm Banks

Banco de Germoplasma de E.T.S. Ingenieros Agrónomos de Madrid.
Jardín Botánico Canario Viera y Clavijo.

Exsiccata (Herbarium Specimens)

Tenerife: Monte de Las Mercedes, 2.5.1973, I. La Serna (TFC 35707); Anaga, 3.1979, E. Valdés & P.L. Pérez de Paz (TFC 34891); Anaga, 16.4.1979, J. Fernández Casas (TFC 20749); Cruz del Carmen, 5.1979, M.C. León Arencibia, C. Gil Rodríguez & M. del Arco Aguilar (TFC duplic. 8663); Agua García, Madre del Agua, 8.4.1980, A. García Gallo & C. Suárez (TFC 10448 duplic.; 10449 duplic.); Agua García, Madre del Agua, 21.5.1980, A. García Gallo, A. Losada Lima &

A. Viera (TFC 10450 + duplic); Chinyero, 1.3.1981, R.M. Arnay (TFC 27199); Proximidades de Pico del Inglés, 13.3.1982, R. Lecuona (TFC 13096 + duplic.); El Escobonal, Bco. de Herques, 300 m.s.n.m., 4.4.1982, O. Rodríguez Delgado (TFC 12937); El Escobonal, Bco. de Herques, 4.4.1982, O. Rodríguez Delgado (TFC 28049); Icod, Monteverde de La Vega, 750 m.s.n.m., 29.4.1984, F. Ardévol González (TFC 19057); Bco. Hondo, Pista de las Raíces, 1150 m.s.n.m., 23.3.1990, M. del Arco Aguilar & M. Salas Pascual (TFC 38414); Anaga, El Moquinal, 17.3.1992, J.R. Acebes (TFC duplic. 34579); El Moquinal, Anaga, 14.3.1994, J.R. Acebes (TFC duplic. 33423); Los Silos, Pista principal del Monte del Agua, 30.5.1996, V. Lucía Sauquillo & M. del Arco Aguilar (TFC 38299); Los Silos, La Cruzada, UTM 28RCS 2138, 29.11.1996, V. Lucía Sauquillo & C. Acosta (TFC 39992 + duplic.); Bco. de Tahodio, cerca del Roque Lombarda, 27.2.1999, S. Becker & T. Himstedt (TFC 41990); Altos de El Sauzal, UTM 28RCS 6245, 1350 m.s.n.m., 11.5.2002, V. Lucía Sauquillo & M.L. Rodríguez Navarro (TFC 44710).

El Hierro: El Golfo, 26.3.1972, E. Beltrán Tejera (TFC 21970); El Cres, Ventejea, 10.4.1979, W. Wildpret, M. del Arco Aguilar, P.L. Pérez de Paz & C. Hernández Padrón (TFC 8094).

Gran Canaria: Riscos de Faneque, 16.4.1976, E. Beltrán tejera, J.R. Acebes, B. Méndez Pérez & M. del Arco Aguilar (TFC 23253 + 2 duplic.); Moya, Bco. Oscuro, 2.5.1979, P.L Pérez Suárez & C. Suaréz (TFC 11041); Andenes, Monte de Inagua, 29.4.1985, M. Nogales (TFC 20212); Bco. del Cernícalo, primavera 1989, S. Ortega González & E. Fraile Bonafonte (TFC 30477); Parque Natural del Bco. de Guayadeque, 10.2.1990, F. Amor & P.L. Pérez de Paz (TFC 30665).

Fuerteventura: Pico de la Zarza, ?, P.L. Pérez de Paz & J.R Acebes Ginovés (TFC 4766).

Lanzarote: Mirador de Haría, 17.2.1978, E. Beltrán Tejera, J.R. Acebes, M.C. León Arencibia & M. del Arco Aguilar (TFC duplic. 26983); Mirador de Haría, 17.1.1992, P.L. Perez de Paz, MMG & ECA (TFC 35297); Haría, Valle del Malpaso, 24.3.1993, J.A. Reyes-Betancort (TFC 39645); Bajo Mirador del Río, 26.12.1993, J.A. Reyes-Betancort (TFC duplic. 38554); Yaiza, Atalaya de Femés, 18.2.1996, J.A. Reyes-Betancort (TFC duplic. 39872); Yaiza, Caldera Gritana, 21.2.1997, J.A. Reyes-Betancort (TFC 40360).

?, W. Wildpret, A. Santos & P.L. Pérez de Paz (TFC duplic. 917).

Uses, Exploitation and/or Medicinal Properties

"*Analgesic, caustic, vesicant, antihemorrhoidal; poor quality fodder. Useful parts: flowers, root, rhizomes. Instructions for use: external use; food: cooked rhizomes in times of scarcity.*" (Translated from: Pérez de Paz and Hernández Padrón 1999)

3.5.80 Rhamnus integrifolia *DC.*

Cat. Pl. Hort. Monsp., 136 (1813)

Rhamnus coriacea Brouss. *Schultz, Syst.*, vol. 5, pág. 291; Nees ab Esenb., *Hor. phys.*, pág. 114, table 22.
Rhamnus coriacea Nees ab Esenb. *Hort. Phys.:* 114, t. 22 (1820)

Endemism, Certain native
Family: *Rhamnaceae*
Common name: Mocanillo o Moralito
Biotype: nanophanerophyte
Flowering: May–June
Fruiting: July–September
Pollination: wind pollination (anemogamy)
Dispersal: animal dispersal (zoochory)
Chromosome No: –

Descripción

Fruticulous nanophanerophyte 60–90 cm tall, very branched in a scattered manner and leafy. Soft branches, dark in colour, knotty, rough-striated, the youngest puberulous, with shorter, spatulate and simple hairs.

Leathery, glabrous, alternate or subopposite **leaves**, elongated-lanceolate, acute, mucronulate, very intact, cartilaginous and glabrous margins, very green, reddened underside, ascending primary veins, the secondary veins perfectly reticulate pellucid. Short petioles, narrowly canaliculate, carinate underside, sulcate on both sides, with the lower leaves soft and puberulous. Stipules linear, acute, wide base, very deciduous.

Flowers hermaphrodite, racemose, the axils of the leaves of the racemes incurved. Base of the calyx cupuliform, on the outside with puberulous pedicels with short, patulous hairs, with five narrowly lanceolate and acute laciniae, inner part three-nerved, with a very small disc. Petals absent. Five stamens, filaments gradually dilated from the base and with a curved back, anthers oblong, always obtuse. Ovary globose, trisulcate. Three styles, wide, base canaliculate, apex of the stigma obtuse, papillose.

Drupe round (Fig. 3.247), large, fleshy, black, with one to three cartilaginous, elongated-ovate, yellowish seeds, with a crustaceous cover, punctate. Embryo briefly arched with abundant perisperm. Cotyledons are fleshy, whitish, round, tetra-angular, with an obcordate apex. Short radicle.

Fig. 3.247 (**a**) Teide National Park towards Vilaflor, June 2006. (Photo credit: Beatriz Hernández) (**b**) Guajara, June 2003. (Photo credit: Victoria Eugenia Martín Osorio)

Distribution

A species with a wide altitudinal distribution, from 500 m above sea level in the ravines of the south-southwest of Tenerife, to 2550 m above sea level (Guajara Mountain). In the Teide National Park it is closely related to the interior and exterior sectors of the Pared de Las Cañadas, and is also found in La Fortaleza and Roques de García. It is also present in the Wolfredo Wildpret Botanical Garden of the Teide National Park (Figs. 3.248 and 3.249).

> "*Endemism from Tenerife distributed throughout large sectors of the Island, although almost always represented by populations with a small number of individuals (…), preferably in the upper part of the course of the southern ravines and Circo de Las Cañadas. It is in this last cacuminal area of the Island where the plant is most abundant, (…).*" (Translated from: Pérez de Paz et al. 1992)

Habitat and Phytosociology

Individuals of this species tend to take refuge in a relict manner in inaccessible escarpments and cliffs of the Thermo, Meso and Supramediterranean (Pérez de Paz et al. 1992). Within the Teide National Park, its character is preferably rupicolous and, occasionally, glericolous.

It is a characteristic species and bioindicator of the ***Greenovio aureae-Aeonietea*** class, which groups comophytic and chasmo-comophytic permanent perennial communities, endemic to the Canary-Madeiran biogeographic subregion.

Remarks:

According to Webb and Berthelot (1836–1850), this species was found in Las Cañadas, in rock fissures and in a very rare and dispersed way.

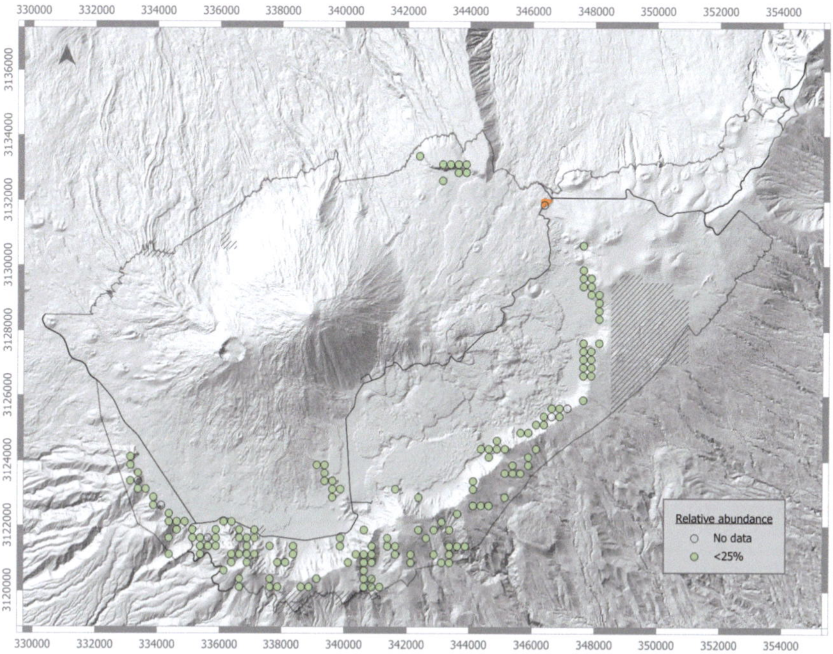

Fig. 3.248 *Rhamnus integrifolia* distribution, N° Subgrids (250 × 250 m): **160**/3741

Fig. 3.249 Endemism of the island of Tenerife

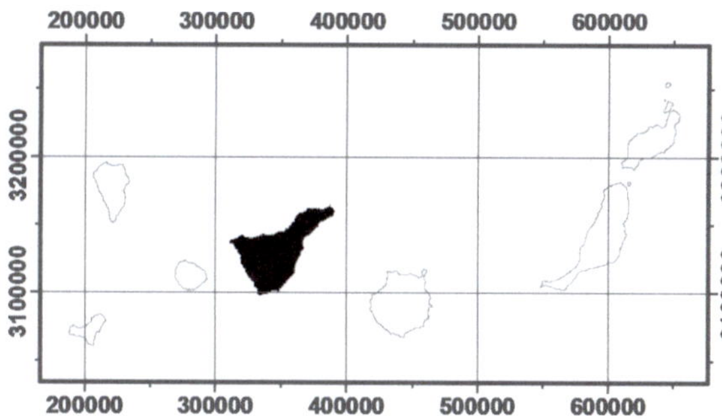

Conservation Status

According to the sampling, in 95% (152) of the sub-grids in which this species was found, no damage of any kind was observed. Only in almost 2% (3), damage was detected (browsing, high mortality without regeneration, etc.), but this did not affect more than 25% of the specimens. Damage was only found on more than 25% of the specimens in just over 1% (2) of the subgrids where this taxon appeared. In the rest (3), this variable was not specified at the time of sampling.

> "*...finding itself in clear recovery in the scree and fissures of the geologically oldest areas of the Teide National Park.*" (Translated from: Pérez de Paz et al. 1992)

Threat Category

R (Barreno et al. 1984)
Anexo II: Protegidas (Gobierno de Canarias 1991)
VU D2 (Bañares et al. 1998)
VU (Bañares et al. 2004)
NT (Moreno 2008)

Germplasm Banks

E.T.S. Ingenieros Agrónomos de Madrid

Exsiccata (Herbarium Specimens)

Tenerife: 23.02.1969, Bramwell 789 (TFC 433); 18.07.1969, Bramwell 1789 (TFC 434); Las Cañadas, 28.07.1971, F. Estévez & A. Acuña (TFC 8406 + Dupl.); Teno, 17.03.1973, P.L. Pérez de Paz, E. Beltrán Tejera, CG & B. Méndez (TFC 3953 + Dupl.); Ibid., 17.03.1973, P.L. Pérez de Paz (TFC 24067 + Dupl.); Ibid., 17.03.1973, E. Beltrán (TFC 4013); La Fortaleza, 19.05.1973, P.L. Pérez de Paz, E. Barquín Díez & E. Beltrán Tejera (TFC 3885); Ibid., 19.05.1973, P.L. Pérez de Paz (TFC 23339, 23346 + Dupl.); Ibid., 19.05.1973, E. Barquín Díez (TFC 33179); Cañada de Diego Hernández, 1.06.1974, J.R. Acebes (TFC 6776); Teno, 17.03.1975, E. Beltrán Tejera (TFC 4013); Teno Alto, 10.06.1978, P.L. Pérez de Paz, C. Hernández Padrón, J.R. Acebes, C. León Arencibia, M. del Arco Aguilar & ERH (TFC 23000 + Dupl.); Bco. de Tamay-Arafo, 9.05.1981, W. Wildpret, M. del Arco Aguilar & P.L. Pérez de Paz (TFC 23718 + Dupl., 27738 + Dupl.); Cañada de Chavao, 17.05.1981, CRP (TFC 24497); Cañada de Diego Hernández-Las Cañadas, 30.05.1981, W. Wildpret, E. Beltrán Tejera, A. Losada Lima & A. García Gallo (TFC 33124, 33131 + Dupl.); La Fortaleza-Las Cañadas, 5.04.1984, C. Rodríguez (TFC 13184); Teno, 10.04.1984, P.L. Pérez de Paz, RML, L. Negrín & C.T. Falcón (TFC 7645 + Dupl.); Roques de García, 9.05.1984, C. Rodríguez (TFC 13183); Degollada de Guajara-Las Cañadas, 20.06.1985, P.L. Pérez de Paz, I. La Serna & IMM (TFC 7655, Dupl. in MPU, G, BM); Ruta Siete Cañadas-Las Cañadas, 20.06.1988, M. del Arco Aguilar & V. Lucía (TFC 30083); Bco. de Chacorde, 19.11.1988, MMG & M. Salas Pascual (TFC 29882 + Dupl.); Bco. del Infierno, 6.03.1989, M. Marrero (TFC 29844); Bco. del Niágara, 24.04.1989, M. Marrero, W. Wildpret & O. Rodríguez Delgado (TFC 29893 + Dupl.); Inmediaciones de Boca Tauce-Cumbres de Tenerife, 2.08.1998, P.L. Pérez de Paz, C. Hernández Padrón & DPH (TFC 41625 + Dupl.). Inventario de la Flora y Vegetación del Parque Nacional del Teide (2001–2003) V.E. Martín Osorio, W. Wildpret & B. Hernández Bolaños: Cañada de Diego Hernández, 12/05/2004, B. Hernández Bolaños, V.E. Martín Osorio & W. Wildpret de la Torre (TFC 45435 + Dupl.).

3.5.81 Rhaponticum canariense *(Webb & Berth.) DC.*

Prodr., vol. 6, pág. 664. 1838
Stemmacantha cynaroides (C. Sm. *in* Buch) Dittrich Candollea, 39(1), 46 (1984)
Serratula canariensis (L. fil.) Webb & Berth. Phyt. Canar., III, 2° Parte (2/C): 370–371
Centaurea cynaroides C. Sm. *in* Abh. Königl. Akad. Wiss. Berlin 1816-1817: 369 (1819) with description: Simple stems, petiolate leaves, tomentose, the lower ones runcinate, the upper ones pinnatifid, lobes incised, very intact; membranous margin, rounded appendage.
Leuzea conifera Spreng. Syst. veg., vol. 3, pág. 382, non DC, cum cl. Sprgl.
Leuzea cynaroides (Link.) Font Quer. Publ. Univ. Barc.: 189 (1956)
Cnicus cynara Lam. *in* Fl. Franç. 2: 14 (1779)

Endemism, Certain native
Family: *Asteraceae*
Common name: Cardo de plata (ES) Tenerife Silver Thistle (EN)
Biotype: hemicryptophyte, geophyte
Flowering: June–July
Fruiting: July–August
Pollination: insect pollination (entomogamy)
Dispersal: short range self or wind-dispersal (atelocory—anemochory)

Description

Hemicryptophyte with herbaceous stems, simple, rigid, monocephalous, 22.5 cm long, with an average diameter of 4 mm at the bottom, cylindrical, sulcate-striated, crossed by a thin medulla; foliate, whitish tomentum with small curly hairs, finally subevanescent. Long pinnatisect leaves, oblong-ovate with great hairiness on the underside. Intermediate leaves 1.6–5 cm, n. 6 succulent, petiolate. Petioles of the lower leaves 5–7.5 cm long, succulent, triangular, two lateral ribs, three dorsal, slightly scarious-dilated towards the base. Leaf blade crossed by a thick rib, oblong-ovate, 7.5 cm long, 1.25–5 cm wide, pinnatipartite, subopposite lobes, the upper ones smaller, ovate-oblong, penninervate, irregular margin serrated-dentate, wavy, obtuse.

Upper leaves of the leaf mass gradually escalate downwards; multifloral, homogamous stem naked and slightly dilated towards the involucre. Involucre hemispherical, (Fig. 3.250) glabrous, imbricate, 2.5 cm or higher, external scales without appendages 6–14 mm long, 3–4 mm wide, ovate-oblong-lanceolate, coriaceous, striate, hardly scarious towards the margin and apex a scarious appendage, ovate, 1.32 mm average diameter, whitish, convex dorsum, concave inside, erose margin broken by fissures; inner scales narrow and elongated 18–22 mm long, 3–4 mm wide, lanceolate-linear, very scarious, appendage small or not terminal, obtuse. Receptacle flat, subosseous, 16 mm average diameter, setose-fimbriliferous, fimbriae approximately 1.25 cm long, very narrowly linear, acute, complanate, very intact.

Fig. 3.250 Llano de Maja, Teide National Park. (Photo credit: Victoria Eugenia Martín Osorio)

Pink-purple **flowers**, 12 mm long, with a very dilated base; bulbous, incrassate style below. The tubular, hermaphrodite, 5-toothed, with narrow, linear, acute teeth; anthers subincluded, oblong-lanceolate appendage, acute terminal cornea, with bidentate base; upper filaments papillose, upper style penicillate-incrassate, puberulent. Achene not completely mature, 6 mm long, approximately 2 mm average diameter, cylindrical-compressed, glabrous, striated, basal areola suboblique, with a great hardness surrounding it, upper margin very prominently acute, crenulate, protected papule; with little nectar. Papule 14 mm or more long, multiseriate, irregular complanate setae. Cypselas cylindrical, compressed, glabrous and striated, with pappus (Fig. 3.251).

Remarks:

> "*Hermaphrodite species. Its productivity can be quite high (>1500 seeds), and its diasporas are easily transported by the wind over short distances, but said production can be affected by the incidence of insects that parasitize the heads.*" (Translated from: Marrero Gómez et al. 2004)

Fig. 3.251 *Rhaponticum canariense*, Llano de Maja. (Photo credit: Victoria Eugenia Martín Osorio)

Distribution

Endemism of the island of Tenerife exclusive to the high summit (2000–2600 m a.s.l.). Classic localities: Montaña Chahorra (Cerro Las Chajoras), Corredor de Mario, Montaña Rajada, Llano de Maja, Izaña, Corral del Niño and La Fortaleza; in these last two no individuals were found according to the sampling carried out, however, some individuals were detected that are outside the limits of the National Park (Figs. 3.252 and 3.253).

Habitat and Phytosociology

It usually settles in glericolic environments, stony lithosols, pyroclastic or formed by alluvial contribution; both in flat areas, and especially, in areas with steep slopes. In environments with a high mobility of the substrate, it usually appears practically isolated, due to the difficulty that other species have in settling there. It is a characteristic species and bioindicator of ***Pterocephalenion lasiospermi***, a suballiance that groups permanent and serial dwarf shrublands associations, which colonize volcanic deposits and soils denuded of pine forests and broom shrublands, dry-subhumid Meso-Supra-Mediterranean, endemic to the island of Tenerife.

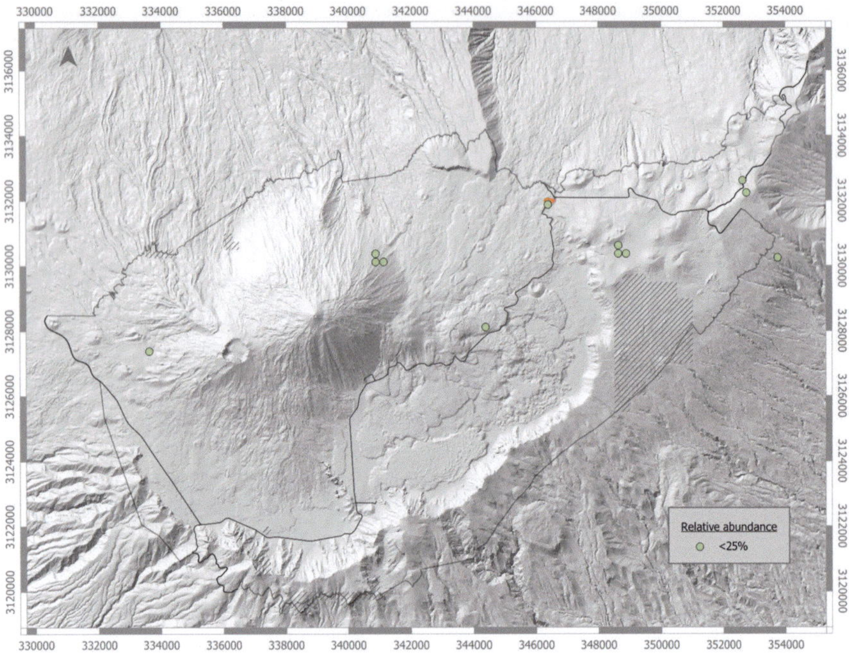

Fig. 3.252 *Rhaponticum canariense* distribution, N° Subgrids (250 × 250 m): **10**/3741

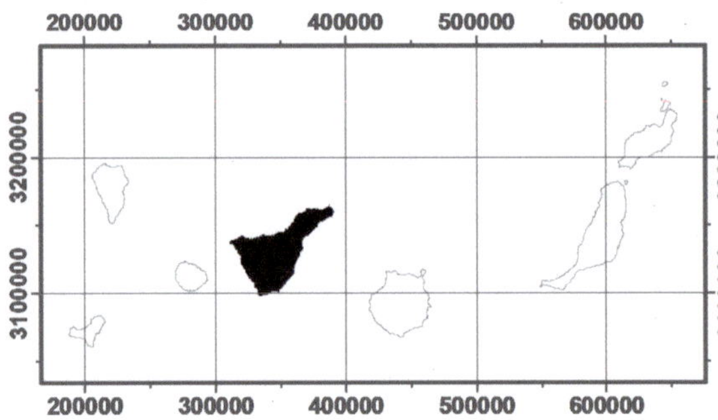

Fig. 3.253 Endemism of the Teide National Park on the island of Tenerife

Conservation Status

Its few numbers are seriously threatened by the incidence of introduced herbivores (rabbits and mouflons). These have a more or less continuous impact, especially by preying on flower heads, which translates into a potential limitation of floral production, a decrease in seed production and a significant mortality rate among adults. In this sense, although it is a notoriously long-lived taxon (>100 years), the survival of its populations is largely dependent on the seed bank in the soil, exhibiting a risk of extinction of over 90% over a period of 50 years. It is also worth highlighting the parasitism carried out by certain insects and collecting, which continues to affect the best-known and most easily accessible populations. Currently only one population maintains an acceptable size (Corredor de Mario).

All individuals are located within the Teide National Park (SCI) and in Habitat of Community Interest. A specific recovery program is being developed in the National Park, which includes measures such as monitoring its population dynamics, reintroductions and the establishment of protective fences. In addition, it is recommended to continue with the reinforcement, the creation of new populations in places suitable for their self-maintenance, to promote the obtaining of material of sexual origin and to send accessions to germplasm banks. Its main threats: predation, parasitism, low ecological plasticity and, in certain populations, collecting (Marrero Gómez et al. 2004).

According to the sampling carried out, in 80% (8) of the subgrids in which this species was found, no damage of any kind was observed. Only in 10% (1), damage was detected (browsing, high mortality without regeneration, etc.), but this did not affect more than 25% of the specimens.

"*A rare endemic species from Las Cañadas del Teide, considered to be in danger of extinction until now; the recent discovery of new populations of this taxon has diminished this fear. The population of its classic locality, Llano de Maja, has been severely punished in the past decades by military maneuvers and unscrupulous collectors. In danger.*" (Translated from: García Gallo et al. 1993)

"*A very rare and very localized plant, considered relict within the flora of Tenerife. Currently only five populations are known, four of them with a very small number of specimens. The historical records that we have always catalogue it as a fairly rare plant. We must point out that even within this critical state, its habitat continues to be threatened by various factors, both biotic and abiotic, including military manoeuvres, the circulation of all-terrain vehicles, unscrupulous collectors, etc., in addition to the potential threat of rabbits and mouflons. Its seeds are also possibly attacked by tripetid diptera.*

Proposed conservation measures: Closing of access roads, suppression of military manoeuvres, cultivation in nurseries and botanical gardens, increased surveillance of localities, control of predators and parasites and study of potential habitat with a view to creating new populations." (Translated from: Gómez Campo et al. 1996)

"*POPULATION Corredor de Mario (La Orotava), 2,600 m a.s.l.: This population, located on the northern slope of Teide, has a remarkably high number of individuals (1000–1250 specimens). The population is dispersed and in an excellent state of conservation. Habitat where the population is located:* **Violetum cheiranthifoliae** Rivas-Martínez et al. 1993 *(of community interest, 732041). Data regarding the habitat.- Naturalness: high; Degree of endemicity: high; Degree of biodiversity: low; Coverage: 26–50%; State of conservation: excellent. Quality of the area: high. Vulnerability of the area: introduced herbivores (mouflons and rabbits). The population is located within the Protected Natural Area of the Canary Islands called the Teide Natural Monument, which in turn is included in the Teide National Park.*

POPULATION Llano de Maja (La Orotava), 2,240 m a.s.l.: Endorheic plain where the number of individuals present is very scarce (9). The population is continuous and presents a medium state of conservation. Data referring to the habitat.- Naturalness: low; Degree of endemicity: high; Degree of biodiversity: low; Coverage: 0–25%; State of conservation: partially degraded. Quality of the area: high. Vulnerability of the area: collecting, past military manoeuvres and introduced herbivores. The population is included in a Protected Natural Area of the Canary Islands (Teide National Park).

POPULATION Izaña-Corral del Niño (La Orotava), 2,200 m a.s.l.: Small slope of pyroclastic material where the number of specimens present is scarce (60). The population is continuous and presents a medium state of conservation. Data referring to the habitat: Naturalness: medium; Degree of endemicity: high; Degree of biodiversity: low; Coverage: 0–25%; State of conservation: moderately conserved. Quality of the area: medium. Vulnerability of the area: forest track, road, hikers. The population is included in a Protected Natural Area of the Canary Islands (Corona Forestal Natural Park).

POPULATION Montaña Rajada (La Orotava), 2,250 m a.s.l.: Valley with little or no slope in the heart of a volcanic lava flow, with a pumice substrate where the present individuals are scarce [24 natural and 44 restored (Plan for the Recovery of Endangered Flora of the Teide National Park)]. The population is dispersed and has a medium state of conservation. Data referring to the habitat: Naturalness: high; Degree of endemicity: high; Degree of biodiversity: low; Coverage: 0–25%; State of conservation: partially degraded. Quality of the area: high. Vulnerability of the area: introduced herbivores (rabbit and mouflon). The population is included in a Protected Natural Area of the Canary Islands (Teide National Park).

POPULATION Cuevas Negras (Guía de Isora), 2,300 m a.s.l.: The number of individuals in the area is very scarce [2 natural and 6 restored (Recovery Plan for Endangered Flora of the Teide National Park)]. The population is continuous and has a medium state of conservation. Data referring to the habitat: Naturalness: medium; Degree of endemicity: high; Degree of biodiversity: low; Coverage: 0–25%; State of conservation: well preserved. Quality of the area: high. Vulnerability of the area: there are no apparent anthropic pressures. The population is included in a Protected Natural Area of the Canary Islands (Teide National Park).

Without a doubt, the locality in which the taxon is best represented is the northern slope of Teide, called Corredor de Mario, where the vigor of the specimens and the reproductive potential make this population the best stable manifestation of the species. The rest of the localities have a low number of individuals and correspond to ephemeral, juvenile or adult populations crushed by human activities." (Translated from: Beltrán Tejera et al. 1999)

"*…a clear decline in numbers has been observed in recent years as a consequence of the incidence of introduced herbivores (mouflons and rabbits). Thus, currently barely 170 breeding specimens survive, all within the Teide National Park (…). With the data obtained from 7 years of monitoring, it was observed, in the absence of predators, a very low mortality rate of adult individuals and a certain reproductive success (…). However, in those situations subject to the incidence of herbivores, a significant reduction in semen production, zero recruitment and a mortality rate close to 50% were observed (…). The main conclusion obtained is the pressing danger of extinction to which the species is subjected.*" (Translated from: Carqué Álamo et al. 2004)

Threat Category

LR pm (Bañares et al. 1998)
Anexo I: Estrictamente protegidas (Gobierno de Canarias 1991)
Anexo II (Consejo de las Comunidades Europeas 1992)
CN (UICN 1994)
EN B1 + 2 cde (Bañares et al. 1998)

E (Walter and Gillett 1998)
EN B1 + 2 cde (VVAA 2000)
En peligro de Extinción (Gobierno de Canarias 2001)
CR C2 a (ii) (Bañares et al. 2004)
CR C2 a (ii) (Moreno 2008)
En peligro de Extinción (Gobierno de Canarias 2010)
En peligro de Extinción (Ministerio de Medio Ambiente, Medio Rural y Marino 2011)

Germplasm Banks

E.T.S. Ingenieros Agrónomos de Madrid
Jardín Botánico Canario Viera y Clavijo

Exsiccata (Herbarium Specimens)

Tenerife: Cañada de Diego Hernández, 2.06.1973, C. Gil (TFC 24087); "Such as *Leuzea canariensis*" Llano de Maja—Las Cañadas, 15.06.1975, P. Pérez, M. del Arco & J. R. Acebes (TFC 4573 + Dupl.).

3.5.82 Rosa cannadas-teydensis *P Vargas, M Nogales & M Luceño*

Flora Montiberica 89:89–96 (2024)

Endemism, Certain native
Family: *Rosaceae*
Common name: Rosa salvaje, Rosa de cumbre, Escaramujo, Rosera (ES); Teide Dogrose (EN)
Biotype: nanophanerophyte
Flowering: spring–summer
Fruiting: summer
Pollination: insect pollination (entomogamy)
Dispersal: animal dispersal (endozoochory)

Description

Shrub with strong straight or slightly curved prickles on the young branches but few, smaller prickles on older branches; leaf rachis hairy, stipitate-glandular; leaflets with uniserrate denticulation (sometimes irregular), with no glands on the tooth tips but sometimes on the lowest teeth; leaflets more or less stipitate-glandular on the midrib; **flowers** numerous (3–8); petals whitish (Fig. 3.254); fruit disc flat (3.5–4 mm) and orifice wide (1.2–1.7 mm); styles lanate covering most of the disc;

Fig. 3.254 El Riachuelo, June 2003. (Photo credit: Juvenal Soto Martín)

pedicels hairy, sometimes glabrous with no stipitate glands; sepals with no glands on the back, deflexed and deciduous at maturity.

Diagnosis

Although numerous Canarian dog-roses usually show leaflets stipitate-glandular beneath, this plant has leaflets eglandular, except for a few glands along midribs. Besides, a unique combination of two more key characters is found: fruit orifice wide (>1.2 mm); and pedicels with hairs.

Discussion

Rosa dumalis **Bescht. subsp.** ***teydensis*** **Weller and Reichter** was described from Cañadas del Teide (Tenerife) (Weller and Reichert 2023). Speciation within the archipelago is also manifested by the combination of morphological characters found in three endemic dog-roses (*R. cannadas-teydensis*, *R. roque-muchachensis*, *R. gran-canariae*), whose origins need further investigation (Weller and Reichert 2023).

Weller and Reichert (2023) report that individuals from the Teide population do not present the characters found in European rose hips in the following points:

1. Using Klášterský (1968) key to species, the population fits into *R. subcanina* better than into *R. dumalis* Bescht. (*R. vosagiaca* Desportes sensu Klášterský) mostly because sepals are deflexed and deciduous at maturity. In any case, the use of *R. dumalis* is still debated (see Bakker et al. 2019; Reichter 2021).
2. There is not a single morphotype in Cañadas del Teide as revealed by a high level of morphological variation mostly in two key characters: leaflet hairiness, stipitate-glandular or glabrous beneath; and pedicels hairiness, stipitate-glandular or glabrous.

3. Although the original description includes new combination of key characters that coincide with some individuals of Cañadas del Teide, the specimen selected as the holotype (TFC 53627) by Weller and Reichert (2023) is closely related to *R. rubiginosa* group because it holds two important key characters for dog-roses: leaflets densely stipitate-glandular beneath and hairy beneath (Klášterský 1968). These characters are found in some other populations at lower altitudes on the island of Tenerife. Indeed, the holotype is quite similar to the herbarium specimen from Vargas (MA 655590).
4. In contrast, the population has two individuals not used as type material by Weller and Reichert (2023) although they display extreme forms with two key characters: leaflets eglandular except for a few glands along midribs; and pedicles hairy (sometimes glabrous), with no glands. These two characters are found neither in the holotype specimen of *Rosa dumalis* subsp. *teydensis* nor the other populations of Tenerife island.

All this leads us to conclude that on the one hand, some specimens display different key characters because of some hybridization or uncertain origin. To reflect the occurrence of the most singular morphotype occurring in the Teide valley (Cañadas del Teide, Tenerife), a new species needs to be described.

Rosa cannadas-teydensis (Vargas and Luceño Garcés 2024).

Holotype:

Spain, Canary Islands, **Santa Cruz de Tenerife**, Tenerife island: 28° 13′ 30.63″ N 16° 36′ 46.76″ W, Cañadas del Teide National Park, Corral del Guanche, 2212 m, sandy soils, 29-X-2023, *Manuel Nogales* 199PV23(2) (MA 962620). **Isotype**: TFC?.

Distribution

In the Teide National Park, it is found more or less continuously, on the escarpments of the Pared de Las Cañadas, from Risco Verde to Roque del Almendro, including the Roques de García; more punctually to the north of Izaña. It is distributed between 2050 and 2600 m a.s.l. (Figs. 3.255 and 3.256).

Remarks:

> "*On the crest of Las Cañadas*." (Translated from: Webb and Berthelot 1836–1850); "*Between Topo la Grieta & el Sombrerito*." (Translated from: Sventenius 1946); "*Vicinity of the houses of the Sanatorio de las Cañadas*." (Translated from: Ceballos and Ortuño 1976)

Habitat and Phytosociology

Naturally, they are found in subxerophytic rock stations in the cedar forests of Pared de Las Cañadas, ***Adenocarpo viscosi-Juniperetum cedri*** Martín Osorio et al. 2007, as well as in hygrophytic alluvial zones of the ***Mentho-Caricetum calderae***

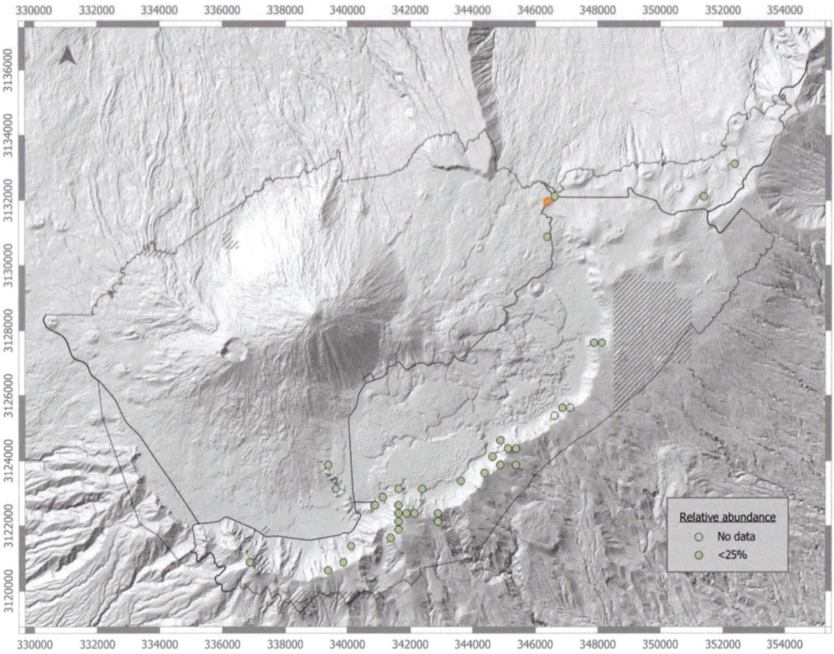

Fig. 3.255 *Rosa cannadas-teydensis* distribution, N° Subgrids (250 × 250 m): **36**/3740

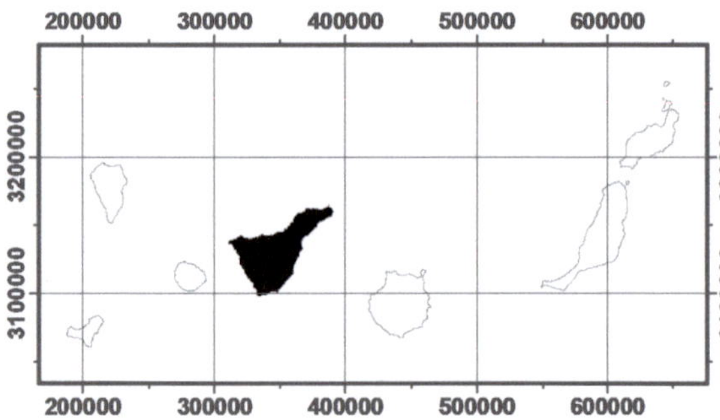

Fig. 3.256 Endemism of the Teide National Park on the island of Tenerife

community Wildpret de la Torre et al. 1987, in the exclusion zone of El Riachuelo. It develops in the dry lower Supra-Oromediterranean bioclimatic zones.

Conservation Status

In most of the subgrids in which this species was found (30), no damage of any kind was observed. Only in three (3) of them, damage was detected (browsing, high mortality without regeneration, etc.), but this did not affect more than 25% of the specimens. Damage was found on more than 25% of the specimens in only two (2). In the remaining subgrid (1), its state of conservation was not determined.

Germplasm Banks

Jardín Botánico Canario Viera y Clavijo.
Real Jardín Botánico de Madrid.

Exsiccata (Herbarium Specimens)

Tenerife: Las Cañadas, 31.V.73, Wildpret de la Torre. W., PPP, M. Galván, H. Metlesics, E. Beltrán Tejera, (TFC 3879); Las Cañadas. Topo de la Grieta, 31.V.7, Wildpret de la Torre, W., E. Beltrán Tejera, P.L Pérez de Paz, MG., LGR, H. Metlesics, (TFC 21912).

Uses, Exploitation and/or Medicinal Properties

> "*Astringent, antiscorbutic, diuretic, toxic, febrifuge, ophthalmic, tonic. Useful part: pseudofruit or rosehip; bedegar; whole plant, leaves, root, petals. Instructions for use: decoction, infusion, jam, fresh fruit.*" (Translated from: Pérez de Paz and Hernández Padrón 1999)

3.5.83 Rumex maderensis *Lowe*

Trans. Camb. Phil. Soc., 6, 12 (1838)

Acetosa maderensis (Lowe) Á.Löve & B.M.Kapoor *in* Taxon 16: 521 (1967)

Macaronesian endemism, Certain native
Family: *Polygonaceae*
New record for Teide National Park
Common name: Agria, Aceda de Madeira (ES); Madeira dock (EN)
Biotype: shrub (chamaeophyte)
Flowering: spring–summer
Fruiting: spring–summer
Pollination: wind pollination (anemogamy)
Dispersal: wind dispersal (anemochory)
Chromosome N°: 2n = 40

Description

Chamaephyte plant. Unisexual flowers. Perianth in 2 groups of 3. **Inflorescence** generally simple; ovate-deltoid leaves, withered at the base. Dense panicle, membranous valves, suborbicular. The fruits are small nuts (Fig. 3.257).

Fig. 3.257 *Rumex maderensis*. (Photo credit: GG Casanova)

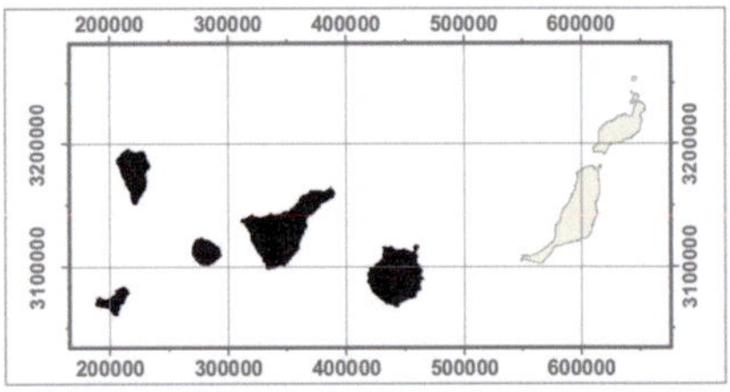

Fig. 3.258 Present on all the Canary Islands except in Lanzarote and Fuerteventura

Distribution

Macaronesian endemism found in Madeira and the Canary Islands. Present throughout the Canary Islands, except in Lanzarote and Fuerteventura (Fig. 3.258).

In the Teide National Park (Fig. 3.259) it has been found occasionally on the western slopes of Montaña de Igueque, southwestern vicinity of La Tarta, at about 2100 m a.s.l.

Habitat and Phytosociology

It usually colonizes stony terrain, rocky, humid and shady slopes, ravines and open formations in the potential domains of green mountain and pine forest, between 600 and 1200 m a.s.l.; Thermo and Mesomediterranean bioclimatic zones.

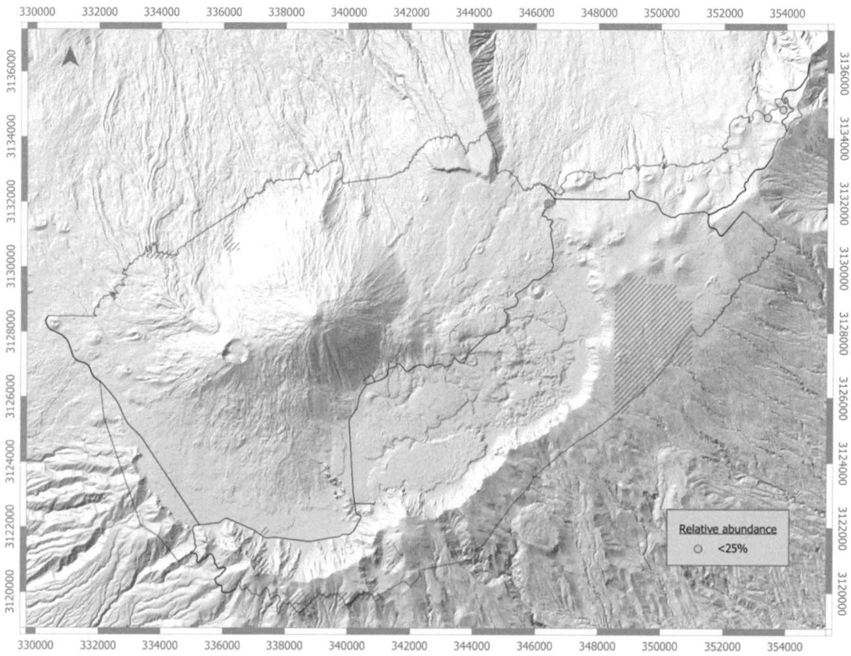

Fig. 3.259 *Rumex maderensis* distribution, N° Subgrids (250 × 250 m): **2**/3740

It is a characteristic species and bioindicator of ***Ranunculo cortusifolii-Geranion canariensis***, an alliance that groups the megaphoric communities of the mesophytic edges of the laurel forests and the humid, Thermo-Mesomediterranean subhumid-humid Canary Island pine forests of the Canary Islands.

Conservation Status

In none of the subgrids in which this species was found (2), damage of any kind was observed.

Threat Category

LR pm (Bañares et al. 1998)

Germplasm Banks

Banco de Germoplasma de E.T.S. Ingenieros Agrónomos de Madrid.
Jardín Botánico Canario Viera y Clavijo.
Real Jardín Botánico de Madrid.

Exsiccata (Herbarium Specimens)

Tenerife: Monte de la Esperanza, 19.5.1973, W. Wildpret, A. García Gallo, C. Gil Rodríguez & E. Beltrán Tejera (TFC 16928) (parasitada); Trebejos, 25.7.1973, E. Beltrán Tejera (TFC 8434); Los Silos, Monte del Agua, 17.8.1973, Anón. (TFC 27122); Güímar, Bco. de Tamay, 17.5.1975, A.D. Benítez (TFC 43025); La Esperanza, 2.6.1979, E. Beltrán Tejera (TFC 7931; 7932); La Orotova, Los Órganos, 26.6.1979, J.R. Acebes (TFC duplic. 20679); La Laguna, 20.6.1986, A. García Gallo & J.M. González (TFC 23115 duplic.; duplic. 23116, duplic. 23117, duplic. 23118, duplic. 23119); Bco. Hondo, límite entre el Término de El Rosario y Candelaria, 1100 m.s.n.m., 9.5.1994, M. del Arco Aguilar (TFC 36583); Los Silos, La Juncía, 15.8.1994, V. Lucía Saquillo & C. Acosta (TFC 38178 + duplic.); Los Silos, UTM 28RCS 2233, 9.7.1995, V. Lucía Saquillo & C. Acosta (TFC 38149 + duplic.); Las Lagunetas, 1430 m.s.n.m., P.L. Pérez de Paz, V. Lucía Saquillo & M.L. Rodríguez Navarro (TFC 44588).

El Hierro: Lomo de Tábano, 28.7.1988, P.L. Pérez de Paz, M. del Arco Aguilar, V. Lucía Sauquillo & M. Salas (TFC duplic. 30033); El Salvador, 29.7.1994, M. del Arco Aguilar & E. Montero (TFC 36864); Las Montañetas, 2.8.1994, E. Montero (TFC duplic. 36857).

?, 2.5.1969, A. Santos (TFC 443); [Como *Rumex* cf. *maderensis*]: ?, 1849, 8.6.1969, Bramwell (TFC 444).

Uses, Exploitation and/or Medicinal Properties

> "*Vulnerary, astringent, diuretic; fodder. Useful parts: leaves, juice; whole plant. Instructions for use: infusion, poultice; browsing, fodder.*" (Translated from: Pérez de Paz and Hernández Padrón 1999)

3.5.84 Sagina procumbens *L. ssp.* stolonifera *Martín Osorio, Rivas-Martínez & Wildpret ssp.* nova

New record for the Teide National Park

Endemism, Certain native
Family: *Caryophyllaceae*
Common name: Morujilla canaria (ES); Procumbent pearlwort (EN)
Biotype: hemicryptophyte
Flowering: July–September
Fruiting: September–October
Pollination: wind or insect pollination (anemogamy—entomogamy)
Dispersal: wind dispersal (anemochory)

Description

Perennial, cespitose plant (Fig. 3.260). Main stem short; lateral stems up to 10 cm, numerous, rooted at the base, ascending and branched. **Leaves** (3) 5–10 (12) mm, glabrous, sometimes with ciliate margin.

Floral pedicels 5–15(25) mm, which arise from the leaf axils, glabrous—very rarely glandular. **Flowers** tetramerous—rarely some pentamerous in the same specimen -, solitary (Fig. 3.261). Sepals (1.5) 2–2.2 (2.5) mm, ovate, obtuse, glabrous—very rarely glandular -, generally patent. Petals absent or smaller than the sepals, rarely subequal. Stamens 4. Capsule (2) 2.4–2.8 (3.5) mm.

It differs from the typical subspecies in its stoloniferous habit and in the size of the main and lateral stems, which are much shorter, up to 10 cm. Summer phenology. Its optimal habitat is the surroundings of the fumaroles that emanate water vapour at 80 °C, from the cacuminal part of the Teide stratovolcano at 3600 m above sea level, although it has also been found in recent lava flows at the base of the volcano. **Holotype** (TFC 45413), Telesforo Bravo trail up to the crater, 24/10/2004, V.E. Martín Osorio, W. Wildpret de la Torre & S. Rivas-Martínez (TFC 45413);

It is worth highlighting the chorological occurrence of the ***Sagina*** taxon, from the high mountains, where they are found both in the Temperate Bioclimate and in the Mediterranean. Montserrat Martí and Montserrat Martí (1990) point out that a heterogeneous group of taxa have been included in ***Sagina saginoides*** that require detailed studies. The specimens that most closely resemble the type are found in the Pyrenees, where there are also other forms of sepals and capsules smaller than those of the type, with erect pedicels, which have been named ***S. pyreniaca*** Rouy, I11. Pl. Eur.: 26 (1895) [***S. fasciculata*** Boiss.- nom. Illeg., nom. subst.; ***S. saginoides* subsp. *pyrenaica*** (Rouy) Font Quer, ***S. procumbens* subsp. *muscosa*** auct.]. ***Sagina saginoides*** (L.) H. Karst. **subsp. *nevadensis*** (Boiss. & Reut.) Greuter & Burdet *in*

Fig. 3.260 *Sagina procumbens* subsp. *stolonifera* near fumaroles at the Pilón de Azúcar—Pico del Teide, Aug 2006. (Photo credit: Victoria Eugenia Martín Osorio)

Fig. 3.261 *Sagina procumbens* subsp. *stolonifera*, near the crater of the Teide stratovolcano at 3600 m above sea level. (Photo credit: Victoria Eugenia Martín Osorio)

Willdenowia 12(2): 189 (1982) [***S. nevadensis*** Boiss. & Reut., Pugill. Pl. Afr. Bor. Hispan.: 22 (1852) basion] refers to populations of densely cespitose plants, with generally sickle-shaped basal leaves, erect pedicels and glandular in their upper third, sepals glandular at their base and capsules of 2.5–3(3.5) mm, typical of Sierra Nevada, although there are transitional forms, with larger capsules, in the Central and Iberian Systems.

On the other hand, Dobignard (1997) describes a new subspecies for the Moroccan Atlas, ***Sagina procumbens* subsp. *atlasica***, (=***S. procumbens* var. *parviflora*** Ball in J. Linn. Soc., Bot. 16: 367. 1877) indicating that it seems very close to ***S. procumbens* subsp. *muscosa***, which even being pentamerous could be included in ***S. saginoides***, but that, in any case, it could be the vicariant of atlasica. An opinion already expressed by Maire (Fl. Afr. M 9:247). In particular, the **subsp. *parviflora*** Litard. & Maire with five stamens, could have a hybridogenic origin between the **subsp. *nevadensis*** with inflorescence and flowers with glandular sepals with ten stamens, and another Moroccan representative of ***S. procumbes*** or ***S. saginoides***. For all the above, we can confirm that ***S. procumbens* subsp. *stolonifera*** would be the vicariant of ***S. procumbens* subsp. *atlasica***, with similar ecology to high mountains and humid soils of an acidic nature between 1600–2800 and 3600 m above sea level on the Teide stratovolcano. Its geographic isolation makes it a member of this group of vicariant subspecific taxa typical of the high Mediterranean mountains of Nevada, Atlas and the Canary Islands, together with the temperate Pyrenees.

Distribution

Endemic subspecies of the islands of La Palma and Tenerife (Teide National Park). In the latter it is found in areas of the Pico del Teide, at about 3650 m above sea level; very specifically, inside the Las Cañadas depression, on recent lava (Figs. 3.262 and 3.263).

Habitat and Phytosociology

It lives in the hot lands of the Pilón de Azúcar de El Teide and in some places on recent volcanic lava.

It belongs to the alliance ***Saginion procumbentis*** Tüxen & Ohba in Géhu, Richard & Tüxen, of the class ***Polygono-Poetea annuae*** Rivas-Martínez and is a characteristic species of the association ***Poo infirmae-Saginetum stoloniferi*** Wildpret de la Torre and Martín Osorio 2005, an oligospecific, thermopedophygrophilous and somewhat sciophilous community that occupies only a few square decimetres of surface, in a microclimatic environment on permanently humid and hot lands (35 °C), in the shade of the rocks of the lava flow that descends from the top of the peak of Teide to the area of La Rambleta (3650–3700 m a.s.l.); Canarian high mountain geomicroseries of Teide, which forms part of the Canarian oromediterranean vegetation complex of the cacuminal zone of the stratovolcano.

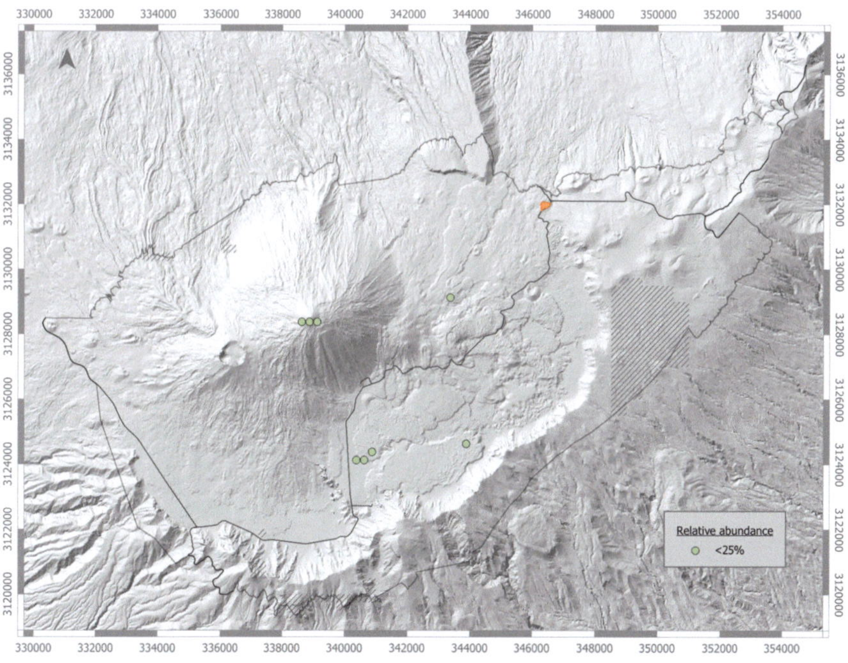

Fig. 3.262 *Sagina procumbens* subsp. *stolonifera* distribution, N° Subgrids (250 × 250 m): **8**/3740

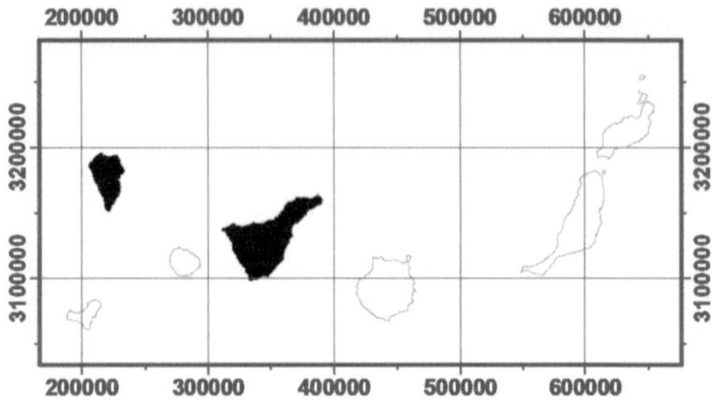

Fig. 3.263 Endemic subspecies of the islands of La Palma and Tenerife (Teide National Park)

Conservation Status

In all the subgrids in which this species was found, no damage of any kind was observed, so its conservation status was determined as high.

Exsiccata (Herbarium Specimens)

Tenerife: Inventario de la Flora y Vegetación del Parque Nacional del Teide (2001–2003) V.E. Martín Osorio, W. Wildpret & B. Hernández Bolaños: Holotipo. Sendero Telesforo Bravo subida al cráter, 24/10/2004, V.E. Martín Osorio, W. Wildpret de la Torre & S. Rivas-Martínez (TFC 45413); Cono del Teide (35B 37 a), 23/04/2003, A. Montañés Rodríguez (TFC 44842).

3.5.85 Salix canariensis *C. Sm.* ex *Link*

In Buch., *Phys. Beschr. Can. Inst.*, pág. 159, 179 (1825).

Salix pedicellata subsp. *canariensis* (C.Sm. *ex* Link) A.K.Skvortsov First published in Willows USSR: 165 (1968) (https://www.worldfloraonline.org/taxon/wfo-0000930946)

Not endemic to the Canary Islands, Certain native
Family: *Salicaceae*
Common name: Sauce canario o Sao (ES); Canary Island Willow (EN)
Biotype: phanerophyte
Flowering: January–April
Fruiting: spring–summer
Pollination: wind or insect pollination (anemogamy–entomogamy)
Dispersal: water or wind dispersal (hydrochory—anemochory)
Chromosome N°: 2n = 38

Description

Dioecious, deciduous **phanerophyte** (Fig. 3.264), 7–10 m tall, with diffuse branches. **Sympodial stem** branched from the base (Fig. 3.265), greyish bark provided with butterfly-like lenticels located on the lower part of the trunk. Gnarled, inflexible branches; the youngest ones have tomentum, which is progressively lost until they become glabrous. Axillary buds of 0.5–0.8 cm are completely covered by a deciduous, tomentose-whitish scale, which turns reddish when mature. Given its water requirements, it develops, apart from its main roots, another very fasciculated root system, with a large number of rootlets that, like a root ball, are installed along the channel where the water flows, a phenomenon that is very easy to observe in specimens that live on cliffs. It can be considered a heliophilous species.

Leaves simple, alternate, stipules; young leaves densely tomentose, adult leaves with green upper surface, glabrous, finely pubescent on the central vein and glaucous tomentose underside; lanceolate leaf blade, attenuated at the base, 4.5–24 × 1.5–5 cm, with smooth, slightly wavy margin, occasionally somewhat serrated and toothed in adult leaves, variable in morphology and size; main vein very prominent, reddish; venation very extended towards the edges of the leaf blade, of the eucamptodrone (secondary nerves are arched) type, although sometimes there are anastomoses between the secondary veins tending towards brachidodrome (the secondary veins do not terminate at the margin); petiole 0.3–3 cm long,

Fig. 3.264 Wolfredo Wildpret Botanical Garden of Teide National Park, June 2015. (Photo credit: Victoria Eugenia Martín Osorio)

Fig. 3.265 El Riachuelo, June 2003. (Photo credit: Juvenal Soto Martín)

pubescent; stipules small, oval, 0.2–0.4 cm long, toothed, submembranous, pubescent and deciduous; inflorescences amentiform, axillary.

Inflorescence shoot covered by a completely hairy, deciduous scale. Catkins arranged laterally, inserted in the stem by a cylindrical peduncle 1.2–3 cm long, often bearing 1.6–8 × 0.6–1.6 cm tomentose-whitish leaves, rachis densely hairy. Male catkins quite dense, erect or barely drooping, 2.5–10.2 × 0.4–1 cm. Male flower bracteate; Bract more or less lanceolate, 0.1–0.3 × 0.03–0.3 cm, dark brown in its apical portion, completely covered with whitish hairs of variable length between 0.03 and 0.2 cm, with a nectary, rarely orange, 0.04–0.2 cm long, free stamens in pairs of two, rarely three, adnate anthers yellow, with longitudinal dehiscence, 0.03–0.1 cm, long filaments, 0.3–1.2 cm long, hairy at the base.

Female catkins not very dense, allowing the rachis to be quite clearly seen, 2.5–8.9 × 0.5–1 cm. Female flower bracteate, bract 3–4 times longer than the nectary, completely hairy, hairs 0.02–0.18 cm long, with a nectary reaching 0.04–0.1 cm in length; gynophore completely glabrous, three times larger than the nectary; ovary ovate-conical, glabrous, 0.2–0.8 cm; style rather short, not exceeding 0.03 cm in length; stigmas two in number, emarginate at the apex, with drooping lobes of 0.02–0.04 cm. Capsule ovoid-conical, glabrous, bicarpellate, unilocular, with parietal placentation, bearing numerous anatropous ovules; longitudinal dehiscence in two valves. Seeds are small, 0.1–0.19 × 0.03–0.09 cm, olive-green in colour, with a tuft of hairs from the placental outgrowth. The number of seeds per capsule varies between 2 and 19.

The pollen grain has an isopolar, radiometric symmetry and shape, and radial symmetry of order 3.

Distribution

Macaronesian endemism (Madeira and the Canary Islands). It is found throughout the Canary Islands, except in Lanzarote and Fuerteventura.

In the Teide National Park, it has been found forming a considerable and continuous population in the Barranco del Río; more punctually, in El Riachuelo and at the foot of the Pared towards Llano Ucanca (Lomo Grande); between 1900 and 2300 m a.s.l. Rodríguez Piñero et al. (1987). There is also a specimen in the Teide Botanical Garden (close to the artificial pond) (Figs. 3.266 and 3.267).

Habitat and Phytosociology

Hydrophile that settles in humid ravine beds and on oozing walls, both in sunny and shady environments. The Canary willow characterizes a gallery forest type formation with dominance in the tree stratum of this species or, occasionally, of ***Myrica faya***.

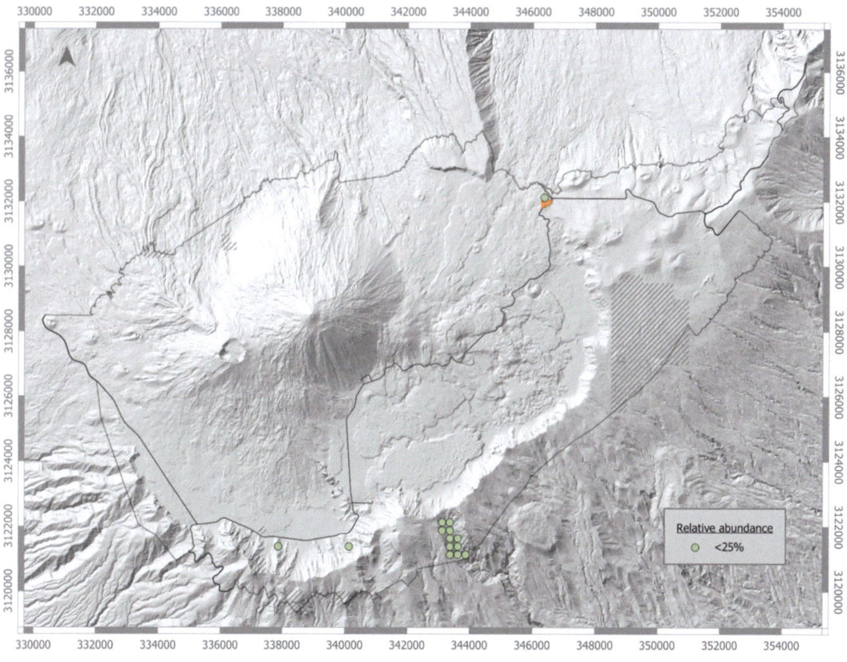

Fig. 3.266 *Salix canariensis* distribution, N° Subgrids (250 × 250 m): **14**/3740

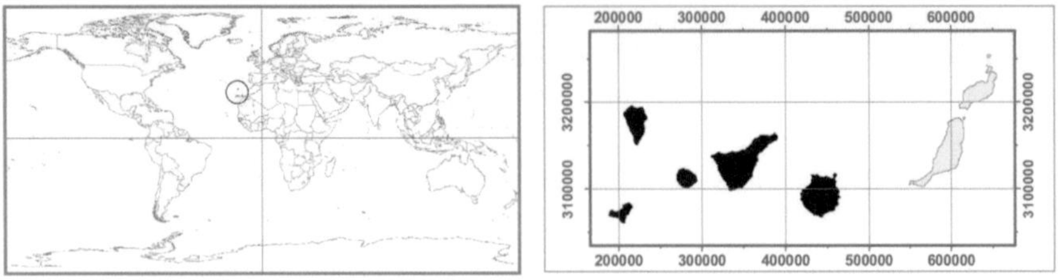

Fig. 3.267 Present on all the Canary Islands except in Lanzarote and Fuerteventura

Characteristic species of the ***Salicion canariensis*** alliance, deciduous broadleaf microforests of Canary willows, Mediterranean pluviseasonal and xeric oceanic, Infra-Thermo-Mediterranean semi-arid-humid, eu-subhyperoceanic, of western Canary and Madeiran distribution.

Conservation Status

In almost all the subgrids in which this species was found (13), no damage of any kind was observed. Only in one (1) of them, damage was detected (browsing, high mortality without regeneration, etc.), but this did not affect more than 25% of the specimens.

Threat Category

Anexo II: Protegidas (Gobierno de Canarias 1991)
VU B1 + 2c (Bañares et al. 1998)
De interés especial (Gobierno de Canarias 2001)
EN (Bañares et al. 2004)
EN B2ac (ii, iv) (Moreno 2008)
Vulnerable (Gobierno de Canarias 2010)

Exsiccata (Herbarium Specimens)

Tenerife: "E. BOURGEAU, Pl. Canarienses (*ex* itinere secundo) 1855, 1502. *Salix canariensis* Chr. Sm.- Phy. Can. III, 270.t. 215 (J. Gay) Teneriffa: *in* sylvaticis regionis mediae Las Mercedes, 26 Januar" (STR); "*Salix canariensis* Chr. Sm. Tenerife: Feuchte Berkhänge welt. Mercedes. 15.4.87." firma illegit (STR), Bco. de San Andrés, 20.XII. 1970, Wildpret, W., L. Gallo & A. Santos (TFC 449); Pedro Álvarez, 28.II.71, Barquín Díez, E., (TFC 21690); Tenerife, 28.II.71, Pérez de Paz, P.L., (TFC 1007); Las Mercedes, 18.VII.1980, Rodríguez, C., (TFC 12300, 12301 y 12302); Ibid., 11.I.1982 (TFC 13078); El Moquinal, 23.III. 1980, Rodríguez, C., (TFC 12328, 12329 y 12330); Bco. Seco. Adeje, 5.IV.80, Suárez, C.L., (TFC 23031); Bco. de Riachuelo, 12.V.84, Charpin, A., W. Wildpret, (TFC 20119); Ibid., 19.X.1985 (TFC 12345, 12346, y 12347); Bco. del Río cerca del canal, 12.II.1972, Gallo, L. & P.L. Pérez de Paz (TFC 787), Ibid., Beltrán, E., R. Afonso & M. del Arco (TFC 4871); Bco. del Río, 2300 m, 15.V.83, Sergio, (TFC 28291); Bco. del

Riachuelo, 2125 m, 3.X.82, (TFC 34251); Bco. Tahodio, 21.II.85, Pérez de Paz. P.L., (TFC Duplic. 13274); Barranco del Río, parte alta y media, 30.X.1985, Rodríguez, C. (TFC 12332, 12333, 12334, 12335, 12336, 12337, 12338, 12339, 12340, 12341, 12342); Bco. del Riachuelo, Las Cañadas, 19.X.1985, Rodríguez, C. (TFC 12343, 12344); Bco. del Infierno, Adeje, 4.IV.1980, C. Rodríguez (TFC 12305, 12306, 12307, 12308, 12309, 12310, 12311, 12312, 12313, 12314, Duplic. 12315, 12317, 12318, 12326 y 12327); Ibid., 24.X.1985, (TFC 12330, 12348, 12349, 12350, 12351 y 12359); Icod, 6.II.86, C.R.P. (TFC 29414, 29415, 29416, 29417, 29418, 29419, 29420, 29421, 29422, 29423, 29424); Monte de Las Mercedes, La Laguna. 15.II.92, Pérez Hernández. Alums., P.L. Pérez de Paz, (TFC 35129); Bco. del Agua. Adeje. 380 m.s.n.m., 28.I.97, ECA & M.M. Gómez, (TFC. Duplic. N 40401); Charcos de Erjos, Los Silos, 12.I.97, Lucía Sauquillo, V., C. Acosta, (TFC 40001); La Isleta, Los Silos, 15.II.97, Lucía Sauquillo, V., (TFC 40578 y Duplicc.); San Mateo. Bco de la Mina, 1350 m, 3.IV.99, Beltrán Tejera, E., J.L. Rodríguez Armas, L.M. Monzón, S.A. Medina (TFC 41932). Inventario de la Flora y Vegetación del Parque Nacional del Teide (2001–2003) V.E. Martín Osorio, W. Wildpret & B. Hernández Bolaños: Cauce del Barranco del Río, 21/05/2003, F. Mesa (TFC 44922); Riachuelo, 01/09/2002, V.E. Martín Osorio & W. Wildpret de la Torre (TFC 44932 + Dupl.)

Uses, Exploitation and/or Medicinal Properties

"*Antipyretic, analgesic, antirheumatic, astringent, antispasmodic; forage: toxic/salicylic. Useful parts: bark, branches, leaves, flowers; pappus of the seeds: help coagulation. Instructions for use: decoction; powder: stamps; herbal tea, enolith, extract; forage: excessively toxic.*" (Translated from: Pérez de Paz and Hernández Padrón 1999)

3.5.86 Scrophularia glabrata *Aiton*

Hort. Kew., II: 341 (1789)

Scrophularia berthelotii Bolle in Verh. K. K. Zool.-Bot. Ges. Wien 11: 202 (1861)
Scrophularia elongata Webb ex Benth. *in* A.P.de Candolle, Prodr. 10: 593 (1846)

Endemism, Certain native
Family: *Scrophulariaceae*
Common name: Yerba de la cumbre o Fistulera de cumbre (ES) Summit figwort (EN)
Biotype: shrub (chamaephyte)
Flowering: May–June
Fruiting: May–July
Pollination: insect pollination (entomogamy)
Dispersal: wind dispersal (anemochory)
Chromosome N°: n = 29; 2n = 56, 58, 60

Description

Perennial species, with a woody base, completely glabrous, short, tetragonous and purplish branches, sometimes narrowly winged.

Leaves ovate-triangular, 10.4 × 0.4–4.7 cm, coriaceous, shiny, with regular serrated-toothed margin. Petiole 0.4–4 cm; blade densely glandulated.

Inflorescence thyrsus-shaped 4–33 cm long, leafless, cymes with up to 16 flowers (37-flowered) (Fig. 3.268), rarely more than once dichotomously branched, usually densely glandular; peduncle 2–24 mm, pedicels 5–18 mm glandular. Bracteoles 1.5–8 × 0.5–1.5 mm, narrowly elliptic to narrowly oblong, acute. Calyx lobes 2–3 × 1.5–2.5(−3) mm, very broadly ovate or oblong to circular, rounded, sometimes dark purple. Corolla 5–8 mm long, deep red to violet-brown. Corolla tube usually striated green. Stamens 4.5–6.5 mm long, exserted. Anthers violet-brown. Staminode 0.9–2 × 0.8–1.7 mm, one to two times as wide as anthers, erect to patent erect, obovate to elliptic transversely, apex bluntly obtuse or emarginate. Style 3–5.5 mm, with the base sometimes persistent, like a mucron in the fruit. Capsule 4–7.5 (−8) × 3–6 (−7) mm, ovoid or sometimes globose, rounded acuminate, with a mucron of 0.1–1 mm. Seeds of 0.7–1 mm.

Remarks:

Dalgaard *(*1979*) indicates as a lectotype chosen by him in BM Orig. coll.: "Tenerife, 1778", Masson s.n. published in Aiton 1789.*

Fig. 3.268 (**a**, **b**) Cañada de Diego Hernández, May 2004. (Photo credit: Beatriz Hernández)

Distribution

Endemism of the islands of Tenerife and La Palma. In Tenerife it has a wide distribution and can occasionally be found from 250 to 2750 m a.s.l. in a punctual manner on the southern slopes of Teide—Pico Viejo. Within the National Park, it is very widespread throughout the Wall and interior areas of the Las Cañadas depression, except on the slopes and lava flows of the Teide-Pico Viejo and Montaña Blanca buildings, and the highest areas of the Ridge (Figs. 3.269 and 3.270).

Habitat and Phytosociology

It preferentially inhabits rocky and stony soils. It is a characteristic species and bioindicator of ***Chamaecytiso-Pinetea canariensis***, a vegetation class that includes coniferous forests, rhetamoid shrublands and colonizing mountain dwarf shrublands serial of Tenerife and La Palma.

It preferably intervenes in communities included in the ***Cisto symphytifolii-Pinion canariensis*** alliance, with greater prominence in the communities of the ***Spartocytision supranubii*** alliance.

It is also presented in the following units:

Adenocarpo viscosi-Chamaecytisetum angustifolii
Adenocarpo viscosi-Juniperetum cedri,
Arrhenathero calderae-Plantaginetum webbii.

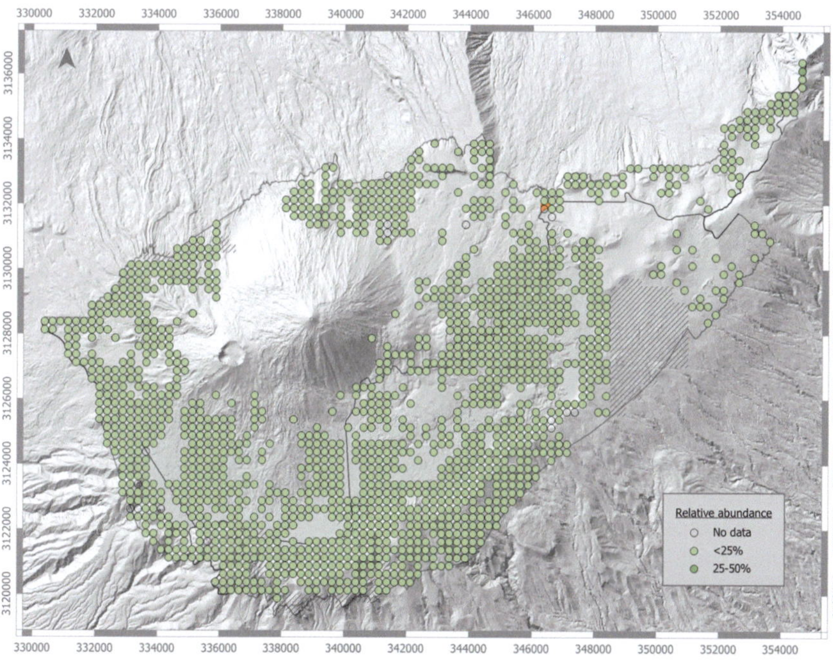

Fig. 3.269 *Scrophularia glabrata* distribution, N° Subgrids (250 × 250 m): **1635**/3741

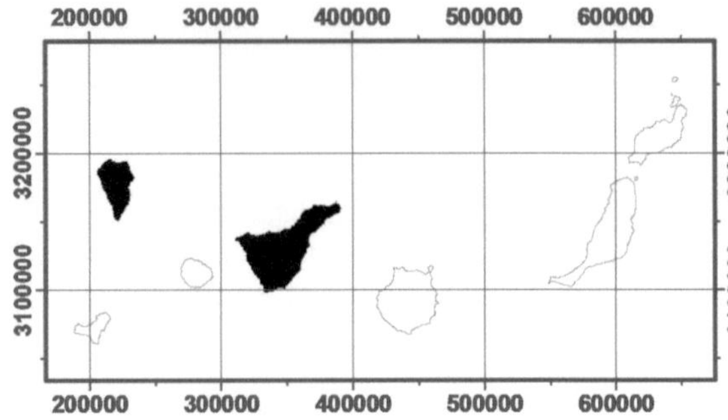

Fig. 3.270 Endemism of the island of Tenerife and La Palma

Cheirolopho teydis-Carlinetum xeranthemoides
Tolpido webbii-Micromerietum lachnophyllae
Arrhenathero calderae-Adenocarpetum viscosi

Conservation Status

In 95% (1551) of the subgrids in which this species was found, no damage of any kind was observed. Only in 4% (68), damage was detected (browsing, high mortality without regeneration, etc.), but this did not affect more than 25% of the specimens. Damage was only found on more than 25% of the specimens in less than 0.5% (7) of the subgrids where this taxon occurred. In the rest (9), this variable was not specified at the time of sampling.

Threat Category

nt (Barreno et al. 1984)
LR pm (Bañares et al. 1998)

Germplasm Banks

E.T.S. Ingenieros Agrónomos de Madrid.
Jardín Botánico Canario Viera y Clavijo.
Real Jardín Botánico de Madrid.

Exsiccata (Herbarium Specimens)

Tenerife: W. Wildpret, L. Gallo, E. Beltrán Tejera, A. Aldrige, C. Gil & P.L. Pérez de Paz (TFC 927); 16.02.1969, Bramwell 702 (TFC 470); Mayo 1969, ejusd. 1313 (TFC 468); 9.06.1969, ejusd. 1872 (TFC 469); 11.06.1969, ejusd. 1932 (TFC 467); 1.09.1969, ejusd. 2038 (TFC 466); Las Cañadas, 28.07.1971, A. Acuña & F. Estévez (TFC 8420); Bco. Orchilla, 20.05.1973, E. Beltrán Tejera (TFC 3810, 3811); La Esperanza, 26.05.1973, I. La Serna (TFC 35779); Risco Azul, 11.04.1974, W. Wildpret de la Torre (TFC 20335); Las Cañadas, 2.05.1975, A.D. Benítez (TFC 43011); Aguamansa, 29.05.1975, ejusd. (TFC 43048); Mña. Bermeja, 19.02.1978,

E. Beltrán Tejera, C. León Arencibia, J.R. Acebes & M. del Arco Aguilar (TFC 29093). Pista Lomo Blanco, 13.05.1979, P.L. Pérez de Paz & C. Hernández Padrón (TFC 20449); Las Cañadas, 28.05.1979, E. Beltrán Tejera (TFC 7942, 7943); Las Arenas-San José de los Llanos-Erjos, 1200 m.s.n.m., 29.08.1979, W. Wildpret, I. La Serna, L. Negrín, A. Losada Lima & A. García Gallo (TFC 24704); Caldera de Pedro Gil-Arafo, 6.05.1981, M. del Arco Aguilar (TFC 24385 + Dupl.); Valle del Volcán de Las Arenas-Cumbres de Arafo, 9.05.1981, W. Wildpret, M. del Arco Aguilar & P.L. Pérez de Paz (TFC 23725 + Dupl.); Ibid., 1700 m.s.n.m., 9.05.1981, ejusd. (TFC 23748 + Dupl.); Cañada de Diego Hernández-Paredón, 30.05.1981, W. Wildpret, E. Beltrán Tejera, A. Losada Lima & A. García Gallo (TFC 33142); La Crucita, 1980 m.s.n.m., 3.05.1984, RLN (TFC 389, 13228); Mña. Samara-Boca Tauce, 3.05.1984, RLN (TFC 358, 13197); El Portillo-La Orotava, 3.05.1984, ejusd. (TFC 357, 13196); Mña. Cangrejo-Chío, 1584 m.s.n.m., 3.05.1984, ejusd. (TFC 343, 13190); Las Cañadas del Teide-Junto al Km. 44 de la carretera, 2200 m.s.n.m., 3.05.1984, ejusd. (TFC 356, 13195); Pista Madre del Agua-Vilaflor, 15.05.1984, P.L. Pérez de Paz, I. La Serna & L. Negrín (TFC 403, 13242); Ctra. Vilaflor, 15.05.1984, ejusd. (TFC 404, 13243 + Dupl.); Pista de la Caldera a los Órganos-Aguamansa, 1135 m.s.n.m., 15.04.1985, RLN & A. García Gallo (TFC 484, 13318); Entrada Galería "El Peral"-Vilaflor, 29.04.1990, C.D. Hernández (TFC 33427); Izaña-Güímar, 1800 m.s.n.m., 31.05.1993, M. del Arco Aguilar (TFC 36052). Inventario de la Flora y Vegetación del Parque Nacional del Teide (2001–2003) V.E. Martín Osorio, W. Wildpret & B. Hernández Bolaños: Roques al Sur del Llano Ucanca (41D 35 d), 14/05/2003, A. Montañés Rodríguez (TFC 44296).

La Palma: Garafía, 6.04.1980, J. Fernández Casas (TFC 20743); Cumbre Nueva, 1000 m.s.n.m., 19.04.1984, P.L. Pérez de Paz & RLN (TFC 390, 13229).

Uses, Exploitation and/or Medicinal Properties

"*Treatment of fistulas, dermatological, vulnerary, antihemorrhoidal, hemostatic, hypoglycemic; fodder. Useful part: leaves, tops. Instructions for use: infusion, decoction, washes, poultice; browsing.*" (Translated from: Pérez de Paz and Hernández Padrón 1999)

3.5.87 Sideritis eriocephala *Marrero Rodr.* ex *Negrín & P. Pérez*

Lagascalia, 15 extra, 295–318 (1988)

Leucophae eriocephala Webb *ex* Pit. & Proust *in* Iles Canaries: 313 (1909), *nom. nud.*

Leucophae eriocephala Webb.mss., E. Bourgeau, n° 918 (P!); nom. *in* schaed (1945), *nomen nudum.*

Leucophae candicans f. *eriocephala* Webb *ex* Burchard. *Bibl. Bot.* 98: 187–188 (1929), *nomen nudum.*

Sideritis candicans v. *eriocephala* Webb *ex* Ceb. et Ort. *Veg. y Fl. Forest. Can. Occ.* p. 418 (1951), *nomen nudum.*

Sideritis cretica v. *eriocephala* (Clos) Mend.—Heuer. *Vieraea*, 3: 136 (1974), *nomen nudum.*

Sideritis eriocephala (Webb *ex* Christ) A. Marrero. *Bot. Mac.*, 14: 41 (1985), *nomen nudum.*

Endemism, Certain native
Family: *Lamiaceae*
Common name: Salvia blanca, Chahorra o Chajorra del Teide (ES); Teide Ironwort (EN)
Biotype: shrub (chamaephyte)
Flowering: June–July
Fruiting: August–September
Pollination: insect pollination (entomogamy)
Dispersal: short range self-dispersal (atelochory)
Chromosome N°: 2n = 34

Description

Clumpy in habit; short stems, profusely branched, densely tomentose, like the leaves and the rachis.

Leaves ovate-lanceolate or narrow-ovate-lanceolate, obtuse, with a rounded or cuneate base, tomentose, underside with sub-manifest nerves despite the density of the tomentum; the basal ones 2.2–7.6 × 0.9–2.3 cm, long-petiolate, petioles 1.7–7 cm, very close together, almost rosette-shaped (internodes 0.2–1 cm); the inhibition ones 1.5–4.8 × 0.4–1.4 cm, the lower ones petiolate, petioles 0.2–3 cm, the upper ones sessile or subsessile (Fig. 3.271, Perez de Paz and Negrín Sosa 1992).

Inflorescences with long rachis up to 80 cm tall, unbranched, ending in terminal spikes of yellowish-white flowers. Verticillasters in number of 7–20, with 9–23 flowers, the 2–4 lower ones more or less distant, the rest agglomerated. Bracts of 0.5–3.5 × 0.2–1.0 cm, the upper ones sessile, the basal ones subsessile (petioles of 0.2–0.5 cm). Bracteoles very scarce, filiform, of 1.0–1.4 × 0.1–0.2 mm. Flowers with pedicels of 1.0–2.3 (3) mm. Calyx of 5.5–8.3 mm, tubular-campanulate, glandular-tomentose on its external face; teeth, occasionally six in number, jagged, hairy on their inner surface of 1–2.5 mm. Corolla of 6.0–7.3 mm barely exserted, tube of 4–6 mm yellowish-white, cylindrical at the base widened at the top; limb violet-purple veined, upper lip of 1–1.8 mm, obovate-triangular; lower lip with obovate lateral lobes of 0.6–1.5 mm, rounded. Stamens with filaments of 1.5–2.3 mm and anthers of 0.7–1.0 mm, the upper ones; the lower ones with filaments of 2.0–2.8 mm and anthers of 0.5–0.8 mm. Gynoecium with style of 2.2–3.9 mm, stigma with laciniae of 0.5–0.8 mm the largest, and 0.4–0.6 mm the smallest. Nutlets 2.3–2.7 × 1.4–1.8 mm, elliptic and obovate, variegated with black and brown.

Fig. 3.271 Teide National Park. (Photo credit: Victoria Eugenia Martín Osorio)

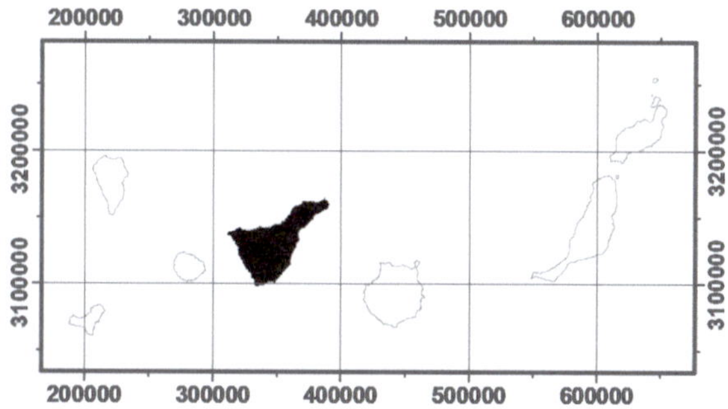

Fig. 3.272 Endemism of the island of Tenerife

Distribution

Endemism of the island of Tenerife (Fig. 3.272), which is usually found at altitudes between 2000 and 2600 m above sea level. In the Teide National Park (Fig. 3.273), its distribution is mainly related to the inner and outer slopes of the Pared de Las Cañadas, from Diego Hernández to Valle de Ucanca and Boca Tauce; to a lesser extent, to Cañada de El Cedro. It is also found, but very sparsely and punctually, in some areas of the northern slopes of Teide—Pico Viejo (Llano Quemón Volcano and Echisere Mountain), Valle de las Piedras Arrancadas and in the Teide Botanical Garden.

Habitat and Phytosociology

It normally appears in the fissures and scree of steep areas, in xerophilous and heliophilous environments.

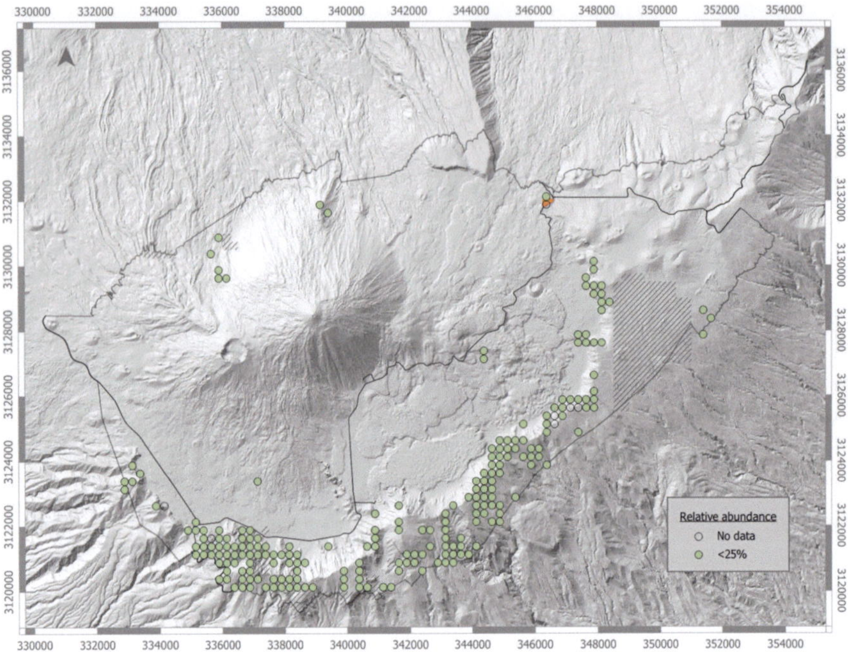

Fig. 3.273 *Sideritis eriocephala* distribution, N° Subgrids (250 × 250 m): **221**/3741

It is a characteristic and bioindicator species of ***Spartocytision supranubii***, an alliance that groups the mountain summit shrublands of brooms, flatpods and sweet brooms, climatophilous or edaphoxerophilous, sometimes catenal or serial of the Canary pine forests; of distribution in La Palma and Tenerife, dry-humid Meso-Oromediterranean.

It also intervenes in the association ***Cheirolopho teydis-Carlinetum xeranthemoides***, substitution scrub, sometimes acting as permanent vegetation, of the summit retamar and Canary pine forests on soils of old materials of the Wall or precaldera building of Teide, denuded and exposed to strong dry-subhumid Meso-Supramediterranean insolation. Its proliferation is frequent after some fires such as the one in the southern area of El Sombrerito.

Conservation Status

In 95% (211) of the subgrids in which this species was found, no damage of any kind was observed. Only in a little less than 3% (6), damage was detected (browsing, high mortality without regeneration, etc.), but this did not affect more than 25% of the specimens. In the rest (4), this variable was not specified at the time of sampling.

"*Species with a small distribution area, locally not too scarce. No direct threat can be identified, other than possible collecting or the geological instability of the land where it is installed.*" (Translated from: Gómez Campo et al. 1996)

Threat Category

VU D2 (Bañares et al. 1998)
VU (Bañares et al. 2004)
EN B2 ab (iii, v) c (iv) (Moreno 2008)

Germplasm Banks

E.T.S. Ingenieros Agrónomos de Madrid.
Jardín Botánico Canario Viera y Clavijo.

Exsiccata (Herbarium Specimens)

Tenerife: Las Cañadas, 28.07.1971, F. Estévez & A. Acuña (TFC 15994); Topo de la Grieta, 5.07.1973, P.L. Pérez de Paz (TFC 17753); Cañada de Diego Hernández, 3.08.1974, W. Wildpret & J.R. Acebes (TFC 4597, 4598, 4599); Carretera de Boca Tauce a Vilaflor, 12.06.1983, L. Negrín (TFC 17369); Inmediaciones del Roque del Almendro-El Sombrerito-Ctra. de Boca Tauce hacia Vilaflor, Julio 1984, S. Socorro, J.M. González & L. Negrín (TFC 17657+ Dupl. in P, MA; 17658, 17659, 17660 + Dupl., 17661 + Dupl., 17662, 17663, 17664, 17665 + Dupl. in BM, TFMC; 17666 + Dupl.); Cañada de Diego Hernández, 2200 m.s.n.m., 20.06.1985, P.L. Pérez & I. La Serna (TFC 17751); Degollada de Guajara, 2400 m.s.n.m., 20.06.1985, ejusd. (TFC 17752 + Dupl.); Pista del Parador Sanatorio-Las Cañadas, 25.04.1987, P.L. Pérez de Paz & W. Wildpret (TFC 28888). Inventario de la Flora y Vegetación del P.N. del Teide (2001–2003) V.E. Martín Osorio, W. Wildpret & B. Hernández Bolaños: Al W de Izaña—Montaña de El Cabezón (31C 1 a), 24.04.2001, L. Afonso, A. Hernández, Y. de Rioja & F. Tapia (TFC 44913); Siete Cañadas (36D 31), 22/06/2001, L. Afonso, A. Hernández, Y. de Rioja & F. Tapia (TFC 45405); Pared de Las Cañadas—Risco Verde, 29/08/2001, W. Wildpret de la Torre (TFC 45421); Cañada de Las Pilas—Pared (36D 4, 14 y 23), 22/06/2001, L. Afonso, A. Hernández, Y. de Rioja & F. Tapia (TFC 44869 + dupl.); Ladera occidental de la Montaña del Cedro (41A 24 c), 07/05/2001, L. Afonso, A. Hernández, Y. de Rioja & F. Tapia (TFC 44886); Por debajo de El Sombrero, hacia el Sur, en Los Retamares (41D 42 c), 29.04.2003, F. Mesa (TFC 44259); Al Oeste de La Hoya del Cedro (35A 7 b), 2211 m.s.n.m., 03.06.2003, Y. de Rioja Santos (TFC 44202); Pico Viejo (35B), 03/06/2003, Y. de Rioja Santos (TFC 44260).

3.5.88 Sideritis oroteneriffae *Negrín & P. Pérez*

Lagascalia, 15: 295–318 (1988)

Leucophae candicans Webb & Berth. Phyt. Can., 3(2/3): 100 (1845).
Sideritis cretica L. s. Mend.—Heuer. Ber. Scheiz. Bot. Ges., 84(4): 261–303 (1975 a).
Sideritis oroteneriffae var. arayae Négrin & P.Pérez in Lagascalia 15(Extra): 313 (1988)

Endemism, Certain native
Family: *Lamiaceae*
Common name: Chahorra, Chagorro o Chajorra de cumbre (ES); Summit iron-wort (EN)
Biotype: shrub (chamaephyte)
Flowering: June–August
Fruiting: July–September
Pollination: insect pollination (entomogamy)
Dispersal: short range self-dispersal (atelochory)
Chromosome N°: 2n = 40

Description

Chamaephyte up to 1 m tall, densely tomentose with a white-cinerate or white-yellowish hue. Leaves ovate or ovate-oblong, crenate or not, with obtuse apex and truncated, rounded, subcordate or more rarely cordate base, whitish to greyish upper surface, white-cinerate underside; the basal leaves more or less close (internodes 0.5–3.0 cm), 2.2–5.0 × 1.1–3.3 cm and petioles 1.3–3.8 cm. The first order para-cladial leaves 2.2–4.1 × 1.4–2.7 cm and petioles 1.8–3.5 cm; First order inhibition bracts are lanceolate and ovate-lanceolate, 1.5–4.6 × 0.6–2.2 cm and petioles are 0.5–2.6 cm (Perez de Paz and Negrín Sosa 1992).

Inflorescences with elongated rachis ending in a lax verticillaster (Fig. 3.274), rarely branched, in which case it usually has a single pair of branches close to the basal leaves, developing or not coverticillasters. Lanceolate bracts of 0.3–3.5 × 0.1–1.1 cm, the lower ones petiolate (petioles of 0.2–1.5 cm), the upper ones with an attenuated base. Filiform bracteoles of 0.7–4.0 × 0.2–0.3 mm. Pedicellate **flowers** (pedicels of 1.2–2.2 mm). Calyx 5.0–7.2 mm, tubular-campanulate, glabrous inside, densely tomentose outside; teeth 0.8–1.7 mm, with small mucro hidden by the tomentum. Corolla 7.0–9.2 mm, yellow with brown limb; tube dilated and barely swollen in the upper third; upper lip 1.2–2.0 mm, lower lip with median lobe 1.0–1.7 mm and lateral lobes 0.7–1.3 mm. Lower stamens with filaments 2.5–3.5 mm and anthers 0.5–0.8 mm; upper stamens with filaments 1.5–2.8 mm and anthers 0.7–1.2 mm. Gynoecium with style measuring 2.0–4.5 mm and stigma with unequal laciniae, the largest measuring 0.6–1.0 mm and the smallest 0.5–0.7 mm. Nutlets measuring 1.8–2.3 × 1.3–1.8 mm, with an elliptical or ovate section, dark brown or black, variegated with light brown.

Fig. 3.274 Teide National Park, June 2005. (Photo credit: Victoria Eugenia Martín Osorio)

Fig. 3.275 Endemism of the Teide National Park on the island of Tenerife

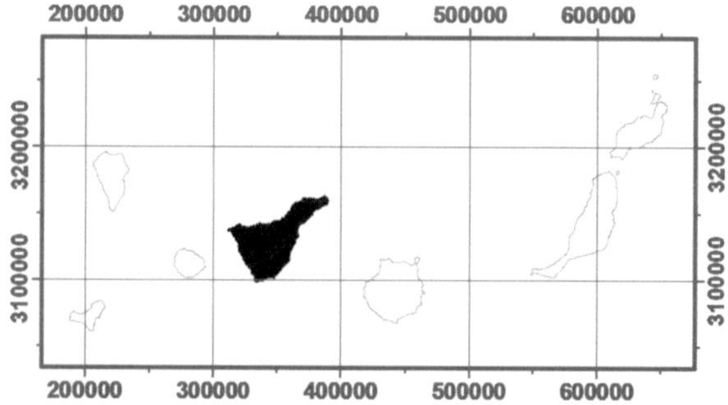

Remarks:

Type: ***Sideritis oroteneriffae*** *L. Negrín et P. Pérez. Habitat in fissuris rupium abruptarum, clives argilloso-glareosus soleatarum et ad pineto soleatis; monte dicto "Montaña Ayosa" circa 2000 m a.s.l. Legit cum flore et fructu die 19 Junii 1982. TFC 17239: holo (Plate 13: 1); ISO. in BM; MA; P.*

Distribution

Endemism of the island of Tenerife (Fig. 3.275), whose most notable populations are distributed in the mountainous escarpments of the NE Ridge from approximately 1300–1500 m a.s.l. (Pine forest domain), and up to 2000–2200 m a.s.l., already integrating into the brooms that border the Circo de Las Cañadas. According to the bibliography consulted, its distribution can be confused with that of ***Sideritis soluta*** Clos (southern peaks and high areas of the Güímar slope).

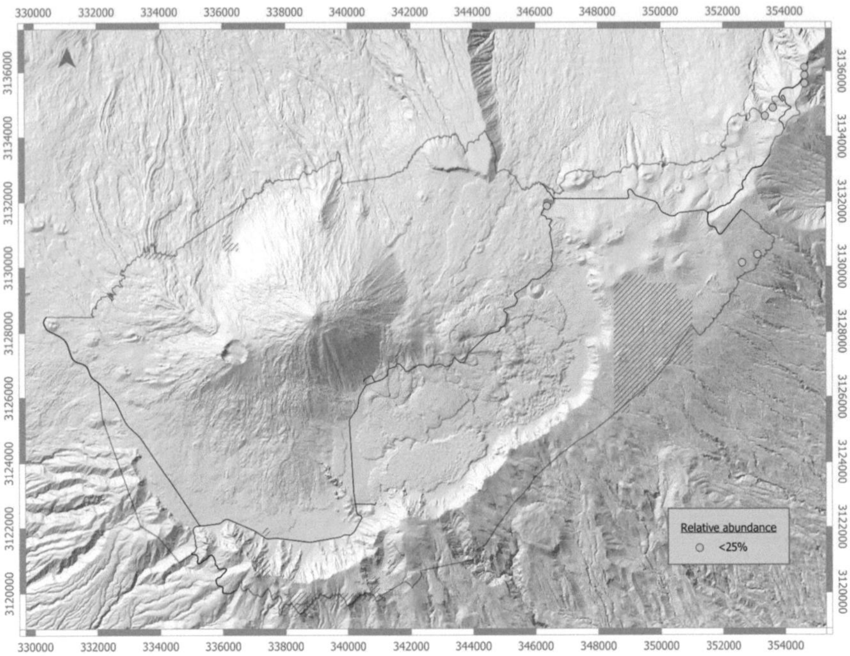

Fig. 3.276 *Sideritis oroteneriffae* distribution, N° Subgrids (250 × 250 m): **8**/3741

In the Teide National Park (Fig. 3.276), it only appears in the surrounding areas of La Crucita, La Tarta and the surrounding areas and southern slopes of Izaña (Altos de Güímar); also in the Teide Botanical Garden.

Habitat and Phytosociology

It is an essentially heliophilous and apparently xerophilous species, although it does not shy away from cliffs exposed to the trade winds on the northern slope of the Ridge. Within its range, it inhabits both rocky escarpments and rocky slopes and partially weathered lapillis of the Meso and Supra-Mediterranean floors. Its presence is very significant in cleared pine forests and broom shrublands (Perez de Paz and Negrín Sosa 1992).

It is a characteristic species and bioindicator of ***Chamaecytiso-Pinetalia canariensis***, an order that represents the natural pyrophilous forests of Canary Island pines. Also included are the microforests of Canary cedars in the midlands and summits, dry-subhumid Meso-Supra-Mediterranean. Within this, it is a typical species of ***Telinetum spachianae***, a dry Meso-Supra-Mediterranean, euoceanic, pluviseasonal, oceanic, Mediterranean, edaphoxerophilous series of the ***Teline stenopetala* subsp. *spachiana*** and ***Sideritis oroteneriffae*** summit shrublands with ***Bystropogon canariensis*** and ***Chamaecytisus proliferus* subsp. *angustifolius***.

Conservation Status

In all the subgrids in which this species was found, no damage of any kind was observed.

Threat Category

Anexo II: Protected (Gobierno de Canarias 1991)
LR pm (Bañares et al. 1998)

Germplasm Banks

E.T.S. Ingenieros Agrónomos de Madrid.
Jardín Botánico Canario Viera y Clavijo.

Exsiccata (Herbarium Specimens)

Tenerife: Hoya Fría-Arafo, 27.06.1971, E. Barquín Díez (TFC 28926); Arafo, 11.04.1973, P.L. Pérez & J.R. Acebes (TFC 4674); Altos de Igueste-Arafo, 1.06.1974, J.R. Acebes, B. Méndez & C. León (TFC 6777+ Dupl.); Fuente de Joco, Junio 1977, P.L. Pérez & C. Hernández (TFC 8386); Bco. de Afoña-Altos de Arafo, 950 m.s.n.m., 21.06.1979, I. La Serna, C. León & E. Beltrán (TFC 10054, 10055, 10056); Valle del Volcán de Las Arenas-Arafo, 1700 m.s.n.m., 9.05.1981, P.L. Pérez, W. Wildpret & M. del Arco (TFC 10057, 10058, 10059, 10060, 10061, 10062, 10063, 10064, 10065, 10066, 10067); Altos de la Caldera de Aguamansa-La Orotava, 1750 m.s.n.m., 23.05.1981, M. del Arco & W. Wildpret (TFC 17220); Los Órganos de La Orotava, 1450 m.s.n.m., 23.05.1981, ejusd. (TFC 17221); Ayosa, 19.06.1982, L. Negrín (TFC 17238, 17239); Hoya Larga-Ctra. dorsal, 19.06.1982, ejusd. (TFC 17240, 17241, 17242, 17243, 17244); El Gaitero-Ctra. Dorsal de La Esperanza, 20.06.1982, ejusd. (TFC 17252, 17253, 17254, 17255, 17256, 17257, 17258); Proximidades Fuente de Joco-Ctra. Dorsal de La Esperanza, 20.06.1982, ejusd. (TFC 17245, 17246 + Dupl., 17247, 17248 + Dupl., 17249 + Dupl., 17250 + Dupl., 17251 + Dupl.); Ctra. de la Cumbre-Entrada por la pista a los Montes de Fayal, Valle y Chafa-Monte público de Candelaria, 1600 m.s.n.m., 19.08.1982, W. Wildpret & M. del Arco (TFC 17370); Cumbres de Arafo, 1200 m.s.n.m., 3.04.1983, P.L. Pérez & C. Hernández (TFC 17329, 17330, 17331, 17332, 17333, 17334, 17402); La Ladera-Güímar-Canal, 1000 m.s.n.m., 3.04.1983, O. Rodríguez Delgado (TFC 28925); Bco. de Erques, 2.04.1984, ejusd. (TFC 28923 + Dupl., 28924 + Dupl.); Güímar-Canal, 1000 m.s.n.m., 4.05.1984, A. Charpin & O, Rodríguez (TFC 28922 + Dupl.); Mirador de Humboldt, 4.06.1984, P.L. Pérez de Paz & L. Negrín (TFC 17627 + Dupl.); Ayosa-Dorsal de La Esperanza, 17.07.1984, L. Negrín, S. Socorro & J.M. Mancebo (TFC 17744, 17745, 17746); Ladera de Güímar, 4.11.1984, P.L. Pérez de Paz & C. Hernández Padrón (TFC 17767); Altos del Escobonal-Güímar, 10.05.1985, P.L. Pérez de Paz & M. del Arco Aguilar (TFC 17736, 17737, 17738, 17739, 17740, 17741, 17742, 17743; Dupl.); Ayosa-Dorsal de La Esperanza, 20.06.1985, P.L. Pérez & I. La Serna (TFC 17747, 17748, 17749, 17750); Arafo, 1550 m.s.n.m., 11.07.1985, M. del Arco Aguilar & J.R. Acebes (TFC 17429 + Dupl., 17430 + Dupl.); Articosa, 12.05.1988, M. Morales (TFC 25250); Güímar, Arafo, 30.05.1988, ejusd. (TFC 25249); Pinar de la Victoria, 29.05.1998, E. Beltrán Tejera, C. León Arencibia & A. García Gallo (TFC 41475). Inventario de la Flora y Vegetación del P.N. del Teide (2001–2003) V.E. Martín Osorio, W. Wildpret & B. Hernández Bolaños: Alrededores de la Tarta, 29/08/2001, W. Wildpret de la Torre (TFC 45417 + dupl.).

Uses, Benefits and/or Medicinal Properties

> "*Anti-catarrhal, stimulant of the circulatory system. Useful parts: leaves, tops. Instructions for use: infusion, foot baths.*" (Translated from: Pérez de Paz and Hernández Padrón 1999)

3.5.89 Sideritis soluta *Clos ssp.* soluta

Ann. Sci. Nat. Bot., sér. IV, 16, 81 (1862)

Leucophae soluta Webb. mss., E. Bougeau n° 921 (P!); nom. *in* schaed (1846), nomen nudum.
Leucophae stricta Webb. mss., E. Bougeau n° 920 (P!); nom. *in* schaed (1846), nomen nudum.
Leucophae penzigii Pit. Pit. et Proust, Iles Can. Fl. Archip. p. 312 (1908).
*Sideritis penzigii (*Pit.) Bornm. *Fedde Rep.*, 19: 278 (1924).
Sideritis dendrochahorra v. *soluta* (Clos) Svent. *Coll. Bot.*, 7: 1152 (1968) p.p.
Leucophae dendrochahorra v. *soluta* (Webb *ex* Clos) Kunkel. *Mon. Biol. Can.*, 3: 65 (1972) p.p.
Sideritis cretica v. *stricta* (Webb) Mend.—Heuer. *Vieraea*, 3: 135 (1974).

Endemism, Certain native
Family: *Lamiaceae*
Common name: Chahorra o Chajorra sureña (ES); Southern ironwort (EN)
Biotype: shrub (chamaephyte)
Flowering: May–July
Fruiting: July–September
Pollination: insect pollination (entomogamy)
Dispersal: short range self-dispersal (atelochory)
Chromosome N°: 2n = 36

Description

Chamaephyte up to 1.5 m tall. Stems, rachis, leaf underside and petioles densely tomentose; tomentum yellowish-white to white-cinerate. Branches elongated and lax with long internodes (0.4–5.8 cm).

Fig. 3.277 Teide National Park, Aug. 2008. ((**a**) Photo credit: Beatriz Hernández); ((**b**) Photo credit: GG Casanova)

Leaves lanceolate (Fig. 3.277), more rarely ovate-lanceolate or ovate, crenate with obtuse apex sometimes acute; base rounded, truncate or cuneate, sometimes cordate or asymmetrical; upper surface greyish-green or yellowish-green, sometimes variegated, in mature leaves densely tomentose. Vegetative leaves 2.5–9.5 × 0.7–4.4 cm and petioles 1–7 cm. First order paracladials leaves 2.2–9 × 0.9–4.5 cm and petioles 1–4.7 cm; The second order paracladials measure 2.3–6.3 × 0.5–2.7 cm and petioles measure 0.8–2.3 cm; the second order inhibition measures 1.1–4.6 × 0.4–1.5 cm and petioles measure 0.4–2.3 cm.

Inflorescence with elongated rachis, with 1–3 pairs of paracladial stems that, in turn, can branch into secondary paracladials. Verticillastres with 1–2 distant basal whrols, the rest approximate, allowing the rachis to be seen. Bracts measure 0.4–2.9 × 0.1–0.8 cm, lanceolate, the basal ones petiolate (petioles 0.1–1.0 cm long), the upper ones attenuated at the base. Filiform bracteoles measure 0.7–5.0 × 0.1–0.3 mm. Flowers with 0.8–2.5 mm petioles. Calyx 5.3–7.7 mm tubular-campanulate, densely tomentose on the outside, glabrous on its inner surface with 0.6–1.6 mm toothed teeth. Corolla 7.2–9.5 mm, yellowish-white with brown limb; tube 6–8 mm, cylindrical at the base, widened and swollen at the top; triangular upper lip 0.7–1.5 mm and lower lip with rounded median lobe 0.8–1.5 mm and ovate lateral lobes 0.6–0.8 mm. Upper stamens with 1.7–2.5 mm filaments and 0.7–1.0 mm anthers; the lower ones with 2.3–3.5 mm filaments and 0.6–0.8 mm anthers. Gynoecium with style of (1.9) 2.3–4.5 mm and stigma with lacinia greater than 0.6–1.0 mm and lesser lacinia of 0.4–0.8 mm.

Nucules 1.8–2.3 × 1.3–1.7 mm with elliptical and ovate-elliptical section, dark brown or black, variegated light brown.

Remarks:

Type: *Leucophae soluta* Webb mss., Teneriffa Tamadaya prope pagum die 19 Junii 1846; E. Bourgeau, *Plantae Canarienses* n° 921 (P!; read; Mend.-Heuer, 1975 a). Lam. 13:4.

Distribution

Endemism of the island of Tenerife (Fig. 3.278), where it has a fairly wide distribution area: between 450 and 2500 m a.s.l.

In the Teide National Park (Fig. 3.279), it is mainly found in two sectors of the southern slope of the Pared de Las Cañadas: southwestern slopes of Topo de la Grieta and southwestern valley of Montaña Guajara (Barranco Eris). Occasionally, it was also found around the road from Vilaflor to Boca Tauce, Alto de Montaña Guajara, Risco Verde and on the southern slopes of Izaña (Altos de Güímar).

Remarks:

According to the bibliography consulted, the most characteristic populations live in the ravines of the central sector of the southern slopes of Tenerife (600–800 m a.s.l.); Those that grow towards the summits of the Pared de Las Cañadas, present more differences with respect to *Typus*.

Habitat and Phytosociology

It usually colonizes xerophilous and heliophilous environments, although it can also be found in more humid and shady ravine areas. It reaches its optimum in areas of potential domain of ***Mayteno canariensis-Juniperion canariensis*** and of the ***Cisto symphytifolii-Pinion canariensis*** alliance, where it appears as a characteristic species, and which groups closed climatophilous and edaphoxerophilous meso-macroforests. Very rarely, it can also be found as an integral part of ***Spartocytision supranubii***.

Conservation Status

In all the subgrids in which this species was found, no damage of any kind was observed.

Threat Category

Anexo II: Protected (Gobierno de Canarias 1991)
LR pm (Bañares et al. 1998)

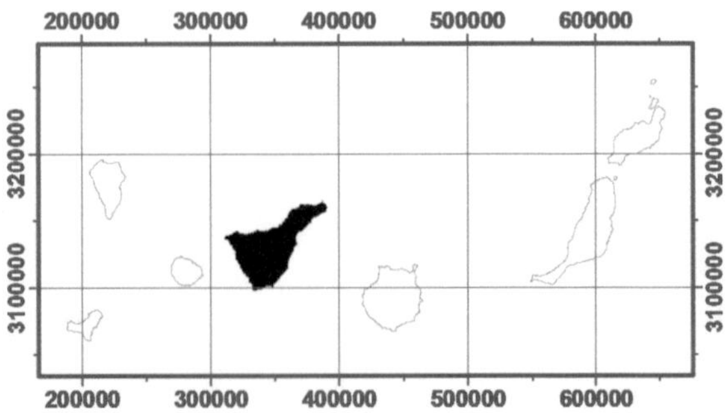

Fig. 3.278 Endemism of the island of Tenerife

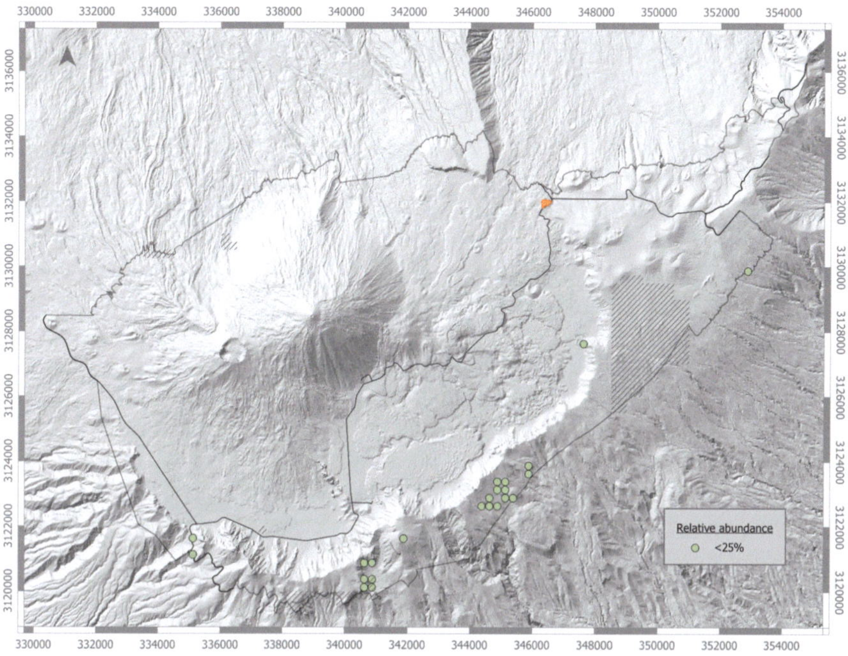

Fig. 3.279 *Sideritis soluta* subsp. *soluta* distribution, N° Subgrids (250 × 250 m): **23**/3741

Germplasm Banks

E.T.S. Ingenieros Agrónomos de Madrid.
Jardín Botánico Canario Viera y Clavijo.

Exsiccata (Herbarium Specimens)

Tenerife: Km. 63—Ctra. Vilaflor, 600–1800 m.s.n.m., 3 de Mayo, M. Fraga (TFC 28930 + Dupl.); El Sombrerito-Alrededor de Vilaflor, 2000 m.s.n.m., Abril 1969, D. Bramwell 1704 (TFC 409); Montaña de Diego Hernández, 2300 m.s.n.m., Mayo 1969, ejusd. 1698 (TFC 491); Bco. de la Orchilla, 24.04.1973, E. Beltrán (TFC 8444); Ladera de Güímar-Canal, 1000 m.s.n.m., 24.01.1975, E. Beltrán Tejera (TFC 2193); Bco. de Tamadaya-Canal-Arico, 16.07.1975, E. Beltrán Tejera, I. La Serna & M. del Arco Aguilar (TFC 8445, 8447, + Dupl.); Arico-La Sabinita, 500 m.s.n.m., 12.08.1977, E. Barquín (TFC 23369); Bco. del Río, 800 m.s.n.m., 16.05.1978, E. Beltrán (TFC 28928); Vilaflor-Ctra. a Las Cañadas, 1450 m.s.n.m., 13.05.1979, P. L. Pérez & C. Hernández (TFC 10056); Bco. de Badajoz-Ladera de Güímar, 16.04.1981, P.L. Pérez de Paz, W. Wildpret & M. del Arco Aguilar (TFC 10050 + Dupl., 10070 + Dupl.); Ladera de Güímar-Trayecto por el canal de los 1050–1100 m.s.n.m.-Cabeceras de los Cuatro Reales del Bco. de Badajoz, 16.04.1981, ejusd. (TFC 10068); Vilaflor, 24.05.1982, R. Lecuona (TFC 17104, 17105, 17106, 17107, 17108, 17109, 17110, 17111); Ctra. de Boca Tauce a Vilaflor, 12.06.1983, L. Negrín (TFC 17359, 17360, 17361, 17362, 17363, 17364, 17365, 17366, 17367, 17368); Vilaflor-Pista Madre del Agua, 15.05.1984, P.L. Pérez, I. La

Serna & L. Negrín (TFC 17582, 17583, 17584, 17585, 17586, 17587, 17588, 17589, 17590, 17591, 17592, 17593, 17594, 17595, 17596, 17597, 17599, 17600, 17601, 17602, 17603, 17604, 17605, 17606, 17607, 17608, 17609, 17610, 17611, 17612, 17613, 28931; Dupl.); Fasnia, 15.05.1984, P.L. Pérez de Paz, I. La Serna & L. Negrín (TFC 17605); Ibid., 450 m.s.n.m., 15.05.1984, ejusd. (TFC 10046 + Dupl.); Degollada de Guajara, 2400 m.s.n.m., 20.06.1984, P.L. Pérez & I. La Serna (TFC 28917, 28918, 28919); Arico-La Sabinita 600 m.s.n.m., 24.05.1985, P.L. Pérez & M. del Arco (TFC 17693, 17695, 17697, 17698, 17699, 17700); Tamadaya, 600–800 m.s.n.m., 24.05.1985, ejusd. (TFC 17728); Ibid., 600–800 m.s.n.m., 24.05.1985, P.L. Pérez de Paz (TFC 10041, 10042, 10043, 10044, 10045, 17728, 17729, 17730, 17731, 17732); Degollada de Guajara, 20.06.1985, P.L. Pérez, I La Serna & I. Medina (TFC 24751); Ladera de Güímar, 16.07.1985, P.L. Pérez de Paz et al. (TFC 17757, 17759); Ladera de Güímar, 16.07.1985, P.L. Pérez de Paz (TFC 10058); Laderas de Guajara-Las Cañadas, 2150 m.s.n.m., 25.04.1987, P.L. Pérez, W. Wildpret & E. Biondi (TFC 17779); Pista Vilaflor-El Río, 25.04.1990, M. Salas Pascual & M. del Arco Aguilar (TFC 37882); Tenerife, Tágara, 1600 m.s.n.m., 3.05.1990, M. del Arco Aguilar & M. Salas Pascual (TFC 38398 + Dupl.); Bcos. del Sur-Desde Chimiche hasta El Bueno, Bco. del Río-Arico, 6.06.1991, ADH, MMG & P.L. Pérez de Paz (TFC 35270). Inventario de la Flora y Vegetación del Parque Nacional del Teide (2001–2003) V.E. Martín Osorio, W. Wildpret de la Torre & B. Hernández Bolaños: Guajara, 19.05.2003, Y. de Rioja Santos (TFC 44235); Las Cañadas, 29.08.2001, W. Wildpret de la Torre (TFC 44891); [Como *Sideritis* cf. *soluta*] Guajara (42C 10 b), 17.06.2003, A. Montañés Rodríguez (TFC 44241); Alto de Guajara (42C 21), 02.06.2003, Y. de Rioja Santos (TFC 44261).

3.5.90 Silene berthelotiana *Webb*

ex Christ, *Bot. Jahrb.* Syst., 9, 100 (1887)

Endemism, Certain native
Family: *Caryophyllaceae*
Common name: Clavel salvaje o Canutillo de Berthelot (ES); Berthelot's Catchfly (EN)
Biotype: shrub (chamaephyte)
Flowering: May–June
Fruiting: June–August
Pollination: –
Dispersal: short range self-dispersal (atelochory)
Chromosome N°: –

Description

Chamaephyte with elongated pubescent stems, lignified from the base, erect or ascending.

Leaves ovate to ovate-lanceolate, puberulent, apiculate or acuminate towards the apex, soft, pubescent, with long petioles and ciliate margins; the highest ones lanceolate or linear.

Inflorescences with a dichasial cyme (below the stem apex, which ends in a flower (Fig. 3.280), two lateral branches also develop, which are also floriferous),

Fig. 3.280 Walls towards Vilaflor near Boca Tauce, June 2006 (**a**, **b**). (Photo credit: Victoria Eugenia Martín Osorio)

sparsely pubescent or subglabrous; linear bracts. Flowers erect or patent, in anthesis, racemose, lax, short, pedicellate (pedicels of 25 mm or more), pubescent, subnutant (pendulum). Calyx folded, oblong to oblong-ovate, 10–12 mm, truncated from the base, glandular-pubescent or puberulent, 10-veined; veins greenish-brown to brown, anastomosed; teeth small, 2 mm, triangular, acute, scarious, obtuse, erose margins. Petals exserted, pinkish-whitish, bifid, blades narrow obovate, with two oblong scales, bi-tridentate or bi-tri-fimbriate. Carpophore very short or almost absent. **Capsule** 8 × 3 mm, ovoid-oblong, with six erect teeth, almost triangular and acute.

Fruits pedicellate; erect pedicels.

Remarks:

Type: Bourgeau—Barranco de Añavingo, Arafo (Tenerife).

Distribution

Endemism of the island of Tenerife (from 2000 m above sea level) and of El Hierro (Fig. 3.281).

In the Teide National Park (Fig. 3.282) it extends mainly along the Pared de Las Cañadas (inner and outer slopes), where it can reach up to 2700 m above sea level, including the Roques de García. It is also found in the vicinity of Montaña Echisere, La Fortaleza, southern slopes of Izaña and around La Crucita.

Habitat and Phytosociology

It preferably colonizes rocky and stony environments. It is a characteristic species of ***Sileno berthelotianae-Cerastietum sventenii***, a rocky association of phonolitic cracks and spurs and trachybasalts oriented to the N and NW of the dry Supra-Mediterranean floor. It is found at altitudes between 2200 and 3000 m, but always in windward situations within the ***Spartocytisetea supranubii***, ***Spartocytision supranubii*** and ***Violion cheiranthifoliae*** communities. This community replaces ***Cheilantho guanchicae-Aeonietum smithii*** at altitudes above 2200 m a.s.l. It is also a characteristic species and bioindicator of the ***Greenovio aureae-Aeonietea*** class.

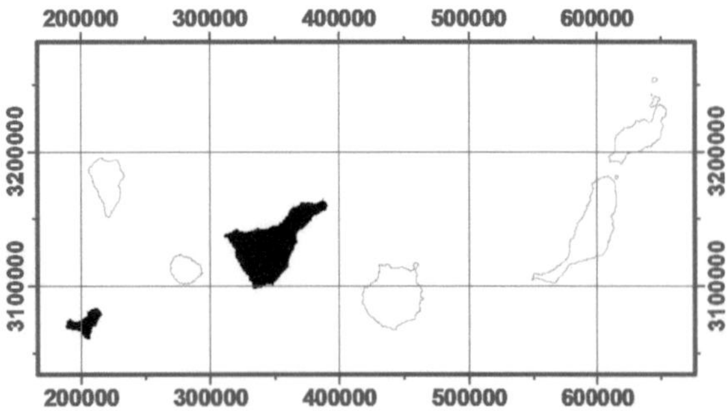

Fig. 3.281 Endemism of the island of Tenerife an El Hierro

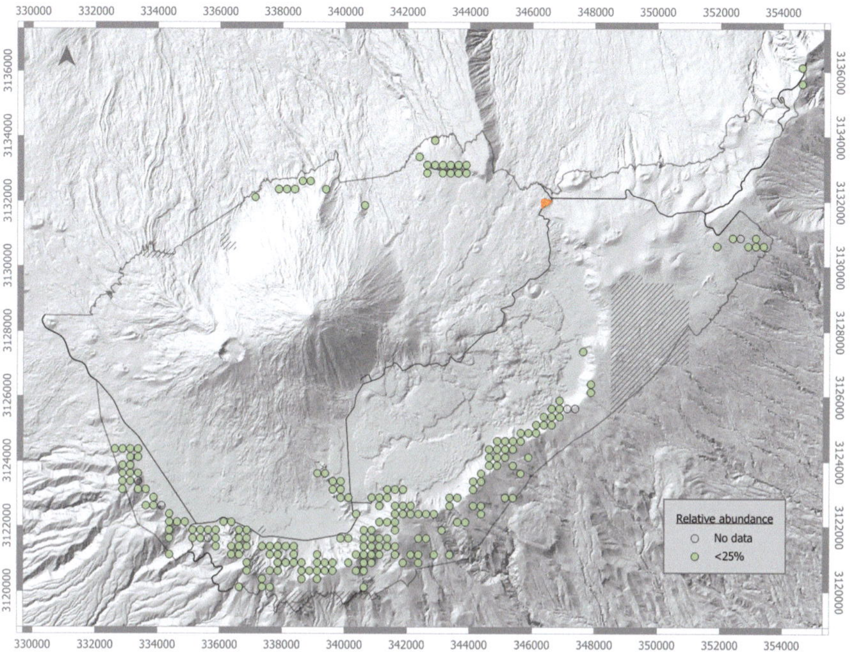

Fig. 3.282 *Silene berthelotiana* distribution, N° Subgrids (250 × 250 m): **218**/3741

Conservation Status

In 62% (136) of the subgrids in which this species was found, no damage of any kind was observed. In almost 30% (64), damage was detected (browsing, high mortality without regeneration, etc.), but this did not affect more than 25% of the specimens. Damage was found on more than 25% of the specimens in just over 6% (14) of the subgrids where this taxon occurred. In the rest (4), this variable was not specified at the time of sampling.

Threat Category

R (Barreno et al. 1984).
Anexo II: Protegidas (Gobierno de Canarias 1991)
VU (Bañares et al. 2004)
VU D2 (Moreno 2008)

Germplasm Banks

E.T.S. Ingenieros Agrónomos de Madrid.
Jardín Botánico Canario Viera y Clavijo.

Exsiccata (Herbarium Specimens)

Tenerife: Bco. de Añavingo-Arafo, 26.05.1973, P.L. Pérez de Paz (TFC 13138); Pista del Lomo Blanco-Acantilados umbrosos del Bco. de Vilaflor, 1600 m.s.n.m., 13.05.1979, P.L. Pérez de Paz & C. Hernández Padrón (TFC 13002 + Dupl.). Tágara-Aldea Blanca, 1700 m.s.n.m., 3.05.1990, M. del Arco Aguilar & M. Salas Pascual (TFC 38400). Inventario de la Flora y Vegetación del P.N. del Teide (2001–2003) V.E. Martín Osorio, W. Wildpret & B. Hernández Bolaños: Pared de Las Cañadas—SE del Llano Ucanca (41D 37), 22/04/2003, Y. de Rioja Santos (TFC 44883); Guajara, cerca de El Capricho (42C 10 c), 17/06/2003, A. Montañés Rodríguez (TFC 44239 + dupl.); Al SE de Guajara (42C 30), 23/06/2003, Y. de Rioja Santos (TFC 44256); Alrededores de La Crucita y La Tarta, 06/11/2004, V.E. Martín Osorio & W. Wildpret de la Torre (TFC 45412).

3.5.91 Silene nocteolens *Webb & Berthel.*

Hist. Nat. Iles Canaries Phyt. Canar., 3(2/1), 141 (1840)

Endemism, Certain native
Family: *Caryophyllaceae*
Common name: Hierba conejera, Clavel salvaje o Canutillo del Teide (ES); Teide Catchfly (EN)
Biotype: hemicryptophyte/geophyte
Flowering: May–June
Fruiting: July–August
Pollination: insect pollination (entomogamy)
Dispersal: wind and self-dispersal (anemo-balistochory)
Chromosome N°: 2n = 24

Description

Large-rooted, perennial, fusiform, whitish, fibrous **herb**. Abundant branching from the base. Large stem root tips, 15–30 cm, ascending, with a purplish-white base; leafy, hairy, white curls densely tomentose.

Leaves 24–30 mm long, 4–6 mm wide, ascending, lanceolate or linear-lanceolate, acute, tomentose.

Abundant **flowers**, white or slightly pink (Fig. 3.283), paniculate, panicle 2–3 parts, nutant, subsecundine (distributed unilaterally on an axis), finally erect, peduncle subviscous-tomentose. Calyx 2.5 cm, 4 mm wide, ovate-cylindrical, tomentose, pale, with 10 green or completely subpurpurescent striae, ovate-subescarious capsular appearance, calyx nails subelongated. Corolla whitish, petals almost laminar, deeply bifid, acute, linear, apex obtuse; the parts of the petals wide, trinervated, laminar parapetals with two individualized apices. Stamens ten, five external opposite to the petals with short aestivations, 5 alternating with the petals, long, filiform,

Fig. 3.283 Montaña Blanca, July 2006. (Photo credit: Victoria Eugenia Martín Osorio)

with long nails, ending in the thick gynophore cupula enclosed from the base; anthers inserted from the middle of the filament, the front ones separated, versatile, closely joined to the filaments, base sagittate; longitudinal dehiscence; thecae obtuse on both sides; spherical pollen. Trilocular ovary with obtuse apex and cylindrical sheaths clearly separated; gynophore hardly duplicated along its length. Ovule ovate, with short, tenuous funiculus and thick subapex. Three weak, filiform styles, with short filaments. Capsule of calyx and corolla dry, scarious, covered and ovate; provided with 6 narrow valves at the apex, acute, dehiscent, trilocular from base to apex; gynophore short in the lower part. Placenta central, columnar, connate to the apex of the capsule, biseriate funiculus, horizontal.

Seed blackish, rough-scrobiculate, cochleate, subreniform. Perisperm mealy with coalescent tegmen. Embryo campylotropic (seminal primordium that rotates arching). Radicle cylindrical, tenuous, obtuse, converse hilum. Cotyledons tenuous, obtuse with subduplicated radicles along them.

Distribution

Endemism of the island of Tenerife (Fig. 3.284), exclusive to the Teide National Park (Fig. 3.285), where it is only found in the central area of Las Cañadas, between 2275 and 3150 m of altitude: Montaña Blanca and Pico Viejo, the latter with greater genetic diversity (González González 2011); more punctually, on the northeastern

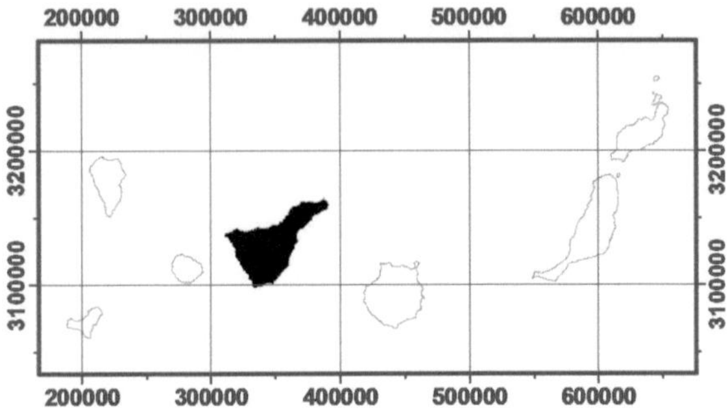

Fig. 3.284 Endemism of the Teide National Park on the island of Tenerife

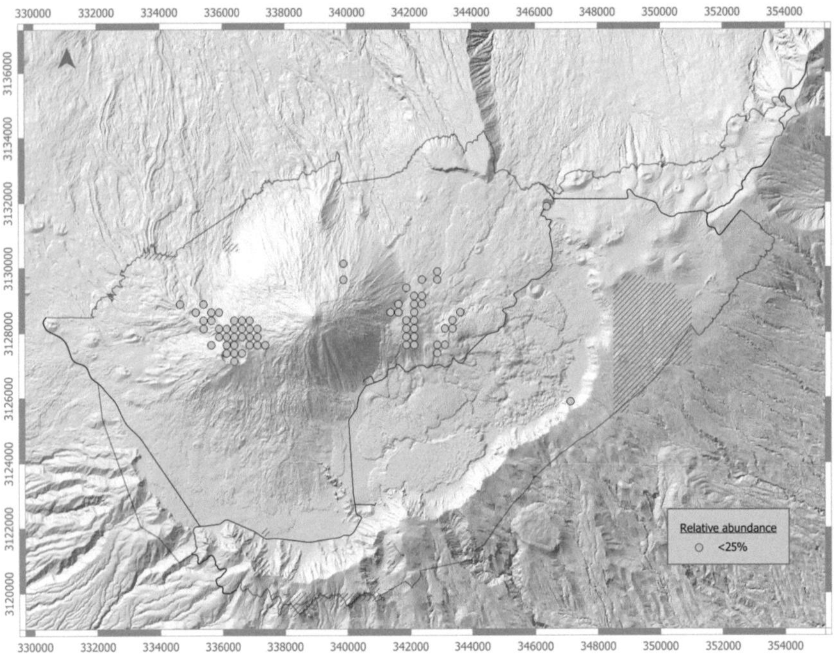

Fig. 3.285 *Silene nocteolens* distribution, N° Subgrids (250 × 250 m): **64**/3741

slopes of Teide. It was also found in Cañada de Las Pilas during the sampling period, although it is thought that it may be a doubtful record.

Habitat and Phytosociology

It lives exclusively on permeable soils, preferably on mobile substrates or pumice pyroclastic material. Characteristic species of the ***Violetum cheiranthifoliae*** association, formation of hemicryptophytes on pumice and glericola lapillis; Geopermaseries of the high summit of Tenerife Mediterranean Pluviseasonal Oceanic Oromediterranean Dry Euoceanic of the open glareose woody scrubs of ***Viola cheiranthifolia*** with ***Argyranthemum tenerifae***.

Conservation Status

In almost 94% (60) of the subgrids in which this species was found, no damage of any kind was observed. Only in almost 5% (3), damage was detected (browsing, high mortality without regeneration, etc.), but this did not affect more than 25% of the specimens. In the rest (1), this variable was not specified at the time of sampling.

Due to the scarcity of existing populations and the low number of specimens that make up the species, this species is in serious danger, which has increased in recent years due to the predatory effect of mouflons that destroy plants in search of their rhizomatous roots. It is poorly known in the scientific world, which is why it was given a much wider distribution and was previously considered out of danger (Marrero Gómez et al. 2003).

Traditionally, the incidence of introduced rabbits and mouflons has been cited as the main threat factor for this species, which may have led to the extinction of some localities cited in the bibliography such as La Rambleta or La Fortaleza. The populations are located close to busy paths, which translates into a clear risk of artificialization of the habitat. Its numbers are limited to two unique populations that contain less than 1000 specimens, and all of them are developed in Habitats of Community Interest and in the Teide National Park (LIC) where a recovery program is being developed that includes sending accessions to germplasm banks (Polytechnic University of Madrid, Viera y Clavijo Botanical Garden). In addition to these actions, it is recommended to reinforce surveillance in the sectors close to the most frequented areas and to continue with the tasks of population reinforcement and to address translocations to optimal sectors for its development (Marrero Gómez et al. 2004).

> "*Very rare in Montaña Blanca and completely extinct in La Fortaleza. Unique locality.*" (Sventenius 1946)

Threat Category

E (Barreno et al. 1984)
Anexo I: Strictly protected (Gobierno de Canarias 1991)
CR (UICN 1994)
CR B1 + 2c (Bañares et al. 1998)
E (Walter and Gillett 1998)
CR B1 + 2E (VVAA 2000)
Sensitive to habitat alteration (Gobierno de Canarias 2001)
CR B2 ab (iii) (Bañares et al. 2004)
CR B2 ab (iii) (Moreno 2008)
Vulnerable (Gobierno de Canarias 2010)
Vulnerable (Ministerio de Medio Ambiente, Medio Rural y Marino 2011)

Germplasm Banks

E.T.S. Ingenieros Agrónomos de Madrid.
Jardín Botánico Canario Viera y Clavijo.

Exsiccata (Herbarium Specimens)

Tenerife: 18.07.1969, Bramwell 1791 (TFC 512); Paredones de Guajara-Las Cañadas, 7.07.1977, SMG & LDldo (TFC 23229 + Dupl.); Cañada del Riachuelo-Las Cañadas, 13.06.1981, CRP (TFC 24502, 24503); Pared N de Guajara-Las Cañadas, 18.07.1982, T. Bravo (TFC 237, 13132); Valle de Ucanca, 2250 m.s.n.m., 7.05.1983, S. Socorro & W. Wildpret (TFC 18461); Pistas Madre del Agua-Vilaflor, 15.05.1984, P.L. Pérez de Paz, I. La Serna & L. Negrín (TFC 361, 13200 + Dupl.); Degollada de Guajara, 20.06.1985, P.L. Pérez de Paz, I. La Serna & IMM (TFC 24753 + Dupl.); Montaña Blanca-Las Cañadas del Teide, 12.06.1994, W. Wildpret de la Torre & V.E. Martín Osorio (TFC 38801); Ibid., 14.06.1995, ejusd. (TFC 41159). Inventario de la Flora y Vegetación del P.N. del Teide (2001–2003) V.E. Martín Osorio, W. Wildpret & B. Hernández Bolaños: Pico Viejo, al Este del cráter (35B 43 b), 16/06/2003, A. Montañés Rodríguez (TFC 44206).

3.5.92 Silene vulgaris *(Moench) Garcke ssp.* commutata *(Guss.) Hayek*

Repert. Spec. Nov. Regni Veg. Beih., 30(1),258 (1924).

Not endemic, Probable native
Family: *Caryophyllaceae*
Common name: Colleja, Conejera, Conejera rillabuey, Arrabuey, Jarrabuey, Rillabuey, Cascabelillo, Cucúbalo, Clavel salvaje (ES); Bladder campion, maiden-stears (EN)
Biotype: therophyte, hemicryptophyte
Flowering: (February) April–October
Fruiting: May
Pollination: insect pollination (entomogamy)
Dispersal: self-dispersal (ballistochory)
Chromosome N°: 2n = 24, 48; n = 12

Description

Plant 25–80 cm, glabrous, glaucous. Root axonomorphic, lignified, not tuberous.

Stems decumbent, medium thick, whitish or greenish in the lower half. **Leaves** up to 60 × 25 mm, herbaceous, obtuse; lower leaves broadly ovate, attenuate on short broad petiole, wavy-margined and usually ciliate-scapitate; mid-stem leaves ovate-elliptic, less than 3 times as long as wide, stems usually prostrate.

Inflorescence usually highly developed. Calyx 11–15 × c. 10 mm, veins sometimes very weakly marked. Petal blades 5–7 mm, broadly elliptic lobes (Fig. 3.286). Carpophore c. 2 mm. Capsule 9–11 × 6–7 mm, subequal to calyx, erect-toothed. **Seeds** tuberculate.

Fig. 3.286 *Silene vulgaris* Feb. 2024. (Photo credit: Victoria Eugenia Martín Osorio)

Remarks:

Logotypic index: "*In glareosis, et rupibus montosis; Busambra, Pizzuta, Favignana*".

Distribution

Native to the Mediterranean region, widespread in Eurasia, North Africa and Macaronesia. Present in all the Canary Islands (Fig. 3.287). ***Silene vulgaris*** ssp. ***commutata*** is considered an introduced species within the Teide National Park (Martín Osorio et al. 2006) (Fig. 3.288). Within the park, it was mainly distributed along the Pared de Las Cañadas, from Risco Verde to Cañada del Cedro, also including Roques de García; in areas of the Dorsal, it was also found in Montaña Los Mallorquines (NW-Izaña) and around Volcanes de Fasnia; between 1900 and 2700 m a.s.l.

Habitat and Phytosociology

The species usually colonises rock crevices and limestone scree, between 200 and 1700 m above sea level. It is characteristic of ***Stellarietea mediae*** communities, an anthropogenic annual nitrophilous vegetation type, consisting of therophytes or small geophytes of an arvense, ruderal, or scionitrophilous character, with a holarctic distribution and cosmopolitan distribution.

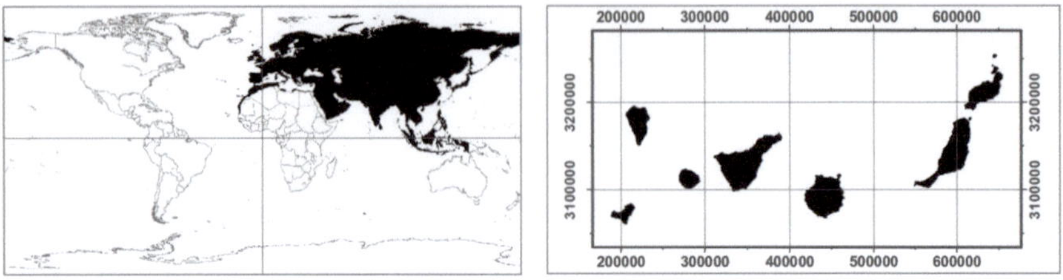

Fig. 3.287 Present on all the Canary Islands

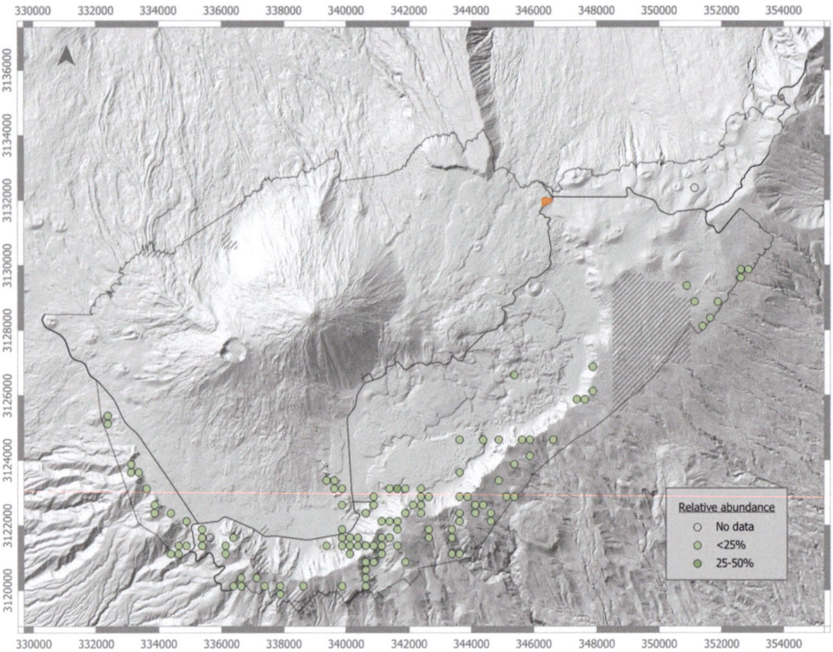

Fig. 3.288 *Silene vulgaris* subsp. *commutata* distribution, N° Subgrids (250 × 250 m): **115**/3740

Conservation Status

In practically all the subgrids in which this species was found (110), no damage of any kind was observed. Damage (browsing, high mortality without regeneration, etc.) was detected in only four (4) of them, but this did not reach more than 25% of the specimens. In the rest (1), this variable was not specified at the time of sampling.

Exsiccata (Herbarium Specimens)

Tenerife: Los Silos, Pinar, 7.8.1991, V. Lucía Sauquillo & C. Acosta (TFC 38161); Los Silos, La Caleta, 27.8.1994, V. Lucía Sauquillo (TFC 38179 + duplic.); Bco. Ijuana, Ladera S, 500 m.s.n.m., 18.3.1996, R. Racca (TFC 39112); Los Silos, Tierra del Trigo, 30.5.1996, V. Lucía Sauquillo (TFC 38063 + duplic.); Los Silos, Erjos, 16.3.1997, V. Lucía Sauquillo (TFC 40048 + duplic.); Las Lagunetas, 1190 m.s.n.m., 5.6.2003, P.L. Pérez de Paz, V. Lucia Sauquillo (TFC 44560); El Sauzal, Palmeral por encima de Ctra. TF-1224, cerca de galería Fuente de los Hinojos, 350 m.s.n.m., UTM 28RCS 592497, 30.3.2002, V. Lucía Sauquillo & M.L. Rodríguez Navarro (TFC 44659). Inventario de la Flora y Vegetación del P.N. del Teide (2001–2003) V.E. Martín Osorio, W. Wildpret & B. Hernández Bolaños: Al W de Izaña—Montaña de El Cabezón (31C 1 a), 24/04/2001, L. Afonso, A. Hernández, Y. de Rioja & F. Tapia (TFC 44916); Al SW de La Tarta (31A 4 a), 09/05/2001, L. Afonso, A. Hernández, Y. de Rioja & F. Tapia (TFC 44887); Por debajo de El Sombrero, hacia el Sur, en Los Retamares (41D 42 a), 29/04/2003, A. Montañés Rodríguez (TFC 44847).

Gran Canaria: ?, 16.4.1977, I. La Serna (TFC 16891); Agüimes, Bco. de Guayadeque, frente a Mña. de Las Tierras, 950 m.s.n.m., en matorrales umbríos, AHIM-CENTUR. VII (2002), 9.4.2002, A. Marrero (TFC 44126).

Lanzarote: Valle de Guiñate, 28.12.1993, J.A. Reyes-Betancort (TFC 38892); Montaña Blanca, Hoya de la Cruz, 31.3.1994, J.A. Reyes-Betancort (TFC duplic. 37370); Monte Guatisea, 26.12.1994, J.A. Reyes-Betancort (TFC 37683); Haría, Carretera de Los Valles, próximo a Peñas del Chache, 25.4.1996, J.A. Reyes-Betancort & C. García Corujo (TFC duplic. 38830).

Uses and/or Medicinal Properties

"*Ophthalmic, antipyretic, stomachic. Useful part: Whole plant. User manual: Infusion, juice, poultice.*" (Translated from: Pérez de Paz and Hernández Padrón 1999)

3.5.93 Spartocytisus supranubius *(L. f.) Christ*

First published in Bot. Jahrb. Syst. 9: 162 (1887)

Spartium supranubium Linn. fil. *Suppl.*, pág. 339.
Cytisus fragans Lam. *Encycl.*, vol. 2, pág. 248.
Spartium nubigena L'Hért. *Hort. Kew.*, ed. 1, vol. 3, fasc. 7, pág. 183, table 85. fig. quoàd habitum plantae bona, sed ramum cultum foliosum et pauciflorum exhibet.
Spartium nubigenum Willd. *Sp. pl.*, vol. 3, pág. 13; *Hort. Kew.*, ed. 2, vol. 4, pág. 256; Buch, *Phys. Beschr. Can. Ins.*, pág. 185.
Cytisus nubigenus (Aiton) Link. Enum. Hort. ber., vol. 2, pág. 240; DC., Prodr., vol. 2, pág. 153.
Spartocytisus supranubius (L.) Web & Berth. *Phyt. Canar.*, 3 (II/A), 50 (1836–1850)

Endemism, Certain native
Family: *Fabaceae*
Common name: Retama del Teide o de cumbre, Retama blanca (ES); Teide or Summit broom (EN)

Biotype: tall shrub (nanophanerophyte)
Flowering: May–June
Fruiting: May–July
Pollination: insect pollination (entomogamy)
Dispersal: self-dispersal (ballistochory)
Chromosome No: 2n = c.48

Distribution

A species with robust, erect, striated, cinereous branches.

Leaves trifoliolate, soon deciduous, with short, thick petioles, lanceolate leaves, densely tomentose.

Abundant **flowers** that appear before the leaves; short, hirsute calyx, rapidly deciduous; wings and vexillum carine-acute, sub-elongated. Calyx briefly campanulate, obsolete bilabiate, upper part convex, 2–3 mm long, with hairs in the central part, sub-narrow, hyaline, wide, shiny, acute; upper lip dark emarginate, the lower one subporrect, briefly tridentate; teeth acute, elongated in the centre, after anthesis the lower part circumcised deciduous. Corolla thick, petals white (Fig. 3.289), sometimes purple-white, entirely devoid of hairs or these are very scattered; vane ovate, subobtuse or dark emarginate, glabrous, very veined, margin sinuate, lamina elevated laterally curved backwards, claw narrow, arched, lower ancipital (flattened stems) part free; wing of vane longitudinal, carina subalongate, oblong-lanceolate or acinaciform, recurved, lower margin revolute, from claw to apex curled, apex rounded, anterior part of claw folded-swollen, the thin lower part of claw with a slightly adherent staminal sheath: carina ovate, pointed, recurved, the anterior part of claw long and folded, pointed auriculate. Stamens 10, monadelphous, elongated in early flowering, filaments elongated at the base, anthers oblong, apex rounded, subsagittal, in early flowering anthers ovate-oblong, obtuse on both sides, fixed to

Fig. 3.289 (**a**) Montaña del Limón, May 2005. (Photo credit: Beatriz Hernández) (**b**) Toward Vilaflor, June 2006. (Photo credit: Victoria Eugenia Martín Osorio)

the back. Ovary linear, complanate, apex extending beyond the style, base attenuate, sessile, multiovulate. Styles filiform, very incurved, apex devoid of hairs, or rarely covered with black papules. Stigma capitate, globuliform, papillate, papillae finally elongated, subinclined downwards.

Legumes 24–36 mm long, 6–8 mm wide, crowded, smooth or rarely hairy, shiny, blackened, temporarily obsolete, reticulate-veined, towards the fleshy seeds, upper and lower suture hardly thick, narrow apex, apiculate, with the lower part with short attenuated gynophore, inner valve blackened cinereous, shiny-sericaceous, subalongated striate, sometimes, towards the seeds, shiny hyaline radiated hairs slightly scattered. Seed ovate, shiny black, with short funiculus, thick caruncle, dilated, galeate, crenate margin, coriaceous tegmen. Embryo oblong-ovate, perispermophore chlamys hyaline, thin, involute. Cotyledons elongated-ovate, rounded apex, attenuate base, sessile. Radicle clavate, acute, incumbent.

Distribution

Endemism of the islands of La Palma and Tenerife (Fig. 3.290). The altitudinal distribution in the latter ranges between 1500 and 3350 m above sea level (SE slopes of Teide). The natural presence of broom on the island of La Palma is doubtful.

It is the taxon with the largest distribution area within the Teide National Park (Fig. 3.291); it extends almost continuously throughout the protected area, even outside its limits, except in some areas of Montaña Blanca, Montaña Rajada, Tabonal Negro, certain sectors of the slopes of Teide, Crater of Pico Viejo, Cerro las Chajoras and coladas de las Narices del Teide.

Habitat and Phytosociology

A characteristic species of the legume scrub of the high Canary mountains, where it lives with greater or lesser abundance on the various substrates characteristic of Las Cañadas del Teide (Tenerife) and La Caldera de Taburiente (La Palma).

It is involved in the formation of the highest altitudinal limits of ***Cisto symphytilolii-Pinion canariensis***, closed climatophilous and edaphoxerophilous meso-macroforests, well characterized by the Canary Island pine, but its main role is characterizing the ***Spartocytision supranubii*** alliance, which groups the mountain

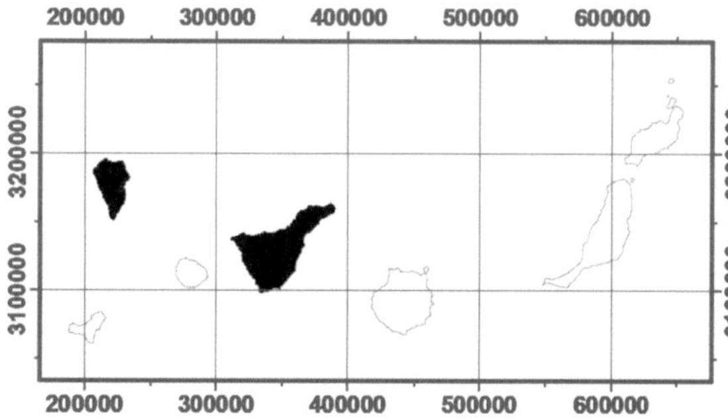

Fig. 3.290 Endemism of the island of Tenerife and La Palma

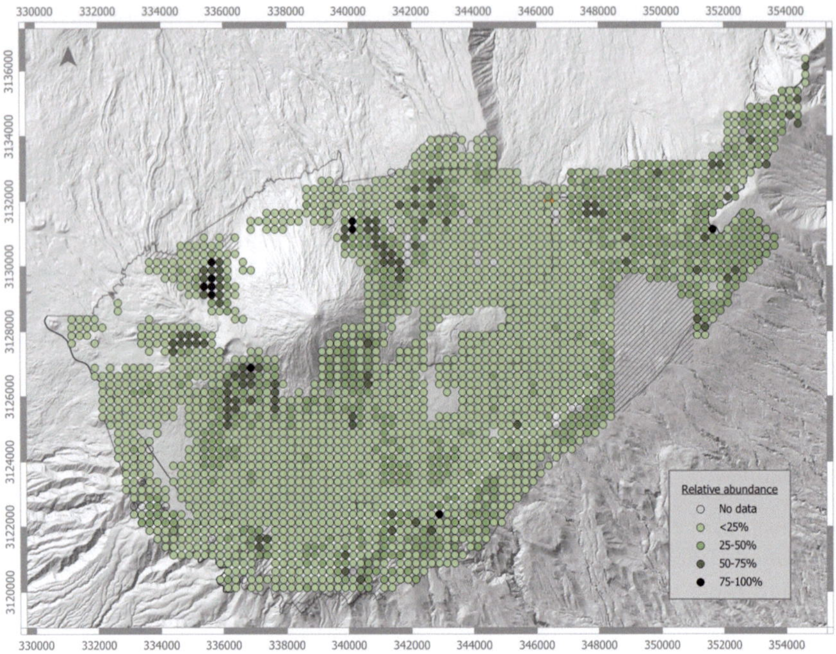

Fig. 3.291 *Spartocytisus supranubius* distribution, N° Subgrids (250 × 250 m): 2689/3741

summit shrublands of brooms, laburnums and summit sweet brooms (*Teline*), climatophilous or edaphoxerophilous, sometimes catenal or serial of the Canary Island pine forests; of distribution in La Palma and Tenerife, dry-humid Meso-Oro-Mediterranean.

It is involved in the following phytosociological communities:

In ***Juniperion cedri***:

Adenocarpo viscosi-Juniperetum cedri, head of the Tenerife dry-subhumid Supra-Mediterranean edaphoxerophilous series of Canary Island cedars.

In ***Spartocytision supranubii***:

Spartocytisetum supranubii, mature stage of the climatophilous series, Tenerife, (Meso)-Supra-Oro-Mediterranean dry-subhumid, of the high shrublands of *Spartocytisus supranubius* with *Adenocarpus viscosus*.

- ***typicum***.
- ***aeonietosum***
- ***echietosum wildpretii***
- ***descurainietosum lemsii***, introgression of the summit brooms into the dry upper Mesomediterranean floor, over recent lapillis and filling cones by the so-called "dorsal effect".
- ***descurainietosum gonzalesii***, dry upper Mesomediterranean subassociation of the brooms of the Precaldera wall or building on its southern slope with *Descurainia gonzalesii*.

- ***facies primocolonizer*** of recent badlands and lapillis, where the broom forms monospecific populations.

Descurainio bourgeauanae-Spartocytisetum supranubii, dry-subhumid Supra-Mediterranean floor, installed on fill materials from after the La Orotava landslide, formed by lava and pyroclastic cones from different eruptive episodes.

- ***typicum***.
- ***erigerontetosum calderae***, spread between the clayey shrinkage cracks originating in the temporarily flooded endorheic deposits.

In ***Plantaginion webbii***:

Arrhenathero calderae-Plantaginetum webbii
Cheirolopho teydis-Carlinetum xeranthemoides
Tolpido webbii-Micromerietum lachnophyllae
Arrhenathero calderae-Adenocarpetum viscosi

Conservation Status

No damage of any kind was observed in only 24% (648) of the subgrids in which this species was found. In 70% (1893), damage was detected (browsing, high mortality without regeneration, etc.), but this did not affect more than 25% of the specimens. Damage was found on more than 25% of the specimens in just under 5% (129) of the sub-grids where this taxon occurred. In the rest (19), this variable was not specified at the time of sampling.

Threat Category

nt (Barreno et al. 1984)
Anexo II: Protegidas (Gobierno de Canarias 1991)
LR pm (Bañares et al. 1998)

Germplasm Banks

E.T.S. Ingenieros Agrónomos de Madrid.
Jardín Botánico Canario Viera y Clavijo.
Jardín de Aclimatación de la Orotava.

Exsiccata (Herbarium Specimens)

Tenerife: 15.04.1969, Bramwell 1292 (TFC 530); 2.06.1969, ejusd. 1925 (TFC 531); Las Cañadas, 28.07.1971, F. Estévez & A. Acuña (TFC 21945); Ibid., 26.05.1973, I. La Serna (TFC 35610); Llano de Maja-Las Cañadas, 15.06.1975, J.R. Acebes, P.L. Pérez de Paz & M. del Arco Aguilar (TFC 4584); Las Cañadas, 2.06.1979, E. Beltrán Tejera (TFC 7958, 7959, 7960, 7961, 7962); Caldera de Pedro Gil-Arafo, 6.05.1981, M. del Arco Aguilar (TFC 24382 + Dupl., 24383 + Dupl.);

Valle del Volcán de las Arenas-Cumbres de Arafo, 9.05.1981, M. del Arco Aguilar, W. Wildpret de la Torre & P.L. Pérez de Paz (TFC 23731, 23752 + Dupl.); Mña. Mostaza-Las Cañadas, 27.06.1982, R. Lecuona (TFC 13123); Teide-Las Cañadas, 12.05.1984, A. Charpin & W. Wildpret de la Torre (TFC 20031); Degollada de Guajara, 20.06.1985, P.L. Pérez de Paz, IMM & I. La Serna (24758 + Dupl.); Alrededores del Corral del Niño-Las Cañadas, 21.05.1995, I. La Serna (TFC 831, 37851 + Dupl.). Inventario de la Flora y Vegetación del P.N. del Teide (2001–2003) V.E. Martín Osorio, W. Wildpret & B. Hernández Bolaños: Cerca del Refugio de los Montañeros (41D 19 a), 14/05/2003, A. Montañés Rodríguez (TFC 44841)

Uses, Exploitation and/or Medicinal Properties

Until it was declared a National Park in 1954, it was used as a fodder plant, as fertilizer in banana cultivation and as a highly valued element for lumberjacks and charcoal burners (natural fuel). From that date on, its use was limited solely to beekeeping and as an ornamental element, as well as decorative in the making of the flower carpets for the Corpus Christi festival in the city of La Orotava.

> "*Nutritious, melliferous; agricultural interest: fodder, livestock bedding; crafts. Useful part: nectar. Instructions for use: honey; firewood, charcoal, manure; crafts: sticks, spears.*" (Translated from: Pérez de Paz and Hernández Padrón 1999)

3.5.94 Spergula pentandra *L.*

Sp. Pl., 440 (1753)

Alsine pentandra (L.) Crantz *in* Inst. Rei Herb. 2: 408 (1766)
Spergula arvensis var. *pentandra* (L.) Wahlenb. *in* Fl. Suecica 1: 291 (1824)
Spergula vernalis Willd. *in* Fl. Berol. Prodr.: 158 (1787), nom. superfl.
Spergula vernalis var. *pentandra* (L.) Samp. *in* Man. Fl. Port.: 285 (1911), nom. illeg.
Spergularia pentandra (L.) G.Don *in* Gen. Hist. 1: 425 (1831)
Stellaria pentandra (L.) Scop. *in* Fl. Carniol., ed. 2, 1: 318 (1771)

Not endemic, Probable native
Family: Caryophyllaceae
Common name: Esparcilla o Esparcilla macho (ES); Five-anther spurrey (EN)
Biotype: therophyte
Flowering: February–May
Fruiting: February–May
Pollination: –
Dispersal: wind dispersal (anemochory)
Chromosome N°: 2n = 18

Description

Stems up to 40 cm, branched from the base, with erect, glabrous or very sparsely puberulent-glandular branches, especially in the inflorescence.

Leaves 5–40 × 0.1–1.2 mm, linear, not grooved on the underside.

Sepals (2–) 2.3–4.5 mm. White petals (Fig. 3.292), shorter or slightly longer than the sepals. Androecium with five stamens. Capsule 3–4.5 (−5) mm.

Seeds 1.6–2.2 mm (including the wing), with a black, non-papillous disk and a wing 0.4–0.7 mm wide, scarious-silver, translucent.

Distribution

Distributed in W, Central and S Europe, N Africa, SW Asia and the Macaronesian region. Present in the Canary Islands on the islands of El Hierro, La Palma, Tenerife and Gran Canaria (Fig. 3.293).

In the Teide National Park (Fig. 3.294) it is found mainly, but dispersed, in inland areas of the eastern Las Cañadas: between Montaña Mostaza and Montaña Rajada, Valle de las Piedras Arrancadas, Cañada Blanca, Cañada del Capricho and Cañada del Montón de Trigo; although it has also been located, outside the depression, on the southeastern slopes of the Topo de la Grieta and on the so-called skirts of the northern slopes of the Teide building; between 1700 and 2450 m a.s.l.

Fig. 3.292 *Spergula pentandra*, Oct. 2015. (Photo credit: Victoria Eugenia Martín Osorio)

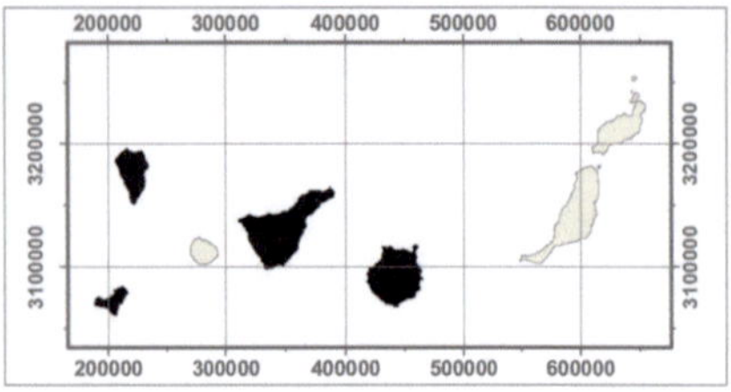

Fig. 3.293 Present in the Canary Islands on the islands of El Hierro, La Palma, Tenerife and Gran Canaria

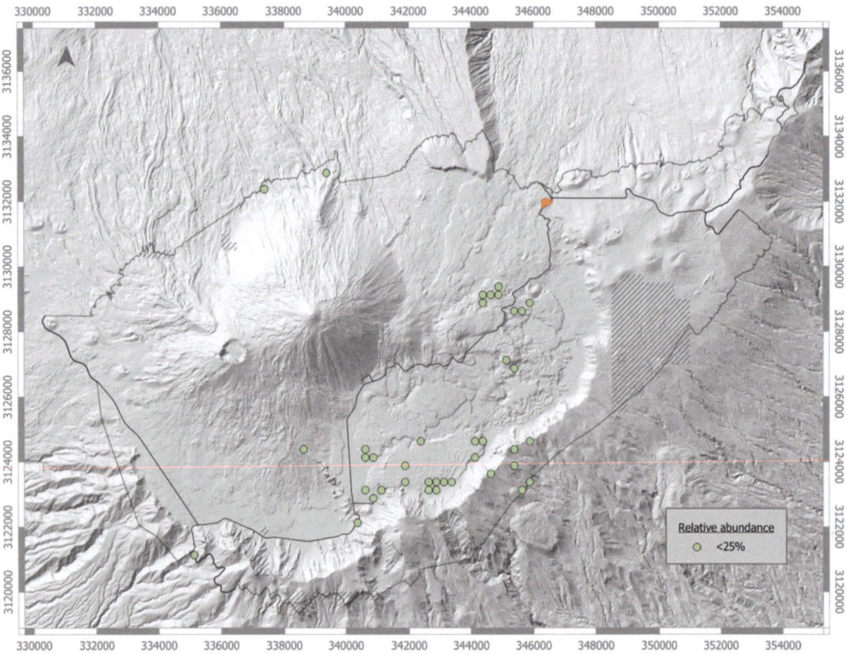

Fig. 3.294 *Spergula pentandra* distribution, N° Subgrids (250 × 250 m): **39**/3740

Habitat and Phytosociology

Very common as a ruderal. It is a characteristic species of the ***Scleranthion alliance***, of the ***Stellarietea mediae*** class, a type of anthropogenic annual nitrophilous vegetation, made up of therophytes or small geophytes of a weedy, ruderal, or scio-nitrophilous character, with a Holarctic distribution and cosmopolitan irradiation.

Conservation Status

In practically all the subgrids in which this species was found (37), no damage of any kind was observed. Only in two (2) of them, the state of conservation was low because the different specimens were completely dry.

3.5.95 Tolpis lagopoda *C. Sm.* in *Buch*

apud Buch., *Phys. Beschr. Can. Ins.*, pág. 147 et 184; DC., p. 7, p. 87. 1828

Aethonia lagopoda D.Don *in* Edinburgh New Philos. J. 6: 309 (1828–1829 publ. 1829)

Schmidtia lagopoda. Sch.Bip. *in* P.B.Webb & S.Berthelot, Hist. Nat. Iles Canaries 3(2; 3): 408 (1847)

Endemism, Certain native
Family: *Asteraceae*
Common name: Lechuguilla montana (ES); Mountain umbrela milkwort (EN)
Biotype: hemicryptophyte
Flowering: May–June
Fruiting: June–July
Pollination: insect pollination (entomogamy)
Dispersal: wind dispersal (anemochory)
Chromosome N°: n = 9; 2n = 18

Description

Hemicryptophyte with perennial, fusiform roots, 2–6 mm in average diameter, rounded, sublignous, dark on the outside, whitish on the inside. Solitary herbaceous stems, about 45 cm tall or more, with a terminal polycephalous corymb, lower part 1.25–2 mm in average diameter, thin leaves, straight below and also folded; incipient middle stem, ascending, protected by a whitish tomentum, with articulated, simple hairs, irregularly braided, evanescent composition; lower part of the stem with very elongated, dense and persistent hairs; adult stems glabrous, greenish, striated, whitish medulla.

Small and thin **leaves**; the lower ones in rosettes; those on the stem in a spiral line and arranged in branches, the 6 or 12 lower ones half as small, enlarging later, being smaller in the branches; elliptic-lanceolate, attenuate on both sides, acute, almost apiculate, rounded corpuscle, curved, sinuate-dentate; teeth 2–6 mm long, triangular-lanceolate, antrorse, acute, sometimes small and more or less flattened and curved, terminal corpuscle. Younger leaves with arachnoid tomentum, the adult ones glabrous, slightly intact and attenuated, the lower ones less tomentose. Lower branches on leaf axis, similar to stems, but less arranged, 21 cm, leaves 5–8 cm long, 5–12 mm wide, subintegrate, acute towards the subsimple corymb, slightly protected or naked; upper branches shorter on scale axis, narrowly linear, naked. Medium capitula multifloral, up to 33, always erect, pedicels naked, with linear, acute and protected scales, those of the central capitula shorter. Involucre 4.5 mm high, campanulate, subimbricate, with a tomentose peduncle at the top, inner scales narrowly linear and acute, the outer ones three times shorter, linear, dark-green dorsal, appressed, approximately eight in number. **Flowers** scarce, golden (Fig. 3.295), dry, somewhat greenish, radiant, involucre divided into parts that exceed them, radial 5 mm, central 3 mm long, pentadentate; triangular-lanceolate teeth; tube

Fig. 3.295 Detail of *Tolpis lagopoda* in its rock habitat November 2004. (Photo credit: Victoria Eugenia Martín Osorio)

2 mm long, short and simple hairs; anthers exserted, bifid styles of erect stylar branches, sometimes arched reflexed. Alveolate receptacle, alveolar margin not very high. **Achene** conformable, 1.5 mm long, columnar-turbinate, lower part of the apex attenuate, margin slightly prominent, 10-costate, glabrous, rough, black, crowned papo, monoserial; between 6 and 10 setae, denticulate and with a sucrase base.

Distribution

Endemism of the islands of La Palma, Tenerife and Gran Canaria (Fig. 3.296). On the island of Tenerife it has a wide distribution: from 800 m above sea level to 2500 m a.s.l., reaching La Pared de Las Cañadas (already within the National Park) (Fig. 3.297), where it appears in a dispersed form, such that it is present very punctually in the surroundings of Montaña del Cerrillar and Montaña de Arenas Negras, Risco Verde, Barranco del Río, Majada Vieja (S-Montaña Guajara), Andén del Gato (Ucanca), western slope of Roque del Almendro and Cañada del Cedro. Outside La Pared, it appears in the Corredor de La Corbata, Malpaís de Las Pilas and Cañada del Descanso de Las Gallinas (S-El Portillo Alto). It is most concentrated along the northwestern slope that goes from La Crucita to Hoya El Peñón, Llano la Rosa and Montaña Limón. It is also found in the Teide Botanical Garden.

Habitat and Phytosociology

It mainly inhabits rocky environments. It is a characteristic species and bioindicator of ***Greenovio aureae-Aeonietea***, a class of perennial vegetation that groups comophytic and chasmo-comophytic permanent perennial communities, endemic to the Canary-Madeiran biogeographic subregion.

Fig. 3.296 Endemism of the La Palma, Tenerife and Gran Canaria

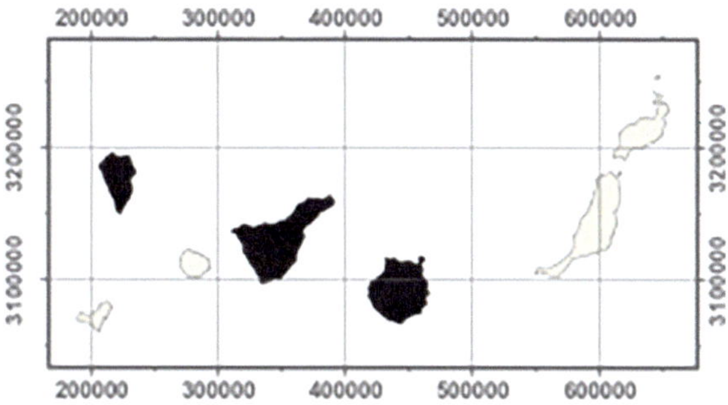

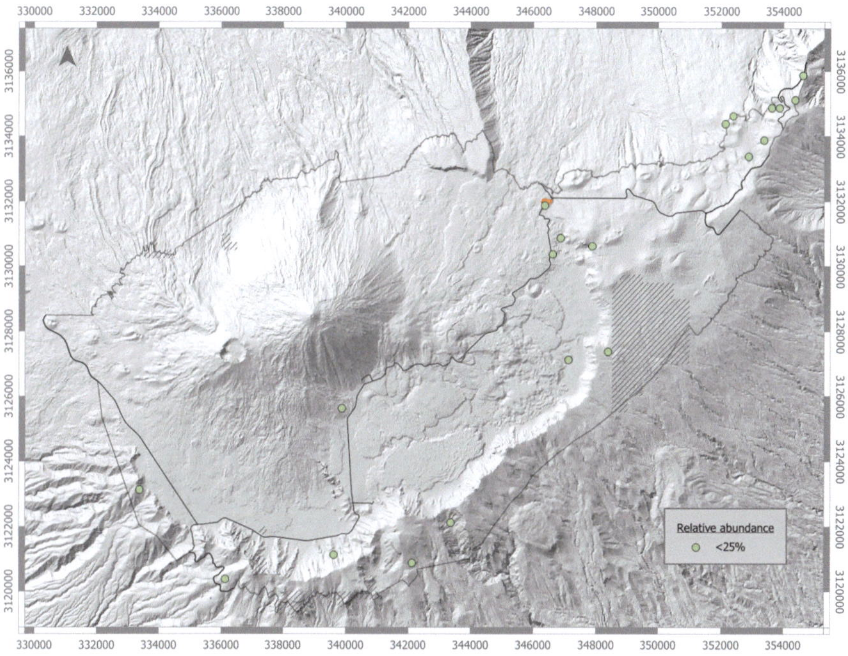

Fig. 3.297 *Tolpis lagopoda* distribution, N° Subgrids (250 × 250 m): **20**/3741

Conservation Status

In all the subgrids in which this species was found, no damage of any kind was observed.

Threat Category

nt (Barreno et al. 1984)
LR ca (Bañares et al. 1998)

Germplasm Banks

E.T.S. Ingenieros Agrónomos de Madrid.
Jardín Botánico Canario Viera y Clavijo.

Exsiccata (Herbarium Specimens)

Tenerife: 20.07.1969, Bramwell 1782 (TFC 553); Octubre 1972, Bramwell (TFC 4361); El Bailadero, 17.10.1984, O. Rodríguez Delgado (TFC 27733); La Ladera-Bajo Mña. Arguazo, 5.11.1984, ejusd. (TFC 27732); Bco. de La Palma-Faneque, 29.07.1989, C.S. Rodríguez (TFC 31793); Pista ppal. Monte del Agua-Los Silos, 22.08.1994, V. Lucía Sauquillo, C. Acosta & J.A. Reyes-Betancort (TFC 40592 + Dupl.); Ibid., 4.06.1995, V. Lucía Sauquillo, M. del Arco Aguilar & J.A. Reyes Betancort (TFC 40591 + Dupl.); La Juncia-Los Silos, 30.05.1996, ejusd. (TFC 40590 + Dupl.); Bco. del Infierno, 29.07.1996, M. León Arencibia & A. García Gallo (TFC 39234 + Dupl.). Inventario de la Flora y Vegetación del P.N. del Teide (2001–2003) V.E. Martín Osorio, W. Wildpret & B. Hernández Bolaños: Alrededores de La Crucita y La Tarta, 06/11/2004, V.E. Martín Osorio & W. Wildpret de la Torre (TFC 45414).

Gran Canaria: Bco. Los Mocanes-Valsequillo, 12.10.1985, C. Suárez & A. Marrero (TFC 31650, 31651, 31653); Ibid., 12.10.1985, C. Suárez (TFC 31652); Bco. del Agua-Base del Roque Sauquillo-Valsequillo, 18.10.1985, ejusd. (TFC 31646, 31647, 31648); Tamadaba, 27.10.1985, ejusd. (TFC 31656, 31658); Bco. del Agua-Valsequillo, 18.06.1986, ejusd. (TFC 31610); Mña. Las Brujas, 1100 m.s.n.m., 15.07.1993, M. Salas Pascual (TFC 37003).

3.5.96 Tolpis webbii *Sch. Bip.* ex *Webb & Berthel.*

In Webb & Berth., Phyt. Canar., 2, 409 (1836–1850). 1849

Schmidtia webbii C.H. Schultz Bip. *in* litt. ad amiciss. Webb aliosque.

Endemism, Certain native
Family: *Asteraceae*
Common name: Lechuguilla de cumbre (ES); Summit umbrela milkwort (EN)
Biotype: hemicryptophyte
Flowering: May–June
Fruiting: June–August
Pollination: insect pollination (entomogamy)
Dispersal: wind dispersal (anemochory)
Chromosome N°: n = 9; 2n = 18

Description

Hemicryptophyte with perennial, fusiform roots, more than 4–6 mm in average diameter, rounded, brown on the outside, whitish on the inside, bitter, with a milky juice that coagulates into a brown mass when it coagulates.

Stems erect, solitary, graceful, rush-like, exceeding 1–2 mm in average diameter, rounded, striated, green, sometimes with a subpurpurescent base, foliate, usually branched from the ground, glabrous, with hairs on the lower part that are poorly protected, short and patent.

Radical **leaves** rosulate (withered in the fruiting plant), numerous, on the upper part in a circumference of 12–16 cm wide, elongated-linear-lanceolate, pinnatifid; Leaflets not equal, slightly twisted, linear-lanceolate, 6 cm long, 12 mm wide, acute, sometimes very toothed; triangular-lanceolate teeth, sometimes more or less flattened and curved; petiole narrow and winged at the bottom; apex of acute teeth and margin of corpuscle whitish, on both sides short, simple, articulated, whitish hairs, rarely hirto-pubescent. Cauline leaves elongate-linear, entire, acute, glabrous, the lower ones rarely toothed. The lower branches are similar to the smaller stems (about 37 cm long), foliate, apex with simple corymb, the leaves of the axis elongate-linear; the upper branches are shorter (about 21 cm), leaves of the axis 2–10 mm long, narrow, linear-subulate. Medium capitula, with 50–60 flowers, always erect, thin pedicels, upper part with some linear-subulate and erect scales and with a smaller, whitish, tomentose-pubescent involucre. Involucre campanulate-hemispherical, 4–4.5 mm high, reflexed, biserial, with 13–21 subequal, narrow, linear and acute inner scales, with a convex back; on the outer part 8–10 narrow, linear-lanceolate, acute, tight scales, less short than the inner ones. Receptacle 1.5 mm in average diameter, convex, foveolate; foveole with short margin, on both sides with slightly raised teeth.

Inflorescence in loose corymb. Yellowish flowers (Fig. 3.298), when dry greenish, radial, pentadentate; 2 mm long, whitish tube, short hairs patent, simple, articulated, protected; ligule on the back; anthers exserted, style with short, erect, slightly hirst branches. Achenes conformable, 1.5 mm long, columnar-turbinate, pentagonal, rounded rib, not striated, glabrous, black, with five separate, narrow and glabrescent valleculate. Persistent whitish monoserial stem, larger setae with 13–16 teeth, lower part fleshy.

Distribution

Endemism exclusive to the island of Tenerife (Fig. 3.299). In the Teide National Park (Fig. 3.300) it is widespread inland and at the Pared de Las Cañadas, reaching altitudes of 3500 m above sea level (base of the Pilón de Azúcar); more dispersed or non-existent in the highlands of the areas most related to the Dorsal, slopes and lava flows of Montaña Blanca and the Teide-Pico Viejo complex, more western areas of the Park (lava of the Narices del Teide and surroundings of Montaña de Samara), Ucanca plain, Cañada Blanca and Tabonal Negro.

Fig. 3.298 (**a**) Surroundings of Black Sands Volcano, June 2006. (**b**) *Guajara*, June 2003. (Photo credit: Victoria Eugenia Martín Osorio)

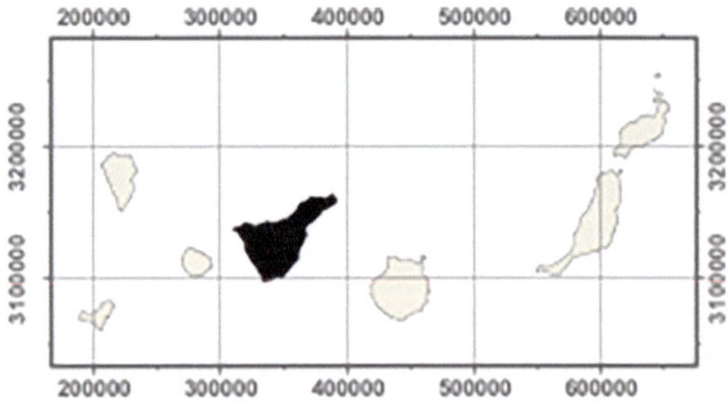

Fig. 3.299 Endemism of the island of Tenerife

Habitat and Phytosociology

It lives in stony and rocky environments. It is present in the communities of the ***Spartocytision supranubii*** alliance, although it is more closely related to the ***Pterocephalenion lasiospermi*** suballiance, which groups permanent and serial dwarf shrubland associations that colonize volcanic deposits and denuded soils of pine forests and broom shrublands, dry-subhumid Meso-Supra-Mediterranean, endemic to the island of Tenerife.

In addition, it characterizes or is present in the following units:

Adenocarpo viscosi-Chamaecytisetum angustifolii
Violo cheiranthifoliae-Adenocarpetum viscosii
Arrhenathero calderae-Plantaginetum webbii
Cheirolopho teydis-Carlinetum xeranthemoides
Tolpido webbii-Micromerietum lachnophyllae
Arrhenathero calderae-Adenocarpetum viscosi

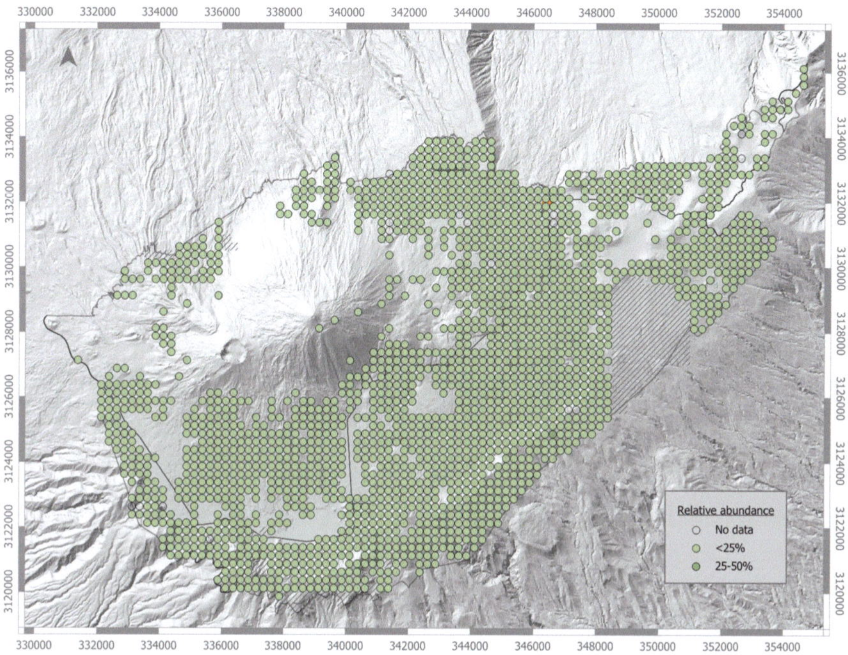

Fig. 3.300 *Tolpis webbii* distribution, N° Subgrids (250 × 250 m): **2074**/3741

Conservation Status

In 57% (1176) of the subgrids in which this species was found, no damage of any kind was observed. In almost 42% (865), damage was detected (browsing, high mortality without regeneration, etc.), but this did not affect more than 25% of the specimens. Damage was found on more than 25% of the specimens in just under 1% (19) of the sub-grids where this taxon occurred. In the rest (14), this variable was not specified at the time of sampling.

Threat Category

nt (Barreno et al. 1984)
LR pm (Bañares et al. 1998)

Germplasm Banks

E.T.S. Ingenieros Agrónomos de Madrid.
Jardín Botánico Canario Viera y Clavijo.

Exsiccata (Herbarium Specimens)

Tenerife: 1.08.1969, Bramwell 2129 (TFC 556); Pista de Lomo Blanco-Vilaflor, 1600 m.s.n.m., 13.05.1979, P.L. Pérez de Paz & C. Hernández Padrón (TFC 30413 + Dupl.); Cañada de Diego Hernández, 20.06.1985, P.L. Pérez de Paz, IMM & I. La Serna (TFC 24755 + Dupl.); Pico de las Flores-Fuente de Joco-Ctra. de Las Cañadas, 13.06.1987, P.L. Pérez de Paz (TFC 25394); El Montañón-Cerca de El Portillo-Las Cañadas, 26.06.1990, P.L. Pérez de Paz & D. Pérez (TFC 32597 + Dupl.); El Portillo-Las Cañadas, 13.07.1994, P.L. Pérez de Paz & DPH (TFC 36783 + Dupl.); Boca Tauce al Portillo-Las Cañadas, 2.08.1998, P.L. Pérez de Paz, C. Hernández Padrón & DPH (TFC 41628 + Dupl., Test muestra fitoquímica); Inventario de la Flora y vegetación del Parque Nacional del Teide (2001–2003) V.E. Martín Osorio, W. Wildpret de la Torre & B. Hernández Bolaños: Cañada del Capricho, 02.03.2003, F. Mesa, A. Montañés & Y. de Rioja (TFC 44871); Por debajo de El Sombrero, hacia el Sur, en Los Retamares (41D 42 a), 29.04.2003, A. Montañés Rodríguez (TFC 44851); Alrededores de la Tarta (25C 4 c), 03.04.2001, L. Afonso, A. Hernández, Y. de Rioja & F. Tapia (TFC 44964).

3.5.97 Tragopogon porrifolius *L.*

Sp. Pl., 789 (1753)

Tragopogon porrifulis ssp. *australis* sensu Richarson. *Fl. Eur.*, 4: 323 (1976), non (Jordan) Nyman (1879).
Tragopogon sinuatus Ave-Lall.

Not endemic, Probable native
Family: Asteraceae
Common name: Barbón (ES); Common salsify, goatsbeard (EN)
Biotype: therophyte
Flowering: April–May
Fruiting: April–May
Pollination: insect pollination (entomogamy)
Dispersal: wind dispersal (anemochory)
Chromosome N°: 2n = 12

Description

Annual or biennial, with thick fusiform root. Stems (15–) 30–60 cm, generally branched. Linear-lanceolate **leaves**, widened at the base, semi-amplexicaul, glabrous; the cauline leaves 15–30 mm wide at the base. Peduncles markedly thickened at the apical part. Involucre with 8–10 bracts. **Involucral** bracts 2 times longer than the ligules. **Ligules** purple (Fig. 3.301). Style arms yellow. Achenes 26–50 mm, longitudinally muricate, attenuated at a 10–36 mm peak, cylindrical and slightly widened at the top. **Pappus** 32–35 mm, feathery.

Fig. 3.301 *Tragopogon porrifolius*, March. 2018. (Photo credit: Victoria Eugenia Martín Osorio)

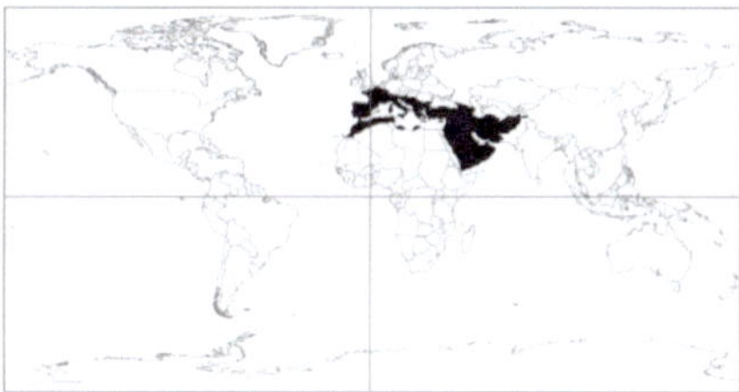

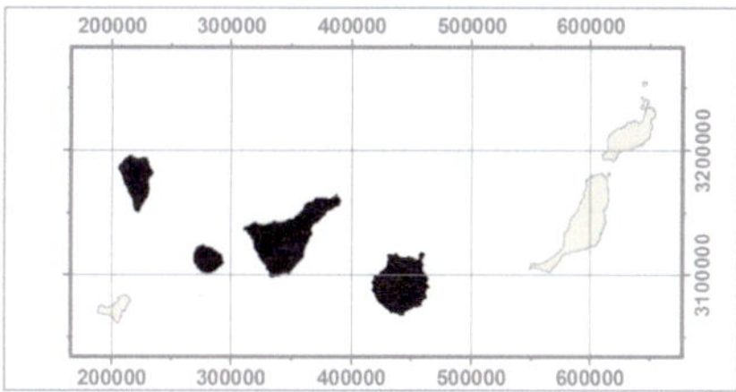

Fig. 3.302 Present on the islands of La Palma, La Gomera, Tenerife and Gran Canaria

Distribution

It is found in the Macaronesian Region and more generally in southern Europe, northern Africa, and southwestern Asia. In the Canary Islands it is only present on the islands of La Palma, La Gomera, Tenerife, and Gran Canaria (Fig. 3.302).

In the Teide National Park (Fig. 3.303) it was found in a very dispersed manner, mostly in areas of the Ridge: La Crucita and the northern and southern slopes of Izaña. Also occasionally in La Fortaleza, Portillo Bajo, Portillo Alto, Cañada del Arrife del Codeso, and Montaña del Palo; between 2000 and 2300 m above sea level.

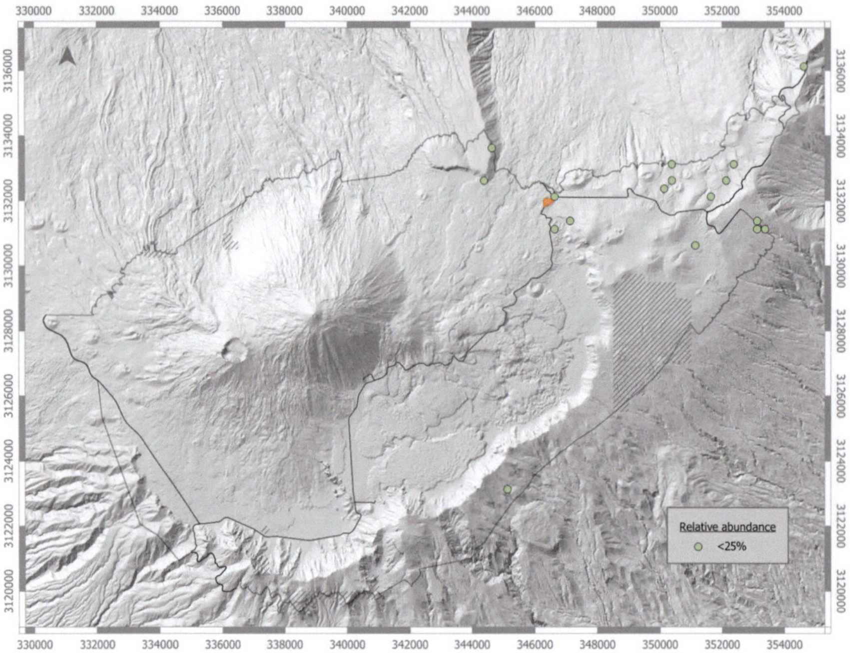

Fig. 3.303 *Tragopogon porrifolius* distribution, N° Subgrids (250 × 250 m): **17**/3740

Habitat and Phytosociology

It colonizes grasslands on basic soils, generally in the range between 200 and 1300 m above sea level.

It is a characteristic species of ***Brachypodietalia phoenicoidis***, of the ***Festuco-Brometea*** class, typical of nutrient-rich and well-drained anthropogenic grasslands, covered by perennial herbs; subhumid-humid oceanic Meso-Supratemperate and non-steppe sub-Mediterranean.

Conservation Status

In all the subgrids in which this species was found (17), no damage of any kind was observed.

Exsiccata (Herbarium Specimens)

Tenerife: W. Wildpret, 4.1969 (TFC 559); 4.1970, J.R. Acebes Ginovés, (TFC 560); W. Wildpret, L. Gallo & A. Santos (TFC 625); Icod, Cueva de las Hiedras, 12.9.1988, F. Ardévol (TFC 19222); La Ladera, junto a camino Real, 19.3.1984, O. Rodríguez Delgado (TFC 27709); Bco. del Río, 15.5.1983, J.S. Socorro (TFC 18458); La Montañeta, 25.2.1986, O. Rodríguez Delgado (TFC duplic. 27675); Bco de Tahodio, Cerca del embalse de Tahodio, 1.4.1999, S. Becker & T Himstedt (TFC 42098); Las Lagunetas, UTM 28RCS 6245, 2.6.2001, M. Rodríguez López &

M.L. Rodríguez Navarro (TFC 44644); Las Lagunetas, Mña. Cabeza de Toro, 1450 m.s.n.m., 362/3144, 12.2.2002, M. Rodríguez López & M.L. Rodríguez Navarro (TFC 44618). Inventario de la Flora y Vegetación del P.N. del Teide (2001–2003) V.E. Martín Osorio, W. Wildpret & B. Hernández Bolaños: Jardín Botánico (P.N. Teide) (I34—30D 2 c), 22/06/2004, B. Hernández Bolaños (TFC 45394).

Gran Canaria: Parque Natural de Bco. de Guayadeque, cauce del Bco. a la altura de Cueva Labrada, 700 m.s.n.m., 3.3.1990, F. Amor (TFC 30829).

3.5.98 Umbilicus gaditanus *Boiss.*

Diagn. Pl. Orient., ser.1, 6, 58 (1846)

Cotyledon gaditana (Boiss.) Pau *in* Visita Bot. Riff 2: vi (1911)
Cotyledon umbilicus-veneris subsp. *gaditana* (Boiss.) Emb. & Maire *in* É.Jahandiez & al., Cat. Pl. Maroc 4: 1019 (1941)
Cotyledon umbilicus-veneris var. *gaditana* (Boiss.) Batt. *in* J.A.Battandier & L.C.Trabut, Fl. Algérie, Dicot.: 329 (1889)
Umbilicus horizontalis subsp. *gaditanus* (Boiss.) Losa & Rivas Goday *in* Arch. Inst. Aclim. Cons. Super. Invest. Ci. 13: 232 (1974)
Umbilicus rupestris subsp. *gaditanus* (Boiss.) O.Bolòs, X.Font & X.Pons *in* Atlas Corol. Fl. Vasc. Països Catalans 4: 583 (1994)
Umbilicus vulgaris subsp. *gaditanus* (Boiss.) Batt. & Trab. *in* Fl. Algérie Tunisie: 133 (1905)

Not endemic, Probable native
Family: *Crassulaceae*
New record for Teide National Park
Common name: Papa, Sombrerillo, Ombliguillo, Ombligo de Venus (ES); Venus navelwort (EN)
Biotype: geophyte
Flowering: April–June
Fruiting: April–June
Pollination: wind pollination (anemogamy)
Dispersal: wind dispersal (anemochory)
Chromosome N°: 2n = 24

Description

Perennial plant (5) 13–70 cm, glabrous. Subspherical stump, with numerous fibrous roots. Stem erect, generally simple, rarely branched.

Fleshy **leaves**; basal ones peltate (Fig. 3.304), with petiole (1.5) 2–7 cm, eccentric or sulateral, lamina 1.5–4.5 cm, subreniform, crenate, slightly concave; the caulines progressively decreasing towards the upper part and varying from subspatulate to lanceolate (never linear), with marked segments.

Fig. 3.304 *Umbilicus gaditanus*, March. 2018. (Photo credit: Victoria Eugenia Martín Osorio)

Inflorescence that occupies 40–60% of the stem length, racemose or, very rarely, paniculate; bracts 4–6 mm, always larger than the pedicel, filiform, ascending arched. **Flowers** with pedicel 0.7–3 mm which sometimes carries a bracteole 0.5–1 mm, usually horizontal or slightly inclined downwards, but not pendulous. Sepals 1–1.8 mm, acute. Corolla 5.5–7.5 (8) mm, dark straw-coloured to light greenish, tubular, with five marked longitudinal striations, with segments 1–2.1 mm, triangular-ovate, acute, which usually do not overlap at their edges or do so very scarcely. Stamens 10. Carpels 5, attenuate in a short style.

Seeds 0.6–0.7 mm, dark brown, ovoid.

Distribution

Distributed throughout Macaronesia, except Cape Verde, and throughout the western Mediterranean. Scattered throughout the S and E of Spain and the Balearic Islands. Present in all the Canary Islands (Fig. 3.305).

In the Teide National Park (Fig. 3.306) it occurs in a very dispersed way along the Pared de Las Cañadas, southeastern slopes of Izaña, Altos de La Orotava and Montaña Echisere or Pico Cabras (N-Teide), between 1750 and 2400 m a.s.l.

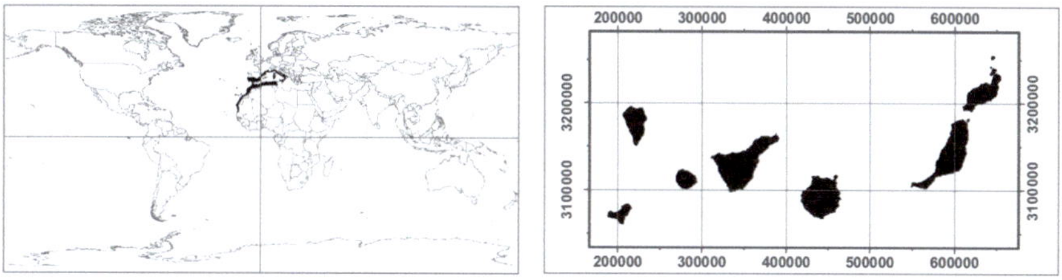

Fig. 3.305 Present on all the Canary Islands

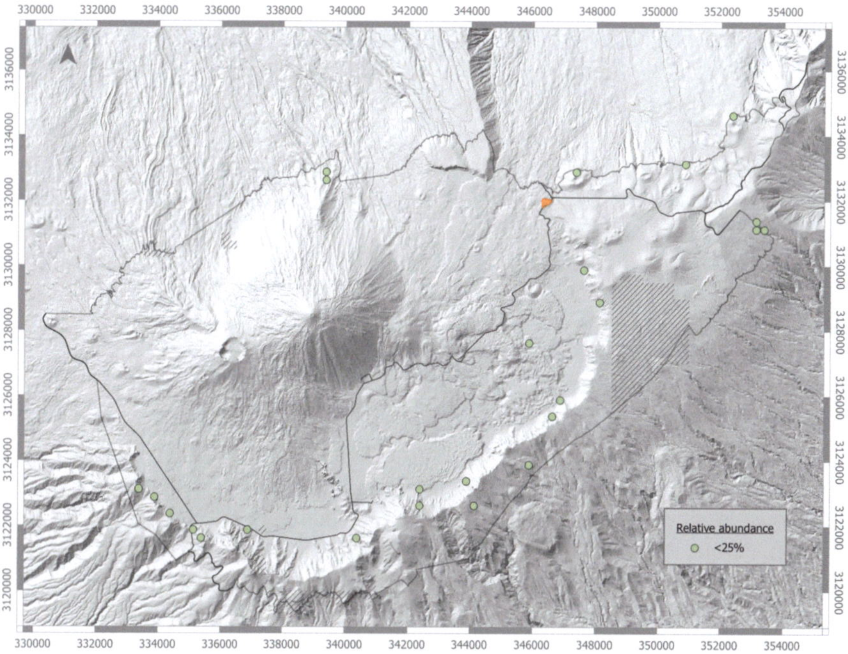

Fig. 3.306 *Umbilicus gaditanus* distribution, N° Subgrids (250 × 250 m): **25**/3740

Habitat and Phytosociology

It often grows on stone walls that delimit orchards, cracks in damp walls and roofs; between 0 and 500 m a.s.l.

It is a characteristic species of the ***Parietarietalia judaicae*** communities, and the ***Parietario judaicae-Centranthion rubri*** alliance, which groups the cosmopolitan extratropical nitrophilous chasmophytic associations of walls, walls and nitrified, non-oozing caves, especially in rural, urban and suburban environments, Infra-Supra-Mediterranean and Infra-Mesotemperate semi-arid-subhumid.

Conservation Status

In practically all the subgrids in which this species was found (23), no damage of any kind was observed. Only in two (2) of them, its conservation status was low, because the specimens were completely dry.

Exsiccata (Herbarium Specimens)

Tenerife: Anaga, 3.1979, E. Valdés, P.L. Pérez de Paz & alumnos (TFC 34927); Ladera de Tamaimo, 2.4.1980, W. Wildpret, J.R. Acebes, M. del Arco Aguilar, M.C. León Arencibia, L. Negrín & AVR (TFC 25135 + duplic.); La Laguna, finca de Las Dominicas, 10.3.1982, W. Wildpret, M. del Arco Aguilar & A. García Gallo (TFC duplic. 22145); Icod, 28.12.1983, F. Ardévol González (TFC 19087); La Laguna, Muro del Nuryana, 7.3.1985, A. García Gallo (TFC 2246; 22147; 22148); Bco. en Pinar de Agache, 1300 m.s.n.m., 20.4.1990, M. Salas Pascual (TFC 37874); La Matanza, Casas de El Caletón, 16.1.1994, F. Ramallo (TFC 37262); Los Silos, Las Canteras, ctra. Vieja, 26.4.1996, V. Lucía Sauquillo & M. del Arco Aguilar (TFC 38074 + duplic.); Bco. Ijuana, Ladera N, 320 m.s.n.m., 18.6.1996, R. Racca (TFC 39153); Los Silos, Bco. Los Cochinos, 22.7.1996, V. Lucía Sauquillo & C. Acosta (TFC 39325); Fasnia, Bco. de La Canal, SICAH, 6.9.1996, GMCT (TFC 39522); La Laguna, Rancho Grande, 11.3.1997, 370/3151 J.R Acebes (TFC 40503 + duplic.); El Sauzal, Punta Pesquero Alto, 2.3.2002, M. Rodríguez López & M.L. Rodríguez Navarro (TFC 44598).

El Hierro: Sabinar, 23.4.1978, W. Wildpret et al. (TFC 8146).

Gran Canaria: Bco. del Cernícalo, primavera 1989, S. Ortega González & E. Fraile Bonafonte (TFC 30483); Parque Natural de Bco. de Guayadeque, cauce del bco. a la altura de Las Cabezadas, 1300 m.s.n.m., 17.6.1989, F. Amor (TFC 30508); Parque Natural de Bco. de Guayadeque, bco. de Guayadeque, Morro Garañón, 31.3.1990, F. Amor (TFC 30623); Parque Natural de Bco. de Guayadeque, Bco. de Guayadeque, sendero Cueva Bermeja, 25.3.1990, F. Amor & P.L. Pérez de Paz (TFC 30723).

Lanzarote: Montaña Clara, 31.3.1983, M.C. León Arencibia, E. Beltrán Tejera & W. Wildpret (TFC 32849); Teguise, Bco. próximo al Cortijo del Rincón, 29.3.1994, J.A. Reyes Betancort (TFC 38883); Montaña Clara, la Caldera, 22.2.1995, P.L. Pérez de Paz, W. Wildpret & J.A. Reyes-Betancort (TFC 41046); San Bartolomé, El Islote, 13.3.1996, J.A. Reyes-Betancort, M.C. León Arencibia & A. Losada Lima (TFC 39680 + duplic.).

Uses, Exploitation and/or Medicinal Properties

> "*Diuretic, refreshing, vulnerary. Useful part: whole plant. Instructions for use: poultice, juice, infusion.*" (Translated from: Pérez de Paz and Hernández Padrón 1999)

3.5.99 Valeriana calcitrapae L.

First published in Sp. Pl.: 31 (1753)

Centranthus calcitrapae (L.) Dufr. *Hist. Nat. Méd. Fam. Valér.*, 39 (1811)
Rittera calcitrapae (L.) Raf. in Autik. Bot.: 88 (1840)
Valeriana laciniata Salisb. in Prodr. Stirp. Chap. Allerton: 56 (1796), nom. superfl.

Not endemic, Probable native
Family: *Valerianaceae*
Common name: Abrojo, Camellera, Cardo estrellado, Lágrimas de la virgen, Valeriana española (ES); Annual Valerian, Pink Valerian (EN)
Biotype: therophyte
Flowering: March–May
Fruiting: March–May
Pollination: insect pollination (entomogamy)
Dispersal: wind dispersal (anemochory (Martín Osorio et al. 2008))
Chromosome N°: 2n = 32

Description

Annual glabrous plant, with stems up to 70 cm long, simple or branched.

Leaves up to 9 × 4 cm; the basal ones petiolate, with obovate to subobicular, toothed to crenulate, rarely pinnatifid; the uppermost ones pinnatifid, with linear or narrowly oblong, toothed or crenulate divisions.

Inflorescences with short (Fig. 3.307), dense branches, extending up to 5 (−12) cm when fructifying. Corolla 1.5–3 (−4.5) mm, pale pink, rarely white; tube 1–3 mm, with a gibbosity or spur of less than 0.5 mm that does not exceed or barely exceeds the base of the corolla.

Fruits (1.7–) 2–2.5 (−3.5) mm, ovoid, glabrous or hirsute, with persistent calyx of c. 0.5 mm; frequently with some fruit with an accrescent calyx formed by 10–14 plumose setae of 3–4 mm.

Distribution

Abundant in the Mediterranean region and in the Macaronesian region (except Cape Verde). Present in all the Canary Islands, except in Fuerteventura and Lanzarote (Fig. 3.308). In the Teide National Park (Fig. 3.309) it is found in a very concentrated form in the vicinity of La Tarta and La Crucita, between 1900 and 2200 m a.s.l.

Habitat and Phytosociology

It usually colonizes preferably, stony terrains. It is a characteristic species of ***Cardamino hirsutae-Geranietea purpurei***, which groups scionitrophilous communities, generally of small size, which thrive under the protection and semi-shade of

Fig. 3.307 Surroundings of La Crucita October 2019. (Photo credit: GG Casanova)

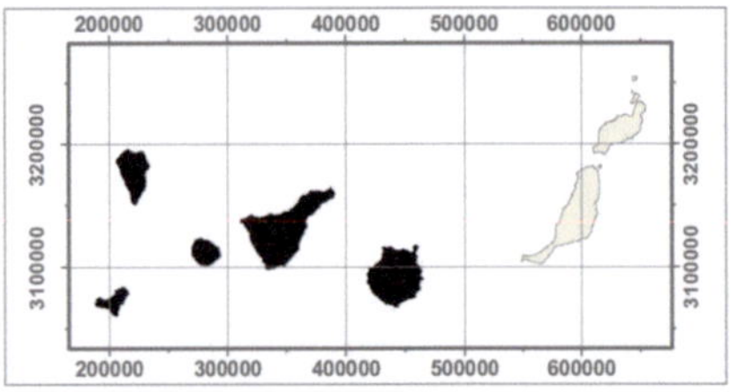

Fig. 3.308 Present on all the Canary Islands except in Fuerteventura and Lanzarote

forests, bushes, walls or rocks, in areas of the Mediterranean region or bordering with a Mediterranean-influenced bioclimate.

Conservation Status

In most of the subgrids in which this species was found (7), no damage of any kind was observed. In the rest (1), this variable was not specified at the time of sampling.

Exsiccata (Herbarium Specimens)

Tenerife: Bco. de las Huertas, 28.2.1973, E. Barquín Díez (TFC 21957 + duplic.); Altos de Güímar, 27.4.1974, J.R. Acebes (TFC 6798); Güímar, W. Wildpret, E. Beltrán Tejera, NA, P.L. Pérez de Paz, B. Méndez, RA & KL (TFC 20339); Bco. de Badajoz, 27.3.1975, M. del Arco Aguilar, C. León Arencibia (TFC 4882);

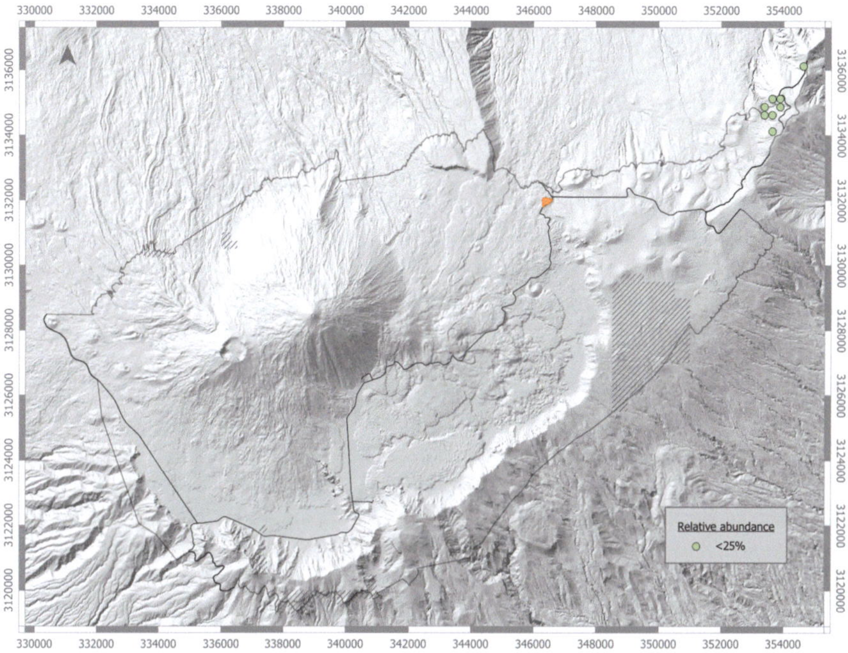

Fig. 3.309 *Valeriana calcitrapae* distribution, N° Subgrids (250 × 250 m): **8**/3740

Güímar, Bco. de Tamay, 17.7.1975, A. Delgado Benítez (TFC 42888); Bco. del Río, 29.3.1977, E. Beltrán Tejera & SM (TFC 6488); El Escobonal, Canal de Anocheza, 1.4.1980, W. Wildpret, J.R. Acebes, M. del Arco Aguilar, C. León Arencibia, L. Negrín & AVR (TFC 25123 + duplic.); Ladera de Güímar, 16.4.1981, P.L. Pérez de Paz, W. Wildpret & M. del Arco Aguilar (TFC 13740); Ibid, 16.1.1981, ejusd. (TFC 13713); Las Cañadas, 30.5.1981, W. Wildpret, E. Beltrán tejera, A. Losada Lima & A. García Gallo (TFC duplic. 33127); Icod, Bco. de Castro, 26.4.1985, F. Ardévol (TFC 19181); Pinares de Adeje, 20.4.1987, P.L. Pérez de Paz, M. del Arco Aguilar & M. Chacana (TFC 24075); Los Silos, Las Moradas, 6.3.1997, V. Lucía Sauquillo, A. Díaz, J.A. Reyes-Betancort & C. Hernández (TFC 40051); Los Silos, Erjos, 23.4.1997, V. Lucía Sauquillo, C. Acosta & A. Díaz (TFC 40597); Los Silos, pista principal del Monte del Agua, 25.5.1997, V. Lucía Sauquillo & C. Acosta (TFC 40805 + duplic.); Cerca del Embalse de Tahodio, 1.4.1999, S. Becher & T. Himstedt (TFC 42092).

Gran Canaria: Bco. de Guayadeque, 16.4.1971, I. La Serna (TFC 6439); San Mateo, Bco. de Antona, 4.4.1987, C. Suárez (TFC 31528).

La Palma: Laderas del Birigoyo, (El Paso) 1.5.1992, P.L. Pérez de Paz (TFC 35183). 9.4.1971, W. Wildpret & A. Santos (TFC 602); (?), 20.5.1972, B. Méndez Pérez (TFC 1014).

Uses, Exploitation and/or Medicinal Properties

"*Sedative. Useful part: whole plant, rhizome. Instructions for use: infusion.*" (Translated from: Pérez de Paz and Hernández Padrón 1999)

3.5.100 Viola cheiranthifolia *Humb. & Bonpl.*

First published in F.W.H.A.von Humboldt & A.J.A.Bonpland, Pl. Aequinoct. 1: 111 (1807)

Mnemion cheiranthifolium Webb & Berth. Phyt. Canar., 3(2/1): 113–115 (1837)
Viola tenerifera longifolia Feuill. Voy. Mss. Bibl. Reg. Par. cum ic (1724)
Viola pumicis! Masson. Mss. Herb. Banks. (1778)
Violette à feuilles un peu allongées! Labill. Voy. Réch. Lapeyr, vol. 1, pág. 21.
Viola teydea Berthel. Mem. Soc. Linn. Par. 4. Livr. pág. 420, table 7.
Viola canariensis Berthel. in Mém. Soc. Linn. Paris 5*:* 420 (1827)

Endemism, Certain native
Family: *Violaceae*
Common name: Violeta del Teide (ES); Teide Violet (EN)
Biotype: hemicryptophyte
Flowering: May–June
Fruiting: May–July
Pollination: entomogamy
Dispersal: short range self-dispersal (atelochory)
Chromosome N°: 2n = c. 64

Description

Perennial, fusiform, white, fleshy, fibrous root, truncated apex, very short stem.

Stems erect; branches very numerous, striated base, white, flaccid; leaf petioles and dry stipules rough, with pale green apex, densely tomentose, subtetragonal.

Leaves small, ascending, apex obtuse, margin entire or rarely sinuate-dentate, ciliate, succulent, completely white-tomentose, hairs very dense, straight, soft, simple, ascending; upper leaves compact, lanceolate or linear-lanceolate, with descending petiole, 24–30 mm long, 4–6 mm wide, conformal, entire and not very short; Lower leaves circular, ovate, stipules short, spatulate-linear, entire, petiolate; the lower ones very short, subviridine (high purity green), rounded, stipules, with broad petiole.

Leaves on the peduncle double, rarely subtriple, elongated, erect, striate, subtetragonal, whitish, weakly hispid, hairs short, patent, from the centre towards the top stipules tomentose, hairy, often shaded, acute apex, base in an elongated pouch.

Showy, purplish-pink **flowers** (Fig. 3.310) with yellow spots, rarely white. Calyx tomentose, sepals linear-lanceolate to the acute apex attenuate, subtrinervated,

Fig. 3.310 Montaña Blanca, July 2006. ((**a**) Photo credit: Beatriz Hernández); ((**b**) Photo credit: Victoria Eugenia Martín Osorio)

10 mm long, 2 mm wide. Corolla 24 mm wide, amethystine-blue, yellow lineolate eyes; Upper petals 12–14 mm long, 8 mm wide, ovate, acute; lateral petals with rounded, narrow apex; lower petals short, cuneate, crenate. Spur with appendage five times the length of the appendage, apex often tomentose. Anthers 2 mm long, ovate, apex attenuate, upper appendage elongate-pyramidal, obtuse, from the base 6 mm long, attenuate, filiform. Ovary rounded-ovate, glabrous. Style very geniculate, short. Stigma subrounded, with large foramen, approximately vertical opening, thick appendage. Ovule ovoid, horizontal. **Capsule** 6 mm long, ovoid, subobtuse, deep to the placenta 6-sulcate, rough, subglabrous, yellow, valve cartilaginous, hard, subcarious margin, whitish placenta. Seed 2 mm long, ovoid, yellow or subobscura, glabrous, large, white caruncle. Whitish perisperm. Embryo 1.5 mm long, greenish. Orbicular cotyledons. Sharp radicle.

Distribution

Endemism in Tenerife (Fig. 3.311), limited specifically to the Teide National Park (Fig. 3.312), between 2275 m a.s.l. (northern flank of Montaña Blanca) and 3700 m a.s.l. (Pico del Teide). We can also find it in the surroundings of the Pico Viejo crater and in the Teide Botanical Garden.

Habitat and Phytosociology

A species well adapted to stony and scoriaceous slopes, sometimes mobile, with pumice material, and at considerable altitudes. Less commonly it appears in rocky terrain.

It is a characteristic species and bioindicator of ***Violion cheiranthifoliae***, an alliance that groups the open permanent glareous woody subshrubs and perennial herbs of Teide violets, colonizers of the recent lapillis and lavas of the Oro-Mediterranean summit areas of the island of Tenerife, with a single association, ***Violetum cheiranthifoliae***; high mountain Geopermaserie of Tenerife Mediterranean pluviseasonal

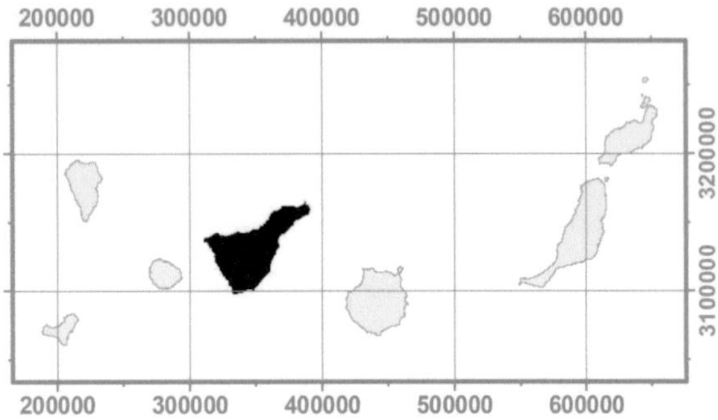

Fig. 3.311 Endemism of the Teide National Park on the island of Tenerife

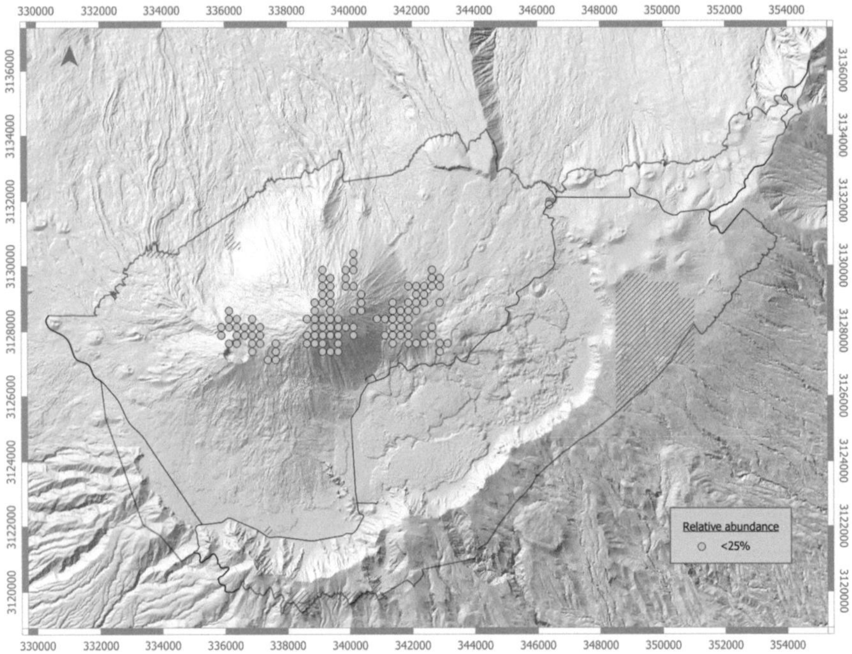

Fig. 3.312 *Viola cheiranthifolia* distribution, N° Subgrids (250 × 250 m): **113**/3740

oceanic Oro-Mediterranean dry euoceanic of the open glareous woody subshrublands of ***Viola cheiranthifolia*** with ***Argyranthemum tenerifae*** and ***Silene nocteolens*** with ***Laphangium teydeum***.

Conservation Status

In 90% (111) of the subgrids in which this species was found, no damage of any kind was observed. Only in 8% (10), damage was detected (browsing, high mortality without regeneration, etc.), but this did not affect more than 25% of the specimens. Damage was only found on more than 25% of the specimens in one (1) of the subgrids where this taxon appeared. In the rest (2), this variable was not specified at the time of sampling. "In danger" (García Gallo et al. 1993).

> "*Most of its populations are highly threatened by the high density of visitors to the Teide peak, the ease of access to its populations, and the introduction of mouflons in the 1970s. No expansion has been observed since the eradication of goat grazing, probably due to the increase in the other risks indicated.*" (Translated from: Gómez Campo et al. 1996)

Threat Category

E (Barreno et al. 1984)
Anexo I: Strictly protected (Gobierno de Canarias 1991)
VU D2 (Bañares et al. 1998)
E (Walter and Gillett 1998)
VU (Bañares et al. 2004)
VU D2 (Moreno 2008)

Germplasm Banks

E.T.S. Ingenieros Agrónomos de Madrid.
Jardín Botánico Canario Viera y Clavijo.

Exsiccata (Herbarium Specimens)

Tenerife: 19.04.1957, T. Bravo (TFC 586); Roque Blanco-W Pico Viejo, 19.04.1957, ejusd. (TFC 682, 25710); Cañada de Diego Hernández, 2.06.1973, M.C. Gil (TFC 24090); Topo de la Grieta, 15.06.1975, P.L. Pérez de Paz, J.R. Acebes & M. del Arco Aguilar (TFC 4582); Montaña Blanca, 12.06.1994, W. Wildpret de la Torre & V.E. Martín Osorio (TFC 38800); Inventario de la Flora y Vegetación del Parque Nacional del Teide (2001–2003) V.E. Martín Osorio, W. Wildpret & B. Hernández Bolaños: Ladera Sur del Teide (35B 49 a), 24/04/2003, A. Montañés Rodríguez (TFC 44843).

3.5.101 Viola guaxarensis *Marrero, Docoito Díaz & Martín Esquivel*

First published in Willdenowia 50: 16 (2020)

Endemism, Certain native
Family: *Violaceae*
Common name: Violeta de Guajara (ES); Guajara Violet (EN)
Biotype: hemicryptophyte
Flowering: May–June
Fruiting: May–July
Pollination: insect pollination (entomogamy)
Dispersal: short range self-dispersal (atelochory)

Description

Perennial cushion-forming herb with a clear overwintering period, to 20 cm tall × 50 cm in diam. Indumentum present on almost all structures of plant, comprising smooth, hyaline trichomes 0.1–0.2 mm long.

Stems 4–5 cm long, subterranean, with green, fistulous, subtetragonous branches. Leaves 3-grouped, greenish, petiolate or subpetiolate, stipulate. Lower leaves (25–) 30–45 × (8–) 10–15 mm, ovate to orbicular-ovate, petiolate, generally subdentate, stipulate; stipules linear, entire and 2–3 × c. 2 mm or bisected with median lobe 3–5 × c. 1 mm.

Upper **leaves** frequently unevenly dentate-crenate with short incisions, lanceolate to linear-lanceolate, (30–) 35–45 × 5–7 mm, base cuneate, apex subacute, usually purple; petiole <5 mm long, shorter than lamina; central leaf normally without stipules; lateral leaves stipulate; external stipules 2- or 3-sect (main lobe 10–15 × 1–1.5 mm); internal stipules linear, 12–17 × 1–1.5 mm.

Flowers axillary, solitary. Peduncle more than 2× length of leaves, 7–10 cm long, hairy, arcuate, subtrigonous; bracteoles purple, 3–6 mm long, subopposite, inserted in upper ½ of peduncle, usually slightly untilte. Sepals subequal, 10–13 × 3–4 mm at base, linear-triangular, acuminate, purple at base and margin; basal appendix 1.5–2 mm long, crenate-sinuate at base, usually with 2 or 3 incipient lobes. Corolla 25–30 mm in diameter, white-cream to violet (Fig. 3.313), occasionally entirely white, petals slightly eroded. Upper petals13–10 × 8–11 mm, glabrous, obovate, apex subacute. Lateral petals 12–10 × 7–9 mm, oval, glabrous, slightly clawed and densely hairy at base of claw, apex rounded. Lower petals

Fig. 3.313 Montaña de Guajara, April 2023. (Photo credit: Wolf Hermann Wildpret Martín)

11–14 × 9–11 mm, lightly coloured, yellow near spur, retuse, slightly conduplicate, glabrous. Spur 8–9 × 2–2.5 mm, arcuate, glabrous, purple, obtuse, protruding 6–7 mm from sepal appendices. Anthers ovate, hairy, with an appendix orange and almost triangular. Style c. 1 mm long, geniculate. Stigma globose.

Capsules subrounded, c. 8 × 7 mm, slightly sulcate, opening at ripening into 3 valves, with 25–30 seeds. Seeds c. 2 × 1 mm, chestnut-brown; elaiosome covering c. 1/5 of seed.

Remarks:

A new variety (var. *albiflora*) has been found, located in the highlands of Montaña Guajara, which we describe as new with the differential characteristic over the typical variety of having all the petals white (Fig. 3.314).

Distribution

Tenerife endemism limited specifically to the Teide National Park, Endemic to the high-mountain zone of Tenerife, Canary Islands, Spain, where it is known from only two localities. Locally common near the summit of Montaña Guajara with nearly 3000 individuals, but scarce (with only a few dozen individuals) at the second locality, Topo de la Grieta 3 km to the northeast (Figs. 3.315 and 3.316). In general, the

Fig. 3.314 Montaña de Guajara, April 2023. (Photo credit: Wolf Hermann Wildpret Martín)

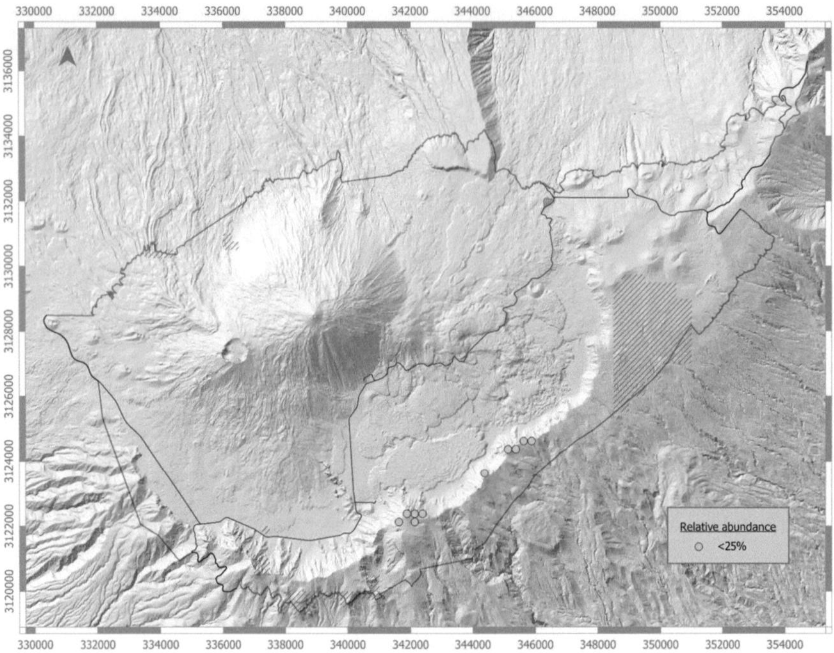

Fig. 3.315 *Viola guaxarensis* distribution, N° Subgrids (250 × 250 m): **10**/3740

Fig. 3.316 Endemism of the Teide National Park on the island of Tenerife

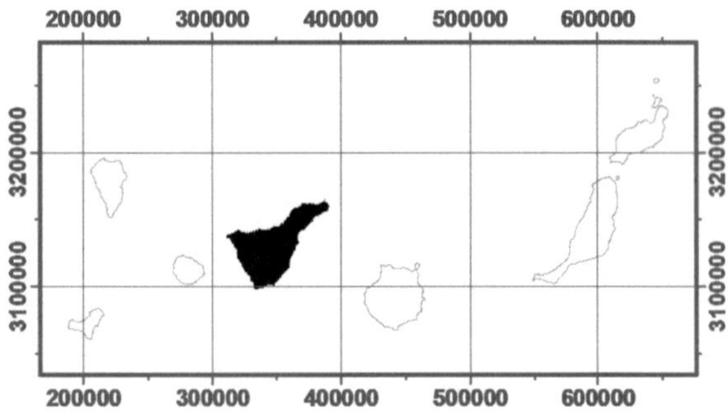

habitat is characterized by low or moderate slopes, with acid and rocky substrates sometimes mixed with pumice. These environments are subject to relatively strong winds (Wildpret de la Torre and Martín Osorio 2005) and frequent winter frosts. Vegetation at the sites is dominated by a low mixed cushion-shrub community with ***Adenocarpus viscosus*** (Willd.) Webb & Berthel. **subsp.** ***viscosus*** (*Fabaceae*) one of the more common species. Other Canary Island endemics present are ***Erysimum scoparium*** (Brouss. *ex* Willd.) Wettst. (*Brassicaceae*), ***Nepeta teydea*** Webb & Berthel. (*Lamiaceae*), ***Spartocytisus supranubius*** (L. f.) Christ *ex* G. Kunkel (*Fabaceae*), and ***Tolpis webbii*** Sch. Bip. *ex* Webb & Berthel. (*Asteraceae*).

Locally common near the summit of Montaña Guajara with nearly 3000 individuals, but scarce (with only a few dozen individuals) at the second locality, Topo de la Grieta 3 km to the northeast (Marrero Gómez et al. 2020).

In August 2020, the known populations of the taxon were visited, proceeding to the counting of specimens. The results obtained indicate the presence of 1382 specimens in Alto de Guajara and 46 in Topo de La Grieta. However, the sampling date was somewhat late, so it is possible that the total figures are underestimated.

Habitat and Phytosociology

A species well adapted to stony and scoriaceous slopes, sometimes mobile, with pumice material, and at considerable altitudes. Less commonly found in rocky terrain, it is well adapted to extreme climatic conditions, especially from strong winds, high on the summit plateau of Montaña Guajara.

Wildpret de la Torre and Martín Osorio (2005) published the syntaxon ***Violo cheiranthifoliae-Adenocarpetum viscosii*** Wildpret de la Torre and Martín Osorio 2005, ***Spartocytision supranubii. Spartocytisetalia supranubii, Chamaecytiso-Pinetea canariensis***, which according to the Code of Phytosociological Nomenclature (Barkman et al. 1986) in its article 43, should be corrected by ***Violo guaxarensis-Adenocarpetum viscosi***

Threat Category

Being a recently published taxon, its conservation status must be studied for inclusion in some threat category.

Germplasm Banks

Instituto Universitario de Estudios Ambientales y Recursos Naturales (IUNAT), Universidad de Las Palmas de Gran Canaria, Campus Universitario de Tafira, 35017 Las Palmas de Gran Canaria, Canary Islands, Spain

Exsiccata (Herbarium Specimens)

Tenerife: Cañada de Diego Hernández, 2.06.1973, M.C. Gil (TFC 24090); Topo de la Grieta, 15.06.1975, P.L. Pérez de Paz, J.R. Acebes & M. del Arco Aguilar (TFC 4582); Guajara-Parque Nacional del Teide, 2700 m.s.n.m., Orientación SW, 27.02.1998, A. Bañares & E. Carqué (TFC 41482 + Dupl.). Guajara, Nacimiento del Barranco del Río (42C 3 b), 15/05/2003, Y. de Rioja Santos (TFC 44211).

Tenerife: type locality 1998, *A. Bañares & E. Carqué* (TFC 41482); ibidem, 2003, *Y. de Rioja* (TFC 42211); ibidem, 2019, *J. R. Docoito* (TFC 53317, TFC 53318);

3.5.102 Wahlenbergia lobelioides *(L.f.) Link ssp.* lobelioides

Handbuch, 1, 632 (1829); *Monogr. Camp.*, 157 (1830)

Campanopsis lobelioides (L.f.) Kuntze *in* Revis. Gen. Pl. 2: 379 (1891)
Campanula lobelioides L.f. *in* Suppl. Pl.: 140 (1782)
Campanula parviflora Salisb. *in* Prodr. Stirp. Chap. Allerton: 126 (1796), nom. illeg.
Cervicina lobelioides (L.f.) Druce *in* Rep. Bot. Soc. Exch. Club Brit. Isles 8: 547 (1927 publ. 1928)
Cervicina pendula Druce *in* Rep. Bot. Soc. Exch. Club Brit. Isles 6: 32 (1920 publ. 1921)
Hecale lobelioides (L.f.) Raf. *in* Fl. Tellur. 2: 79 (1837)
Wahlenbergia pendula Schrad. *in* Blumenbachia: 38 (1827), nom. superfl.

Not endemic, Certain native
Family: *Campanulaceae*
Common name: Almirón o Escarchalagua (ES); Bellflower (EN)
Biotype: therophyte
Flowering: March–June
Fruiting: March–June
Pollination: insect pollination (entomogamy)
Dispersal: short range self-dispersal (autochory)
Chromosome N°: 2n = 18

Description

Annual plant, up to 20 (−30) cm, erect. Stem frequently branched at the base, with short hairs, scattered in the lower half or only at the base, glabrous on the upper part.

Leaves located mainly on the lower part of the stem, alternate, the lower ones subopposite, attenuated on the very short or almost non-existent petiole; blade 20–70 × 5–15 mm, elliptic-oblong, serrulate to subentire, ciliate.

Flowers solitary (Fig. 3.317), with pedicel 5–40 mm. Calyx with (4) 5 teeth of 1–3 mm, erect, narrowly triangular. Corolla with (4) 5 lobes, white, pale blue or mauve, campanulate-infundibuliform; tube 2–5 mm; lobes 2–2.5 mm, more or less erect-patent. Stamens (4) 5. Style even, with very small glands at the base of the stigma; stigmas 3.

Capsule 5–12 × 3–4 mm, with 3 locules, ellipsoid to obovoid, with 10 prominent nerves. Seeds 0.5–0.6 × 0.2–0.3 mm, smooth.

Distribution

It can be found in the Macaronesian regions (Canary Islands, Cape Verde and Madeira) and the Mediterranean (in Spain: coast of Almeria and Valencia); from NW Africa to Egypt, Sudan, Ethiopia, Socotra and the Arabian Peninsula. Present in all the Canary Islands (Fig. 3.318).

Fig. 3.317 *Wahlenbergia lobeliodes*. (Photo credit: GG Casanova)

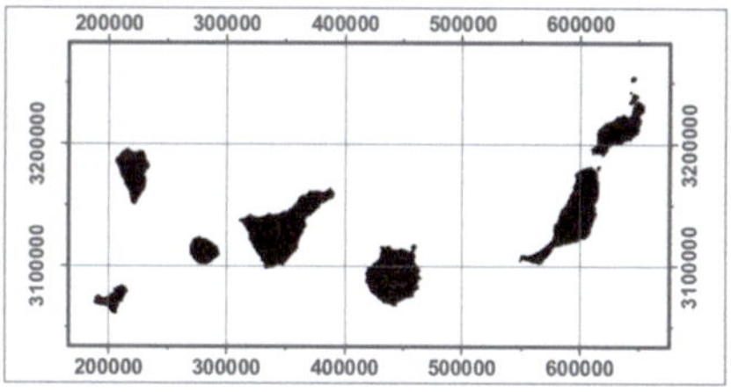

Fig. 3.318 Present on all the Canary Islands

In the Teide National Park (Fig. 3.319), it is widely distributed between 1600 and 2750 m a.s.l. It is found practically throughout the entire protected area, except for the slopes and lava flows of the Pico Viejo-Teide-Montaña Blanca complex, sectors of the Pared de Las Cañadas and areas of La Dorsal.

Remarks:

W. lobelioides is a species widely distributed in tropical Africa and Macaronesia. Taxa closely related to it live in northern Africa, southern Asia, Australia and America. Three subspecies are generally recognized, one of them being the type subspecies, from Madeira, the Canary Islands and Cape Verde.

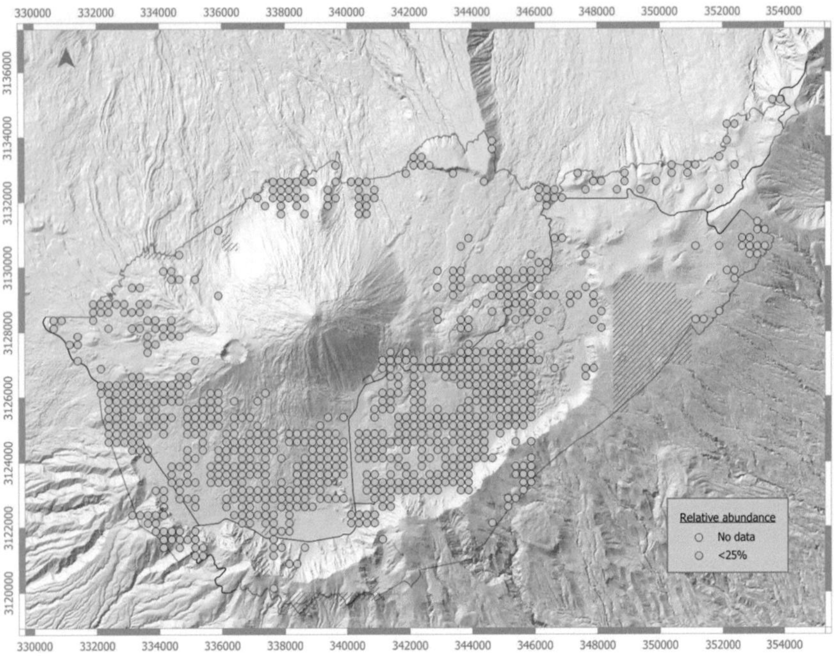

Fig. 3.319 *Wahlenbergia lobelioides* distribution, N° Subgrids (250 × 250 m): **819**/3740

Habitat and Phytosociology

It generally grows in siliceous sands and rocky places, between 0 and 400 m a.s.l. Widely distributed from the sub-Mediterranean to the Supra-Mediterranean bioclimatic zone.

It is involved in the ***Bromo madritensis-Piptatherion miliacei*** alliance, which groups the associations of perennial grasses-herbs, structured by mijeras that develop on deep soils in ruderal-nitrophilous environments and abandoned crops; of circum-Mediterranean distribution with Cantabrian-Atlantic and Canary-Madeiran disjunctions Infra-Mesomediterranean semi-arid-subhumid.

Conservation Status

In most of the subgrids in which this species was found (802), no damage of any kind was observed. Only in four (4) of them, damage was detected (browsing, high mortality without regeneration, etc.), but this did not affect more than 25% of the specimens. And damage was only found on more than 25% of the specimens in nine (9). In the rest (4), this variable was not specified at the time of sampling.

Threat Category

LR pm (Bañares et al. 1998)

Germplasm Banks

Banco de Germoplasma de E.T.S. Ingenieros Agrónomos de Madrid.

Exsiccata (Herbarium Specimens)

Tenerife: Güímar, Bco. de Badajoz, 31.5.1973, I. La Serna (TFC 35734); San Andrés, 15.4.1975, M.C. Godoy (TFC 33515); Salvaje Grande, 2.1976, P.L. Pérez de Paz & J.R Acebes (TFC 4941); Bco. de La Orchilla, 20.5.1973, E. Beltrán Tejera (TFC 3805; 3806); Bco. de San Andrés, 24.3.1979, P.L. Pérez de Paz & E. Rodríguez (TFC 20454); Punta del Hidalgo, 17.4.1980, I. La Serna (TFC 7455; 7456; 7457); Las Cañadas, 30.5.1981, W. Wildpret, E. Beltrán Tejera, A. Losada Lima & A. García Gallo (TFC 33122); Icod, La Punta, 15.6.1983, E. Beltrán Tejera & L. Gallo (TFC 34011); Bco. de Herques, 450 m.s.n.m., 14.5.1984, O. Rodríguez Delgado (TFC 27618); La Laguna, 6.3.1985, W. Wildpret & A. García Gallo (TFC 22001); La Laguna, 22.5.1985, I. Zaldivar, W. Wildpret & A. García Gallo (TFC 22002; 22003); Guía de Isora, Chío, 20.4.1993, P.L. Pérez de Paz (TFC duplic. 35662); Candelaria, Altos de la Mesa, 1993, W. Wildpret, MMG, O. Rodríguez Delgado & E. Carqué Álamo, 362–3140, (TFC 36136); La Pardilla, 500 m.s.n.m., 18.3.1996, R. Racca (TFC 39129); Bco. Anosma, 330 m.s.n.m., 20.3.1996, R. Racca (TFC 39130); *Inventario de la Flora y Vegetación del P.N. del Teide (2001–2003) V.E. Martín Osorio, W. Wildpret & B. Hernández Bolaños*: Cañada del Montón de Trigo, junto a la laguna, 01.01.2003, W. Wildpret (TFC 44940); Ladera Sur del Teide, hacia Pico Viejo (35D 33 b), 04/06/2003, A. Montañés Rodríguez (TFC 44243); Jardín Botánico (P.N. Teide) (Ñ25), 17/06/2004, B. Hernández Bolaños (TFC 45398); La Fortaleza, 06/06/2004, V.E. Martín Osorio & W. Wildpret de la Torre (TFC 45430).

El Hierro: Sabinar, 10.4.1979, W. Wildpret, P.L. Pérez de Paz, M. del Arco Aguilar & C. Hernández (TFC 8182); Los Jables, 22.3.1998, P.L. Pérez de Paz & GCT (TFC 42933) (Termófilos 206–201/97); Malpaís bajo Los Llanitos, 23.3.1998, ejusd. (TFC 42477) (Observ. Ver Inventario n. 928 en la libreta n.18 de P.L. Pérez).

La Gomera: Valle Gran Rey, Playa del Inglés, 20.3.1976, E. Beltrán Tejera (TFC 24166); Valle Gran Rey, Bco. Argaga, 13.4.1981, B. Méndez Pérez & J.R. Acebes (TFC 13621 + duplic.).

Gran Canaria: Teror, Carretera de San Isidro, 30.5.1986, C. Suárez & G. Rodríguez (TFC 31328); Parque Natural Bco. de Guayadeque, Bco. de Guayadeque, cauce a la altura del cruce ctra. de Ingenio-Agüimes, 350 m.s.n.m., 8.3.1990, F. Amor, P.L. Pérez de Paz (TFC 30838); Parque Natural Bco. de Guayadeque, Bco. de Guayadeque, cauce a la altura de la Sª, 450 m.s.n.m., 8.3.1990, F. Amor, P.L. Pérez de Paz (TFC 30850); Parque Natural Bco. de Guayadeque, Camino Real de Pajonales, exposición N, 650 m.s.n.m., 25.3.1990, F. Amor, P.L. Pérez de Paz (TFC 30842).

Lanzarote: Malpaís de la Corona, 27.2.1994, J.A. Reyes-Betancort (TFC 40340 + duplic.); San Bartolomé, 23.2.1995, J.A. Reyes-Betancort (TFC 37165); San Bartolomé, Vega Quintero, 24.2.1995, ejusd. (TFC 37378); Ibid, 23.2.1995, ejusd. (TFC duplic. 37464); Hoya de la Pila, próximo al Volcán de La Corona, 26.4.1996, ejusd. (TFC 38762); Tías, Rompimiento, 7.3.1997, ejusd. (P-TFC 860) (TFC 40393).

Fuerteventura: Macizo de Betancuria, 18.4.1984, E. Beltrán Tejera (TFC 45015); [Como *Wahlenbergia* cf. *lobelioides*]: Jandía, 4.5.1980, W. Wildpret, M.C. León Arencibia, M. del Arco Aguilar & I. La Serna, CRP & A. García Gallo (TFC 20641). ?, 9.3.1969 W. Wildpret (TFC 591).

Uses, Benefits and/or Medicinal Properties

"*Treatment of urinary tract. Useful part: whole plant. Instructions for use: Infusion.*" (Translated from: Pérez de Paz and Hernández Padrón 1999)

References

Acebes Ginovés JR, León Arencibia MC, Rodríguez Navarro ML, del Arco Aguilar M, García Gallo A, Pérez de Paz PL, Rodríguez Delgado O, Martín Osorio VE, Wildpret de la Torre W (2010) División Pteridophyta, Spermatophyta. In: Arechavaleta M, Rodríguez S, Zurita N, García A (eds) Lista de Especies Silvestres de Canarias. Hongos, plantas y animales terrestres. Gobierno de Canarias, pp 119–172

Bañares Á, Romero P, Rodríguez C (1992) Adiciones corológicas de algunos endemismos canarios en peligro de extinción. Notas corológicas 12-27. Bot Macaronésica 19(20):141–150

Bañares Á, Marrero M, Carqué E (1998) The application of the Revised IUCN Red List Categories to the National Parks Flora of the Canary Islands. In: Synge H, Akeroyd J (eds) Planta Europa. Proceedings of the second European conference on the conservation of wild plants. Swedish Threatened Species Unit, Upsala, pp 198–204

Bañares Á, Blanca G, Güemes J, Moreno JC, Ortiz S (eds) (2004) Atlas y Libro Rojo de la Flora Vascular Amenazada de España. Dirección General de Conservación de la Naturaleza, Madrid

Bañares , *et al.* Atlas y Libro Rojo de la flora vascular amenazada de España. Adenda 2008. , https://www.miteco.gob.es/es/biodiversidad/temas/inventarios-nacionales/inventario-especies-terrestres/inventario-nacional-de-biodiversidad/ieet_flora_vasc_adenda_2008.html (2010).

Bañares Á, Blanca G, Güemes J, Moreno JC, Ortiz S (2008) Lista Roja de la Flora Vascular Española. https://bibdigital.rjb.csic.es/records/item/1526095-lista-roja-2008-de-la-flora-vascular-espanola?offset=6

Barbero M, Quezel P (1975) Les Forets de sapin sur le pourtour mediterranèen. Anal Inst Bot Cavanilles 32(2):1245–1289

Barkman JJ, Moravec J, Rauschert S (1986) International code of phytosociological nomenclature. 2rd edition. Vegetatio 67(66):145–195. Trad. Izco J, Del Arco M (1988) Código de nomenclatura fitosociológica. Opusc Bot Pharm Complutensis 4:5–74

Barreno E et al (1984) Listado de plantas endémicas, raras o amenazadas de España. Información Ambiental 3:48–71

Bakker P, Maes B, Maskew R & Stace C (2019) Dog-Roses (Rosa Sect. Caninae): towards a consensus taxonomy. Bristish & Irish Botany 1 (1): 7–19.

Batista F, Bañares A, Caujapé-Catells J, Carqué E, Marrero-Gómez M & Sosa P A (2001) Allozyme diversity in three endemic species of Cistus (Cistaceae) from the Canary Islands: intraspecific and interspecific comparisons and implications for genetic conservation. American Journal of Botany 88: 1582–1592.

Beltrán Tejera E, Wildpret de la Torre W, León Arencibia MC, García Gallo A, Reyes Hernández J (1999) Libro Rojo de la Flora Canaria contenida en la Directiva-Hábitats Europea. La Laguna, Organismo Autónomo Parques Nacionales, Tenerife, 694 pp

Bolle C (1860) Addenda ad floram Atlantidis, praecipue insularum Canariensium Gorgadumque. Bonplandia 8:279–287

Bramwell D (1973) New species of *Cruciferae* from the Canary Islands. Cuad Bot Canar XVII:19–26

Carqué Álamo E, Durbán Villalonga M, Marrero Gómez M, Bañares Baudet Á (2004) Influencia de los herbívoros introducidos en la supervivencia de *Stemmacantha cynaroides* (Asteraceae) Una especie amenazada de las islas Canarias. Vieraea 32:97–105

Castroviejo S (ed) (1986) Flora Ibérica. Plantas vasculares de la Península Ibérica e Islas Baleares, vol I. Real Jardín Botánico de Madrid, Servicio de Publicaciones del C.S.I.C., Madrid, 575 pp

Ceballos L, Ortuño F (1976) Estudio sobre la Vegetación y Flora Forestal de las Canarias Occidentales. Excmo. Cabildo Insular de Tenerife, Santa Cruz de Tenerife

Christ DH (1888) Spicilegium canariense. Engl Bot Jahrb 9:86–172

Dalgaard V (1979) Biosystematics of the Macaronesian species of *Scrophularia*. Opera Bot 51:1–64. Stockholm

Demoly J-P, Marrero MV, Bañares Á (2006) Contribution à la connaissance des *Cistus* de la section *Macrostylia* Willk. (*Cistus* L., Cistaceae). J Bot Soc Bot 36:13–38

Dickson JH, Rodríguez JC, Machado A (1987) Invading plants at high altitudes on Tenerife especially in the Teide National Park. Bot J Linn Soc 95:155–179

Directiva 92/43/CEE del Consejo, de 21 de mayo de 1992, relativa a la conservación de los hábitats naturales y de la fauna , https://www.boe.es/buscar/doc.php?id=DOUE-L-1992-81200 (1992).

Dobignard A (1997) Nouvelles observations sur la flore du Maroc. Candollea 52(1):119

Dukes J, Money H (2004) Disruption of ecosystem processes in western North America by invasive species. Rev Chil Hist Nat 77:411–437

García Gallo A, Acebes Ginovés JR, Vera Galván MA, Marrero Gómez M, Pérez de Paz PL (1993) Avance del atlas cartográfico de los endemismos canarios. Itinera Geobotanica 7:405–436

García Gallo A, Acebes Ginovés JR, Martín Osorio VE, Wildpret de la Torre W (2003) La Flora Vascular de Gran Canaria. In: Rodríguez Delgado O (ed) Apuntes sobre flora y vegetación de Gran Canaria. Cabildo Insular de Gran Canaria, pp 41–70

Gobierno de Canarias (1991) Orden de 20 de febrero de 1991, de la Consejería de Política Territorial, sobre protección de especies de la flora vascular silvestre de la Comunidad Autónoma de Canarias. BOC, n° 35, de 18 de marzo de 1991

Gobierno de Canarias (2001) Decreto 151/2001, de 23 de julio, por el que se crea el Catálogo de Especies Amenazadas de Canarias. https://www.gobiernodecanarias.org/boc/2001/097/004.html

Gobierno de Canarias (2010) Ley 4/2010, de 4 de junio, del Catálogo Canario de Especies Protegidas. BOE, n°150, de 21 de junio de 2010 y BOC, n° 112, de 9 de junio de 2010.

Gómez Campo C et al (1996) Libro Rojo de especies vegetales amenazadas de las Islas Canarias. Consejería de Política Territorial. Viceconsejería de Medio Ambiente. Gobierno de Canarias, Santa Cruz de Tenerife

González González EA (2011) Análisis de la diversidad genética en poblaciones naturales de especies vegetales amenazadas: *Ilex perado* ssp. *lopezlillo* (Aquifoliaceae), *Silene nocteolens* (Caryophyllaceae) y *Sorbus aria* (Rosaceae). Tesis doctoral inédita, Universidad de Las Palmas de Gran Canaria. http://hdl.handle.net/10553/6709

Hedberg O. 1964. Features of afroalpine plant ecology; avec un résumé en français. Acta Phytogeographica Suecica 49: 1–144.

Humphries CJ (1976) A revision of the Macaronesian genus *Argyranthemun* Webb *ex* Schultz Bip. (Compositae-Anthemidae). Bull Br Mus Nat Hist (Bot) 5(4):147–240

Jalas J (1966) *Cerastium sventenii* Jalas, sp. nova, and the related Macaronesian taxa. Ann Bot Fenn 3:129–139

Klášterský L (1968) *Rosa* L. In: Tutin TG, Heywood VH, Burgess NA, Moore DM, Valentine DH, Valters SM, Webb DA (eds) Flora Europaea, 2nd edn. Cambridge University Press

Lindinger L (1926) Beiträge zur Kenntnis von Vegetation und Flora der kanarischen Inseln. Hamburgische Universität, Hamburg, 350 pp

Lid J (1968) Contributions to the flora of the Canary Islands. Skr. Norske Videnk. - Akad. Oslo, 1. Mat.-Naturv Kl., n.s. 23 (1967): 212 pp

Lowe R & Mösch W (1964), De Cerastiis Africae septentrionalis, *Mem. Soc. Broter. 17, 5 — 121.*

Marrero Gómez MV, Bañares Baudet Á, Carqué Álamo E, Padilla Cubas A (1999) Size structure in populations of the two threatened endemic plant species of the Canary Islands: *Cistus osbaeckiaefolius* and *Helianthemum juliae*. Nat Areas J 19(1):79–86

Marrero Gómez MV, Bañares Baudet Á, Carqué Álamo E (2003) Plant resource conservation planning in protected natural areas: an example from the Canary Islands, Spain. Biol Conserv 113:399–410

Marrero Gómez MV, Carqué Álamo E, Bañares Baudet Á (2004) *Gnaphalium teydeum* Knapp. In: Bañares A, Blanca G, Güemes J, Moreno JC, Ortiz S (eds) Atlas y Libro Rojo de la Flora Vascular Amenazada de España. Dirección General de Conservación de la Naturaleza, Madrid, pp 706–707

Marrero Gómez M, Martínez Esquivel JL, Docoito Díaz JR, Suarez Izquierdo M (2020) *Viola guaxarensis* (*Violaceae*): a new *Viola* from Tenerife, Canary Islands, Spain. Willdenowia 50(1):13–21. https://doi.org/10.3372/wi.50.50102

Marrero M, Carqué E, Bañares Á, Palomares A (2004) *Bencomia exstipulata* Svent. In: Bañares A, Blanca G, Güemes J, Moreno JC, Ortiz S (eds) Atlas y Libro Rojo de la Flora Vascular amenazada de España. Dirección General de Conservación de la Naturaleza. Ministerio de Medio Ambiente, Madrid, p 881

Martín Osorio VE, Wildpret de la Torre W, Hernández Bolaños B, De Rioja Santos Y (2004) Atlas y Libro Rojo de la Flora Amenazada de España (*Dactylis metlesicsii*). In: Bañares A, Blanca G, Güemes J, Moreno JC, Ortiz S (eds) Atlas y Libro Rojo de la Flora Vascular Amenazada de España. Dirección General de Conservación de la Naturaleza. Ministerio de Medio Ambiente, Madrid, pp 668–669, 1072 pp

Martín Osorio VE, Wildpret de la Torre W, Hernández Bolaños B (2006) Estudio de las plantas vasculares exóticas invasoras en el Parque Nacional del Teide, Tenerife (Islas Canarias), mediante un Sistema de Información Geobotánica (S.I.G.). In: GEIB Grupo Especialista en Invasiones Biológicas (ed) Invasiones Biológicas: un factor del cambio global. EEI 2006 Actualización de conocimientos, 2° Congreso Nacional sobre Especies Exóticas Invasoras “EEI 2006”, Serie Técnica n° 3. GEIB, pp 142–153, 280 pp

Martín Osorio VE, Wildpret de la Torre W, Del Arco Aguilar M, Pérez de Paz PL, Hernández Bolaños B, Rodríguez O, Acebes JR, García Gallo A (2007) Estudio Bioclimático y Fitocenótico comparativo de la alta cumbre canaria: Tenerife—La Palma. Islas Canarias. Phytocoenologia 37(3–4):663–697. Berlín–Stuttgart

Martín Osorio VE, Wildpret de la Torre W, Ojeda Land E, Jardim R, Silva L (2008) *Centranthus ruber* (L.) DC. In: Silva L, Ojeda Land E, Rodríguez Luengo JL (eds) Flora y Fauna Terrestre Invasora en la Macaronesia. TOP 100 in Azores, Madeira y Canarias. ARENA, Ponta Delgada, pp 391–393

Martín Osorio VE, Wildpret de la Torre W, Alcántara Vernet E (2010) *Cistus osbeckiifolius* Webb *ex* Christ subsp. *osbeckiifolius*. In: Bañares A, Blanca G, Güemes J, Moreno JC, Ortiz S (eds) Atlas y Libro Rojo de la Flora Vascular Amenazada de España. Adenda 2010. Dirección General de Medio Natural y Política Forestal (Ministerio de Medio Ambiente, y Medio Rural y Marino) – Sociedad Española de Biología de la Conservación de Plantas, Madrid, pp 66–67

Martín Osorio VE, Wildpret de la Torre W, Hernández Bolaños B (2015) Los bosques de cedros de la alta cumbre canaria. Makaronesia 17:102–119

Mesa Coello R, Acevedo A, Oval de la Rosa JP, Hernández A, Marrero M, Carqué E (2004) *Cerastium sventenii* Jalas. In: Bañares Á, Blanca G, Güemes J, Moreno JC, Ortiz S (eds) Atlas y Libro Rojo de la Flora Vascular amenazada de España. Dirección General de Conservación de la Naturaleza. Ministerio de Medio Ambiente, Madrid, pp 634–635

Ministerio de Medio Ambiente, Medio Rural y Marino (2011) Real Decreto 139/2011, de 4 de febrero, para el desarrollo del Listado de Especies Silvestres en Régimen de Protección Especial y del Catálogo Español de Especies Amenazadas

Montserrat Martí G, Montserrat Martí JM (1990) *Sagina* L. In: Castroviejo S et al (eds) Fl. Iberica, 2nd edn. CSIC, Madrid, pp 295–296

Moreno JC (ed) (2008) Lista Roja 2008 de la flora vascular española. Dirección General de Medio Natural y Política Forestal (Ministerios de Medio Ambiente y Medio Rural y Marino, y Sociedad Española de Biología de la Conservación de Plantas), Madrid

Nyffeler R (1992) A taxonomic revision of the genus *Monanthes* Haworth (*Crassulaceae*). Bradleya 10:49–82

Pérez de Paz PL (1978) Check-list del gén. Micromeria Bentham en la Macaronesia. I, Seminario de *Lamiaceae* endémicas de Canarias. Opúscula del Dpto. de Botánica, Fac. Biología Universidad de La Laguna

Pérez de Paz PL, Hernández Padrón CE (1999) Plantas Medicinales o útiles en la Flora Canaria. *Aplicaciones populares*. Francisco Lemus Editor, Arafo

Perez de Paz PL, Negrín Sosa ML (1992) Revisión taxonómica de *Sideritis* L subgénero Marrubiastrum (Moench) Mend.-Heuer (endemismo macaronésico). J Cramer, 58pp

Pérez de Paz PL, La Serna Ramos IE, Hernández Padrón CE (1992) El género *Rhamnus* L. en las Islas Canarias. Stud Bot 10:91–107

Pitard J & Proust L (1908). Les Iles Canaries. Flore de l'archipel. Librairie des Sciences Naturelles Paul Klincksieck, Paris (reprint: Otto Koeltz, Koenigstein 1973).

Praeger RL (1967) An account of the Sempervivum group. In Praeger, R.L. [reedición] (1929) *Semperviva* of the Canary Islands área. Proc Roy Irish Acad 38(Sect. B):454–499, como Plant Monograph vol. 1, Lahre.

Puppo P (2015) Diversification patterns of Micromeria Benth. (*Lamiaceae*) in the Canary Islands: genetic and morphological implications of inter-island colonization. PhD Thesis, inédita, University of Porto, Portugal

Reichter H (2021) Erneute Klarstellung: Rosa dumalis und nicht Rosa vosagiaca ist der korrekte Name für die Blaugrüne Rose (Vogesen-Rose). Kochia 14:51–59.

Rivas Goday S, Borja J (1961) Estudio de la vegetación y Flórula del Macizo de Gúdar y Jabalambre. Anales Inst Bot Cavanilles 19:328

Rivas-Martínez S, Wildpret W, del Arco Aguilar M, Rodríguez Delgado O, Pérez de Paz PL, García Gallo A, Acebes JR, Díaz TE, Fernández-González F (1993) Las comunidades vegetales de la Isla de Tenerife (Islas Canarias). Itinera Geobotanica 7:169–374

Rivas-Martínez S, Díaz TE, Fernández-González F, Izco J, Loidi J, Lousa M, Penas A (2002) Vascular plant communities of Spain and Portugal. Addenda to the Syntaxonomical checklist of 2001. Itinera Geobotanica 15(1):5–432. Servicio de Publicaciones de la Universidad de León

Rivas-Martínez S et al (2011) Mapa de series, geoseries y geopermaseries de vegetación de España [Memoria del mapa de vegetación potencial de España]. Itinera Geobotanica 18(1 & 2):5–800

Rodríguez Piñero JC, Wildpret de la Torre W, Del Arco-Aguilar M (1987) Contribución al estudio biosistemático de *Salix canariensis* (Salicaceae). Vieraea 17:121–142

Rodríguez ML, Fariña B, Rodríguez M, Escolástico Moreno J (2009) *Arrhenatherum calderae* A. Hansen (Poaceae) y *Petrorhagia nanteuilii* (Burnat) P.W. Ball & Heywood (Caryophyllaceae) en el Cono del Volcán Teide, Parque Nacional del Teide (Tenerife, Islas Canarias). Rev Acad Canar Cienc XXI(3–4):45–48

Rumeu B, García C, Nogales M (2015) El sistema de dispersión de semillas del cedro canario (*Juniperus cedrus*) en el Parque Nacional del Teide: aproximación ecológico y molecular. In:

Proyectos de investigación en Parques Nacionales 2010-2013. Organismo Autónomo Parques Nacionales (2014), pp 307–324

Salvo Tierra E (1990) Guía de Helechos de la Península Ibérica y Baleares. Ediciones Pirámide S.A, 377 pp

Sangüensa-Barreda G, García-Cervigón A, García Hidalgo M, Rozas V, Martín-Esquivel JL, Martín-Carbajal J, Martínez R, Olano JM (2022) Vertical cliffs harbor millennia-old junipers in the Canary Islands. Ecology 103:e3633. https://esajournals.onlinelibrary.wiley.com/doi/epdf/10.1002/ecy.3633

Schenck H (1907). Beiträge zur Kenntnis der Vegetation der Canarischen Inseln. Wiss. Ergebn. Dtsch. Tiefsee-Exped. Valdivia 1898-1899. 2. Bd., 1:2/3, 225-406

Socorro JS (1985) Contribución al conocimiento de la Flora y Vegetación del Piso Bioclimático Supracanario: Vegetación higro-hidrófila de Las Cañadas del Teide. Memoria de Licenciatura inédita, Departamento de Botánica, Facultad de Biología, Universidad de La Laguna, 92 páginas

Sventenius ERS (1946) Notas sobre la Flora de Las Cañadas de Tenerife. Bol Inst Nac Invest Agronómicas 15(78):148–171

Sventenius ERS (1960) Additamentum ad floram canariensem I. Instituto Nacional de Investigaciones Agronómicas. Ministerio de Agricultura, Madrid, 25 pp

UICN (1994) Categorías de las Listas Rojas de la UICN. Preparadas por la Comisión de Supervivencia de Especies de la UICN. Adoptadas por la 40ª Reunión del Consejo de la UICN, Gland, Suiza

Vargas P, Luceño Garcés M (2024) On the genus *Rosa* (*Rosaceae*) in the Canary Islands: proposal of three new species. Flora Montiberica 89:89–96

VVAA (2000) Lista Roja de Flora Vascular Española (valoración según categorías UICN). Conservación Vegetal 6(extra):11–38

Walter KS, Gillett HJ (eds) (1998) 1997 IUCN Red List of Threatened Plants. Compiled by The World Conservation Monitoring Centre. IUCN – The World Conservation Union, Gland/Cambridge, UK

Webb P, Berthelot S (1836–1850) Histoire naturelle des Iles canaries. Phytographia Canariensis. Béthune, Paris

Weller AA, Reichert H (2023) On the identity of the Teyde dog-rose (*Rosaceae*): evidence for a new endemic taxon from Tenerife, Spain. Phytotaxa 578:261–274

Wildpret de la Torre W, Martín Osorio VE (2005) Dos nuevas asociaciones para la vegetación del Parque Nacional del Teide: *Poo infirmae – Saginetum stoloniferi* ass. nova y *Violo cheiranthifoliae – Adenocarpetum viscossii* ass. nova. Vieraea 33:359–365. Santa Cruz de Tenerife

Wildpret de la Torre W, Socorro JS, Arencibia MC (1987) *Mentho-Caricetum calderae*, comunidad higro-hidrófila del piso supracanario de Tenerife (Islas Canarias). Ser Informes 22:35–40

Chapter 4
Vegetation of Teide National Park

4.1 Vegetation Belts/Series

For the preparation of the syntaxa, the groups of plant communities in a landscape, catalogue, the 500 × 500 m digital grid network of the Teide National Park (Fig. 4.1) was taken as a reference base, determining for each of them and for each plant entity within them, a set of information through a database in Access format (Microsoft) that brings together the following details:

At the 500 × 500 m grid level:

- UTM coordinates of the vertices and central point of the grid.
- Catalogue of vegetation units or syntaxa recognized within the grid.
- Phytosociological inventories of vegetation units recognized according to grid.
- List of grids where a certain taxon appears inventoried within a vegetation unit.

The georeferenced systematization of the information package collected according to the 500 × 500 m grid networks or 250 × 250 m subgrids has been carried out by connecting the table structure of the *Fichas Florísticas y Fitocenoticas*.mdb database with a Geographic Information Systems (GIS) software that, when applied to the phytosociological methodology, has given rise to the implementation of a new concept, the **Geobotanical Information System** (Martín Osorio, Wildpret de la Torre, Hernández Bolaños 2005) that has been developed using the open source free software QGIS v. 2.6.1 (Brighton). The basic sampling unit was established at the association level, although it was also detailed down to the sub-association level and, in some cases, also to the facies level.

V. E. Martín Osorio et al., *Flora and Vegetation of The Teide National Park*,
https://doi.org/10.1007/978-3-031-88400-9_4

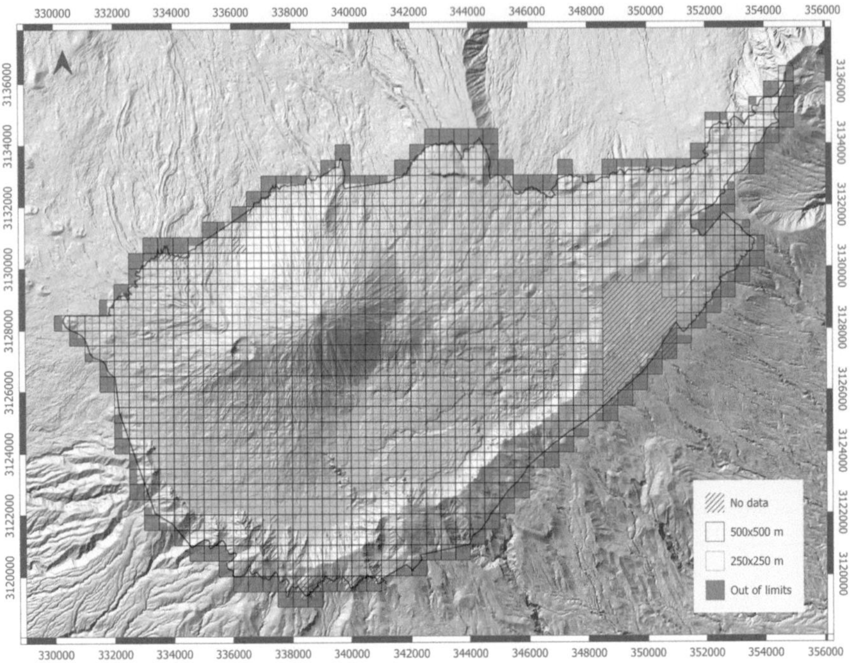

Fig. 4.1 Grid network in Teide National Park

Vegetation Units

I. ANOMODONTO-POLYPODIETEA Rivas Martínez 1975

Chasmocomophytic, epiphytic and scree vegetation consisting of medium or small sized bryophytes and ferns. Thermo, Meso, Supra-Mediterranean mesophytic with a dry to hyperhumid ombrotype.

***ANOMODONTO-POLYPODIETALIA** O. Bolós & Vives *in* O. Bolós 1957

*******Selaginello denticulatae-Anogrammion leptophyllae*** Rivas-Martínez, Fernández-González & Loidi 1999

Bryo-pteridophytic and ephemeral herbaceous communities on shady slopes and walls rich in organic matter with small sciophilous terrestrial comophytes that thrive in pluviseasonal Thermo-Mesomediterranean bioclimates.

Characteristic on the island of Tenerife: *Anogramma leptophylla*

Community of *Anogramma leptophylla*

Community that develops in small earthly cavities facing N-NW from 2000 to 2200 m above sea level, where *Anogramma leptophylla* (Fig. 4.2) is frequently found accompanied by lichens and bryophytes.

Fig. 4.2 *Anogramma leptophylla*. (Photo credit: Victoria Eugenia Martín Osorio)

II. CHAMAECYTISO ANGUSTIFOLII-PINETEA CANARIENSIS Rivas Goday & Esteve *ex* Esteve 1969

This Phytosociological Class includes the Canarian coniferous forests, retamoid shrubland (escobonales, codesares or retamares) and the colonizing and serial micro shrubland native of Tenerife and La Palma, as well as the oromediterranean sub-shrub of the Teide volcanic cone, climatophilous, edaphoxerophilous, pyrophilous or serial, endemic to the midlands and summits of the Central-Western Canary Islands, typical of dry-humid Meso-Oromediterranean bioclimates, of the Western Canary Islands biogeographic subprovince (Rivas-Martínez 2009).

Within this strictly western Canarian class, two orders are recognised, *Chamaecytiso angustifolii—Pinetalia canariensis* Rivas Goday & Esteve *ex* Esteve 1969 and *Spartocytisetalia supranubii* Schönfelder & Voggenreiter 1994. The first includes two alliances, *Cisto-Pinion canariensis* Rivas Goday & Esteve *ex* Esteve 1969, poorly represented in the studied territory and *Juniperion cedri* Martín Osorio et al. 2007. The second order brings together three alliances, *Spartocytision supranubii* Oberdorfer ex Esteve 1973, *Violion cheiranthifoliae* Voggenreiter *ex* Rivas Martínez, Wildpret & Martín Osorio 2007 and *Plantaginion webbii* Martín Osorio, Wildpret & Rivas Martínez 2007.

*CHAMAECYTISO ANGUSTIFOLII-PINETALIA CANARIENSIS** Rivas Goday & Esteve *ex* Esteve 1969

Cisto- Pinion canariensis Rivas Goday & Esteve *ex* Esteve 1969

Alliance that groups climatophilous and edaphoxerophilous meso-macroforests composed of Canary Island pines, brooms and flatpods, essentially dry-humid Meso-Supra-Mediterranean forests with a western Canary Islands distribution.

II.1. ***Sideritido solutae- Pinetum canariensis*** Esteve 1973

Head of series association of the Tenerife Mediterranean pluviseasonal oceanic Thermo-Supra-Mediterranean dry-subhumid semi-hyperoceanic climatophilous and edaphoxerophilous Series of *Pinus canariensis* forests and *Sideritis soluta.*

II.1.1 ***typicum***

II.1.2 ***cistetosum osbaeckiaefolii*** Rivas- Martínez et al., 1993 present in the areas of Fortaleza and Boca Tauce, developed on siliceous soils with a poor base, acidic phonolites of the dry Mesomediterranean belt.

II.1.3 ***ericetosum canariense*** (del Arco, P. Pérez & Wildpret 1987) Rivas-Martínez et al. 1993. Subhumid Thermo-Meso-Mediterranean with northern slopes.

II.1.4 ***spartocytisetosum supranubii*** Rivas-Martínez et al. 1993 of shady areas and ecotonic zones with the dry mesophytic Supra-Mediterranean association of Teide broom.

II.2. ***Adenocarpo viscosi-Chamaecytisetum angustifolii*** Martín Osorio, Hernández Bolaños & Wildpret *ass. nova*

Type Relevé N° 1, Table 4.1 Codesar-escobonal that forms the heliophilous fringe of dry-humid Meso-Supra-Mediterranean pine forests on the island of Tenerife. In the Cañadas del Teide it is located in an arc almost bordering the protected area (Figs. 4.3 and 4.4).

Fig. 4.3 Adenocarpo *viscosi-Chamaecytisetum angustifolii* near Boca Tauce. June 2006. (Photo credit: Victoria Eugenia Martín Osori)

Table 4.1 *Adenocarpo viscosi-Chamaecytisetum angustifolii* Martín Osorio, Hernández Bolaños & Wildpret *ass. nova Cisto-Pinion canariensis, Chamaecytiso-Pinetea*

N° Relevé	**1**	**2**	**3**	**4**	**5**	**6**	**7**	**8**	**9**	**10**	**11**	**12**
Altitude	1993	1875	1780	1820	1980	2100	1860	2025	2084	2218	2047	2016
Surface	100	100	100	100	100	100	100	100	400	400	225	100
Slope	–	40	15	15	60	15	15	10	10	20	–	20
Exposure	–	N	E	N	N	N	N	N	W	S	–	W
Coverage	30	70	70	60	80	70	70	70	40	60	30	5
N° Species	7	6	5	5	5	5	4	4	4	4	4	4
UTM	334874—3130647	338875—3132375	335875—3131375	335875—3131125	335625—3130625	333875—3130125	335875—3130875	334596—3130449	333186—3129440	333667—3129304	333692—3129914	332077—3128561
Date	26/6/03	18/6/03	25/6/03	25/6/03	25/6/03	11/6/03	25/6/03	26/6/03	24/6/02	24/6/02	11/6/03	21/6/02
Characteristics species of Association												
Adenocarpus viscosus subsp. viscosus	2	1	2	+	3	.	3	.	.	.	2	2
Chamaecytisus proliferus subsp. angustifolius	1	1	+	+	.	1	.	3	1	1	1	+
Characteristics of Chamaecytiso Pinetea												
Pinus canariensis	2	+	2	2	3	+	3	3	3	3	2	2
Scrophularia glabrata	1	.	.	.	.	+	.	1	.	.	+	.
Tolpis webbii	.	+	.	.	.	+	.	1	.	.	.	.
Erysimum scoparium	.	.	.	.	.	.	.	.	1	1	.	.
Characteristics of Cisto-Pinion												
Cistus symphytifolius	+	.	1	+	+	.	+	.	.	.	.	.

(continued)

Table 4.1 (continued)

N° Relevé	1	2	3	4	5	6	7	8	9	10	11	12
Altitude	1993	1875	1780	1820	1980	2100	1860	2025	2084	2218	2047	2016
Surface	100	100	100	100	100	100	100	100	400	400	225	100
Slope	–	40	15	15	60	15	15	10	10	20	–	20
Exposure	–	N	E	N	N	N	N	N	W	S	–	W
Coverage	30	70	70	60	80	70	70	70	40	60	30	5
N° Species	7	6	5	5	5	5	4	4	4	4	4	4
UTM	334874—3130647	338875—3132375	335875—3131375	335875—3131125	335625—3130625	333875—3130125	335875—3130875	334596—3130449	333186—3129440	333667—3129304	333692—3129914	332077—3128561
Date	26/6/03	18/6/03	25/6/03	25/6/03	25/6/03	11/6/03	25/6/03	26/6/03	24/6/02	24/6/02	11/6/03	21/6/02
Bystropogon origanifolius var. origanifolius	+	1	1	2	1	.	1	.	.	.	.	.
Characteristics of Spartocytision supranubii												
Argyranthemum tenerifae	.	.	.	.	.	.	.	.	1	1	.	.
Characteristics of Pterocephalenion lasiospermi												
Pterocephalus lasiospermus	+	1	.	.	+	+	+	.	.	.	.	.
Other species												
Wahlenbergia lobelioides subsp. lobelioides	.	.	.	.	.	.	.	.	.	.	.	1

Localities: 1: Asientos de las Colmenas (Teide National Park boundary on the northern flanks of Cerro las Chajoras—Pico Viejo). 2: Northern slopes of Teide (NW-Echisere). 3, 4, 5: Los Asientos de Luis González (Western ridge of the Teide Ravine). 6: North of Los Frontones (Coladas del Cerro de las Chajoras). 7: Los Asientos de Luis González (Western ridge of the Teide Ravine). 8: Asientos de las Colmenas (Teide National Park boundary on the northern flanks of Cerro las Chajoras—Pico Viejo). 9, 10, 11: Los Frontones (NW slopes of the Cerro Las Chajoras-Pico Viejo). 12: Cerro de los Roques Blancos (N-Montaña Reventada)

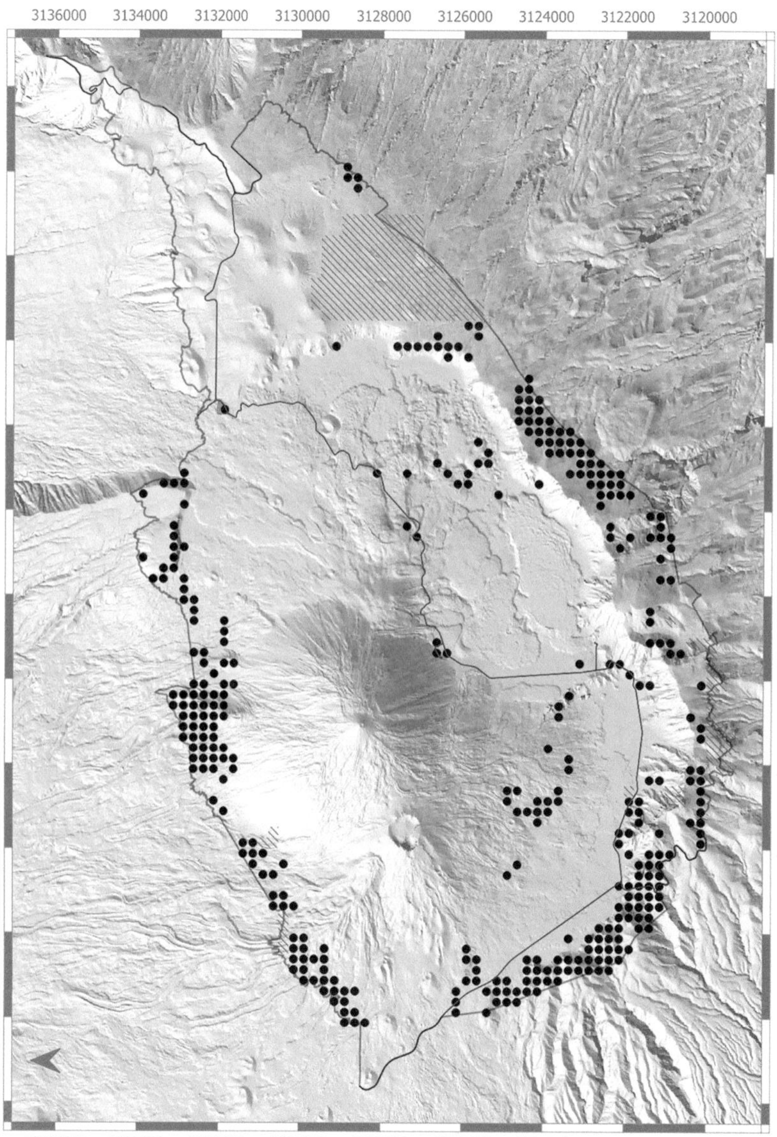

Fig. 4.4 Distribution of Codesar-Escobonal *Adenocarpo viscosi-Chamaecytisetum angustifolii*

Juniperion cedri Martín Osorio, Wildpret & Rivas-Martínez 2007

Alliance that groups open edaphoxerophilous microforests of Canary cedars *Juniperus cedrus* that thrive in the leptosols of the escarpments of the Cañadas del Teide and the Caldera de Taburiente from the upper horizon of the Mesomediterranean belt to the upper dry-humid Supramediterranean belt, out of the reach of trade wind fogs.

II.3. *Adenocarpo viscosi-Juniperetum cedri* Martín Osorio, Wildpret & Rivas-Martínez 2007

Head of association of the Tenerife edaphoxerophilous Series of Mediterranean litho-solic, pluviseasonal, oceanic, Supra-Mediterranean, dry-subhumid, euoceanic microforests of *Juniperus cedrus* and *Adenocarpus viscosus* with *Echium wildpretii* (Tables 4.2a and 4.2b) (Fig. 4.5).

Fig. 4.5 Open Forest of *Juniperus cedrus* cedars in the highland areas of La Fortaleza. November 2011. (Photo credit: Victoria Eugenia Martín Osorio)

Table 4.2a *Adenocarpo viscosi-Juniperetum cedri* Martín Osorio, Wildpret & Hernández Bolaños
Juniperion cedri **Martín Osorio, Wildpret & Rivas-Martínez**
Chamaecytiso angustifolii-Pinetalia canariensis. Chamaecytisetea-Pinetea canariensis

N° Relevé	**1**	**2**	**3**	**4**	**5**	**6**	**7**	**8**	**9**	**10**
Altitude	2343	2070	2350	2300	2350	2346	2350	2300	2050	2325
Exposure	SSW	S	NO	NNE	NO	SW	NNE	S	S	W
Slope	50	10	30	35	10	5	0–10	20	100–50	30
Coverage	60	70	50	50	50	40	60	50	60	60
Surface	25	100	50	400	25	25	200	300	100	25
N° Species	11	11	10	10	10	10	9	8	9	9
UTM	342750—3132750	342750—3133250	343750—3133250	343750—3133250	343750—3133250	342750—3132750	343750—3133250	343750—3133250	342750—3133250	348250—3127750
Characteristics species of Association										
Juniperus cedrus	1	+	2	2	2	1	3	2	2	2
Adenocarpus viscosus subsp. viscosus	1	.	+	+	+	1	2	1	+	3
Characteristics of Spartocytision supranubii										
Spartocytisus supranubius	3	3	1	2	2	2	2	2	2	+
Echium wildpretii subsp. wildpretii	.	.	.	.	.	.	.	.	+	.

(continued)

Table 4.2a (continued)

N° Relevé	**1**	**2**	**3**	**4**	**5**	**6**	**7**	**8**	**9**	**10**
Altitude	2343	2070	2350	2300	2350	2346	2350	2300	2050	2325
Exposure	SSW	S	NO	NNE	NO	SW	NNE	S	S	W
Slope	50	10	30	35	10	5	0–10	20	100–50	30
Coverage	60	70	50	50	50	40	60	50	60	60
Surface	25	100	50	400	25	25	200	300	100	25
N° Species	11	11	10	10	10	10	9	8	9	9
UTM	342750—3132750	342750—3133250	343750—3133250	343750—3133250	343750—3133250	342750—3132750	343750—3133250	343750—3133250	342750—3133250	348250—3127750
Characteristics of Plantaginion webbii, Pterocephalenion lasiospermi										
Pterocephalus lasiospermus	3	1	.	3	2	2	2	1	1	3
Micromeria lacnophylla	1	.	2	3	2	.	3	3	.	2
Tolpis webii	1	.	+	+	1	.	1	.	.	.
Erysimum scoparium	.	1	+	+	1	.	+	1	1	.
Carlina xeranthemoides	+	.	2	+	2	1	.	3	.	.
Silene berthelotiana	.	.	.	+	+	.	.	.	.	.
Ferula linkii	.	1	+	+	.	+	.	.	+	.

Altitude	2343	2070	2350	2300	2350	2346	2350	2300	2050	2325
Exposure	SSW	S	NO	NNE	NO	SW	NNE	S	S	W
Slope	50	10	30	35	10	5	0–10	20	100–50	30
Coverage	60	70	50	50	50	40	60	50	60	60
Surface	25	100	50	400	25	25	200	300	100	25
N° Species	11	11	10	10	10	10	9	8	9	9
UTM	342750—3132750	342750—3133250	343750—3133250	343750—3133250	343750—3133250	342750—3132750	343750—3133250	343750—3133250	342750—3133250	348250—3127750
Lotus campylocladus	.	.	+	.	+	.	.	2	.	.
Arrhenatherum calderae	.	.	+	.	.	.	.	.	.	.
Argyranthemum tenerifae	.	.	.	.	.	.	+	.	.	.
Erica canariensis	.	.	.	.	.	.	+	.	.	.
Descurainia bourgeauana	.	1	.	.	.	.	.	.	.	.
Cheirolophus teydis	2	.	.	.	.	2	.	.	.	1
Characteristics of Chamaecytiso-Pinetea										
Pinus canariensis	.	.	.	.	.	.	.	.	2	.
Chamaecytisus proliferus subsp. angustifolius	.	1	.	.	.	+	.	.	.	.
Sideritis oroteneriffae	2	.	.	.	.	.	.	.	.	+

(continued)

Table 4.2a (continued)

N° Relevé	**1**	**2**	**3**	**4**	**5**	**6**	**7**	**8**	**9**	**10**
Altitude	2343	2070	2350	2300	2350	2346	2350	2300	2050	2325
Exposure	SSW	S	NO	NNE	NO	SW	NNE	S	S	W
Slope	50	10	30	35	10	5	0–10	20	100–50	30
Coverage	60	70	50	50	50	40	60	50	60	60
Surface	25	100	50	400	25	25	200	300	100	25
N° Species	11	11	10	10	10	10	9	8	9	9
UTM	342750—3132750	342750—3133250	343750—3133250	343750—3133250	343750—3133250	342750—3132750	343750—3133250	343750—3133250	342750—3133250	348250—3127750
Scrophularia glabrata	+	1	.	.	.	.	.	.	.	.
Pimpinella cumbrae	+	.	.	.	.	1	.	.	.	2
Nepeta teydea	.	.	.	.	.	1	.	.	.	+
Bufonia paniculata	.	.	.	.	.	.	.	.	.	.
Cistus osbeckiifolius	.	1	.	.	.	.	.	.	.	.
Other species										
Rhamnus integrifolia	.		.	.	.	.	.	.	1	.
Festuca agustinii	.	1	.	.	.	.	.	.	.	.
Bituminaria bituminosa	.	1	.	.	.	.	.	.	.	.

Localities: From 1 to 10: La Fortaleza

Table 4.2b *Adenocarpo viscosi-Juniperetum cedri* Martín Osorio, Wildpret & Hernández Bolaños
Juniperion cedri **Martín Osorio, Wildpret & Rivas-Martínez**
Chamaecytiso angustifolii-Pinetalia canariensis. Chamaecytisetea-Pinetea canariensis

N° Relevé	**11**	**12**	**13**	**14**	**15**	**16**	**17**	**18**	**19**	**20**
Altitude	2200	2350	2350	2200	2392	2330	2200	2348	2320	2200
Exposure	SO	NNE	S	NNE	W	NW	NNE	E	NNO	S
Slope	40	10	30	5	20	20	10	10	0–5	40
Coverage	40	50	50	50	60	50	50	50	60	40
Surface	200	25	25	50	25	25	25	50	100	200
N° Species	7	7	7	7	7	6	6	6	5	5
UTM	347750—3127250	343750—3133250	343750—3133250	347750—3127250	347750—3127750	347750—3127750	347750—3127750	348250—3127750	348250—3127750	347750—3127750
Characteristics species of Association										
Juniperus cedrus	2	1	1	2	3	3	2	3	3	2
Adenocarpus viscosus subsp. viscosus	1	.	1	1	3	.	.	+	2	.
Characteristics of Spartocytision supranubii										
Spartocytisus supranubius	1	+	+	2	+	2	2	3	2	+
Echium wildpretii subsp. wildpretii	.	.	.	.	.	.	.	.	.	.
Characteristics of Plantaginion webbii, Pterocephalenion lasiospermi										
Pterocephalus lasiospermus	.	+	.	3	.	1	2	1	2	.

(continued)

Table 4.2b (continued)

N° Relevé	**11**	**12**	**13**	**14**	**15**	**16**	**17**	**18**	**19**	**20**
Altitude	2200	2350	2350	2200	2392	2330	2200	2348	2320	2200
Exposure	SO	NNE	S	NNE	W	NW	NNE	E	NNO	S
Slope	40	10	30	5	20	20	10	10	0–5	40
Coverage	40	50	50	50	60	50	50	50	60	40
Surface	200	25	25	50	25	25	25	50	100	200
N° Species	7	7	7	7	7	6	6	6	5	5
UTM	347750—3127250	343750—3133250	343750—3133250	347750—3127250	347750—3127750	347750—3127750	347750—3127750	348250—3127750	348250—3127750	347750—3127750
Micromeria lacnophylla	1	2	1	2	.	.	2	.	2	.
Tolpis webii	.	.	.	1	.	+	1	.	.	.
Erysimum scoparium	.	.	.	1	.	+	+	.	.	.
Carlina xeranthemoides	1	2	3	.	.	.	.	.	.	1
Silene berthelotiana	.	.	.	.	.	.	.	.	.	.
Ferula linkii	.	.	.	.	.	.	.	.	.	.
Lotus campylocladus	1	2	1	.	.	.	.	.	.	.
Arrhenatherum calderae	.	.	.	.	.	.	.	.	.	.
Argyranthemum tenerifae	.	.	.	.	.	.	.	.	.	.
Erica canariensis	.	.	.	.	.	.	.	.	.	.
Descurainia bourgeauana	.	.	.	.	.	.	.	.	.	.
Cheirolophus teydis	.	.	.	.	2	.	.	2	.	.
Characteristics of Chamaecytiso-Pinetea										
Pinus canariensis	.	.	.	.	.	.	.	.	.	.
Chamaecytisus proliferus subsp. angustifolius	.	.	.	.	.	.	.	.	.	2
Sideritis oroteneriffae	.	.	.	.	+	.	.	.	.	.

Altitude	2200	2350	2350	2200	2392	2330	2200	2348	2320	2200
Exposure	SO	NNE	S	NNE	W	NW	NNE	E	NNO	S
Slope	40	10	30	5	20	20	10	10	0–5	40
Coverage	40	50	50	50	60	50	50	50	60	40
Surface	200	25	25	50	25	25	25	50	100	200
N° Species	7	7	7	7	7	6	6	6	5	5
UTM	347750—3127250	343750—3133250	343750—3133250	347750—3127250	347750—3127750	347750—3127750	347750—3127750	348250—3127750	348250—3127750	347750—3127750
Scrophularia glabrata	.	.	.	.	.	.	.	.	.	.
Pimpinella cumbrae	.	.	.	.	2	3	.	2	.	.
Nepeta teydea	.	.	.	.	.	.	.	.	.	.
Bufonia paniculata	.	.	.	.	1	.	.	.	.	.
Cistus osbeckiifolius	3	.	+	.	.	.	.	.	.	2
Other species										
Rhamnus integrifolia	.	1	.	.	.	.	.	.	.	.
Festuca agustinii	.	.	.	.	.	.	.	.	.	.
Bituminaria bituminosa	.	.	.	.	.	.	.	.	.	.

Localities: From 11 to 13 and 20: La Fortaleza. From 14 to 19: Risco

***SPARTOCYTISETALIA SUPRANUBII** Schönfelder & Voggenreiter 1994
Order representing the retamoid tall shrublands, dwarf shrublands and mountain and high mountain subshrub, climatophilous, edaphoxerophilous and serial, presided over by Teide brooms (*Spartocytisus supranubius*), the primocolonizing, permanent or serial dwarf shrublands of *Plantago webbii* and the subshrub of Teide violets (*Viola cheirantifolia* & *Viola guaxarensis*), dry-humid Meso-Oro-Mediterranean.

****Spartocytision supranubii** Oberdorfer *ex* Esteve 1973
Alliance that groups the high mountain growing of Teide brooms, rockroses and summit grasses, climatophilous or edaphoxerophilous, sometimes catenales or serials of the Canary pine forests, with a dry-humid Meso-Oro-Mediterranean distribution in La Palma and Tenerife. In the cone of Pico del Teide these communities reach 3000 m above sea level.

II.4. ***Spartocytisetum supranubii*** Oberdorfer *ex* Esteve 1973
Head association of the Tenerife Mediterranean climatophilous and edaphoxerophilous Series of the upper Mesomediterranean-dry-subhumid oceanic Mesomediterranean-oromediterranean euceanic altifruticedas of *Spartocytisus supranubius* Table 4.3.

II.4.1 ***typicum* ***

II.4.2 ***echietosum wildpretii*** P. Schönfelder *in* P. Schönfelder & Voggenreiter 1994
In scree with a considerable slope, such as on the slopes of the Ucanca Valley (Fig. 4.6) La Cortada del Diablo, El Riachuelo, Risco Verde, Topo de la Grieta, small bottoms of temporary channels after the melting of the snow or during the storms of autumn and winter.

II.4.3 ***descurainietosum lemsii*** Martín Osorio & Wildpret 2007
(Meso)-Supra-Mediterranean subassociation of summit broom with *Descurainia lemsii* in the vicinity of La Crucita Mountain and the summit of Pedro Gil at around 2200 m a.s.l. It behaves as an introgression of the summit broom into the dry upper Mesomediterranean belt on the eastern borders of the National Park. The volcanic materials are represented by recent lapillis and cones filled by the so-called "dorsal effect" of continuity in the formation of the La Esperanza Ridge. These are non-welded pyroclastic deposits of an ignimbrite nature (Martí 2003) that could range between 260,000 and 179,000 years old.

Fig. 4.6 *Spartocytisetum supranubii echietosum wildpretii,* slopes Valle Ucanca. May 2011. (Photo credit: Victoria Eugenia Martín Osorio)

II.4.4 ***descurainietosum gonzalesii*** Martín Osorio & Wildpret 2007
Dry upper Mesomediterranean subassociation of the broom bushes of the Pre-caldera wall or building on its southern slope with *Descurainia gonzalesii*. It is also found punctually in Los Roques de García in a westerly Exposure, in Risco Verde and in the vicinity of Montaña del Cerrillar and Arenas Negras, in the western limits of the National Park in the surroundings of Boca Tauce and in Cañada del Cedro. The materials in which it is found belong to the Lower Group of the Cañadas Building with an age between 3 and 2 million years and made up of mostly welded monolithic pyroclasts that give the appearance of flows, but which in reality are pyroclastic deposits that were welded together when they fell due to the high temperature.
II.4.5 Primocolonizing faciation of recent badlands and lapillis.
This faciation is frequently found on very loose lapillis materials, at the bottom of small ravines and on almost uncolonized badlands where the summit broom forms monospecific populations. It is found dispersed throughout the Teide National Park, especially on the southern slopes of the Teide cone, the southern slopes of Pico Viejo and Roque de los Almendros and is sometimes accompanied by *Adenocarpus viscosus* subsp. *viscosus*.

II.5. *Descurainio bourgeauanae-Spartocytisetum supranubii* (Esteve 1973) Martín Osorio, Wildpret & Hernández Bolaños 2007 (*Spartocytisetum supranubii descurainietosum bourgeauanae* Esteve 1973) (Table 4.3).

Association that extends from the beginning of the National Park limits in La Crucita to Portillo, colonizing from here and up to the Roques de García the pumice substrates and recent materials after the Cañadas landslide. On hard materials and before the landslide, the *Spartocytisetum supranubii* association is established. The latter is an endemic association of the island of Tenerife and represents the mature stage of the summit (Meso)-Supra-Oro-Mediterranean summit brooms of Las Cañadas del Teide, which is established preferably on volcanic materials belonging to the Upper and Lower Cañadas Building. The new association *Descurainio bourgeauanae-Spartocytisetum supranubii* was considered by Esteve Chueca (1973) as a nitrophilous sub-association of the typical brooms without validating it.

Following the research work carried out in recent years, we have considered that it represents a community with its own geographical entity and restricted to the dry-subhumid Supra-Mediterranean belts. It is installed on post-landslide fill materials of La Orotava formed by lava and pyroclastic accumulation cones (Navarro Latorre 2000) (Fig. 4.7) that are very mobile and belong to several volcanic episodes. This lapilli-picón material is a good hygroscopic material that captures and maintains moisture in the soil for a long period of time. From the bioclimatic point of view, it corresponds to the so-called cold Cañadas, periodically affected by the rise of the sea of clouds that reaches over 2100 m above sea level and dissipates at the height of the Montaña de la Mostaza. Outside this area of accumulation cones, this community settled on pumice materials from Montaña Blanca and other emission centres, probably from the northern slope of Teide, which has now disappeared due to a process of gravitational sliding towards the bottom of the sea, which, in the form of superficial pyroclastic rain, buried a large part of the territory (Fig. 4.8).

Fig. 4.7 *Descurainio bourgeauanae-Spartocytisetum supranubii.* On post-landslide volcanic cones, north orientation. 31 May 2009. (Photo credit: Victoria Eugenia Martín Osorio)

Fig. 4.8 *Spartocytisus supranubius* with *Descurainia bourgeauana,* south orientation with pumices stones. 31 May 2009. (Photo credit: Victoria Eugenia Martín Osorio)

II.5.1. *typicum.*
II.5.2. *erigerontetosum calderae* Martín Osorio & Wildpret 2007 (Table 4.3).

In sandy environments formed by sediments that temporarily suffer from waterlogging due to the accumulation of runoff water from surrounding lava flows composed of harder and more impermeable materials, as observed in the plains of Siete Cañadas, or in the Llano de Maja, a sub-association of broom and grass develops, where *Erigeron calderae* is the characteristic species. It is a community that extends between the clayey retraction cracks originating in the temporarily flooded endorheic deposits of Siete Cañadas, Llanos deMaja,the dividing line between Montaña Blanca and Montaña Rajada, the base of the Altavista Trail andthebaseoftheFasnia-Volcano.Itisnecessarytomentiontheinfluence of snow melting on these biotopes. The mild presence of periglacial phenomena in the National Park is responsible for various typesofcryonivalformsthat superficially model the volcanic substrates. One of these forms is called "periglacial garlands" (Martínez de Pisón and Quirantes 1981) which specificallyinthe vicinity of Montaña Rajada originate because of the melting of winter snow on platforms located on agentleslopeofpumice. Theyappeargeneratinglobularfigures with compacted fine-grained central materials closed downwards by arches of coarser and looser elements by dragging these surface materials infavorofthe aforementioned slope. The detrital materials rich in fine elements, such as those found in the depression of Llano de Maja, or Siete Cañadas,retainsurfacewater either from melting snow, thawing, rain (scarce) or by segregation attract groundwater at the time of frost to return it to the surfacewhenthawingoccurs;this is the optimal biotope for this sub-association.

Synthetic table of the associations and subassociations of the summit brooms Spartocytision supranubii N° of order1234N° Relevé182722Spartocytisus supranubiusV+VVDescurainia bourgeauanaVV..erigerontetosum calderae Erigeron calderae.V..descurainietosum gonzalesii Descurainia gonzalesii..V.descurainietosum lemsii Descurainia lemsii...VCharacteristic of Spartocytision supranubii Nepeta teydea var. teydeaIIIVIIIIArgyranthemum tenerifaeIII.II.Tolpis webbiiII.IIIICheirolophus teydis..II.Echium wildpretii subsp. wildpretii..II.Characteristic of Pterocephalenion lasiospermi Pterocephalus lasiospermusIIVVIArrhenatherum calderaeIV.III.Erysimum scoparium.. IIIScrophularia glabrata...ICharacteristic of Chamaecytiso-Pinetea Pinus canariensis..IIMicromeria lachnophylla...ICarlina xeranthemoides..IV+Adenocarpus viscosus subsp. viscosus..IIIAndryala pinnatifidaI...Bethencourtia palmensis..IV.Cistus symphytifolius..I.Sideritis soluta..I.Other species Bromus tectorumI...Galium aparine...IIBystropogon canariensis...+Aeonium smithii..+.Tolpis lagopoda..+.Festuca myuros..+.1: Descurainio bourgeauanae-Spartocytisetum supranubii 2: Descurainio bourgeauanae-Spartocytisetum supranubii erigerontetosum calderae3: Spartocytisetum supranubii descurainietosum gonzalesii 4: Spartocytisetum supranubii descurainietosum lemsii Origin of Relevé1–4: Prepared by the authors based on table 3 of Martín Osorio et al., 2007.

Table 4.3 Synthetic table of the associations and subassociations of the summit brooms

Spartocytision supranubii				
N° of order	**1**	**2**	**3**	**4**
N° Relevé	**18**	**2**	**7**	**22**
Spartocytisus supranubius	V	+	V	V
Descurainia bourgeauana	V	V	.	.
erigerontetosum calderae				
Erigeron calderae	.	V	.	.
descurainietosum gonzalesii				
Descurainia gonzalesii	.	.	V	.
descurainietosum lemsii				
Descurainia lemsii	.	.	.	V
Characteristic of Spartocytision supranubii				
Nepeta teydea var. teydea	III	V	I	III
Argyranthemum tenerifae	III	.	II	.
Tolpis webbii	II	.	III	I
Cheirolophus teydis	.	.	II	.
Echium wildpretii subsp. wildpretii	.	.	II	.
Characteristic of Pterocephalenion lasiospermi				
Pterocephalus lasiospermus	II	V	V	I
Arrhenatherum calderae	IV	.	III	.
Erysimum scoparium	.	.	II	I
Scrophularia glabrata	.	.	.	I
Characteristic of Chamaecytiso-Pinetea				
Pinus canariensis	.	.	I	I
Micromeria lachnophylla	.	.	.	I
Carlina xeranthemoides	.	.	IV	+
Adenocarpus viscosus subsp. viscosus	.	.	II	I
Andryala pinnatifida	I	.	.	.
Bethencourtia palmensis	.	.	IV	.
Cistus symphytifolius	.	.	I	.
Sideritis soluta	.	.	I	.
Other species				
Bromus tectorum	I	.	.	.
Galium aparine	.	.	.	II
Bystropogon canariensis	.	.	.	+
Aeonium smithii	.	.	+	.
Tolpis lagopoda	.	.	+	.
Festuca myuros	.	.	+	.
1: *Descurainio bourgeauanae-Spartocytisetum supranubii*				
2: *Descurainio bourgeauanae-Spartocytisetum supranubii erigerontetosum calderae*				
3: *Spartocytisetum supranubii descurainietosum gonzalesii*				
4: *Spartocytisetum supranubii descurainietosum lemsii*				
Origin of Relevé 1–4: Prepared by the authors based on table 3 of Martín Osorio et al., 2007				

Monospecific community that colonizes the grassy meadow: ***Descurainia bourgeauana***

There are populations of grassy meadow that colonize recent volcanic cones, after the landslide, such as in the Montaña Mostaza and around the Montaña Negra de Fasnia (Fig. 4.9).

Fig. 4.9 *Descurainia bourgeauana* near the Black Mountain of Fasnia. (Photo credit: Victoria Eugenia Martín Osorio)

II.6. ***Violo guaxarensis-Adenocarpetum viscosi*** Wildpret & Martín Osorio 2005 *em. prop.*

On a wide plain, formed by a rocky stony substrate of phonolitic origin, occasionally covered by pumice, located at about 2690–2700 m above sea level, southwest of the summit of Montaña Guajara (2715 m above sea level) with a gentle slope of about 5° and whose surface area exceeds just over 1 hectare, there is a dense "codesar" of *Adenocarpus viscosus* subsp. *viscosus*, very attached to the substrate and very branched, about 15–30 cm high due to extreme weather conditions (Fig. 4.10), especially the intensity of the wind, which comes from the southwest and circulates through this summit space.

Very few plants of the Teidean vegetation find refuge inside the branches of the codeso. *Viola guaxarensis* stands out for its abundance (Fig. 4.11). This floral combination is found scattered throughout the "codesar". In this sense, nine inventories of about 2 m^2 and one of 50 m^2 have been made, integrated in the attached table (Table 4.4). Only on two occasions have some specimens of *Erysimum scoparium* and small specimens of *Spartocytisus supranubius* been inventoried.

Fig. 4.10 Aspect of the Summits' codesar Guajara 17 Jun. 2003. (Photo credit: Victoria Eugenia Martín Osorio)

Table 4.4 *Violo guaxarensis-Adenocarpetum viscosii* Wildpret & Martín Osorio, 2005 *em. prop Spartocytision supranubii. Spartocytisetalia supranubii. Chamaecytiso-Pinetea canariensis*

Nª Relevé	**1**	**2**	**3**	**4**	**5**	**6**	**7**	**8**	**9**	**10**
Surface (m²)	2	2	2	2	2	2	2	2	2	50
Slope (%)	0	5	5	10	5	2	5	2	5	5
Altitude (m)	2660	2680	2670	2685	2690	2695	2700	2690	2685	2677
Coverage (%)	90	90	90	90	90	100	95	90	95	50
Exposure	SW	SW	SW	SW	SW	SW	SW	SW	SW	W
UTM X	341709	341709	341709	341709	341709	341709	341709	341709	341709	341697
UTM Y	3122262	3122262	3122262	3122262	3122262	3122262	3122262	3122262	3122262	3122091
N° Species	2	2	2	2	4	3	3	2	3	3
Date	17-6-03	17-6-03	17-6-03	17-6-03	17-6-03	7-6-04	7-6-04	7-6-04	7-6-04	15-4-03
Characteristic species of association										
Adenocarpus viscosus subsp. viscosus	5	5	5	5	5	5	5	5	5	4
Viola guaxarensis	1	+	1	+	1	1	2	1	1	1
Erysimum scoparium	.	.	.	.	+	.	+	.	.	.
Nepeta teydea	.	.	.	.	+	.	.	.	.	.
Tolpis webbii	.	.	.	.	.	.	.	.	.	2
Characteristic of higher units										
Spartocytisus supranubius	.	.	.	.	.	+	.	.	+	.

Localities: Guajara Peaks, Teide National Park

Fig. 4.11 Detail of *Viola guaxarensis* y *Adenocarpus viscosus* subsp. *viscosus*. 17 Jun. 2003. (Photo credit: Victoria Eugenia Martín Osorio)

****Violion cheiranthifoliae** Voggenreiter *ex* Rivas-Martínez, Wildpret & Martín Osorio 2007

Phytosociological alliance that groups the shrubland and perennial herbaceous plants of open permanent stands of Violeta del Teide (*Viola cheiranthifolia*), colonizers of the lapillis and recent lavas of the Oro-Mediterranean summit areas of the island of Tenerife.

II.7. *Violetum cheiranthifoliae* Rivas-Martínez, Wildpret, Del Arco, Rodríguez, Pérez de Paz, García Gallo, Acebes, Díaz & Fernández González 1993

Supra-Oromediterranean Association of Pilón de Azúcar del Teide and Montaña Blanca (Fig. 4.12), on lapilli deposits in scree environments between 2400 and 3500 m a.s.l.

It constitutes the Tenerife Mediterranean pluviseasonal oceanic oromediterranean dry euoceanic Geopermaseries of the open gravel-dwelling dwarf shrublands of *Viola cheiranthifolia* and *Silene nocteolens.*

Fig. 4.12 *Viola cheiranthifolia* on pumice in Montaña Blanca 3 July 2006. (Photo credit: Victoria Eugenia Martín Osorio)

****Plantaginion webii** Martín Osorio, Wildpret & Rivas Martínez 2007
Alliance that groups the pioneer microshrubland of "white pinillos", *Plantago webbii*, of lava flows, lapillis and other volcanic deposits, also behaves as serial communities on eroded soils of Canary pine forests, cedar forests or Teide broom forests, dry-humid Meso-Supra-Mediterranean.

*****Pterocephalenion lasiospermi** Martín Osorio, Wildpret & Rivas-Martínez 2007
Phytosociological suballiance that groups together permanent and serial dwarf shrubland associations that colonize volcanic deposits and denuded soils of pine forests, cedar forests and broom forests, dry-subhumid Meso-Supra-Mediterranean, endemic to the island of Tenerife.

II.8. *Echietum auberiani* P. Schönfelder *in* Schönfelder & Voggenreiter 1994
A community almost monospecific, poor in species and with a cover of 3%–5%, present preferably in open, disturbed pumices soils. It usually forms a vegetation complex with *Descurainea bourgeauana* (Fig. 4.13).

Fig. 4.13 *Echium auberianum*, Tajinaste picante o azul 12 Jun. 2006. (Photo credit: Victoria Eugenia Martín Osorio)

II.9. *Erysimo scoparii-Pterocephaletum lasiospermi* Rivas-Martínez et al. 1993
Formations of “rosalillo”, sometimes with *Erysimum scoparium* and sometimes in facies with Teide broom (*Spartocytisus supranubius*). It is a community that acts as a first colonizer or as a stage of degradation of Supramediterranean broom shrublands and Mesomediterranean pine forests on lithosols or stony ground.

II.10. *Arrhenathero calderae-Plantaginetum webbii* Martín Osorio & Wildpret 2003. Tables 4.5a and 4.5b
“Cerrillar-Crespar”, an endemic community that is the first colonizer of scree and gelifracted (criogenic soil) slopes of the ancient materials of the Pared de las Cañadas (Fig. 4.14), in the surroundings of the Cerrillar Mountain, Los Roques and La Fortaleza, subjected to gelifluction and geliremoval phenomena.

Fig. 4.14 *Arrhenathero calderae-Plantaginetum webbii* Cañada del Capricho Nov. 2003. (Photo credit: Victoria Eugenia Martín Osorio)

Table 4.5a *Arrhenathero calderae-Plantaginetum webbii* Martín Osorio & Wildpret de la Torre
Plantaginion webbii, Spartocytisetalia supranubii, Chamaecytiso-Pinetea canariensis
Scree grassland, pioneer community of gelifract (criogenic soil) slopes. Cerrillar-Crespar de Las Cañadas

N° Relevé	**1**	**2**	**3**	**4**	**5**	**6**	**7**	**8**	**9**	**10**	**11**	**12**
Surface (m^2)	50	50	50	50	25	50	50	25	25	50	25	50
Slope (%)	45	25	30	45	45	45	45	20	20	20	20	30
Altitude (m)	2150	2138	2150	2350	2350	2150	2150	2100	2150	2150	2150	2150
Coverage (%)	70	80	75	70	70	70	30	60	60	30	70	75
Exposure	W	NW	NW	W	W	W	NW	NW	NW	NW	NW	NW
N° Species	14	12	11	11	11	10	10	10	10	9	9	9
Characteristic species of association												
Arrhenatherum calderae	1	2	2	2	3	2	3	3	1	3	3	+
Plantago webbii	3	4	3	3	4	4	4	2	4	3	4	3
Characteristic of higher units												
Pterocephalus lasiospermus	1	3	2	1	1	1	2	1	+	2	2	1
Tolpis webbii	+	.	+	+	1	+	+	1	.	1	1	1
Argyranthemum tenerifae	+	2	+	1	+	.	3	+	+	.	+	.
Spartocytisus supranubius	+	.	+	+	+	+	+	+	1	+	+	+
Scrophularia glabrata	+	+	.	+	+	+	1	+	1	.	1	.
Nepeta teydea	+	+	.	+	1	.	1	+	+	+	+	.
Erysimum scoparium	+	1	.	+	.	+	+	+	+	.	.	.
Echium wildpretii	+	.	+	.	.	+	.	.	+	.	.	.
Carlina xeranthemoides	1	.	.	.	.	+	.	.	.	.	.	.
Melica canariensis	4	.	.	3	3	4	.	.	.	.	.	.
Descurainia gonzalesii	.	.	1	.	.	.	.	+	+	.	.	1
Andryala pinnatifida subsp. *teydensis*	.	+	+	.	+	.	.	.	.	.	.	.

(continued)

Table 4.5a (continued)

N° Relevé	**1**	**2**	**3**	**4**	**5**	**6**	**7**	**8**	**9**	**10**	**11**	**12**
Surface (m²)	50	50	50	50	25	50	50	25	25	50	25	50
Slope (%)	45	25	30	45	45	45	45	20	20	20	20	30
Altitude (m)	2150	2138	2150	2350	2350	2150	2150	2100	2150	2150	2150	2150
Coverage (%)	70	80	75	70	70	70	30	60	60	30	70	75
Exposure	W	NW	NW	W	W	W	NW	NW	NW	NW	NW	NW
N° Species	14	12	11	11	11	10	10	10	10	9	9	9
Polycarpaea tenuis	.	.	.	.	.	.	+	.	.	.	.	.
Cheirolophus teydis	.	+	.	.	.	.	.	.	.	.	.	1
Pimpinella cumbrae	.	1	.	.	.	.	.	.	.	.	+	3
Micromeria lachnophylla	.	.	+	.	.	.	.	.	.	1	.	3
Pinus canariensis	+	.	.	1	+	.	.	.	.	.	.	.
Bufonia paniculata	.	.	.	.	.	.	.	.	.	1	.	.
Adenocarpus viscosus	.	.	.	.	.	.	.	.	.	+	.	.
Descurainia bourgeauana	+	.	.	.	.	.	.	.	.	.	.	.
Other species												
Ferula linkii	.	+	.	.	.	.	.	.	.	.	.	.
Bromus tectorum	.	+	.	.	.	.	.	.	.	.	.	.
Festuca myuros	.	.	+	.	.	.	.	.	.	.	.	.

Localities:
1, 4, 5, 6 y 11: Cañada del Capricho
2, 3, 8, 9 y 12: Cañada de Diego Hernández
7 y 10: Queen's Shoe Rock

Table 4.5b *Arrhenathero calderae-Plantaginetum webbii* Martín Osorio & Wildpret de la Torre
Plantaginion webbii, Spartocytisetalia supranubii, Chamaecytiso-Pinetea canariensis
Scree grassland, pioneer community of gelifract (criogenic soil) slopes. Cerrillar-Crespar de Las Cañadas

N° Relevé	**13**	**14**	**15**	**16**	**17**	**18**	**19**	**20**	**21**	**22**	**23**	**24**
Surface (m^2)	50	50	50	50	100	50	50	25	200	50	100	100
Slope (%)	45	30	10	25	45	20	10	25	5	30	5	0
Altitude (m)	2150	2100	2100	2138	2200	2100	2100	2150	2100	2150	2100	2100
Coverage (%)	50	75	50	80	50	90	50	70	75	50	75	90
Exposure	SW	W	W	NW	NW	SW	W	NW	SW	SW	W	SW
N° Species	9	9	8	7	7	7	7	6	6	5	3	2
Characteristic species of association												
Arrhenatherum calderae	+	+	1	1	+	1	+	2	1	+	1	.
Plantago webbii	3	4	3	4	3	3	3	4	5	4	4	5
Characteristic of higher units												
Pterocephalus lasiospermus	2	2	2	2	+	1	2	3	3	.	3	3
Tolpis webbii	2	2	2	.	2	1	+	1	.	1	.	.
Argyranthemum tenerifae	+	.	3	.	+	1	1	.	.	+	.	.
Spartocytisus supranubius	.	+	.	.	+	.	.	1	.	.	.	.
Scrophularia glabrata	.	+	.	.	1	.	.	+	+	.	.	.
Nepeta teydea	+	2	.	.	.	.	.	.	.	+	.	.
Erysimum scoparium	.	.	+	+	.	.	.	.	.	.	.	.
Echium wildpretii	.	.	+	2	.	.	+	.	+	.	.	.
Carlina xeranthemoides	1	2	1	.	.	2	1	.	.	.	.	.
Melica canariensis	+	.	.	.	.	.	.	.	+	.	.	.
Descurainia gonzalesii	.	.	.	.	.	.	.	.	.	.	.	.
Andryala pinnatifida subsp. teydensis	.	.	.	1	.	.	.	.	.	.	.	.

(continued)

Table 4.5b (continued)

N° Relevé	**13**	**14**	**15**	**16**	**17**	**18**	**19**	**20**	**21**	**22**	**23**	**24**
Surface (m^2)	50	50	50	50	100	50	50	25	200	50	100	100
Slope (%)	45	30	10	25	45	20	10	25	5	30	5	0
Altitude (m)	2150	2100	2100	2138	2200	2100	2100	2150	2100	2150	2100	2100
Coverage (%)	50	75	50	80	50	90	50	70	75	50	75	90
Exposure	SW	W	W	NW	NW	SW	W	NW	SW	SW	W	SW
N° Species	9	9	8	7	7	7	7	6	6	5	3	2
Polycarpaea tenuis	.	.	.	.	.	.	.	.	.	.	.	.
Cheirolophus teydis	.	.	.	1	.	.	.	.	.	.	.	.
Pimpinella cumbrae	.	.	.	.	.	.	.	.	.	.	.	.
Micromeria lachnophylla	.	.	.	.	.	.	.	.	.	.	.	.
Pinus canariensis	.	.	.	.	.	.	.	.	.	.	.	.
Bufonia paniculata	.	.	.	.	.	.	.	.	.	.	.	.
Adenocarpus viscosus	2	+	.	.	.	3	.	.	.	.	.	.
Descurainia bourgeauana	.	.	.	.	.	.	.	.	.	.	.	.
Other species												
Ferula linkii	.	.	.	.	.	.	.	.	.	.	.	.
Bromus tectorum	.	.	.	.	.	.	.	.	.	.	.	.
Festuca myuros	.	.	.	.	.	.	.	.	.	.	.	.

Localities:
12 y 16: Cañada de Diego Hernández
13, 14, 15, 17, 18, 19 y 22: Vértice geodésico Cañada del Capricho
20: Cañada del Capricho
21, 23 y 24: Base de la Cañada del Capricho

II.11. *Cheirolopho teydis-Carlinetum xeranthemoides* Martín Osorio, Hernández Bolaños & Wildpret *ass. nova*

Relevé type n°8, Tables 4.6a and 4.6b. Figure 4.16 Substitute scrub, sometimes acting as permanent vegetation, of the summit broom and Canary Island pine forests on soils made of old materials from the Wall or pre-caldera building of Las Cañadas del Teide, denuded and exposed to strong sunlight in dry-subhumid Meso-supra-Mediterranean bioclimate. Its proliferation is frequent after some fires such as the one in the southern area of El Sombrerito. Characteristic species are the "cabezón" Cheirolophus teydis (Fig. 4.15) and the "malpica" Carlina xeranthemoides (Fig. 4.16).

Fig. 4.15 *Cheirolopho teydis-Carlinetum xeranthemoides* Western slopes of the National Park, April 2004. (Photo credit: Victoria Eugenia Martín Osorio)

Table 4.6a *Cheirolopho teydis-Carlinetum xeranthemoides* Martín Osorio, Hernández Bolaños & Wildpret *ass. nova.* Type Relevé n°8 *Pterocephalenion lasiospermi, Plantaginion webbii, Chamaecytiso-Pinetea canariensis*

N° Relevé	**1**	**2**	**3**	**4**	**5**	**6**	**7**
Altitude (m)	2140	2309	2384	2416	2185	2092	2077
Surface (m²)	49	100	100	100	100	25	25
Slope (%)	–	30	50	50	45	90	–
Exposure	E	E	S	W	SSW	N	NE
Coverage (%)	3	20	80	70	60	2	5
N° Species	9	9	7	7	7	7	7
UTM	336596—3121802	336574—3121365	336224—3121073	336266—3121127	347514—3130618	336785—3121799	336719—3121830
Date	16/4/03	16/4/03	7/5/03	7/5/03	29/5/01	15/4/03	16/4/03
Characteristic species of association							
Carlina xeranthemoides	+	.	5	2	2	+	+
Cheirolophus teydis	+	2	1	2	2	.	.
Characteristic Chamaecytiso-Pinetea canariensis							
Adenocarpus viscosus subsp. viscosus	.	+	+	1	+	.	.
Pterocephalus lasiospermus	.	+	1	1	.	+	+
Tolpis webbii	1	.	1	1	.	1	1
Erysimum scoparium	.	.	1	1	.	.	1
Micromeria lachnophylla	.	1	.	.	2	1	.
Pimpinella cumbrae	.	1	.	.	.	.	.
Scrophularia glabrata	.	.	.	.	.	.	.
Characteristic Plantaginion webbii							
Arrhenatherum calderae	+	1	.	.	1	1	1
Plantago webbii	.	.	.	.	.	.	.

Altitude (m)	2140	2309	2384	2416	2185	2092	2077
Surface (m²)	49	100	100	100	100	25	25
Slope (%)	–	30	50	50	45	90	–
Exposure	E	E	S	W	SSW	N	NE
Coverage (%)	3	20	80	70	60	2	5
N° Species	9	9	7	7	7	7	7
UTM	336596—3121802	336574—3121365	336224—3121073	336266—3121127	347514—3130618	336785—3121799	336719—3121830
Date	16/4/03	16/4/03	7/5/03	7/5/03	29/5/01	15/4/03	16/4/03
Characteristic Cisto-Pinion							
Chamaecytisus proliferus subsp. angustifolius	.	.	.	.	.	.	.
Argyranthemum tenerifae	.	.	.	.	.	.	.
Characteristic Spartocytision supranubii							
Nepeta teydea var. teydea	+	+	.	.	.	+	+
Spartocytisus supranubius	.	1	.	1	1	.	.
Sideritis eriocephala	.	+	+	.	.	.	.
Characteristic Juniperion cedri							
Echium wildpretii subsp. wildpretii	.	.	.	.	.	.	.
Other species							
Silene berthelotiana	+	.	.	.	.	+	1
Ferula linkii	+	.	.	.	1	.	.
Rhamnus integrifolia	1	.	.	.	.	.	.
Bethencourtia palmensis	+	.	.	.	.	.	.

Localities: 1, 6 y 7: Morros Los Codesos (SW-Llano Ucanca, at the foot of the wall). 2: Near Roque El Pinito (SW-Llano Ucanca, wall height). 3 y 4: Southwestern slopes of La Sombrera. 5: Risco de Las Ventanas/Ravine of the Llano de Maja

Table 4.6b *Cheirolopho teydis-Carlinetum xeranthemoides* Martín Osorio, Hernández Bolaños & Wildpret *ass. nova.* Type Relevé n°8 *Pterocephalenion lasiospermi, Plantaginion webbii, Chamaecytiso-Pinetea canariensis*

N° Relevé	**8**	**9**	**10**	**11**	**12**	**13**	**14**
Altitude (m)	2250	2450	2100	2181	2150	2500	2254
Surface (m²)	100	100	100	100	100	100	100
Slope (%)	60	40	50	45	25	0	15
Exposure	NW	SE	E	S	W	–	W
Coverage (%)	60	60	80	70	60	50	35
N° Species	6	5	5	5	5	4	4
UTM	336125—3121625	337625—3120375	346125—3123875	335464—3121118	348080—3128750	336875—3120625	335861—3122102
Date	16/4/03	7/5/03	6/6/03	16/4/02	13/6/01	7/5/03	6/7/01
Characteristic species of association							
Carlina xeranthemoides	2	+	.	4	2	.	2
Cheirolophus teydis	3	3	3	1	1	4	2
Characteristic Chamaecytiso-Pinetea canariensis							
Adenocarpus viscosus subsp. viscosus	+	+	+	+	.	+	.
Pterocephalus lasiospermus	.	.	1	+	1	.	1
Tolpis webbii	.	.	.	.	.	1	1
Erysimum scoparium	+	.	.	.	.	1	.
Micromeria lachnophylla	.	1	.	.	.	.	.
Pimpinella cumbrae	.	.	.	.	.	.	.
Scrophularia glabrata	+	.	.	.	.	.	.
Characteristic Plantaginion webbii							
Arrhenatherum calderae	.	1	.	.	.	.	.
Plantago webbii	.	.	.	.	1	.	.

Altitude (m)	2250	2450	2100	2181	2150	2500	2254
Surface (m^2)	100	100	100	100	100	100	100
Slope (%)	60	40	50	45	25	0	15
Exposure	NW	SE	E	S	W	–	W
Coverage (%)	60	60	80	70	60	50	35
N° Species	6	5	5	5	5	4	4
UTM	336125—3121625	337625—3120375	346125—3123875	335464—3121118	348080—3128750	336875—3120625	335861—3122102
Date	16/4/03	7/5/03	6/6/03	16/4/02	13/6/01	7/5/03	6/7/01
Characteristic Cisto-Pinion							
Chamaecytisus proliferus subsp. angustifolius	.	.	+	1	.	.	.
Argyranthemum tenerifae	+	.	.	.	.	.	.
Characteristic Spartocytision supranubii							
Nepeta teydea var. teydea	.	.	.	.	.	.	.
Spartocytisus supranubius	.	.	.	.	1	.	.
Sideritis eriocephala	.	.	.	.	.	.	.
Characteristic Juniperion cedri							
Echium wildpretii subsp. wildpretii	.	.	1	.	.	.	.
Other species							
Silene berthelotiana	.	.	.	.	.	.	.
Ferula linkii	.	.	.	.	.	.	.
Rhamnus integrifolia	.	.	.	.	.	.	.
Bethencourtia palmensis	.	.	.	.	.	.	.

Localities: 8: Las Escaleras (NW de La Sombrera). 9: Roque Bermejo (Vertiente S de Pared). 10: Llanos de Yarosa/Lomo Fuente Gabriel. 11 y 14: Valle/Barranco Peguero (W-La Sombrera). 12: Cañada de Diego Hernández/Risco Los Cuchillos. 13: Roque del Almendro (Vertiente S de Pared)

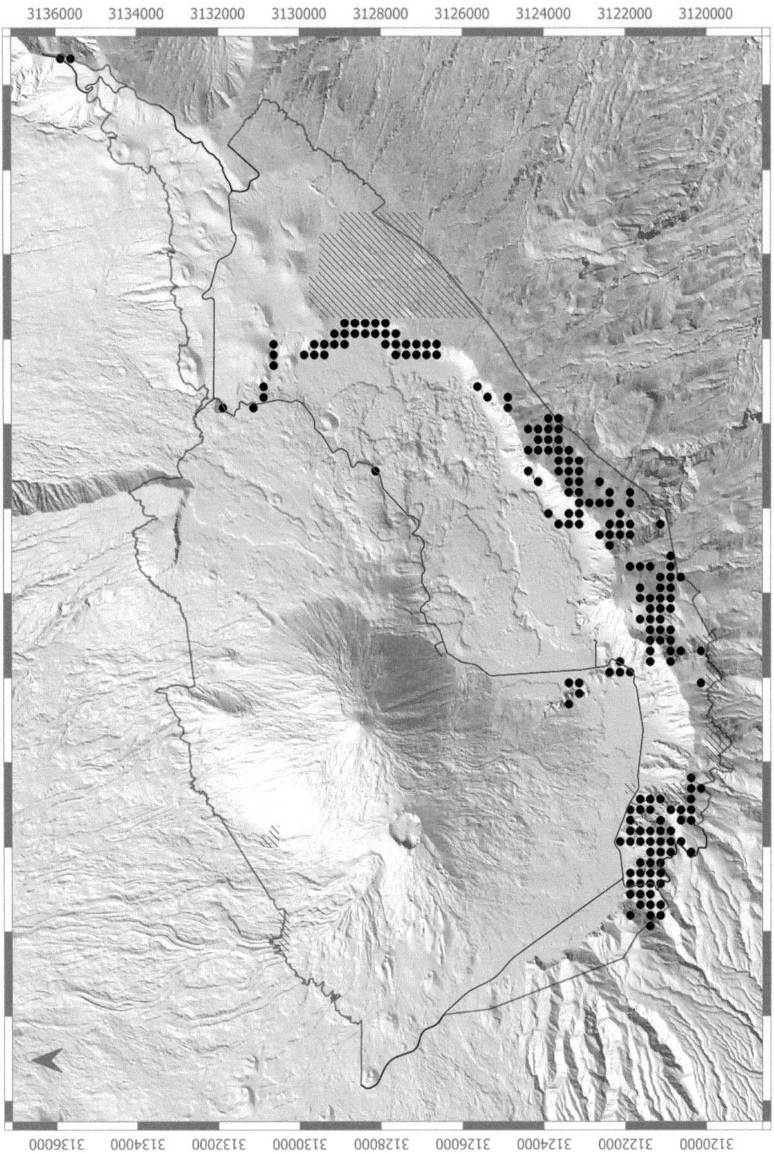

Fig. 4.16 Distribution of the plant community *Cheirolopho teydis-Carlinetum xeranthemoides*

II.12. ***Tolpido webbii-Micromerietum lachnophyllae*** Martín Osorio, Hernández Bolaños & Wildpret *ass. nova.*

Relevé Type No. 13, Tables 4.7a and 4.7b. Figure 4.18. Thyme substitution scrub that colonizes volcanic deposits and denuded soils of pine forests, cedar forests and broom shrublands, dry-subhumid Meso-Supra-Mediterranean, endemic to the island of Tenerife between 1930 and 2400 m a.s.l. Characteristic species are the "cañadas thyme" Micromeria lachnophylla (Fig. 4.17) and the "summit lechuguilla" Tolpis webbii. It is common in the upper part of La Fortaleza and old stony areas of the Pared, or Pre-caldera Building of Siete Cañadas (Fig. 4.18).

Fig. 4.17 *Tolpido webbii-Micromerietum lachnophyllae* between scree of La Fortaleza November 2009. (Photo credit: Victoria Eugenia Martín Osorio)

Table 4.7a *Tolpido webbii-Micromerietum lachnophyllae* Martín Osorio, Hernández Bolaños & Wildpret *ass. nova*
Type Relevé n°13
Pterocephalenion lasiospermi, Plantaginion webbii, Chamaecytiso-Pinetea canariensis

N° Relevé	**1**	**2**	**3**	**4**	**5**	**6**	**7**	**8**	**9**
Altitude (m)	2114	2075	2040	2055	2024	2134	2130	2022	2143
Surface (m^2)	50	50	100	100	100	60	100	50	100
Slope (%)	90	35	30	50	0	90	50	10	0
Exposure	NW	E	S	S	–		SW	NW	–
Coverage (%)	20	45	45	60	80	40	45	20	40
N° Species	16	14	13	11	11	11	11	10	10
UTM	354606—3135691	336674—3121854	343235—3132998	343703—3132988	354706—3136121	346734—3129582	347910—3129004	352043—3133583	345441—3131159
Date	2/4/01	16/04/03	9/7/01	9/7/01	2/4/01	7/6/01	11/6/01	9/4/01	11/6/01
Characteristic species of association									
Tolpis webbii	1	1	.	+	+	1	1	1	1
Micromeria lachnophylla	1	1	1	2	1	1	1	1	1
Characteristic Pterocephalenion lasiospermi									
Pterocephalus lasiospermus	1	.	1	1	2	1	2	2	2
Descurainia bourgeauana	.	.	.	.	.	1	.	.	1
Characteristic Chamaecytiso-Pinetea canariensis									
Erysimum scoparium	1	.	.	+	1	1	1	1	1
Adenocarpus viscosus subsp. viscosus	1	.	+	.	.	.	+	.	1

Altitude (m)	2114	2075	2040	2055	2024	2134	2130	2022	2143
Surface (m²)	50	50	100	100	100	60	100	50	100
Slope (%)	90	35	30	50	0	90	50	10	0
Exposure	NW	E	S	S	–		SW	NW	–
Coverage (%)	20	45	45	60	80	40	45	20	40
N° Species	16	14	13	11	11	11	11	10	10
UTM	354606—3135691	336674—3121854	343235—3132998	343703—3132988	354706—3136121	346734—3129582	347910—3129004	352043—3133583	345441—3131159
Date	2/4/01	16/04/03	9/7/01	9/7/01	2/4/01	7/6/01	11/6/01	9/4/01	11/6/01
Pinus canariensis	.	.	.	.	+	.	.	.	.
Carlina xeranthemoides	.	1	+	3	+	.	1	.	.
Scrophularia glabrata	1	1	.	.	1	1	1	1	.
Nepeta teydea var. *teydea*	.	1	+	.	.	1	.	.	+
Characteristic Sparticytision									
Spartocytisus supranubius	1	1	1	2	4	2	+	1	2
Cheirolophus teydis	.	1	.	.	.	.	2	.	.
Argyranthemum tenerifae	.	.	.	.	.	1	.	.	1
Characteristic Plantaginion webbii									
Arrhenatherum calderae	1	1	1	1	.	1	.	1	1
Plantago webbii	.	.	.	.	.	.	2	.	.
Polycarpaea tenuis	.	.	.	.	.	.	1	.	.
Other species									
Pimpinella cumbrae	.	1	.	.	.	.	.	.	.
Bromus tectorum	.	1	.	.	.	.	.	.	.
Piptatherum coerulescens	.	.	1	1	.	.	.	.	.

(continued)

Table 4.7a (continued)

N° Relevé	**1**	**2**	**3**	**4**	**5**	**6**	**7**	**8**	**9**
Altitude (m)	2114	2075	2040	2055	2024	2134	2130	2022	2143
Surface (m²)	50	50	100	100	100	60	100	50	100
Slope (%)	90	35	30	50	0	90	50	10	0
Exposure	NW	E	S	S	–		SW	NW	–
Coverage (%)	20	45	45	60	80	40	45	20	40
N° Species	16	14	13	11	11	11	11	10	10
UTM	354606—3135691	336674—3121854	343235—3132998	343703—3132988	354706—3136121	346734—3129582	347910—3129004	352043—3133583	345441—3131159
Date	2/4/01	16/04/03	9/7/01	9/7/01	2/4/01	7/6/01	11/6/01	9/4/01	11/6/01
Silene berthelotiana	.	.	.	1	.	.	.	.	.
Echium wildpretii subsp. wildpretii	.	1	1	.	.	.	.	.	.
Ferula linkii	.	.	1	1	.	.	.	.	.
Anogramma leptophylla	1	.	.	.	.	.	.	+	.
Descurainia lemsii	1	.	.	.	.	.	.	1	.
Sideritis eriocephala	.	1	.	.	.	.	.	.	.
Cistus osbeckiifolius	.	.	2	.	.	.	.	.	.
Lotus campylocladus	.	.	1	.	.	.	.	.	.
Bufonia paniculata	.	1	.	.	.	.	.	.	.
Rhamnus integrifolia	.	.	.	+	.	.	.	.	.
Wahlenbergia lobelioides subsp. lobelioides	.	.	.	.	.	.	.	.	.
Piptatherum miliaceum	.	1	.	.	.	.	.	.	.
Aira caryophyllea	1	.	.	.	.	.	.	.	.
Hemionitis pteridioides	.	.	.	.	.	.	.	.	.
Galium aparine	.	.	.	.	1	.	.	.	.

Altitude (m)	2114	2075	2040	2055	2024	2134	2130	2022	2143
Surface (m²)	50	50	100	100	100	60	100	50	100
Slope (%)	90	35	30	50	0	90	50	10	0
Exposure	NW	E	S	S	–		SW	NW	–
Coverage (%)	20	45	45	60	80	40	45	20	40
N° Species	16	14	13	11	11	11	11	10	10
UTM	354606—3135691	336674—3121854	343235—3132998	343703—3132988	354706—3136121	346734—3129582	347910—3129004	352043—3133583	345441—3131159
Date	2/4/01	16/04/03	9/7/01	9/7/01	2/4/01	7/6/01	11/6/01	9/4/01	11/6/01
Bituminaria bituminosa	.	.	1	.	.	.	.	.	.
Gastridium phleoides	.	.	.	.	.	1	.	.	.
Gnaphalium sp.	.	.	.	.	.	.	.	.	.
Aeonium aizoon	1	.	.	.	.	.	.	.	.
Hypericum grandifolium	1	.	.	.	.	.	.	.	.
Monanthes brachycaulos var. nivata	1	.	.	.	.	.	.	.	.
Andryala pinnatifida subsp. teydensis	.	.	.	.	.	.	.	+	.
Bystropogon canariensis	.	.	.	.	+	.	.	.	.
Cystopteris fragilis	+	.	.	.	.	.	.	.	.
Ranunculus cortusifolius	+	.	.	.	.	.	.	.	.

Localities: 1: Morra La Negrita (E-La Tarta). 2: Eastern slopes towards Ucanca from Montaña La Sombrera. 3 y 4: La Fortaleza. 5: La Crucita. 6: Cañada de los Chupaderos (NE-Montaña Mostaza. 7: Filling materials in Cañada de Diego Hernández. 8: SW-Montaña Limón and W-Llano La Rosa. 9: Cuesta La Burra (W-El Portillo)

Table 4.7b *Tolpido webbii-Micromerietum lachnophyllae* Martín Osorio, Hernández Bolaños & Wildpret *ass. nova*
Type Relevé n°13
Pterocephalenion lasiospermi, Plantaginion webbii, Chamaecytiso-Pinetea canariensis

N° Relevé	**10**	**11**	**12**	**13**	**14**	**15**	**16**	**17**	**18**
Altitude (m)	2035	2115	2135	2163	2107	2110	2170	2092	2417
Surface (m²)	100	100	100	100	400	100	100	25	25
Slope (%)	40	35	0	30	25	45	60	90	
Exposure	W-N-E	SE	–	SE	W	SE	NNE	N	
Coverage (%)	10	60	55	35	80	35	35	2	50
N° Species	10	9	9	8	8	8	8	7	7
UTM	352150—3134385	346645—3130863	346706—3130913	344368—3130850	345841—3132026	344667—3131297	335700—3121511	336785—3121799	341679—3121924
Date	6/4/01	31/5/01	31/5/01	17/7/01	18/7/01	16/7/01	6/7/02	16/4/03	19/5/03
Characteristic species of association									
Tolpis webbii	1	1	1	1	1	1	1	1	.
Micromeria lachnophylla	1	1	1	2	1	1	1	1	4
Characteristic Pterocephalenion lasiospermi									
Pterocephalus lasiospermus	1	3	3	1	1	1	+	+	1
Descurainia bourgeauana	.	.	+	.	1	.	.	.	.
Characteristic Chamaecytiso-Pinetea canariensis									
Erysimum scoparium	1	1	1	.	.	1	1	.	1
Adenocarpus viscosus subsp. viscosus	1	2	1	1	+	1	1	.	.

Altitude (m)	2035	2115	2135	2163	2107	2110	2170	2092	2417
Surface (m^2)	100	100	100	100	400	100	100	25	25
Slope (%)	40	35	0	30	25	45	60	90	
Exposure	W-N-E	SE	–	SE	W	SE	NNE	N	
Coverage (%)	10	60	55	35	80	35	35	2	50
N° Species	10	9	9	8	8	8	8	7	7
UTM	352150—3134385	346645—3130863	346706—3130913	344368—3130850	345841—3132026	344667—3131297	335700—3121511	336785—3121799	341679—3121924
Date	6/4/01	31/5/01	31/5/01	17/7/01	18/7/01	16/7/01	6/7/02	16/4/03	19/5/03
Pinus canariensis	.	.	.	.	3	.	.	.	.
Carlina xeranthemoides	.	+	.	.	.	.	.	+	.
Scrophularia glabrata	.	.	.	.	.	.	.	.	.
Nepeta teydea var. *teydea*	.	1	+	.	.	.	.	+	.
Characteristic Spartocytision									
Spartocytisus supranubius	1	1	1	1	1	1	1	.	.
Cheirolophus teydis	.	.	.	.	.	.	1	.	.
Argyranthemum tenerifae	.	.	.	.	.	.	.	.	1
Characteristic Plantaginion webbii									
Arrhenatherum calderae	.	1	1	1	1	1	.	1	+
Plantago webbii	.	.	.	.	.	.	.	.	.
Polycarpaea tenuis	.	.	.	+	.	.	.	.	.
Other species									
Pimpinella cumbrae	.	.	.	.	.	.	1	.	1

(continued)

Table 4.7b (continued)

N° Relevé	**10**	**11**	**12**	**13**	**14**	**15**	**16**	**17**	**18**
Altitude (m)	2035	2115	2135	2163	2107	2110	2170	2092	2417
Surface (m²)	100	100	100	100	400	100	100	25	25
Slope (%)	40	35	0	30	25	45	60	90	
Exposure	W-N-E	SE	–	SE	W	SE	NNE	N	
Coverage (%)	10	60	55	35	80	35	35	2	50
N° Species	10	9	9	8	8	8	8	7	7
UTM	352150—3134385	346645—3130863	346706—3130913	344368—3130850	345841—3132026	344667—3131297	335700—3121511	336785—3121799	341679—3121924
Date	6/4/01	31/5/01	31/5/01	17/7/01	18/7/01	16/7/01	6/7/02	16/4/03	19/5/03
Bromus tectorum	1	.	.	.	.	.	.	.	.
Piptatherum coerulescens	.	.	.	.	.	.	.	.	.
Silene berthelotiana	.	.	.	.	.	.	.	+	2
Wahlenbergia lobelioides subsp. lobelioides	1	.	.	.	.	.	.	.	.
Piptatherum miliaceum	.	.	.	.	.	.	.	.	.
Aira caryophyllea	.	.	.	.	.	.	.	.	.
Hemionitis pteridioides	1	.	.	.	.	.	.	.	.
Galium aparine	.	.	.	.	.	.	.	.	.
Bituminaria bituminosa	.	.	.	.	.	.	.	.	.
Gastridium phleoides	.	.	.	.	.	.	.	.	.
Laphangium luteoalbum	1	.	.	.	.	.	.	.	.

Localities: 9: Cuesta La Burra (W-El Portillo). 10: Montaña Limón. 11 y 12: El Portillo Alto. 13: Montaña Los Tomillos. 14: Corral Nuevo (NW—El Portillo Bajo). 15: Estancia La Cera. 16: Degollada Peguero (E-Boca Tauce). 17: Eastern slopes towards Ucanca of Montaña La Sombrera. 18: Highs of Montaña Guajara

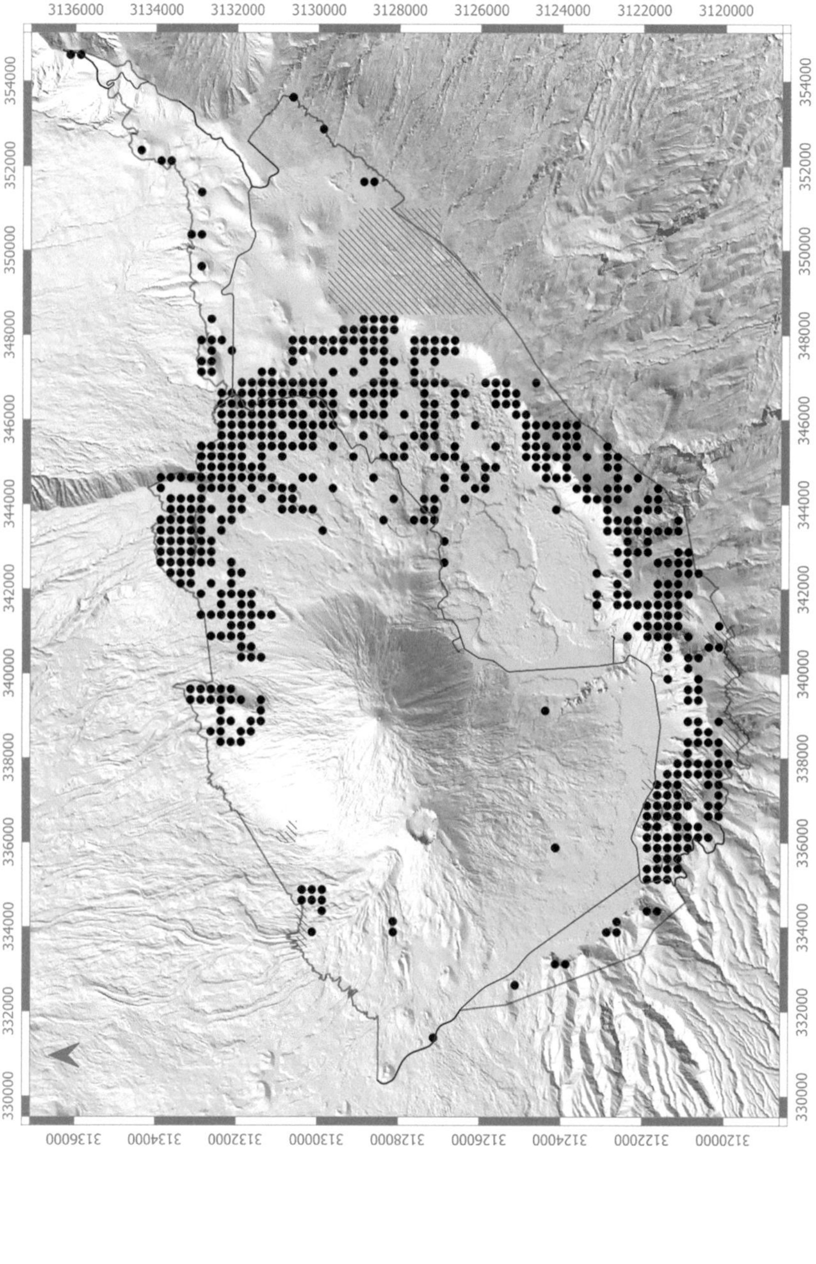

Fig. 4.18 Distribution of *Tolpido webbii-Micromerietum lachnophyllae* association

II.13. *Arrhenathero calderae-Adenocarpetum viscosi* Martín Osorio, Hernández Bolaños & Wildpret *ass. nova.*

Relevé Type n° 2, Tables 4.8a and 4.8b. Figure 4.20. A first-colonizing community of recent badlands with potential for pine forests, cedar forests and broom shrublands, dry-subhumid Meso-Supra-Mediterranean, endemic to the island of Tenerife. It is common on the southern slopes of the Teide Natural Monument, Montaña Escoronada or of La Corona and near Sombrero de Chasna. The shrubland is found in almost monospecific situations with large extensions of Adenocarpus viscosus or accompanied by the "cerrillo de cumbre" Arrhenatherum calderae (Figs. 4.19 and 4.20).

Fig. 4.19 *Arrhenathero calderae-Adenocarpetum viscosi* Trail of Pico Viejo August 2005. (Photo credit: Victoria Eugenia Martín Osorio)

Table 4.8a *Arrhenathero calderae-Adenocarpetum viscosi* Martín Osorio, Hernández Bolaños & Wildpret *ass. nova*
Type Relevé n° 2 *Pterocephalenion lasiospermi, Plantaginion webbii, Chamaecytiso-Pinetea canariensis*

N° Relevé	**1**	**2**	**3**	**4**	**5**	**6**	**7**	**8**	**9**
Altitude (m)	2575	2275	2275	2425	2475	2326	2310	2252	2256
Surface (m²)	100	100	100	100	100	100	100	100	100
Slope (%)	40	5	5	30	10	?	45	45	30
Exposure	S	SW	S	S	S	–	S	S	S
Coverage (%)	80	60	75	70	80	40	40	35	30
N° Species	4	7	8	5	5	2	3	5	4
UTM	338875—3125625	340125—3125625	340375—3125625	336875—3125125	338375—3125375	339196—3125085	339086—3125360	339963—3125363	339822—3125302
Date	13/5/03	28/4/03	28/4/03	4/6/03	13/5/03	30/4/03	30/4/03	30/4/03	30/4/03
Characteristic species of association									
Arrhenatherum calderae	1	1	2	1	1	3	1	1	.
Adenocarpus viscosus subsp. viscosus	2	3	3	1	3	2	3	1	3
Characteristic Spartocytision supranubii									
Spartocytisus supranubius	+	.	1	1	+	.	.	.	+
Echium wildpretii subsp. wildpretii	.	.	.	.	.	.	.	.	.
Argyranthemum tenerifae	.	.	.	1	.	.	.	1	.
Characteristic Pterocephalenion lasiospermi									
Descurainia bourgeauana	.	1	1	.	.	.	.	.	.

(continued)

Table 4.8a (continued)

N° Relevé	1	2	3	4	5	6	7	8	9
Altitude (m)	2575	2275	2275	2425	2475	2326	2310	2252	2256
Surface (m²)	100	100	100	100	100	100	100	100	100
Slope (%)	40	5	5	30	10	?	45	45	30
Exposure	S	SW	S	S	S	–	S	S	S
Coverage (%)	80	60	75	70	80	40	40	35	30
N° Species	4	7	8	5	5	2	3	5	4
UTM	338875—3125625	340125—3125625	340375—3125625	336875—3125125	338375—3125375	339196—3125085	339086—3125360	339963—3125363	339822—3125302
Date	13/5/03	28/4/03	28/4/03	4/6/03	13/5/03	30/4/03	30/4/03	30/4/03	30/4/03
Pterocephalus lasiospermus	.	+	+	.	.	.	.	1	.
Tolpis webbii	1	1	.	1	1	.	+	.	1
Characteristic Chamaecytiso Pinetea									
Andryala pinnatifida subsp. teydensis	.	.	1	.	.	.	.	.	.
Erysimum scoparium	.	.	.	.	.	.	.	.	.
Nepeta teydea var. teydea	.	+	+	.	+	.	.	1	+
Scrophularia glabrata	.	.	.	.	.	.	.	.	.
Other species									
Bromus tectorum	.	+	+	.	.	.	.	.	.
Wahlenbergia lobelioides	.	.	.	.	.	.	.	.	.

Localities: 1,4,5,6,7,8: Southern slopes of the Stratovolcano towards Llano de Ucanca (Corridor of La Corbata). 2 y 3: Corridor of La Corbata (around the road). 9: Southern slopes of the stratovolcano towards Llano de Ucanca (S-Regatones Negros)

Table 4.8b *Arrhenathero calderae-Adenocarpetum viscosi* Martín Osorio, Hernández Bolaños & Wildpret *ass. nova Pterocephalenion lasiospermi, Plantaginion webbii, Chamaecytiso-Pinetea canariensis*

N° Relevé	**10**	**11**	**12**	**13**	**14**	**15**	**16**	**17**	**18**
Altitude (m)	2400	2326	2269	2222	2328	2146	2220	2245	2498
Surface (m^2)	200	100	100	100	100	100	100	100	100
Slope (%)	20	0	0	0	10	5	10	0	10
Exposure	S	–	–	–	S	S	S	–	S
Coverage (%)	70	65	60	50	55	40	60	25	35
N° Species	6	3	2	7	3	6	4	2	3
UTM	342625—3127125	337820—3124311	338661—3124340	339405—3124423	336828—3124852	336428—3123761	337417—3123781	337693—3123721	339418—3120482
Date	12/6/02	27/5/02	23/5/02	21/5/02	5/6/02	3/6/02	29/5/02	27/5/02	21/4/03
Characteristic species of association									
Arrhenatherum calderae	3	2	1	1	1	1	1	1	1
Adenocarpus viscosus subsp. viscosus	4	3	4	3	3	1	3	2	+
Characteristic Spartocytision supranubii									
Spartocytisus supranubius	1	1	.	+	1	1	1	.	.
Echium wildpretii subsp. wildpretii	.	.	.	1	.	.	.	.	.
Argyranthemum tenerifae	1	.	.	.	.	.	1	.	.
Characteristic Pterocephalenion lasiospermi									
Descurainia bourgeauana	.	.	.	.	.	.	.	.	.

(continued)

Table 4.8b (continued)

N° Relevé	**10**	**11**	**12**	**13**	**14**	**15**	**16**	**17**	**18**
Altitude (m)	2400	2326	2269	2222	2328	2146	2220	2245	2498
Surface (m²)	200	100	100	100	100	100	100	100	100
Slope (%)	20	0	0	0	10	5	10	0	10
Exposure	S	–	–	–	S	S	S	–	S
Coverage (%)	70	65	60	50	55	40	60	25	35
N° Species	6	3	2	7	3	6	4	2	3
UTM	342625—3127125	337820—3124311	338661—3124340	339405—3124423	336828—3124852	336428—3123761	337417—3123781	337693—3123721	339418—3120482
Date	12/6/02	27/5/02	23/5/02	21/5/02	5/6/02	3/6/02	29/5/02	27/5/02	21/4/03
Pterocephalus lasiospermus	1	.	.	1	.	1	.	.	.
Tolpis webbii	.	.	.	.	.	1	.	.	.
Characteristic Chamaecytiso-Pinetea									
Erysimum scoparium	.	.	.	.	.	.	.	.	3
Nepeta teydea var. teydea	.	.	.	1	.	+	.	.	.
Other species									
Bromus tectorum	.	.	.	.	.	.	.	.	.
Wahlenbergia lobelioides subsp. lobelioides	.	.	.	1	.	.	.	.	.

Localities: 10: El Culatón/Vueltas del Carnero. 11, 12 y 13: Southern slopes of the stratovolcano towards Llano de Ucanca Llano de Ucanca (Chafarí/Llanos de la Escoronada). 14 y 17: Southern slopes of the stratovolcano towards Llano de Ucanca (Corridor of La Corbata). 15: Southern slopes of the stratovolcano towards Llano de Ucanca (S-Regatones Negros), 16: Mesa de Ucanca (SW-Cumbre del Riachuelo), 18: Colorado de Guajara (western ridge of Barranco del Río)

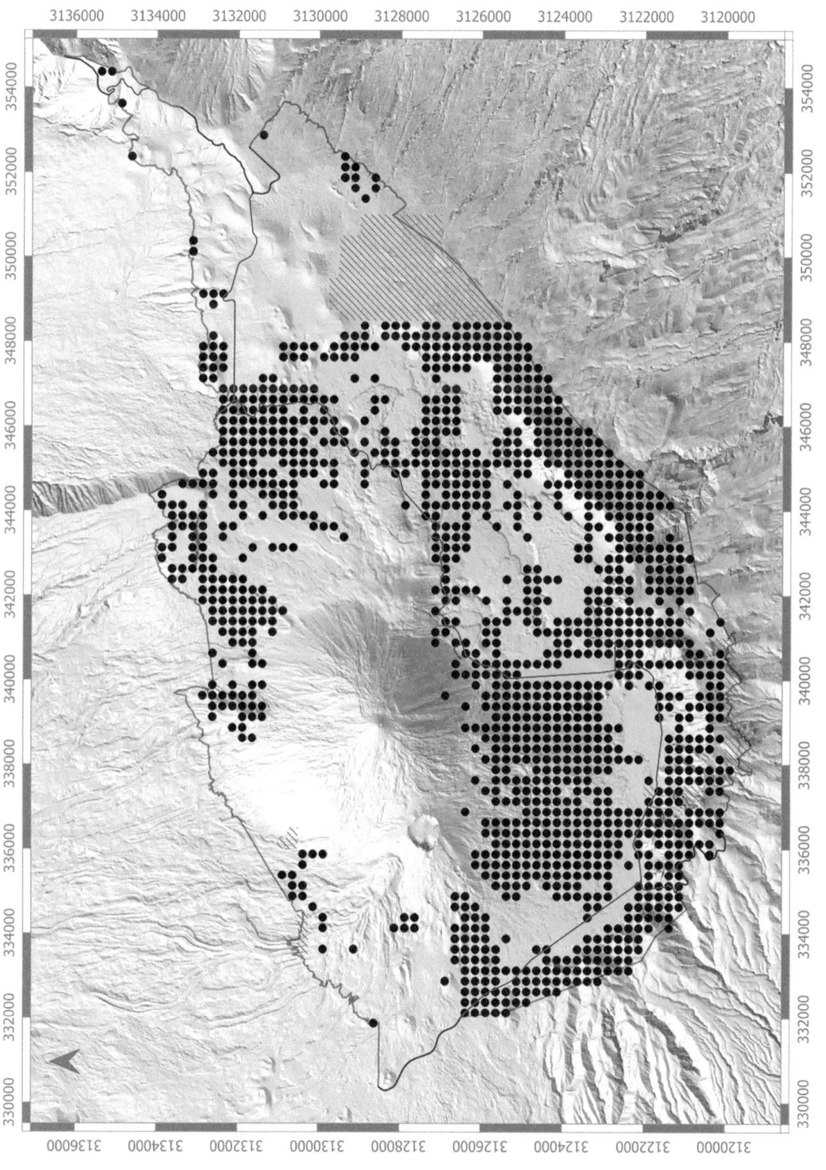

Fig. 4.20 Distribution of *Arrhenathero calderae-Adenocarpetum viscosii* association

III. GREENOVIO –AEONIETEA A. Santos 1976

Phytosociological class of perennial vegetation that groups the comophytic and chasmocomophytic perennial vegetation of the Canary-Madeiran subregion, in which various types of marcescent pinkish succulent chamephytes predominate, which colonize cliffs, recent badlands and other rocky surfaces.

The communities of the *Greenovio-Aeonietea* class are endemic to the Canary-Madeiran biogeographic subregion of the hyperarid-humid Infra-Supra-Mediterranean belts.

Characteristic species in the Teide National Park: *Aeonium smithii, Monanthes brachycaulos, Rhamnus integrifolia, Bethencourtia palmensis, Silene bertheloti-ana, Athamanta cervariifolia, Tolpis lagopoda.*

***SONCHO ACAULIS-AEONIETALIA** Rivas Goday & Esteve *ex* Sunding 1972

In the National Park a single alliance *Greenovion aureae* and four communities are represented.

***Greenovion aureae A. Santos ex* Rivas-Martínez et al. 1993

Phytosociological alliance that groups the endemic summit growing comophytic and chasmocomophytic associations of the western Canary Islands, which show their optimum in mesophytic Thermo-Meso-Supra-Mediterranean environments. They colonize fissures of generally cohesive volcanic rocks.

Characteristics in the Teide National Park: *Arabis caucasica, Festuca agustinii, Aeonium (=Greenovia) aizoon, Aeonium (=Greenovia) aureum, Monanthes niphophyla.*

III.1. ***Cheilantho guanchicae- Aeonietum smithii*** Rivas-Martínez et al. 1993
Endemic association of the mountains surrounding the Circo de las Cañadas del Teide in the dry mesophytic Meso and Supra-Mediterranean belts. It develops in fissures of basaltic and phonolitic vertical rocks, especially those facing north.

III.2. ***Greenovietum aizoi*** Voggenreiter *ex* Rivas-Martínez et al. 1993
Endemic association of Tenerife, Güímar Heights and surroundings of La Cru-cita. From the upper Thermo-Mediterranean horizon to the dry mesophytic Supra-Mediterranean. On almost vertical basalts exposed to the windward side (Fig. 4.21)

Fig. 4.21 *Aeonium aizoon* on the wall around La Crucita. (Photo credit: Victoria Eugenia Martín Osorio)

III.3. ***Greenovietum aureae*** Rivas-Martínez et al. 1993

Association typical of elevated areas (900–2000 m a.s.l.), northern and eastern parts of the island of Tenerife. Exposed to the north on steep basaltic rocks and cliffs. In pine forest environments with Erica canariensis *Sideritido-Pinetum canariensis ericetosum canariense*.

III.4. ***Sileno berthelotianae-Cerastietum sventenii*** Martín Osorio, Hernández Bolaños & Wildpret *ass. nova.*

Relevé Type n° 1, Tables 4.9a and 4.9b (Fig. 4.22). Rupicolous association of phonolitic cracks and spurs and trachybasalts oriented to N and NW of the dry Supra-Mediterranean belts. It is located at altitudes between 2200 and 3000 m but always in windward situations within the scope of the communities of Spartocytisetea supranubii, Spartocytision y Violion cheiranthifoliae.This community replaces Cheilantho guanchicae- Aeonietum smithii at altitudes above 2200 m a.s.l.

Table 4.9a *Sileno berthelothianae-Cerastietum sventenii* Martín Osorio, Hernández Bolaños & Wildpret *ass. nova* Relevé Type n° 1
Greenovion aureae, Greenovio-Aeonietea

N° Relevé	**1**	**2**	**3**	**4**	**5**	**6**
Altitude (m)	2627	2591	2462	2227	2569	2980
Surface (m^2)	15	15	9	4	4	4
Slope (%)	–	–	–	–	–	90
Exposure	N	N	–	E	–	NW
Coverage (%)	25	30	15	3	–	10
N° Species	7	6	6	4	2	2
UTM	342097—3122357	342015—3122265	341861—3122371	339234—3120662	341732—3122245	339875—3129875
Date	15/5/03	15/5/03	15/4/03	22/4/03	19/5/03	27/5/03
Characteristic species of association						
Cerastium sventenii	2	2	2	+	1	1
Silene berthelotiana	1	+	+	1	.	.
Characteristic Greenovio-Aeonietea						
Cheilanthes guanchica	.	.	.	+	.	.
Arabis caucasica	2	1	.	1	.	.
Characteristic Chamaecytiso-Pinetea						
Pimpinella cumbrae	1	+	+	.	.	.
Micromeria lachnophylla	.	.	2	.	3	.
Other species						
Piptatherum caerulescens	1	.	.	.	.	.
Piptatherum miliaceum	.	1	.	.	.	.
Rosa cannadas-teydensis	+	+	.	.	.	.
Argyranthemum tenerifae	.	.	1	.	.	+
Pterocephalus lasiospermus	.	.	+	.	.	.
Erysimum scoparium	+	.	.	.	.	.
Cystopteris fragilis	.	.	.	.	.	.
Viola cheiranthifolia	.	.	.	.	.	.

Localities: 1–3: Chozas (Montaña Guajara). 4: Cortada del Diablo (Las Cañadas Wall towards Llano de Ucanca). 5: Sol de los Muertos (Montaña Guajara). 6: Northeastern slopes of the Teide Building

Table 4.9b *Sileno berthelothianae-Cerastietum sventenii* Martín Osorio, Hernández Bolaños & Wildpret *ass. nova*
Relevé Type n° 1
Greenovion aureae, Greenovio-Aeonietea

N° Relevé	**7**	**8**	**9**	**10**	**11**	**12**	**13**
Altitude (m)	2976	2725	2820	3000	3000	2886	2935
Surface (m^2)	9	100	100	1 × 1	100	1	25
Slope (%)	–	60	60	90	70	90	–
Exposure	N	W	W	W	N	NW	W
Coverage (%)	5	1	1	2	1	15	5
N° Species	2	2	2	2	1	1	1
UTM	336532—3127130	335875—3128375	335875—3128125	339875—3129875	340375—3129625	339875—3130125	336164—3127961
Date	4/6/03	3/6/03	3/6/03	27/5/03	28/5/03	27/5/03	3/6/03
Characteristic species of association							
Cerastium sventenii	1	+	+	+	1	2	1
Silene berthelotiana	.	.	.	.	.	.	.
Characteristic Greenovio- Aeonietea							
Arabis caucasica	+	.	.	.	.	.	.
Other species							
Argyranthemum tenerifae	.	.	.	+	.	.	.
Erysimum scoparium	.	+	.	.	.	.	.
Cystopteris fragilis	1	.	.	.	.	.	.
Viola cheiranthifolia	.	.	+	.	.	.	.

Localities: 7: La Mesita de la Escoronada (Crater of Pico Viejo). 8 y 9: Cerro las Chajoras (NW- Crater of Pico Viejo). 10, 11 y 12: Northeastern slopes of the Teide Building. 13: La Escoronada (Crater of Pico Viejo)

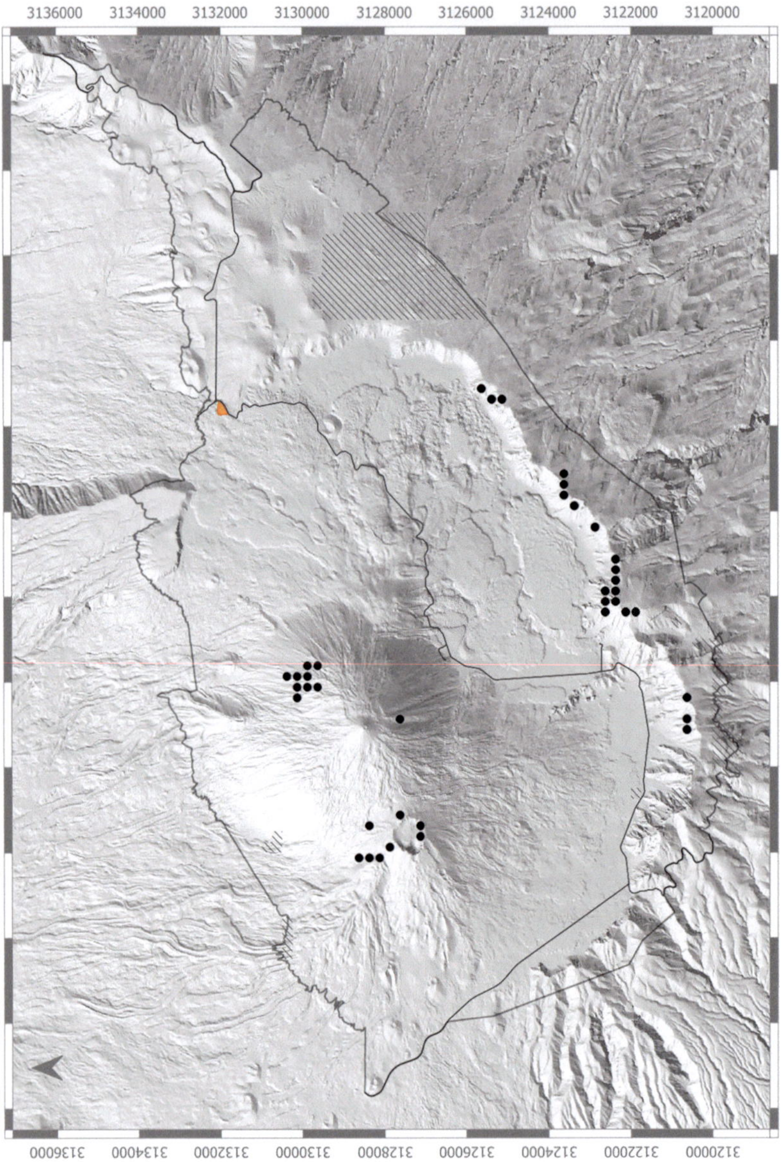

Fig. 4.22 Distribution of *Sileno berthelothianae-Cerastietum sventenii*

IV. *PHRAGMITO-MAGNOCARICETEA* *Klika in* Klika & Novak 1941

Helophytic vegetation typical of river or lake banks bathed by mesoeutrophic or brackish waters, as well as certain hygro-peaty soils that remain waterlogged for much of the year. Its distribution is cosmopolitan.

***PHRAGMITETALIA** W. Koch 1926

Large or medium-sized helophytic vegetation, linked to fresh or subsaline waters.

****Magnocaricion elatae** (W. Koch 1926) Br.-Bl. 1947

Associations of large tufted sedges (*Carex panniculata*) and other medium-sized or tall heliophiles, typical of permanent hydrophiles stations subject to variable temporary flooding. Their distribution is Mediterranean and Euro-Siberian.

IV.1. *Mentho-Caricetum calderae* Wildpret, Socorro & León 1987

Associations that occur around springs or stream banks on the Supra-Mediterranean level of the island of Tenerife (Fig. 4.23).

Fig. 4.23 *Mentho-Caricetum* El Riachuelo. (Photo credit: Victoria Eugenia Martín Osorio)

V. *POLIGONO-POETEA ANNUAE* *Rivas-Martínez* 1975

Nitrophilous, roadside and first-colonizing vegetation, consisting of small, largely prostrate therophytes.

***POLIGONO ARENASTRI-POETALIA ANNUAE** Tüxen *in* Gëhu et al., 1972 *corr.* Rivas-Martínez, Báscones, T.E. Díaz, Fernández-González & Loidi 1991
****Saginion procumbentis** Tüxen *in* Gëhu et al., 1972

V.1. *Poo infirmae-Saginetum stoloniferae* Martín Osorio & Wildpret 2005 *em. prop.*

Road community and first colonizer of the fumaroles and trail of Telesforo Bravo of the Cono or Pilón de Azúcar (Table 4.10, Fig. 4.24). The presence of a small perennial population of pulvinular-looking *Sagina procumbens* subsp. *stolonifera* in its natural environment at about 3650 m a.s.l. was detected for the first time on June 5, 1995 in the company of Professors Pott and Hüppe from the University of Hannover. On July 7, 1999, in the company of Professor Salvador Rivas-Martínez, Universidad Complutense de Madrid, the community was inventoried and its syntaxonomic affinities were discussed. Over the last few years, repeated field observations in different hot and fumarolic land environments have shown an expansion of the population associated with two bryophytes of the genera *Bryum* and *Amphidium* and the grasses *Poa annua* and *Poa infirma* respectively. Phenologically, an expansive autumn population growth has been observed, which is reduced to the warmest environments during the winter frost period.

This oligospecific, thermopedophygrophilous, somewhat sciophilous community occupies only a few square decimetres of surface area, preferably in a microclimatic environment on permanently humid and warm lands at 35 °C in the shade of the rocks of the lava flow that descends from the top of the peak to the Rambleta. Occasionally, the community invades the trampled interstices of the Telesforo Bravo trail built on the lava flow that leads to the top of the Peak. To date, it has only been observed extending between 3650 m a.s.l and 3700 m shortly before entering the crater of the cone, exceeding the altitudinal limit of *Violetum cheirantifoliae* Rivas-Martínez et al. 1993, where the Teidean endemics *Viola cheiranthifolia* and *Argyranthemum tenerifae*, established on the unstable rocky areas of the steep slopes of the cone of Pico de Teide, reach their maximum altitude.

Table 4.10 *Poo infirmae-Saginetum stoloniferae* Martín Osorio & Wildpret
Polygono-Poetea annuae, Polygono arenastri-Poetalia annuae, Saginion procumbentis
Plant community on hot lands of the Teide fumaroles

N° Relevé	**1**	**2**	**3**	**4**	**5**	**6**	**7**	**8**
Altitude (m)	3640	3650	3660	3675	3700	3610	3610	3650
Slope (%)	20	20	20	20	20	20	20	50
Surface (m²)	10	2	5	5	3	3	3	3
Coverage (%)	60	40	35	35	40	60	40	30
Exposure	SW	SW	SW	SW	SW	SW	SW	SE
UTMx	339000	339000	339000	339000	339000	339000	339000	339000
UTMy	3128000	3128000	3128000	3128000	3128000	3128000	3128000	3128000
N° Species	2	2	2	2	2	3	4	3
Date	4-6-05	4-6-05	4-6-05	4-6-05	7-7-99	24-10-04	24-10-04	23-4-03
Characteristic species of association								
Sagina procumbens subsp. *stolonifera*	3	3	3	3	2	3	3	2
Poa infirma	3	.	1	+	1	2	1	1
Bryum sp.	.	3	.	.	.	.	.	.
Amphidium sp.	.	.	.	.	.	.	+	.
Characteristic Polygono Poetea								
Poa annua	.	.	.	.	.	1	+	+

Localidades: Hot lands near the Telesforo Bravo trail, Teide National Park

Fig. 4.24 *Poo infirmae-Saginetum stoloniferae* Teide fumaroles. (Photo credit: Victoria Eugenia Martín Osorio)

VI. PRUNO HIXAE-LAURETEA NOVOCANARIENSIS Oberdorfer 1965 *corr.* Rivas Martínez, Díaz, Fernández González, Izco, Loidi, Lousa & Penas 2002.

Vegetation class that includes the climatophilous laurel cloud forests and heminebulous forests of the Canary Islands and Madeira and represents the serial ericoid or retamoid microforests and summit shrublands and the deciduous rivular planifolia microforests and bramble bushes. In the Canary Islands, they are typical of dry-humid hyperoceanic Infra-Meso-Mediterranean cloudy bioclimates.

***RUBO-SALICETALIA CANARIENSIS**

Riparian deciduous planifolia microforests of Canary willows and brambles, of western Canarian and Infra-Thermo-Mesomediterranean Madeiran distribution.

*******Salicion canariensis*** Rivas-Martínez, Wildpret, Del Arco, Rodríguez, Pérez de Paz, García Gallo, Acebes, Díaz & Fernández González ex Rivas-Martínez, Fernández González & Loidi 1999

Alliance that brings together the deciduous riparian microforests of Canary Island willows of western Canary Islands and Infra-Thermo-Mesomediterranean Madeiran distribution.

VI.1. ***Rubo ulmifolii-Salicetum canariensis*** J.C. Rodríguez, Arco & Wildpret 1987

Willow groves representing the western Canary Mediterranean pluviestational oceanic and xeric Infra-Meso-Mediterranean semi-arid-humid eusubhyperoceanic edaphophilous soil and geoseries of the microforests of *Salix canariensis* and *Rubus ulmifolius*. The main localities where this community is found are El Riachuelo and Barranco del Río.

VII. STELLARIETEA MEDIAE Tüxen, Lohmeyer & Preising *in* Tüxen 1950

Anthropogenic annual nitrophilous vegetation, consisting of therophytes or small geophytes of a weedy, ruderal or scionitrophilous nature.

***BROMETALIA RUBENTI-TECTORI** (Rivas Goday & Rivas–Martínez 1963) Rivas-Martínez & Izco 1977

Subnitrophilous therophytic vegetation of the western Mediterranean, Ibero-Atlantic and Tingitan biogeographic territory optimum, dry, sub-humid, humid, which shows some Cantabrian-Atlantic and Canarian irradiations.

Community of *Bromus tectorum*

VIII. TUBERARIETEA GUTTATAE (Br.-Bl. *in* Br.-Bl. Roussine y Nêgre 1952) Rivas Goday & Rivas-Martínez 1963 em. Rivas-Martínez et al. 2002

Xerophilous and ephemeral therophytic grasslands that colonize incipient or degraded soils, poorly structured, poor in organic matter or nitrogen compounds and devoid of hydromorphic phenomena.

***TUBERARIETALIA GUTTATAE** Br.-Bl. in Br.-Bl. Molinier & Wagner 1940 em. Rivas-Martínez et al. 2002

Surface-rooted therophytic grasslands that colonize oligotrophic skeletal soils with a gravelly, silty or sandy texture, but with a certain cohesion.

****Tuberarion guttati** Br.-Bl. *in* Br.-Bl. Molinier & Wagner 1940 em. Rivas-Martínez et al. 2002

VIII.1. ***Festuco myuri-Laphangietum teydei*** Wildpret, Rodríguez & León *in* Rivas-Martínez et al. 1993 corr.

Therophytic micrograsslands of the Teide fumaroles made up of *Laphangium teydeum*, (Fig. 4.25) Teide edelweiss, the grass *Festuca myurus* and various species of bryophytes that form a cushioned stratum close to the fumaroles, which are usually at around 80 °C. It develops above 2000 m a.s.l., in humid places (springs), which include the fumaroles at the top of Pico del Teide and the hot lands of La Rambleta.

Fig. 4.25 *Festuco myuri-Laphangietum teydei*. (Photo credit: Victoria Eugenia Martín Osorio)

Vegetation Series of the Teide National Park

In this chapter, the set of plant communities that make up the plant landscape of the Teide National Park are presented through the Vegetation Series (Martín Osorio et al. 2007; Rivas-Martínez et al. 2011) representative of the phytocenotic diversity. A Vegetation Series is a sequence of plant communities that occur over time in a specific geographic area, characterized by a particular set of environmental conditions. These Series represent the natural evolution of vegetation towards its climax or mature state, influenced by factors such as climate, soil, topography and human action.

1. **D**iagnosis: Tenerife Climatophilic and Edaphoxerophilous Series Mediterranean pluviseasonal oceanic Thermo-Supramediterranean dry-subhumid semi-hyperoceanic forests of *Pinus canariensis* and *Sideritis soluta* with *Chamaecytisus angustifolius*

[**S**ideritido solutae-**P**ino canariensis sigmetum]

- Typical dry Mesomediterranean faciation of *Pinus canariensis*.
- Thermo-Mesomediterranean subnebular faciation of *Erica canariensis*.
- Lower Supramediterranean faciation of *Spartocytisus supranubius*.
- Edaphoxerophilous faciation of recent badlands.

Nomenclatural reference syntaxon: *Siderito solutae-Pinetum canariensis.*
Common name: Series of the Canary Island pine forests of Tenerife with southern chajorras and brooms with magarzas of Tenerife.
Integrative Alliance: *Cisto symphytifolii-Pinion canariensis.*
Succession:

1. *Adenocarpo viscosi-Chamaecytisetum angustifolii.*
2. *Erysimo scoparii-Pterocephaletum lasiospermi.*
3. *Tolpido webbii-Micromerietum lachnophyllae.*
4. *Arrhenathero calderae-Plantaginetum webbii.*

2. **D**iagnosis: Tenerife Edaphoxerophilous Series of Mediterranean, pluviseasonal, oceanic, Supra-Mediterranean, dry-subhumid, Euoceanic microforests of *Juniperus cedrus* and *Adenocarpus viscosus* with *Echium wildpretii* and *Pterocephalus lasiospermus.*

[**A**denocarpo viscosi-**J**unipero cedri sigmetum]
Nomenclatural reference syntaxon: *Adenocarpo viscosi-Juniperetum cedri.*
Common Name: Series of Tenerife Canary cedars with summit laburnums and red taginastes, with summit rosalito.
Integrative Alliance: *Juniperion cedri.*
Succession:

1. *Erysimo scoparii-Pterocephaletum lasiospermi.*
2. *Tolpido webbii-Micromerietum lachnophyllae.*
3. *Arrhenathero calderae-Plantaginetum webbii.*

3. **D**iagnosis: Tenerife Climatophilic and Edaphoxerophilous Series Mediterranean pluviseasonal oceanic Mesomediterranean upper-Oromediterranean dry-subhumid euoceanic of the summit shrublands of *Spartocytisus supranubius* and *Adenocarpus viscosus* with *Descurainia bourgeauana* and *Nepeta teydea.*

[Spartocytiso supranubii sigmetum]

- Typical Supra-Oromediterranean faciation of *Spartocytisus supranubius.*
- Edaphoxerophilous faciation of badlands and recent lapillis of *Descurainia bourgeauana.*
- Dry upper faciation of *Descurainia gonzalesii.*
- Anemogenic phonolitic Meso-Suprameditterranean faciation of *Viola guaxarensis.*

Nomenclatural reference syntaxon: *Spartocytisetum supranubii.*
COMMON NAME: Series of Teide brooms from Tenerife with summit laburnums and summit grasses with *Nepeta teydea.*
Integrative Alliance: *Spartocytision supranubii.*
Succession:

1. *Erysimo scoparii-Pterocephaletum lasiospermi*
2. *Cheirolopho teydis-Carlinetum xeranthemoides*
3. *Arrhenathero calderae-Adenocarpetum viscosi* (primocolonizer)

4. *Tolpido webbii-Micromerietum lachnophyllae*
5. *Arrhenathero calderae-Plantaginetum webbii*

4. **D**iagnosis: Tenerife High Mountain Geopermaseries Mediterranean Bioclimate pluviseasonal oceanic Oromediterranean dry Euoceanic of the open glareose subshrub of *Viola cheiranthifolia* with *Argyranthemum tenerifae* and *Silene nocteolens* with *Laphangium teydeum.*

[**V**iolo cheiranthifoliae geopermasigmetum]

Nomenclatural reference syntaxon: *Violetum cheiranthifoliae.*

COMMON NAME: High Mountain Geopermaseries of Tenerife of open glareose subshrubs of Teide violets with Teide daisy and Teide catchfly with Teide edelweiss.

Integrative **A**lliance: *Violion cheiranthifoliae.*

5. **D**iagnosis: Series and Edaphohygrophilous Geoseries rivular western Canarian Mediterranean pluviseasonal oceanic and xeric oceanic Infra-Meso-Supramediterranean semiarid-humid eu-subhyperoceanic of the microforests of *Salix canariensis* and *Rubus ulmifolius*. Faciation and geofaciation typical of Tenerife.

[**R**ubo-**S**alici canariensis sigmetum et geosimetum]

Nomenclatural reference syntaxon: *Rubo ulmifolii-Salicetum canariensis.*

Common **N**ame: Series and rivular geoseries of the western Canary Islands willow trees with common brambles and "fayas" with "matoespumas"

Integrative **A**lliance: *Salicion canariensis.*

4.2 Syntaxonomic Scheme of Teide National Park

A syntaxonomic scheme is a hierarchical classification system that organizes plant communities into units called syntaxa, based on their floristic composition and ecology. It is a fundamental tool in phytosociology, the science that studies plant communities.

Syntaxa are organized in a hierarchy similar to that used in the classification of living beings (kingdom, phylum, class, order, family, genus, species), although with its own terminology:

1. **Class (Classis):** Groups orders with common general characteristics. Ex: *Chamaecytiso angustifolii-Pinetea canariensis*
2. **Order (Ordo):** Brings together alliances with shared floristic and ecological characteristics. Ex: *Chamaecytiso angustifolii-Pinetalia canariensis*
3. **Alliance (Alliance):** Groups associations with a more specific floristic composition. Ex: *Cisto-Pinion canariensis*
4. **Association (Associatio):** The basic unit of syntaxonomy, defined by a characteristic set of species (characteristic species). Ex: *Adenocarpo viscosi-Juniperetum cedri*

5. **Subassociation (Subassociatio):** Subdivides the association into smaller units with minor floristic and ecological variations. Ex: *Spartocytisetum supranubii echietosum wildpretii*
6. **Variant (Varietas):** Differentiations within the subassociation.
7. **Facies:** Very local variations, often related to specific environmental factors. Ex: ***Spartocytisetum supranubii*** Primocolonizing faciation of recent badlands and lapillis.

I. ANOMODONTO-POLYPODIETEA Rivas Martínez 1975

***ANOMODONTO-POLYPODIETALIA** O. Bolós & Vives *in* O. Bolós 1957
****Selaginello denticulatae-Anogrammion leptophyllae** Rivas-Martínez, Fernández-González & Loidi 1999
Community of ***Anogramma leptophylla***

II. CHAMAECYTISO ANGUSTIFOLII-PINETEA CANARIENSIS Rivas Goday & Esteve *ex* Esteve 1969

*CHAMAECYTISO ANGUSTIFOLII-PINETALIA CANARIENSIS Rivas Goday & Esteve *ex* Esteve 1969
**Cisto-Pinion canariensis Rivas Goday & Esteve *ex* Esteve 1969

II.1. ***Sideritido solutae- Pinetum canariensis*** Esteve 1973

II.1.1 ***typicum***
II.1.2 ***cistetosum osbaeckiaefolii*** Rivas-Martínez et al. 1993
II.1.3 ***ericetosum canariensis*** (del Arco, P. Pérez & Wildpret 1987) Rivas-Martínez et al 1993. *em. prop.*
II.1.4 ***spartocytisetosum supranubii*** Rivas-Martínez et al. 1993

II.2. ***Adenocarpo viscosi-Chamaecytisetum angustifolii*** Martín Osorio, Hernández Bolaños & Wildpret *ass. nova.*

****Juniperion cedri** Martín Osorio, Wildpret & Rivas-Martínez in Martín Osorio et al. 2007

II.3. ***Adenocarpo viscosi-Juniperetum cedri*** Martín Osorio, Wildpret & Rivas-Martínez in Martín Osorio et al. 2007

***SPARTOCYTISETALIA SUPRANUBII** Schönfelder & Voggenreiter 1994
****Spartocytision supranubii** Oberdorfer *ex* Esteve 1973

II.4. ***Spartocytisetum supranubii*** Oberdorfer *ex* Esteve 1973

II.4.1 ***typicum***
II.4.2 ***echietosum wildpretii*** P. Schönfelder *in* P. Schönfelder & Voggenreiter 1994
II.4.3 ***descurainietosum lemsii*** Martín Osorio & Wildpret in Martín Osorio et al. 2007
II.4.4 ***descurainietosum gonzalesii*** Martín Osorio & Wildpret in Martín Osorio et al. 2007

II.4.5 **Primocolonizing faciation** of recent badlands and lapillis

II.5. ***Descurainio bourgeauanae-Spartocytisetum supranubii*** (Esteve 1973) Martín Osorio, Wildpret & Hernández Bolaños 2007 *stat. nov*. (Spartocytisetum supranubii descurainietosum bourgeauanae Esteve 1973)

II.5.1. ***typicum***
II.5.2. ***erigerontetosum calderae*** Martín Osorio & Wildpret in Martín Osorio et al. 2007

Monospecific pioneer community of "pajonera": ***Descurainia bourgeauana.***
II.6. ***Violo guaxarensis-Adenocarpetum viscosi*** Wildpret & Martín Osorio 2005 *em. prop.*

****Violion cheiranthifoliae** Voggenreiter *ex* Rivas-Martínez, Wildpret & Martín Osorio 2007

II.7. ***Violetum cheiranthifoliae*** Rivas-Martínez, Wildpret, Del Arco, Rodríguez, Pérez de Paz, García Gallo, Acebes, Díaz & Fernández González 1993

****Plantaginion webii** Martín Osorio, Wildpret & Rivas-Martínez in Martín Osorio et al. 2007
*****Pterocephalenion lasiospermi** Martín Osorio, Wildpret & Rivas-Martínez in Martín Osorio et al. 2007

II.8. ***Echietum auberiani*** P. Schönfelder *in* Schönfelder & Voggenreiter 1994
II.9. ***Erysimo scoparii-Pterocephaletum lasiospermi*** Rivas-Martínez et al. 1993
II.10. ***Arrhenathero calderae-Plantaginetum webbii*** Martín Osorio & Wildpret 2003
II.11. ***Cheirolopho teydis-Carlinetum xeranthemoides*** Martín Osorio, Hernández Bolaños & Wildpret *ass. nova.*
II.12. ***Tolpido webbii-Micromerietum lachnophyllae*** Martín Osorio, Hernández Bolaños & Wildpret *ass. nova.*
II.13. ***Arrhenathero calderae-Adenocarpetum viscosi*** Martín Osorio, Hernández Bolaños & Wildpret *ass. nova*

III. GREENOVIO –AEONIETEA A. Santos 1976

***SONCHO ACAULIS-AEONIETALIA** Rivas Goday & Esteve *ex* Sunding 1972
****Greenovion aureae** A. Santos *ex* Rivas-Martínez et al. 1993

III.1. ***Cheilantho guanchicae-Aeonietum smithii*** Rivas-Martínez et al. 1993
III.2. ***Greenovietum aizoi*** Voggenreiter *ex* Rivas-Martínez et al. 1993
III.3. ***Greenovietum aureae*** Rivas-Martínez et al. 1993
III.4. ***Sileno berthelotianae-Cerastietum sventenii*** Martín Osorio, Hernández Bolaños & Wildpret *ass. nova.*

IV. PHRAGMITO-MAGNOCARICETEA Klika in Klika & Novak 1941

***PHRAGMITETALIA** W. Koch 1926
****Magnocaricion elatae** (W. Koch 1926) Br.-Bl. 1947

IV.1. ***Mentho-Caricetum calderae*** Wildpret, Socorro & León 1987

V. POLIGONO-POETEA ANNUAE Rivas-Martínez 1975

***POLIGONO ARENASTRI-POETALIA ANNUAE** Tüxen *in* Gëhu et al. 1972 corr. Rivas-Martínez, Báscones, T.E. Díaz, Fernández-González & Loidi 1991
****Saginion procumbentis** Tüxen *in* Gëhu et al. 1972

V.1. ***Poo infirmae-Saginetum stoloniferae*** Martín Osorio & Wildpret 2005

VI. PRUNO HIXAE-LAURETEA NOVOCANARIENSIS

***RUBO-SALICETALIA CANARIENSIS**
*******Salicion canariensis*** Rivas-Martínez, Wildpret, Del Arco, Rodríguez, Pérez de Paz, García Gallo, Acebes, Díaz & Fernández González ex Rivas-Martínez, Fernández González & Loidi 1999

VI.1. ***Rubo ulmifolii-Salicetum canariensis*** J.C. Rodríguez, Arco & Wildpret 1987

VII. STELLARIETEA MEDIAE Tüxen, Lohmeyer & Preising *in* Tüxen 1950

CHENOPODIENEA MURALIS Rivas-Martínez, Báscones, T.E. Díaz, Fern. Gonz. & Loidi *ex* Theurillat et al. 1995

***BROMETALIA RUBENTI-TECTORI** (Rivas Goday & Rivas–Martínez 1963) Rivas-Martínez & Izco 1977

Community of ***Bromus tectorum.***

VIII. TUBERARIETEA GUTTATAE (Br.-Bl. *in* Br.-Bl. Roussine y Nêgre 1952) Rivas Goday & Rivas-Martínez 1963 em. Rivas-Martínez et al. 2002

***TUBERARIETALIA GUTTATAE** Br.-Bl. in Br.-Bl., Molinier & Wagner 1940 *em.* Rivas-Martínez et al. 2002
****Tuberarion guttati** Br.-Bl. *in* Br.-Bl., Molinier & Wagner 1940 em. Rivas-Martínez et al. 2002

VIII.1. ***Festuco myuri-Laphangietum teydei*** Wildpret, Rodríguez & León *in* Rivas-Martínez et al. 1993 *em. prop.*

4.3 Plant Landscape Cartography

Through Vegetation Cartography we can learn about the distribution of different types of vegetation in a given area. It is a fundamental discipline for understanding biodiversity, ecosystems and ecological processes. It is also useful for environmental management, territorial planning and scientific research. The most commonly used techniques in vegetation cartography are:

- **Field inventories**: Compilation of data on vegetation through direct sampling in the field.
- **Satellite images**: They allow obtaining large-scale information on vegetation cover and changes in it.
- **Photointerpretation**: Visual analysis of aerial and satellite photographs to identify types of vegetation.
- **Geographic Information Systems (GIS)**: Computer tools that allow capturing, storing, analysing and visualizing geographic data related to vegetation, transforming once the data from the Physical Environment and the Biological Environment are integrated into a **Geobotanical Information System** (Martín Osorio et al. 2005).

Based on photointerpretation work on orthophotos at a scale of 1:5000 and revision tasks developed over the last few years, the following cartography has been produced: **Potential Vegetation Map, Current Vegetation Map and Vegetation Cover Map of the Teide National Park**. Potential vegetation is a fundamental concept in ecology and environmental restoration and refers to the Vegetation Units that, under optimal environmental conditions and without the influence of human disturbances, could be established in a given place. In other words, it is the natural vegetation that an ecosystem could have if the limitations imposed by human activities were eliminated.

4.3.1 Potential Vegetation

Potential Vegetation Map E: 1/75,000 (Fig. 4.26) prepared using the Geobotanical Information System connecting information from the Physical Environment (Geology, Soil Science and Bioclimatology) and the Biological Environment (Flora and Vegetation), based on the layout of the cartographic vegetation units recognized in the National Park and according to Rivas-Martínez et al. (2011). It represents the vegetation characterized by the following **Vegetation Series and Geoseries**:

- *Sideritido solutae-Pino canariensis* sigmetum
- *Adenocarpo viscosi-Junipero cedri* sigmetum
- *Spartocytiso supranubii* sigmetum
- *Violo cheiranthifoliae* geopermasigmetum
- *Rubo-Salici canariensis* sigmetum et geosigmetum.

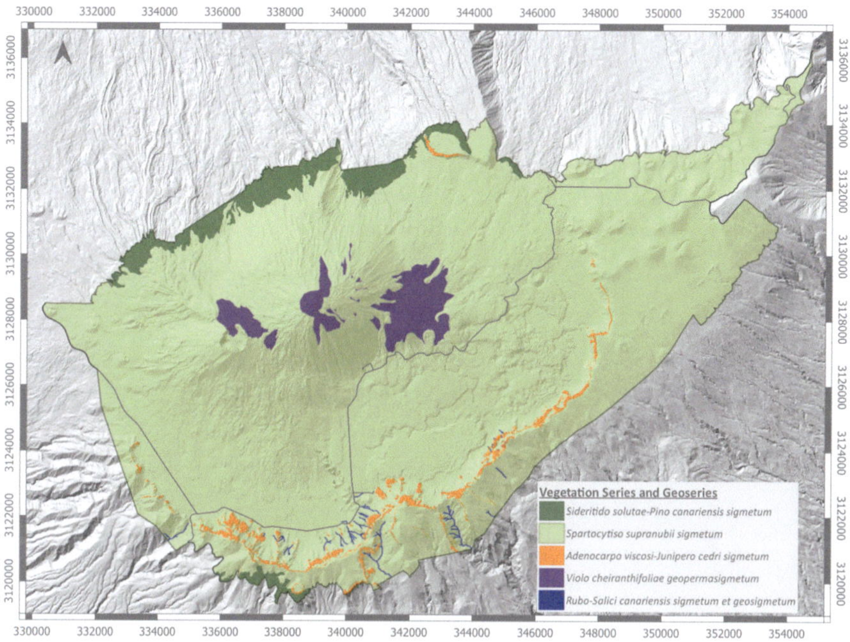

Fig. 4.26 Map of the Potential vegetation of the Teide National Park

4.3.2 Current Vegetation

Current Vegetation Map E: 1/75,000 Where the vegetation complexes in which the complex of vegetation units has been unified according to the 500 × 500 m grid network are represented (Fig. 4.27). Due to the scale of work, it is often difficult to cartographically delimit some syntaxa from others, for this reason in the legend of the vegetation units they are presented as vegetation mosaics, the first of which is the most physiognomically representative (Fig. 4.28).

A total of 41 syntaxonomic units have been mapped, and their representation on the map has been carried out through a total of 1638 polygons or spatial entities. It is worth noting that, of these, some 615 were identified as "non-significant vegetation", distributed to a greater or lesser extent throughout the entire extension of the protected area.

A first approximation of the cartographic units that have the greatest spatial representation within the National Park is the following:

- *Descurainio bourgeauanae-Spartocytisetum supranubii.*
- *Spartocytisetum supranubii.*
- *Erysimo scoparii-Pterocephaletum lasiospermi + Spartocytisetum supranubii + Descurainio bourgeauanae-Spartocytisetum supranubii.*
- *Arrhenathero calderae-Adenocarpetum viscosii.*

- *Sideritido solutae-Pinetum canariensis + Adenocarpo viscosi-Chamaecytisetum angustifolii.*
- *Spartocytisetum supranubii Faciación primocolonizadora de malpaíses y lapillis recientes.*

The unit that seems to present a more significant spatial fragmentation, apart from the complex of rupicolous/special vegetation given as *Greenovion aureae*, is the one corresponding to *Adenocarpo viscosi-Juniperetum cedri*.

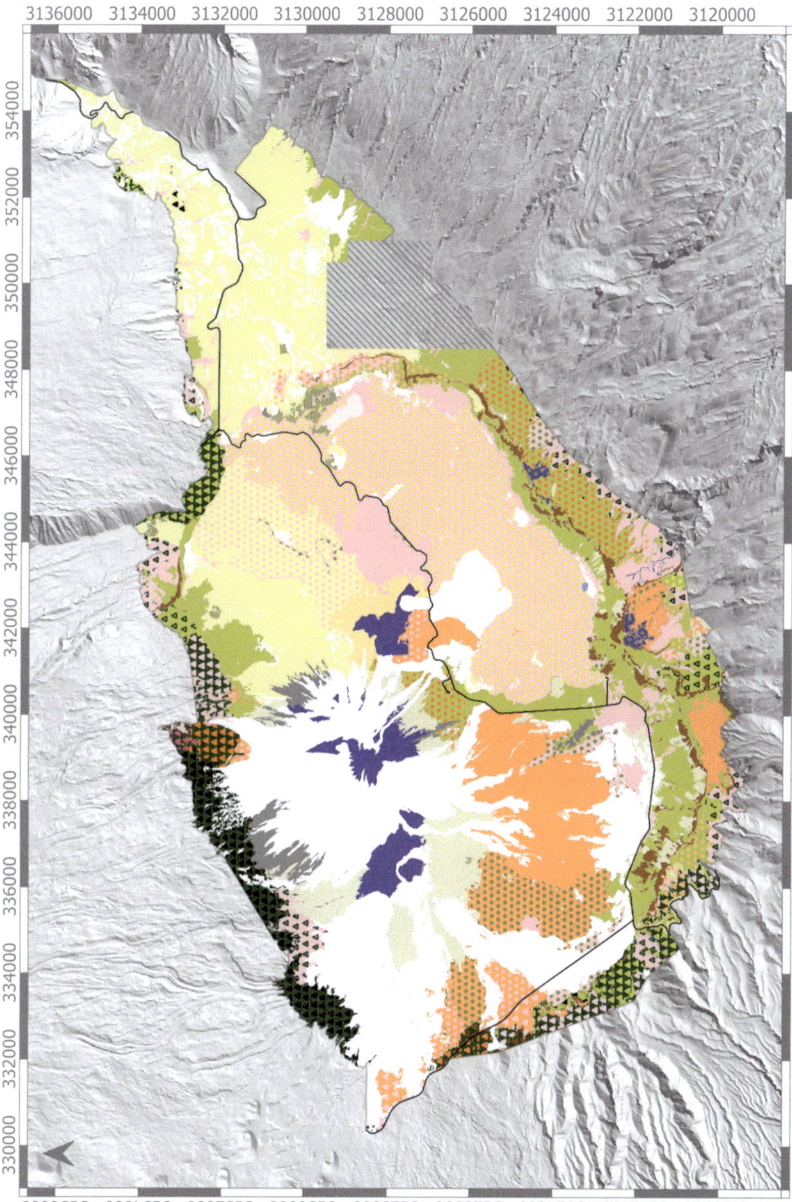

Fig. 4.27 Map of Current Vegetation of the Teide National Park

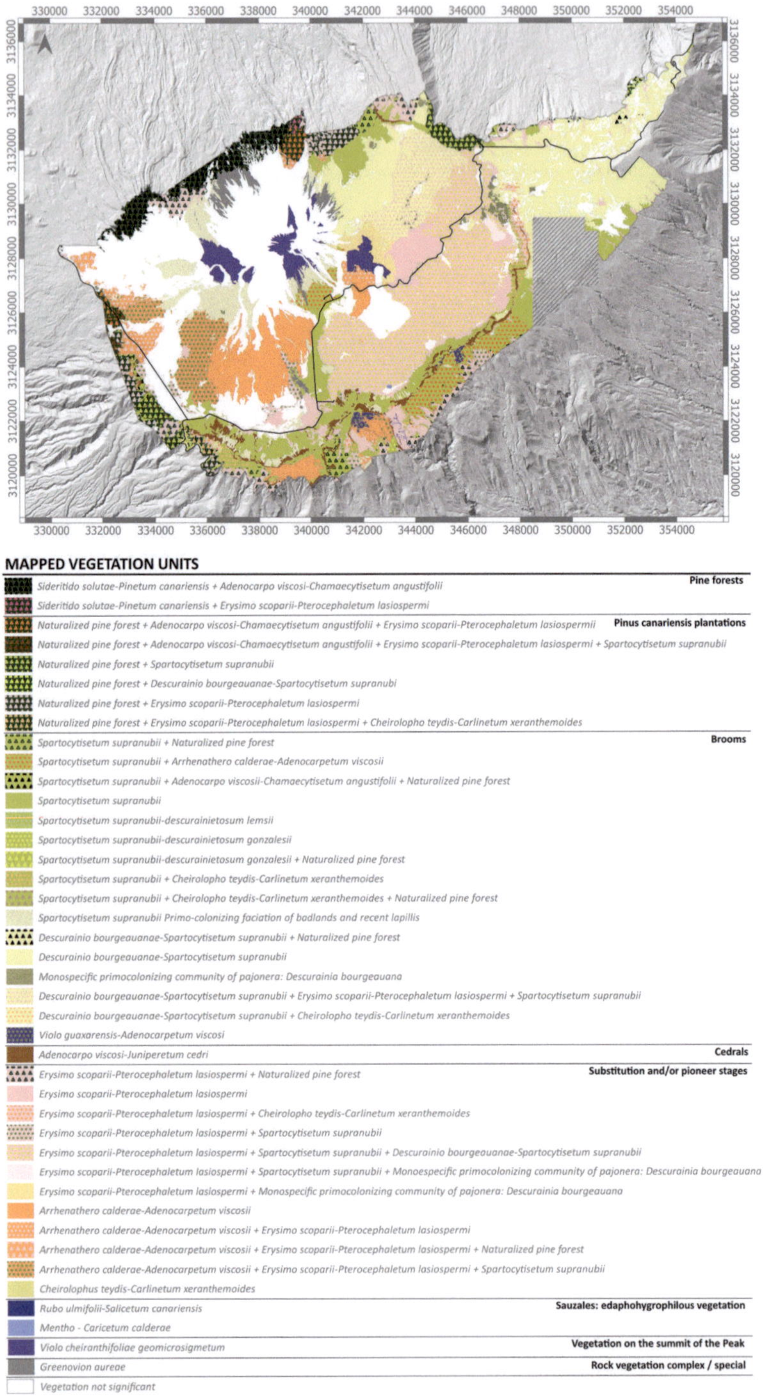

Fig. 4.28 Legend of the Map of Current Vegetation of the Teide National Park (1:75.000), with details on the vegetation complexes mapped

4.3.3 *Phytocenotic Richness*

Vegetation Cover Map E: 1/75,000 prepared based on the photo-interpretation work carried out on the series of orthophotos. A total of 2504 polygons were categorized, taking as a reference those of the cartographic vegetation units recognized in the previous map (Fig. 4.29).

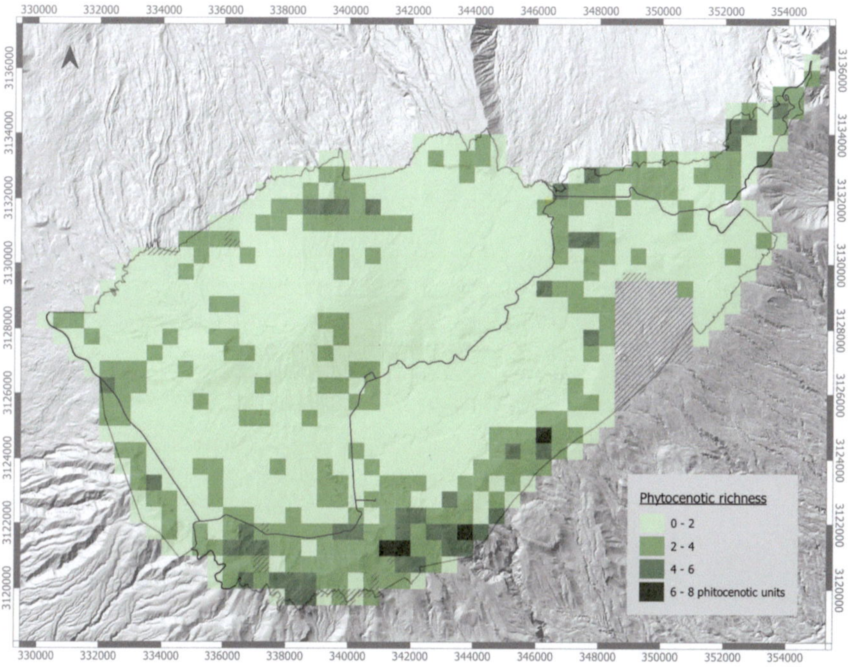

Fig. 4.29 Phytocenotic richness

Annex

Some images of the volcanic landscape of the Teide National Park are presented here, which due to their uniqueness we have preferred to offer together as a gallery of images. It is worth highlighting the collection of *Juniperus cedrus* made by the Naturalist and Professional Climber Javier Martín-Carbajal, whom we thank for his generosity in providing us with the photos (Figs. 4.30, 4.31, 4.32, 4.33, 4.34, 4.35, 4.36, 4.37, 4.38, 4.39, 4.40, 4.41, 4.42, 4.43, 4.44, 4.45, 4.46, 4.47, 4.48, 4.49, 4.50, and 4.51).

Fig. 4.30 Sunset in September 2004. (Photo credit: Victoria Eugenia Martín Osorio)

Fig. 4.31 "Sea of Clouds" and Canarian pine forest in Oct 2024. (Photo credit: Victoria Eugenia Martín Osorio)

Fig. 4.32 Precaldera Building November 2024. (Photo credit: Victoria Eugenia Martín Osorio)

Fig. 4.33 Corded Lava August 2005 and Prof. Wildpret. (Photo credit: Victoria Eugenia Martín Osorio)

Fig. 4.34 Valle de Ucanca May 2011. (Photo credit: Victoria Eugenia Martín Osorio)

Fig. 4.35 El Sombrerito 2532 m a.s.l. and pine forest April 2004. (Photo credit: Victoria Eugenia Martín Osorio)

Fig. 4.36 La Catedral 2011. (Photo credit: Wolf-Hermann Wildpret Martín)

Fig. 4.37 Montaña Cerrillar Sept 2024. (Photo credit: Victoria Eugenia Martín Osorio)

Fig. 4.38 Fasnia Volcano erupted from December 31, 1704 to March 27, 1705, May 2009. (Photo credit: Victoria Eugenia Martín Osorio)

Fig. 4.39 Researchers group with Prof. Pott Institute of Geobotany, Leibniz University Hannover and Universidad La Laguna, Tenerife, 6 Jun. 2004. (Photo credit: Victoria Eugenia Martín Osorio)

Fig. 4.40 *Juniperus cedrus* burned after a fire and Prof. Salvador Rivas-Martínez Nov 2009. (Photo credit: Victoria Eugenia Martín Osorio)

Fig. 4.41 Guajara, 2715 m a.s.l. May 2009. (Photo credit: Victoria Eugenia Martín Osorio)

Fig. 4.42 Pine forest of *Pinus canariensis* and wallflowers Cañada del Cedro Sept. 2024. (Photo credit: Victoria Eugenia Martín Osorio)

Fig. 4.43 The snowy Teide Volcano. (Photo credit: J Martín Carbajal)

Fig. 4.44 Black lava at the top of Teide Volcano. (Photo credit: J Martín Carbajal)

Fig. 4.45 *Echium wildpretii* and Prof. Wolfredo Wildpret in Tabonal Negro, Jun 2006. (Photo credit: Victoria Eugenia Martín Osorio)

Fig. 4.46 Pollinating Bee of *Echium wildpretii*, Cañada del Cedro 2014. (Photo credit: Wolf-Hermann Wildpret Martín)

Fig. 4.47 *Adenocarpo-Juniperetum cedri* Cedral Risco Verde Oct 2024. (Photo credit: J Martín Carbajal)

Fig. 4.48 *Juniperus cedrus* The 1100 year old Matriarch. (Photo credit: J Martín Carbajal)

Fig. 4.49 Barbol, *Juniperus cedrus* The oldest tree in Europe 1480 years old. (Photo credit: J Martín Carbajal)

Fig. 4.50 Barbol, *Juniperus cedrus* surviving between cracks. (Photo credit: J Martín Carbajal)

Fig. 4.51 *Juniperus cedrus,* At 1230 years old, the second oldest. (Photo credit: J Martín Carbajal)

References

Bolós O (1957) De vegetatione valentina, I. Collect Bot (Barcelona) 5(2):527–596

Braun-Blanquet J (1947) The International Station of Mediterranean and Alpine Geobotany at Montpellier 1944–1946, 12 pp

Braun-Blanquet J (1964) Pflanzensoziologie, 3rd edn. Springer Vienna, Wien

Braun-Blanquet J, Molinier R & Wagner H (1940) Prodrome des groupements végétaux. Prodromus der Pflanzengesellschaften. Cisto-Lavanduletea (landes siliceuses à cistes et lavandes). 53 pp. Montpellier

Braun-Blanquet J, Roussine N and Nègre R (1952) The vegetation of the Mediterranean France. National Center of Scientific Research and Service of North Africa Vegetation Map, Vaucluse, 297 p.

Del Arco Aguilar, M. J., P. L. Pérez de Paz y W. Wildpret de la Torre, 1987: Contribución al conocimiento de los pinares de la isla de Tenerife. Lazaroa 7: 67–84

Díaz González T (2004) Pasado, presente y futuro de la Fitosociología española. Lazaroa 25:3–13

Esteve Chueca F (1969) Estudio de las alianzas y asociaciones del orden Cytiso-Pinetalia en las canarias orientales. Bol R Soc esp Hist Nat (Biol) 67:77–104

Esteve Chueca F (1973) Estudio de las asociaciones *Spartocytisetum nubigeni* (Oberd. 1965) Emend. y *Sideriti-Pinetum canariensis* (ass. nova) en las Islas Canarias. Trab Dept Bot Univ Granada 2(1):3–9

Gehú JM, Richard JL & Tüxen R (1972) Compte rendu de l'excursion de l'Association Internationale de Phytosociologie dans le Jura, juine-1967 (1° partie). Doc. Phytosociol. 2: 1–44

Klika J, Novák V (1941) Praktikum rostlinné sociologie, půdoznalství, klimatologie a ekologie;·Vysokoškolské rukověti. Řada V. spisů technickýc, Melantrich, Praha, 385 pp

Koch W (1926) Die Vegetationseinheiten der Linthebene unterBerücksichtigung der Verhältnisse in der Nordostschweiz. Sys-tematisch-kritische Studie. Jahrbuch der St Gallischen naturwissenschaftlichen Gesellchaft 61(II):1–144

Martí J (2003) El volcanismo explosivo de Tenerife. Libro de Apuntes. (Inédito) Curso de Volcanismo del Museo de Ciencias Naturales de Santa Cruz de Tenerife. Instituto de Ciencias de la Tierra "Jaume Almera", CSIC, Barcelona, 99 pp

Martín Osorio VE, Hernández Bolaños B (2003) Comunidad primocolonizadora de taludes de derrubios gelifractos en el Parque Nacional del Teide (Tenerife, Islas Canarias). Vieraea 31:281–292. Santa Cruz de Tenerife

Martín Osorio VE, Wildpret de la Torre W, Hernández Bolaños B (2005) Avances significativos en la elaboración de la base de datos georreferenciada de Flora y Vegetación del Parque Nacional del Teide, Tenerife, Islas Canarias, mediante un Sistema de Información Geobotánica. Vieraea 33:345–358. Santa Cruz de Tenerife

Martín Osorio VE, Wildpret de la Torre W, Del Arco Aguilar M, Pérez de Paz PL, Hernández Bolaños B, Rodríguez O, Acebes JR, García Gallo A (2007) Estudio Bioclimático y Fitocenótico comparativo de la alta cumbre canaria: Tenerife—La Palma. Islas Canarias. Phytocoenologia 37(3–4):663–697. Berlín – Stuttgart

Martínez de Pisón E, Quirantes F (1981) El Teide. Estudio Geográfico. Ed. Interinsular Canaria, S.A. Departamento de Geografía—Universidad de La Laguna, Santa Cruz de Tenerife, 187 pp

Navarro Latorre JM (2000) Geología. En V.V.A.A. Parque Nacional del Teide. Ed. Esfagnos

Oberdorfer E (1965) Pflanzensoziologische Studien auf Teneriffa und Gomera (Kanarische Inseln). Beitr. Naturk. Forsk. SW_Deutschl. 24(1): 47–104

Rivas Martínez S (1975) Sobre la nueva clase Polygono-Poetea annuae. Phytocoenologia 2:123–114

Rivas-Martínez S (2009) Ensayo geobotánico global sobre la Macaronesia. In: Beltrán Tejera E, Afonso-Carrillo J, García Gallo A, Rodríguez Delgado O (eds) Homenaje al Profesor Dr. Wolfredo Wildpret de la Torre. Instituto de Estudios Canarios, Tenerife, La Laguna, pp 255–296. Monografía LXXVIII

Rivas Goday S & Rivas–Martínez S (1963) Estudio y clasificación de los pastizales españoles. Publ. Ministerio de Agricultura. Madrid

Rivas-Martínez S, Coautores (2011) Mapa de series, geoseries y geopermaseries de vegetación de España [Memoria del mapa de vegetación potencial de España]. Itinera Geobot 18(1 y 2):5–800

Rivas-Martínez S, Izco J (1977) Sobre la vegetacion terofitica subnitrofila mediterranea (*Brometalia rubenti-tectori*). Anales Inst Bot Cavanilles 34:355–381

Rivas-Martínez S, Báscones JC, Díaz TE, Fernández-González F, Loidi J (1991) Vegetación del Pirineo occidental y Navarra. VI Excursión Internacional de Fitosociología (AEFA). Itinera Geobot 5:5–456. ISSN 0213-8530

Rivas-Martínez S, Wildpret W, del Arco AM, Rodríguez Delgado O, Pérez de Paz PL, García Gallo A, Acebes JR, Díaz TE, Fernández-González F (1993) Las comunidades vegetales de la Isla de Tenerife (Islas Canarias). Itinera Geobot 7:169–374

Rivas-Martínez S, Fernández-González F, Loidi J (1999) Checklist of plant communities of Iberian Peninsula, Balearic and Canary Islands to suballiance level. Itinera Geobot 13:353–451

Rivas-Martínez S, Díaz TE, Fernández-González F, Izco J, Loidi J, Lousa M, Penas A (2002) Vascular plant communities of Spain and Portugal. Addenda to the syntaxonomical checklist of 2001. Itinera Geobot 15(2):433–922. Servicio de Publicaciones de la Universidad de León

Rodríguez Piñero JC, Wildpret de la Torre W, Del Arco-Aguilar M (1987) Contribución al estudio biosistemático de *Salix canariensis* (Salicaceae). Vieraea 17:121–142

Santos A (1976) Notas sobre la vegetación potencial de la isla de El Hierro. Anal Inst Bot Cavanilles 33:249–261

Schönfelder P, Voggenreiter V (1994) Zur Abgrenzung und Gliederung der Klassen *Spartocytisetea supranubii* cl. nov. und *Cytiso-Pinetea canariensis* auf Tenerife/Kanarische Inseln. Phytocoenologia 24:461–493. Berlin – Stuttgart

Sunding P (1972) The Vegetation of Gran Canaria. *Skrifter Norske Vidensk*. Akademi i Oslo, series I, Ny Serie, vol.29. 1 col. foldg. vegetation map (1: 200 000). 186, LIII p. gr8vo. Paper bd

Theurillat et al. (1995) The higher vegetation units of the Alps. Colloq. Phytosociol. 23:189–239

Tüxen R (1950) Grundriss einer Systematik der nitrophilen Unkrautgesellschaften in der Eurosibirischen Region Europas. Mitt Flor Soz Arb Gem, NF 2:94–175

Wildpret de la Torre W, Martín Osorio VE (2005) Dos nuevas asociaciones para la vegetación del Parque Nacional del Teide: *Poo infirmae—Saginetum stoloniferi* ass. nova y *Violo cheiranthifoliae—Adenocarpetum viscossii* ass. *nova*. Vieraea 33:359–365. Santa Cruz de Tenerife

Wildpret de la Torre W, Socorro JS, León Arencibia MC (1987) *Mentho-Caricetum calderae*, comunidad higro-hidrófila del piso supracanario de Tenerife (Islas Canarias). Ser Informes 22:35–40

MIX
Papier aus verantwortungsvollen Quellen
Paper from responsible sources
FSC® C105338

If you have any concerns about our products,
you can contact us on
ProductSafety@springernature.com

In case Publisher is established outside the EU,
the EU authorized representative is:
Springer Nature Customer Service Center GmbH
Europaplatz 3, 69115 Heidelberg, Germany

Printed by Libri Plureos GmbH
in Hamburg, Germany